RISING TO THE CHALLENGE

U.S. Innovation Policy for the Global Economy

Committee on
Comparative National Innovation Policies:
Best Practice for the 21st Century

Board on Science, Technology, and Economic Policy

Policy and Global Affairs

Charles W. Wessner and Alan Wm. Wolff, Editors

NATIONAL RESEARCH COUNCIL
OF THE NATIONAL ACADEMIES

THE NATIONAL ACADEMIES PRESS
Washington, D.C.
www.nap.edu

THE NATIONAL ACADEMIES PRESS   500 Fifth Street, N.W.   Washington, DC 20001

NOTICE: The project that is the subject of this report was approved by the Governing Board of the National Research Council, whose members are drawn from the councils of the National Academy of Sciences, the National Academy of Engineering, and the Institute of Medicine. The members of the committee responsible for the report were chosen for their special competences and with regard for appropriate balance.

This study was supported by Contract/Grant No. N01-OD-4-2139, TO #245, between the National Academy of Sciences and the National Institutes of Health; Contract/Grant No. DE-PI0000010, TO #15, between the National Academy of Sciences and the U.S. Department of Energy; Contract/Grant No. SB1341-03-C-0032 between the National Academy of Sciences and the National Institute of Standards and Technology of the U.S. Department of Commerce; Contract/Grant No. OFED-858931 between the National Academy of Sciences and Sandia National Laboratories; and Contract/Grant No. NAVY-N00014-05-G-0288, DO #2, between the National Academy of Sciences and the Office of Naval Research. Any opinions, findings, conclusions, or recommendations expressed in this publication are those of the author(s) and do not necessarily reflect the views of the organizations or agencies that provided support for the project.

International Standard Book Number 13: 978-0-309-25551-6
International Standard Book Number 10: 0-309-25551-1

Additional copies of this report are available for sale from the National Academies Press, 500 Fifth Street, N.W., Keck 360, Washington, DC 20001; (800) 624-6242 or (202) 334-3313; Internet, http://www.nap.edu/ .

Copyright 2012 by the National Academy of Sciences. All rights reserved.

Printed in the United States of America

THE NATIONAL ACADEMIES
Advisers to the Nation on Science, Engineering, and Medicine

The **National Academy of Sciences** is a private, nonprofit, self-perpetuating society of distinguished scholars engaged in scientific and engineering research, dedicated to the furtherance of science and technology and to their use for the general welfare. Upon the authority of the charter granted to it by the Congress in 1863, the Academy has a mandate that requires it to advise the federal government on scientific and technical matters. Dr. Ralph J. Cicerone is president of the National Academy of Sciences.

The **National Academy of Engineering** was established in 1964, under the charter of the National Academy of Sciences, as a parallel organization of outstanding engineers. It is autonomous in its administration and in the selection of its members, sharing with the National Academy of Sciences the responsibility for advising the federal government. The National Academy of Engineering also sponsors engineering programs aimed at meeting national needs, encourages education and research, and recognizes the superior achievements of engineers. Dr. Charles M. Vest is president of the National Academy of Engineering.

The **Institute of Medicine** was established in 1970 by the National Academy of Sciences to secure the services of eminent members of appropriate professions in the examination of policy matters pertaining to the health of the public. The Institute acts under the responsibility given to the National Academy of Sciences by its congressional charter to be an adviser to the federal government and, upon its own initiative, to identify issues of medical care, research, and education. Dr. Harvey V. Fineberg is president of the Institute of Medicine.

The **National Research Council** was organized by the National Academy of Sciences in 1916 to associate the broad community of science and technology with the Academy's purposes of furthering knowledge and advising the federal government. Functioning in accordance with general policies determined by the Academy, the Council has become the principal operating agency of both the National Academy of Sciences and the National Academy of Engineering in providing services to the government, the public, and the scientific and engineering communities. The Council is administered jointly by both Academies and the Institute of Medicine. Dr. Ralph J. Cicerone and Dr. Charles M. Vest are chair and vice chair, respectively, of the National Research Council.

www.national-academies.org

Committee on
Comparative National Innovation Policies:
Best Practice for the 21st Century*

Alan Wm. Wolff, *Chair*
Senior Counsel
McKenna Long & Aldridge LLP
and
STEP Board

Michael G. Borrus
Founding General Partner
X/Seed Capital Management

Gail H. Cassell (IOM)
Visiting Professor
Department of Global Health
 and Social Medicine
Harvard Medical School

Carl J. Dahlman
Henry R. Luce Associate Professor
Edmund A. Walsh School
 of Foreign Service
Georgetown University

Charles K. Ebinger
Director, Energy Security Initiative
Senior Fellow, Foreign Policy
The Brookings Institution

Mary L. Good (NAE), *Vice Chair*
Dean Emeritus, Donaghey College
 of Engineering and Information
 Technology
Special Advisor to the Chancellor
 for Economic Development
University of Arkansas
 at Little Rock
and
STEP Board

Kent H. Hughes
Director
Program on America
 and the Global Economy
Woodrow Wilson International
 Center for Scholars

Gregory Kats
President
Capital E

*As of June 2012.

Project Staff

Charles W. Wessner
Study Director

McAlister T. Clabaugh
Program Officer

David S. Dawson
Senior Program Assistant

David E. Dierksheide
Program Officer

Sujai J. Shivakumar
Senior Program Officer

Peter J. Engardio
Consultant

Thomas R. Howell
Consultant

William A. Noellert
Consultant

For the National Research Council (NRC), this project was overseen by the Board on Science, Technology and Economic Policy (STEP), a standing board of the NRC established by the National Academies of Sciences and Engineering and the Institute of Medicine in 1991. The mandate of the Board on Science, Technology, and Economic Policy is to advise federal, state, and local governments and inform the public about economic and related public policies to promote the creation, diffusion, and application of new scientific and technical knowledge to enhance the productivity and competitiveness of the U.S. economy and foster economic prosperity for all Americans. The STEP Board and its committees marshal research and the expertise of scholars, industrial managers, investors, and former public officials in a wide range of policy areas that affect the speed and direction of scientific and technological change and their contributions to the growth of the U.S. and global economies. Results are communicated through reports, conferences, workshops, briefings, and electronic media subject to the procedures of the National Academies to ensure their authoritativeness, independence, and objectivity. The members of the STEP Board* and the NRC staff are listed below:

Paul L. Joskow, *Chair*
President
Alfred P. Sloan Foundation

Ernst R. Berndt
Louis E. Seley Professor
  in Applied Economics
Massachusetts Institute
  of Technology

John Donovan
Chief Technology Officer
AT&T Inc.

Alan M. Garber (IOM)
Provost
Harvard University

Ralph E. Gomory (NAS/NAE)
Research Professor
Stern School of Business
New York University

Mary L. Good (NAE)
Dean Emeritus, Donaghey College
  of Engineering and Information
  Technology
Special Advisor to the Chancellor
  for Economic Development
University of Arkansas
  at Little Rock

William H. Janeway
Partner
Warburg Pincus, LLC

Richard K. Lester
Japan Steel Industry Professor
Head, Nuclear Science
  and Engineering
Founding Director, Industrial
  Performance Center
Massachusetts Institute
  of Technology

continued

*As of June 2012.

William F. Meehan III
Lecturer in Strategic Management
Raccoon Partners Lecturer
  in Management
Graduate School of Business
Stanford University
and
Director Emeritus
McKinsey and Co., Inc.

David T. Morgenthaler
Founding Partner
Morgenthaler Ventures

Arati Prabhakar
Atherton, CA

Luis M. Proenza
President
The University of Akron

William J. Raduchel
Chairman
Opera Software ASA

Kathryn L. Shaw
Ernest C. Arbuckle Professor
  of Economics
Graduate School of Business
Stanford University

Laura D'Andrea Tyson
S.K. and Angela Chan Professor
  of Global Management
Haas School of Business
University of California, Berkeley

Harold R. Varian
Chief Economist
Google, Inc.

Alan Wm. Wolff
Senior Counsel
McKenna Long & Aldridge LLP

STEP Staff

Stephen A. Merrill
Executive Director

Paul T. Beaton
Program Officer

McAlister T. Clabaugh
Program Officer

Aqila A. Coulthurst
Program Coordinator

Charles W. Wessner
Program Director

David S. Dawson
Senior Program Assistant

David E. Dierksheide
Program Officer

Sujai J. Shivakumar
Senior Program Officer

Contents

PREFACE   xiii

OVERVIEW   1

PART I: THE INNOVATION CHALLENGE   15

  Chapter 1: The Innovation Challenge   17

    AMERICA'S INNOVATION CHALLENGES, 18

      Capturing the Economic Value of Innovation, 19
      Coping with the Growth of New Competitors, 21

    NEW TRENDS IN GLOBAL INNOVATION, 23

      Strong Policy Focus on Innovation, 23
      Rapid Growth in R&D Spending, 25
      21st Century Mercantilism, 30
      The Search for Talent, 36
      The Growth of Foreign Research Centers
         of U.S. Multinationals, 37
      The Rise of Open Innovation, 39
      Growth of Innovative Regions Around the World, 40

    THE PILLARS OF U.S. INNOVATIVE STRENGTH, 41

      Strong Protection of Intellectual Property, 43
      Federal Funding of Research, 44
      Research Universities, 44
      National Laboratories, 45
      The Private Sector, 46
      Public-Private Partnerships, 46

    RESPONDING TO THE INNOVATION CHALLENGE, 49

      Policies to Capture the Value of Innovation
         in Some Leading Countries and Regions, 51
      The Growing U.S. Response: Federal Government, 53
      The Growing U.S. Response: State and Regional Initiatives, 56
      Looking Ahead, 59

Chapter 2: Sustaining Leadership in Innovation   61

   IMPROVING FRAMEWORK CONDITIONS, 61
   SUBSTANTIALLY INCREASING R&D FUNDING, 65
   INSTITUTIONAL SUPPORT FOR APPLIED RESEARCH, 72

      Germany's Fraunhofer, 72
      Taiwan's ITRI, 73
      South Korea's ETRI, 73
      U.S. Applied Engineering Programs, 74
      State Programs, 76
      Lessons and Calls for New U.S. Institutions, 77
      Recent Initiatives, 78

   STRENGTHENING MANUFACTURING, 79

      The Link between Manufacturing and Innovation, 84
      Support for Manufacturing Overseas, 86
      U.S. Support for Manufacturing, 88
      Why Manufacturing Matters for the U.S., 95

   PROVIDING EARLY STAGE FINANCE, 97

   DEVELOPING TWENTY-FIRST CENTURY UNIVERSITIES, 102

      U.S. Universities Face Financial Challenges, 103
      Growing Investments in Universities Abroad, 105
      New Models of 21st Century Universities, 108
      New Opportunities for Global Collaboration, 109
      Collaboration with Industry, 110

   INVESTING IN MODERN S&T PARKS, 111

   GROWING INNOVATION CLUSTERS, 113

      International Cluster Initiatives, 114
      U.S. Regional Cluster Initiatives, 116
      The U.S. Federal Role, 116

   HUNTING FOR GLOBAL TALENT, 118
   THE WAY FORWARD, 121

Chapter 3: Findings   127

Chapter 4: Recommendations   163

CONTENTS   xi

PART II: GLOBAL INNOVATION POLICIES   199

Chapter 5: The New Global Competitive Environment   201

    EMERGING POWERS, 203

        China's Rapid Rise, 203
        India's Changing Innovation System, 239

    NEWLY INDUSTRIALIZED ECONOMIES, 255

        Taiwan, 255
        Singapore's Focus, 265

    INDUSTRIALIZED NATION CASE STUDIES, 271

        Germany, 271
        Flanders, 289
        Finland, 298
        Canada, 302
        Japan, 311

Chapter 6: National Support for Emerging Industries   321

    SEMICONDUCTORS, 324

        The Strategic Importance of Semiconductors, 328
        A New Set of Challenges, 329
        Industry Growth and U.S. Policy, 331
        The Role of U.S. Trade Policy, 335
        New U.S. Research Consortia, 337
        Today's Competitive Challenges, 339
        Lessons, 352

    THE PHOTOVOLTAIC INDUSTRY, 353

        Solar Power's Strategic Importance, 359
        The Industry's Origins, 360
        Competing Technologies, 362
        U.S. Advantages, 363
        The New U.S. Solar Policy Thrust, 364
        The Challenges Ahead, 367
        Photovoltaic Policy Questions for the United States, 378
        Conclusion, 383

ADVANCED BATTERIES, 383

Opportunities to Catch Up, 391
The Growing Federal Role, 393
The Military's Electrification Drive, 395
Future Policy Priorities, 396
Conclusion, 399

PHARMACEUTICALS AND BIOPHARMACEUTICALS, 399

Strategic Importance, 400
Evolution of the Industry, 401
Challenges, 412
Looking Ahead, 425

IN CLOSING, 429

Chapter 7: Clusters and Regional Initiatives   431

THE INNOVATION CHALLENGE, 431
POLICIES TO FOSTER INNOVATION, 433
REGIONAL INNOVATION CLUSTERS, 433

Cluster Dynamics, 437
An Emerging U.S. Cluster Strategy, 438
Why Clusters are Relevant Now, 443
State and Regional Case Studies, 445
Policy Lessons for U.S. Innovation Clusters, 462

TWENTY-FIRST CENTURY RESEARCH AND INDUSTRY PARKS, 463

Research Parks Around the World, 466
A New Generation of U.S. Research Parks, 477
Observations on Factors in the Success of Research Parks, 485

IN CLOSING, 487

APPENDIXES   489

A   List of Workshops and Symposia for the Study of *Comparative National Innovation Policies: Best Practice for the 21st Century*   491

B   Bibliography   493

Preface

The ability to combine theory, creativity and engineering was a great achievement of postwar America. For 50 years, economic growth and job creation were propelled by transistors, lasers and other discoveries that came from the willingness to nurture theoretical research in conjunction with applied science and manufacturing skills. But these days, manufacturing is being outsourced, and funding for pure sciences is being curtailed. With Bell Labs and other such idea factories disappearing, and with government research money endangered, what will propel innovation and job creation for the next 50 years?

Walter Isaacson
New York Times, April 8, 2012[1]

The capacity to innovate is fast becoming the most important determinant of economic growth and a nation's ability to compete and prosper in the 21st century global economy. Innovation encompasses not only research and the creation of new ideas, but the development and effective implementation of the technology into competitive products and services. Governments around the world now recognize that innovation, not just inputs such as capital and labor, is critical to sustaining economic growth, creating good jobs, and fulfilling national needs. Industrialized nations and emerging powers alike have boosted spending on research and development and unveiled comprehensive national strategies to build innovation-led economies. Indeed, just as the global

---

[1] Walter Isaacson writing "Inventing the Future" a review in the *New York Times* of April 8, 2012, of Jon Gertner's book *The Idea Factory – Bell Labs and the Great Age of American Innovation*.

movement toward freer markets in the 1990s became known as the Washington Consensus, the second decade of the 21st century is witnessing the emergence of what may be called the Innovation Consensus.

At the same time that the rest of the world is investing aggressively to advance its innovation capacity, the pillars of America's innovation system are in peril. America's public research universities are facing severe financial constraints. High budget deficits and public debt are exerting extraordinary pressure on federal and state lawmakers to cut spending on the very things that made the United States the world's innovation leader in the post-war era—and that are needed to keep the U.S. economy competitive and productive.

Policymakers are being forced to make painful choices about funding for universities, applied-research programs, help for small business, and new energy technologies. While other nations race to build state-of-the-art transportation systems and ubiquitous high-speed broadband networks, America's critical infrastructure suffers from a lack of sustained investment needed to match rising world standards. Failure to invest in these areas threatens to inflict long-term damage to America's innovation ecosystem, and therefore to its economy and security.

Formulating policy to shore up competitiveness is complicated by the fact that the United States is one of the few industrialized nations whose policymakers have traditionally not thought strategically about the composition of the nation's economy. America's international competitiveness is based on its capacity to innovate and manufacture new services and high-technology products. While innovation is often thought to result from the operation of a free market, in fact the government plays an instrumental role through its investments in R&D, as well as through policies that foster the commercialization of new ideas.

Since World War II, U.S. science and technology policy has been conducted under the assumption that federally funded basic research will be translated by the private sector into commercial products and new U.S. industries. Indeed, sometimes this transfer to the private sector does occur as expected. In many other cases, such as with nuclear power, computers, semiconductors, and aerospace, early government support and procurement has proved critical to the development of new industries. But the popular mythology that the American economy has thrived for decades under solely a *laissez-faire* tradition and linear approach to innovation policy tends to discount both the complexity of innovation and the vigorous government role in the development and deployment of new technologies. It is not just policies directly addressing the development and deployment of new technologies but also policies concerning tax, trade, intellectual property, education and training, and immigration, among others that play a role in innovation. In an age where Internet content is increasingly important to the economy, a broad range of skills is needed to secure American capabilities in innovation and competitiveness.

Whatever its source, America's preeminence no longer can be taken for granted. New players that regard innovation as a matter of strategic importance

PREFACE

are on the rise. Many governments are seeking to adapt the best features of America's innovation ecosystem, such as close collaboration between universities and business, public and private pools of risk capital, and programs that encourage researchers to start up their own companies.

Most other industrialized nations also are taking strong measures to bolster industries in which they are or wish to be competitive and to gain the benefits of jobs and growth afforded by established or emerging high-tech industries. In this highly competitive environment, the U.S. needs, once again, to devote policy attention and resources to the process of innovation because our future competitiveness as a nation is at stake. This commitment is needed if high paying jobs in sufficient numbers are to be created and if America's security is to be assured. The U.S. must understand and urgently address the underlying factors that may be weakening industries in which we might well compete.[2] The world of innovation is undergoing rapid and significant change, and America must change with it if the nation is to continue to prosper.

But what exactly should a national innovation policy look like and aim to achieve? In its essence, innovation is the alchemy of transforming ideas into new goods, services, and processes. Fortunately, the United States remains very strong in innovation as it is generally referred to—having ideas that have economic value to the inventors and in many cases other social value. Yet to create substantial value for the U.S. economy, policy must seek to achieve more than to encourage discovery and invention. America's tremendous investments in research and development cannot just be seen as a global public good. The fruits of innovation should translate into new marketable products, companies, industries, and jobs—and better living standards for Americans. There was a time when the proximity of U.S. companies' production to U.S. researchers was sufficient to give U.S. companies a big advantage that made speed less critical. Modern information and communications technologies have greatly reduced the significance of proximity, and many countries are taking actions to increase the pace of innovation.

Understanding how this process works—and how it can be advanced with public policy—is no simple task. The transformation of ideas into economic value occurs within adaptive networks of people and institutions that interact in complex, often ad-hoc ways. National "innovation ecosystems" typically include universities, private enterprises, public agencies, pools of investors, and national laboratories. Cultural norms and policy frameworks condition and shape interactions within and among these organizations. What's more, the innovation process can no longer be confined within geographic boundaries. Globalization has ushered in a swiftly evolving new paradigm of

---

[2] Chapter 6 of this report addresses America's global competitive standing and policy approach in emerging high-technology industries including advanced batteries, next-generation photovoltaics, flexible electronics, and pharmaceutical and bio-medical products.

borderless collaboration among researchers, developers, institutions, and entrepreneurs spanning the world.

Many nations and regions have developed strategies to commercialize and industrialize technological advances. These efforts demand attention from American policymakers. By investing in extensive applied technology programs, for example, Germany and Taiwan have remained successful export manufacturers in advanced industries despite relatively high labor costs. European nations such as Finland and Belgium have demonstrated the power of public-private partnerships. Through its steady investments in education and infrastructure, Singapore is seeking to raise the bar of what it takes to compete in knowledge industries. India is demonstrating how to drive economic growth and exploit its intellectual capital by becoming an integral node in international innovation networks—largely through creating the necessary human resource base and avoiding excessive regulation of this entrepreneurial activity. The sheer ambition and scale of China's investments in science, technology, and next-generation industries, as well as its less laudable interventions, seek to redraw the map of the global economy.

STATEMENT OF TASK

The global economy is characterized by increasing locational competition to attract the resources necessary to develop leading-edge technologies as drivers of regional and national growth. One means of facilitating such growth and improving national competitiveness is to improve the operation of the national innovation system. This involves national technology development and innovation programs designed to support research on new technologies, enhance the commercial return on national research, and facilitate the production of globally competitive products. The Board on Science, Technology, and Economic Policy (STEP) proposes to study selected foreign innovation programs and compare them with major U.S. programs. The analysis, carried out under the direction of an ad hoc Committee, will include a review of the goals, concept, structure, operation, funding levels, and evaluation of foreign programs similar to major U.S. programs, e.g., innovation awards, S&T parks, and consortia. This analysis will focus on key areas of future growth, such as renewable energy, among others, to generate case-specific recommendations where appropriate. The Committee will assess foreign programs using a standard template, convene a series of meetings to gather data from responsible officials and program managers, and encourage a systematic dissemination of information and analysis as a means of better understanding the transition of research into products and of improving the operation of U.S. programs.

PREFACE

The first step toward understanding the implications for public policy of these global trends is to inform ourselves about the new nature of global competition for human and financial capital—not only between and within companies but also between governments.[3] To this end, the Committee on Comparative National Innovation Policies (CIP) of the National Research Council's Board on Science, Technology, and Economic Policy (STEP) convened a series of symposia from 2006 through 2011 examining select innovation policies and programs of different nations and comparing them to those of the United States. These conferences brought together leading government officials, industrialists, academics, researchers, and economists from advanced and emerging nations. The mission was to learn about national strategies designed to meet the new competitive challenges of the 21st century global economy and to identify best practices of private and public programs to strengthen industries, advance new technologies, and meet critical national needs.[4] It is important to note that the Committee did not seek to quantify the impact of these national strategies and programs. Nor did it seek to directly compare them with each other, recognizing that these policies and programs combine different levels of resources and organizational forms to seek different sets of outcomes within the contexts of different national innovation systems.

Participants at these conferences addressed topics that included the future of the solar power and advanced battery industries, the issues and opportunities associated with the rise of China and India, successful applied-technology and commercialization programs in Europe and Asia, regional innovation cluster strategies, and the role of such early-stage finance programs as the U.S. government's Small Business Innovation Research (SBIR) program.

The National Research Council has recently conducted a number of studies of U.S. competitiveness. Of particular note are the 2007 report *Rising Above the Gathering Storm*[5] and a follow-up report published in 2010.[6] The *Gathering Storm* reports focused heavily on the inputs into America's innovation system, such as K-12 science and math instruction, the supply of scientists and engineers, and federal research funding. The report also included a series of recommendations to address these deficiencies.

---

[3] In multinational companies such as IBM, American workers often compete against Indian, Chinese, and other employees that work in their offshore R&D and manufacturing facilities.
[4] The National Academies Board on Science, Technology, and Innovation (STEP) has underway a study examining Best Practices in State and Regional Innovation Systems across the United States. The study is reviewing the practices and policies of particular regions as well as the synergies between federal, state, and regional efforts to build high tech clusters of competency and growth.
[5] National Academy of Sciences, National Academy of Engineering, and the Institute of Medicine, *Rising Above the Gathering Storm: Energizing and Employing America for a Brighter Economic future*, Washington, DC: The National Academies Press, 2007.
[6] National Academy of Sciences, National Academy of Engineering, and Institute of Medicine, *Rising Above the Gathering Storm, Revisited: Rapidly Approach Category 5,* Washington, DC: The National Academies Press, 2010.

This report—the product of a series of international conferences, review of the work of the National Academies and similar institutions, and extensive discussion within the Committee on Comparative National Innovation Policies—by contrast, focuses on the outputs of the innovation process. This volume seeks to increase the understanding of the challenges the U.S. faces in converting new ideas into new commercial products, companies, industries, and jobs. While it endorses the findings of the *Gathering Storm* reports, the emphasis is on policies and programs that can generate more economic value out of the discoveries and inventions that flow from American taxpayers' substantial investments in research.

LIMITATIONS OF THIS STUDY

A report of this nature necessarily has limits to its scope. Recognizing this early on, the Committee chose to focus on a limited set of countries and an illustrative set of industries in its review. No single report can cover the full range of issues and technologies on this complex topic.

Choice of Countries and Regions: As noted in the Statement of Task, the purpose of the study is to take a selective review of important (notably China, India, and Germany) as well as noteworthy policy initiatives (e.g., Flanders) to develop national innovation capacity and industrial competitiveness. The intent is not to present an all encompassing overview such as those produced by the OECD but to highlight major developments and national strategies and consider their implications for the United States. The selection of countries was also driven by the willingness of leading policymakers, industrialists, and academics in these countries to engage with the Committee in an in-depth dialogue on these issues.

Choice of Sectors: The Committee also could not look at all sectors in adequate depth, within the necessarily limited scope of the study. It chose to focus on advanced manufacturing because it serves to illustrate a broad set of major challenges facing the U.S. in a highly globally competitive sector. We are aware that the report does not provide an in-depth discussion of very large and important sectors such as bio-medicine, aeronautics, and services, where the U.S. continues to set the technological pace.

OVERVIEW OF THE BOOK

This volume draws together our findings from this extensive study while also drawing upon existing research concerning the global competitiveness challenge and the policies and programs that drive it. The report is in two parts. Part I describes the role of innovation in addressing the competitiveness challenge and highlights key policies and programs that leading nations and regions are undertaking to address this challenge. Part I concludes with the Committee's consensus findings and recommendations. Part II of this report provides supporting data, including in-depth case studies of policies and programs being promulgated in leading nations and regions of the world to accelerate innovation, grow new industries, and foster knowledge-based economic growth. The Overview at the front of this volume draws together the key points.

Part I: The Innovation Challenge

Chapter 1 describes the policies implemented around the world and the rapidly changing competitive landscape, reviewing the challenges they present to America's technological leadership and our ability to convert research and invention into economic value in the form of new products, companies, industries, and jobs.

Chapter 2 reviews the wide range of innovation policies adopted by other nations and regions, as well as by U.S. states, to attract, retain, and nurture the innovative industries of today and tomorrow. It identifies key trends in foreign programs and contrasts them with the erosion of existing U.S. strengths.

Chapter 3 sets out the Findings of the Committee.

Chapter 4 sets out the Recommendations, the consensus view of the Committee concerning steps the U.S. needs to take to address the challenges and opportunities in research and innovation that the United States faces in the 21st Century.

Part II: Global Innovation Policies

Chapter 5 provides case studies on several major emerging markets (China and India), successful industrializing nations and regions (Singapore and Taiwan), and more mature industrialized nations (Germany, Japan, the Flanders region of Belgium, Finland, and Canada). Despite their wide differences in terms of economic models and levels of development, the striking commonality among the strategies adopted by these nations is that they have adopted national innovation policies that often reflect the influence of U.S. practices, such as greater encouragement for universities to work with industry and incentives to spin off companies.

Chapter 6 of this report addresses America's global competitive standing and policy approach in emerging high-tech industries. Our case studies are of advanced batteries, next-generation photovoltaic cells, semiconductor manufacturing, and pharmaceutical and bio-medical products. In each of these sectors, the U.S. has been at or near the forefront in terms of innovation and/or the creation of promising start-ups. Translating this advantage into globally competitive industries that create high-paying jobs and drive economic growth, however, is a challenge that the United States must effectively address. The case of semiconductors illustrates that U.S. policy can play a role in restoring and preserving the competitiveness of a critical innovation-intensive industry. The studies of the advanced-batteries and photovoltaic products assess policy strategies and options for bolstering U.S. competitiveness in these promising industries.

Chapter 7 addresses the policy instruments adopted by countries and regions around the world and across the U.S. to rise to the challenges of building innovation-led economies. One method is through the research parks with universities or national laboratories at their nucleus. The chapter explains how new research parks in the U.S. and abroad are adapting to the demands and opportunities of the 21st century global economy. The second part of this chapter analyzes regional innovation cluster initiatives around the U.S. It also explains the evolving role of the federal government in advancing regional innovation clusters. Case studies include bold and innovative initiatives in upstate New York, southeast Michigan, northern Ohio, South Carolina, West Virginia, and New Mexico.

Caveat: A few words are in order on the nature of this report. Our purpose in looking at other countries' innovation systems was to draw some useful lessons for the shaping of U.S. policy. Our intended audience is Congress, Executive Branch agencies, and all those interested in shaping U.S. policies that affect innovation.

Each country examined is markedly different from the United States—for example, Germany is the about the size of one and a half California's, China and India are at very different stages of development—but each offers insights into the thinking of policymakers as to what they think will be most effective to spur innovation. It is through observation of other's policies in this globalized world that the Committee members have informed their views as to what adjustments should be considered in U.S. policies.

The challenges and opportunities being created by the worldwide drive for innovation have never been greater in terms of jobs, income distribution, and ultimately competitive strength and the health of the U.S. economy. There is no single program or legislative enactment that will assure complete success; indeed, there is no panacea. But we are able to identify a series of steps necessary to improving the country's outlook in these regards. It has been said that the right thing to do is often hard but seldom surprising.[7] America has great competitive strengths. It is our conviction that if the steps outlined in this report were adopted, our country's future would indeed be brighter.

The responsibility lies fully with the Committee for the recommendations contained in this report.

ACKNOWLEDGEMENTS

This Report would not have been possible without the collaboration with numerous scientific academies, scholars, technology company executives,

---

[7] Adam Gopnik in the *New Yorker*, April 9, 2012, on Albert Camus, in an essay entitled "Facing History" about in part editorials that Camus wrote for *Combat* a resistance newsletter.

and public officials at home and abroad. It also depended heavily on the volunteer efforts of CIP Committee members over an extended period of time. In particular, special appreciation is due to Bill Spencer, the chair of the Committee during the initial conferences. Special recognition is due to Pete Engardio, formerly of *Businessweek*, for his drafting and reportorial skills. His ability to synthesize a vast amount of material was essential for a report of this scope. William Noellert deserves our special thanks for his review of data and economic analysis, as does Thomas Howell for his many substantive contributions. Both were important to the scope and quality of this report. This project would not have occurred nor have been brought to its final report stage without the leadership, knowledge of national and international programs, and organizational skills of Dr. Charles Wessner. The commitment and support of his team at the National Academies, including in particular Dr. Sujai Shivakumar, has been central to the production of this report, as have the efforts of David Dierkshiede, McAlister Clabaugh, and David Dawson for the many international conferences that characterized this effort. This report would also not have been possible without its initiation by the STEP Board and the encouragement given to it by the Board and the National Academies, as well as the prior and ongoing work of the Academies on which this report builds.

Sponsors

The National Academies Board on Science, Technology, and Economic Policy would like to express its appreciation for the sustained support of the following agencies and departments: Office of Naval Research, National Institute of Standards and Technology, National Cancer Institute, Department of Energy, Defense Advanced Research Projects Agency, Sandia National Laboratories, and National Science Foundation Their contributions of time, expertise, and financial support were essential to the success of the project.

We would also like to express our appreciation for the active participation and contributions of the following companies and organizations, who share a desire to better understand the changing competitive landscape of the 21st Century: Intel Corporation; International Business Machines; Cisco Systems; Volkswagen Group of America; Palo Alto Research Center; M Square, the University of Maryland Research Park; and Association of University Research Parks.

Acknowledgement of Reviewers

This report has been reviewed in draft form by individuals chosen for their diverse perspectives and technical expertise, in accordance with procedures approved by the National Academies' Report Review Committee. The purpose of this independent review is to provide candid and critical comments that will assist the institution in making its published report as sound as possible and to ensure that the report meets institutional standards for objectivity, evidence, and responsiveness to the study charge. The review comments and draft manuscript remain confidential to protect the process.

We wish to thank the following individuals for their review of this report: William Bonvillian, Massachusetts Institute of Technology; James Burns, Sanofi/Genzyme; Erica Fuchs, Carnegie Mellon University; Howard Gobstein, Association of Public and Land-grant Universities; Manuel Heitor, Technical University of Lisbon, Portugal; Ron Hira, Rochester Institute of Technology; Krisztina Holly, University of Southern California; Mark Kryder, Carnegie Mellon University; Richard Lester, Massachusetts Institute of Technology; Raghunath Mashelkar, National Innovation Council of India; Richard Nelson, Columbia University; Charles Phelps, University of Rochester; Clyde Prestowitz, Jr., Economic Strategy Institute; Mu Rongping, Chinese Academy of Sciences; Denis Simon, Arizona State University; James Stevens, Dow Chemical Company; James Turner, Association of Public and Land-grant Universities; and Richard Van Atta, Institute for Defense Analysis.

Although the reviewers listed above have provided many constructive comments and suggestions, they were not asked to endorse the conclusions or recommendations, nor did they see the final draft of the report before its release. The review of this report was overseen by Christopher Hill, George Mason University, and Granger Morgan, Carnegie Mellon University. Appointed by the National Academies, they were responsible for making certain that an independent examination of this report was carried out in accordance with institutional procedures and that all review comments were carefully considered. Responsibility for the final content of this report rests entirely with the authoring committee and the institution.

FUTURE WORK PROGRAM

The international competition in innovation is increasing. Globalization has accelerated the pace of change. There is much to be learned from and about foreign measures and policies that will shape the U.S. economy, the nation's security and the well-being of the U.S. workforce. Best practices should be considered for adoption. Measures of foreign governments and entities that distort international competition must be examined and responses crafted. There is much to be gained from international cooperation with respect to global challenges in energy, climate, and health, among others. It is the strongest recommendation of the Committee that that an ongoing work program to address these needs and opportunities be put into place.

To this end, the National Academies Board on Science, Technology, and Economic Policy will establish a new Innovation Policy Forum. The purpose of this forum is to act as a focal point for national and international dialogue on innovation policy. The Forum will bring together representatives from government, industry, national laboratories, research institutes, and universities—foreign and domestic—to exchange views on current challenges and opportunities for U.S. innovation policy and to learn about the goals, instruments, funding levels, and results of national and regional programs and discuss their lessons for U.S. policy and potential impact on the composition of the economy.

                              Alan Wm. Wolff
                              Chair, Committee on
                              Comparative National Innovation Policies

Overview

America's position as the source of much of the world's global innovation has been the foundation of its economic vitality and military power in the post-War era. No longer is U.S. pre-eminence assured as a place to turn laboratory discoveries into new commercial products, companies, industries, and high-paying jobs. As the pillars of the U.S. innovation system erode through wavering financial and policy support, the rest of the world is racing to improve its capacity to generate new technologies and products, attract and grow existing industries, and build positions in the high technology industries of tomorrow.

Sustaining global leadership in the commercialization of innovation is vital to America's security, its role as a world power, and the welfare of its people. Even in a climate of severe budgetary constraint, the United States cannot afford to neglect investing in its future. These are investments, moreover, that will pay for themselves many times over.

The second decade of the 21st century is witnessing the rise of a global competition that is based on innovative advantage. To this end, both advanced as well as emerging nations are developing and pursuing policies and programs that are in many cases less constrained by ideological limitations on the role of government and the concept of free market economics. Not only have these nations placed massive bets on research and higher education, they have also unveiled comprehensive national strategies to build innovation-led economies. Governments everywhere are adopting, adapting, and in some cases improving aspects of America's innovation ecosystem that have long been the envy of the world, such as close collaboration between universities and business, deep pools of risk capital, and effective programs that encourage researchers to start up their own companies. Going beyond, some countries are pursuing a highly interventionist and essentially mercantilist set of innovation policies and programs.

The rapid transformation of the global innovation landscape presents tremendous challenges as well as important opportunities for the United States. Emerging powers such as China and India have critical masses of highly educated scientists and engineers, rising R&D spending, and large, rapidly growing domestic markets for high-tech products. Innovation hubs such as

Silicon Valley, greater Boston, San Diego and Austin that have been magnets for the world's brightest and most visionary innovators, technology entrepreneurs, and investors face greater competition from dynamic new commercialization zones, such as Taipei, Shanghai, Helsinki, Tel Aviv, and Bangalore.

The world of innovation itself is undergoing radical change, calling into question America's ability to benefit fully from U.S. science and technology leadership. In today's world, knowledge, money, and people flow across borders with ever-greater speed and ease, often through open collaborative innovation networks linking corporations, researchers, investors and institutions. The good news is that this opens genuine opportunities for international collaboration that can help solve global health, environment, and energy challenges, as well as enable companies to accelerate product development.

But the globalization of innovation capacity is also undermining traditional assumptions that have guided U.S. policymaking for the past six decades. In particular, it no longer follows that discoveries and inventions flowing from research conducted by America's universities, corporations and national laboratories will naturally lead to products that are commercialized and industrialized on U.S. shores. Although the U.S. federal government remains the biggest sponsor of basic research, spending some $148 billion on public R&D in 2011, traditional trading partners and emerging economies are concentrating their energies on translating new technologies from every available source into industrial applications and job-generating industries. In some cases, nations are using the resources of the state to induce U.S. companies to manufacture their innovations locally and transfer proprietary technologies while giving homegrown champions privileged access to their domestic markets. In other cases, companies produce offshore because they conclude the United States simply lacks the supply chain capacity, technical skills, and the right investment climate for high-volume manufacturing. As a result, the U.S. is finding it increasingly difficult to capture the economic value generated by its tremendous public and private investments in R&D.

The United States urgently needs to adjust to the new great game [or challenge] of 21st century global competition. Just as the 2007 National Academies report *Rising Above the Gathering Storm* was a call to arms that urged the U.S. to increase investment in R&D, education, and other inputs into the innovation system, this report argues that far more vigorous attention be paid to capturing the outputs of innovation -- the commercial products, the industries, and particularly high-quality jobs to restore full employment. America's economic and national security future depends on our succeeding in this endeavor.

THE NEW INNOVATION LANDSCAPE

The search for a new U.S. innovation policy should begin with an understanding of America's changing competitive position as compared with the rest of the world. Over the past several years, the Board on Science, Technology, and Economic Policy of the National Academies has engaged in an extensive dialogue on science, technology and innovation policy with countries that place a high priority on innovation. America's competitive challenge comes into clearer focus when the strong measures taken by other nations to improve their innovation capacity are contrasted directly with the flagging U.S. commitment in many of the same areas. For example:

Support for the Pillars of Innovation:

- **R&D Investment:** The U.S. is losing its once-overwhelming advantage in research. The U.S. share of global R&D spending dropped from 39 percent in 1999 to 34.4 percent in 2010. This is still very substantial, but trends suggest the U.S. share will continue to shrink. While American R&D spending has risen 3.2 percent a year on average for the past decade, for example, growth in South Korea has averaged 8 percent annually and China has averaged 20 percent. Brazil nearly tripled R&D spending between 2000 and 2008, and Singapore plans to triple spending between 2010 and 2015. U.S. federal spending on basic research as a percentage of GDP, which is critical to future technological progress, has virtually stagnated for the past 20 years and risks actual decline in the face of current fiscal pressures.
- **University Funding**: Research universities—the engines of the U.S. innovation system—are suffering severe cutbacks across the U.S. due to state budgetary constraints. Other nations and regions are dramatically increasing funding to upgrade, expand, and open new research universities. China is spending billions to make 39 universities world leaders. India's five-year plan calls for 1,500 new universities and a number of new elite technology institutes. And Taiwan plans to invest $1.7 billion to develop world-class universities.
- **Early-Stage Finance:** Funding from angel investors and venture capitalists, another pillar of America's innovation ecosystem, has fallen sharply since 2000 (albeit a peak year), and venture capital investors have grown steadily more risk-averse, putting less funding in the early-stage investments. But successful U.S. programs, such as the Small Business Innovation Research (SBIR) program, that are important sources of early-stage funding have struggled for reauthorizations. Others, such as NIST's Advanced Technology Program, now the Technology Innovation Program (TIP), have struggled for renewed funding. Meanwhile, other nations have launched large funds to support

start-ups. Japan, Brazil, the United Kingdom, Sweden, India, the Netherlands, Germany, and other nations have adopted programs that often are modeled directly on SBIR or other U.S. policies and address the early-stage funding challenge in the innovation chain.
- **Talent**: Singapore, Canada, and China are among the nations that are attracting star scientists from around the world to their universities and research institutes by offering high salaries and opportunities to run well-funded programs. In the U.S., foreign-born U.S. science and technology graduates and entrepreneurs often face great difficulty obtaining U.S. residency visas and citizenship. Others are investing more in their existing workforce. Germany, for example, is a pathfinder in high-skilled worker training and retention, including dealing with the both the challenge and opportunities presented by an aging population. By contrast, the U.S. lacks any systematic worker-retraining program in an age of drastic technological change.

Efforts to Capture Economic Value:

- **Manufacturing.** U.S. is losing competitiveness as a location for new investment in advanced manufacturing capacity, even in industries where the U.S. is at the technological forefront, driven in part by national policies. This continued erosion of America's high-tech manufacturing base threatens to undermine U.S. leadership in next-generation technologies. Major U.S. trading partners understand that a domestic industrial base that can produce advanced products in high volumes is integral to maintaining global competitiveness in innovation and next-generation technologies. Nations and regions as diverse as Germany, Japan, Taiwan, and South Korea are showing it is possible to remain successful exporters in advanced manufacturing despite relatively high labor costs. The U.S. high-tech manufacturing base, by contrast, has deteriorated to the point that it is sometimes difficult to manufacture in high volumes the products that are invented in the United States —even when labor costs are not a major factor. While many other nations support high-volume manufacturing with tax holidays, grants and credit, U.S. federal incentive programs have short time horizons, limited scope, and uncertain future funding prospects.
- **Translational and Applied Research**: In a time of intense technological change, large, well-funded public-private partnerships such as Germany's Fraunhofer-Gesellschaft, Korea's Electronics and Telecommunications Research Institute, Taiwan's ITRI, and Finland's Tekes have proven remarkably successful at helping domestic manufacturers translate new technologies into products and production processes. Although the U.S. has many applied-research programs, we lack a systematic institutional focus on developing manufacturing

industries at scale for new technology products and or to reinforcing and stimulating the growth of broad industrial clusters.
- **Cluster Development**: Governments around the world are investing aggressively in comprehensive strategies to foster regional innovation clusters. Prominent government-supported successes include the semiconductor, digital display, and notebook PC clusters in Taiwan; telecommunications in Finland; biomedical research in Singapore; micro-electronics in Grenoble, France; and life sciences and information technology in Shanghai's Pudong district. Many promising innovation-cluster initiatives have been launched by U.S. state and local governments, including nano-electronics in upstate New York, advanced batteries in Michigan, flexible electronics in northern Ohio, and biometrics in West Virginia. Unlike in other nations, however, many of these initiatives receive little federal policy or financial support—and new federal initiatives are often small.

Efforts to Enhance National Advantage:

- **Framework Conditions**: The United States still offers one of the world's best environments for commercializing products and launching companies, including strong protection of intellectual property rights, temperate bankruptcy laws, well-developed capital markets, and extensive worker mobility. But the U.S. has not stayed abreast of other nations in areas as diverse as tax policy, regulatory costs, and state-of-the-art infrastructure.
- **Rising Neo-Mercantilism:** Countries such as China and South Korea employ a powerful combination of state subsidies, national standards, preferential government procurement for national firms, and requirements for technology transfer to drive the growth of nationally-based innovation. They also encourage state- owned or –supported enterprises to compete globally in strategic emerging industries with the help of low-cost loans—often with little concern for near-term return on investment or overcapacity. In the United States, trade and investment policy is predicated on the faith that open markets foster innovation. What's more, U.S. trade policy is ill-equipped to avert the serious damage neo-mercantilism inflicts on U.S. industries until it is too late, such as when heavily subsidized competition of a given product forces American manufacturers to shut domestic production. Often, U.S. companies hesitate to seek redress from the federal government because they fear damaging their access to foreign markets. By depriving U.S. companies of the ability to reap the commercial rewards of their significant investments in innovation both at home and abroad, neo-mercantilism poses serious long-term consequences for the U.S. economy and defense capabilities.

RISING TO THE CHALLENGE

In this dramatically more competitive world, the United States cannot return to a path of sustainably strong growth, much less maintain global leadership, by living off past investments in its capacity for innovation. By failing to make the immediate as well as long-term investments needed to ensure that the U.S. remains a dominant location for producing technology-intensive goods and services, we are sacrificing jobs, economic growth, living standards, and national security. Nor can the U.S. compete on the basis of a policy approach that is the legacy of an era when American advantages were overwhelming and innovative activity tended to remain within our borders.

Since publication of *The Gathering Storm*, Congress and the White House have taken a number of measures to shore up U.S. competitiveness in science, technology, and economic policy, though many have lacked adequate follow-through. The reauthorization of the America COMPETES Act, signed into law Jan. 6, 2011, called for sharp increases in the research budgets of federal agencies and federal funding for K-12 science, technology, and mathematics education. However, Congress has not followed up this call with funding and the Obama Administration has proposed flat science budgets below the levels proposed in the legislation. The original America Competes Act also established the Advanced Research Projects Agency-Energy (ARPA-E), which received funding only following the passage of the American Renewal and Reinvestment Act of 2009. In addition to funding ARPA-E, this Stimulus Bill eased immigration rules for skilled talent, and extended billions of dollars in grants and loans to renewable-energy, electricity-transmission, and advanced-battery manufacturing projects, but this was a one-time event. The Obama Administration has unveiled a national innovation strategy that calls for increasing U.S. investments in R&D, higher education, and information-technology and transportation infrastructure along with many other more-targeted innovation programs, such as the National Manufacturing Initiative.

As encouraging as these actions are, they are not enough. Many of the major proposals aimed at boosting U.S. competitiveness and reaping more of the economic value from U.S. innovation have not been enacted into law. Most of the new pro-innovation programs have short time horizons and may well lack sustainable long-term funding. Federal programs also lack the scale and comprehensive approach needed to enable America to rise to meet the acute competitive challenges posed by the rapidly evolving global innovation landscape. We therefore recommend the following strategy to start putting the United States on a clear path to meeting these challenges:

In a dramatically more competitive world, the United States needs to reinforce the traditional pillars of its economic strength and innovation capacity. (Recommendation 2.)

- **Boost R&D investment**: The U.S. should fund R&D at the higher levels authorized under the America COMPETES Act and sustain these levels in the future as part of a plan to boost private and public R&D expenditure to a level of 3% of GDP by 2020. (Recommendation 2a.)
- **Sustain University Research**: Funding for university research should be stabilized at the state and federal level and then increased. Our capacity to train students in science, engineering and mathematics, and in the broad range of future demands for talent, is dependent on well-funded universities and colleges. Funding options should include targeted business tax incentives from dedicated sources of tax revenue as well as incentives for private donations. The government also should reform regulations that make it increasingly expensive for universities to conduct research. (Recommendation 2b.)
- **Help Small Business:** Innovative small businesses are a major source of new job creation. However, many small firms and struggle to raise the funds needed to develop promising new technologies because their commercial potential is often too uncertain to attract needed private venture capital. Proven programs such as SBIR and ATP (or its successor, the Technology Innovation Program), which provide small competitively based innovation awards to small firms or consortia, should be sustained, expanded, and adequately funded. Government agencies should also be encouraged to experiment with and evaluate new initiatives, including prizes for technological advance. The U.S. government should explore offering policy support for angel funds and venture capital. (Recommendation 2c.)
- **Train Workers:** The federal government should expand support for successful state and regional workforce-development programs for advanced industries. It also could provide companies with vouchers to cover training costs for new employees. Programs in community colleges that provide such training need to be reinforced. To encourage experienced talent to remain in the workforce longer, the U.S. should remove tax disincentives for staying employed past age 65. (Recommendation 2d.)
- **Support higher education.** Federal and State governments should make sure that education in all fields, and particularly science, technology, engineering and math, are made affordable and available to all eligible applicants. The land grant colleges were the backbone of the talent infrastructure for the building of America, and the Federal role should not be abandoned now. (Recommendation 2b-i.)
- **Attract Foreign Talent:** Immigration laws should be reformed to attract foreign scientists, engineers, and entrepreneurs to live and work in the U.S. and facilitate their permanent residency and U.S. citizenship. (Recommendation 2d-v)

The United States needs to adopt specific policy measures to capture greater economic value from its public investments in research. (Recommendation 5.)

The America COMPETES Act provides for crucial inputs into the U.S. innovation system. But a similarly comprehensive effort needs to be made to exploit the results of these investments in science, technology, and education into more innovative products and well-paying jobs.

- **Support Advanced Manufacturing:** A 2004 report of the President's Council of Advisors on Science and Technology warned that "with manufacturing leaving the country, the United States runs the risk of losing the strength of its innovation infrastructure of design, research and development and the creation of new products and industries." Many U.S. companies with important technologies cannot develop the full infrastructure and make the high-risk, long-term investments required to support job-creating advanced manufacturing at home. To help stem this erosion of the nation's manufacturing base, current manufacturing tax credits and loan-guarantee programs should be made permanent and expanded in scope. Manufacturing technical assistance and other programs aimed at accelerating commercialization of new technologies should be expanded. In particular, the recent proposal to set up a network of Manufacturing Innovation Institutes should be fully funded. (Recommendation 5d.)
- **Leverage government procurement:** Federal agencies can use their purchasing power to help drive domestic commercialization of emerging technologies. The U.S. government has done this many times previously in industries such as semiconductors, computers, and aerospace. Federal and state agencies can help build domestic markets for important new technologies for electric-drive vehicles, energy-efficient buildings, solid-state lighting, and next-generation photovoltaic cells. Procurement rules of Federal agencies and armed forces should be reformed to put more emphasis on providing incentives for spurring innovation in products and processes that result in continuous performance improvements and lower long-term life-cycle costs (vs. up-front costs). Government agencies also should accelerate innovation by providing early-stage financial support for small companies that can address national needs. (Recommendation 5j.)
- **Foster Clusters:** Recent pilot programs by federal agencies to align current economic development programs with specific regional innovation cluster initiatives by state and local organizations should be assessed and, where appropriate, expanded geographically. The U.S. also needs to assess and draw policy lessons from successful cluster efforts and communicate best practices to those managing regional initiatives.

The Federal government should award competitive grants to support state and regional efforts to develop and sustain modern science parks and also technology development implementation centers that are focused on manufacturing. (Recommendation 5i)
- **Strengthen University Links to the Market:** University seed funds and incubators can help start-ups spun off from research projects. Early-stage funding programs should be expanded to support commercialization of university research. New centers of excellence should be established to foster university-industry-government collaboration on commercial and industrial applications of emerging technologies. (Recommendation 5a.)
- **Promote Public-Private Partnerships:** The U.S. needs to expand successful partnership programs and consider adopting and adapting successful models from abroad, such as Taiwan's ITRI and Germany's Fraunhofer Institutes. The U.S. also should assist in establishing new public-private research and development consortia aimed at fostering the implementation and production in the U.S. of emerging technologies in sectors such as flexible electronics, solid-state lighting, and medical devices. (Recommendation 5c.)

Provide a Competitive Corporate Environment: The United States should assure that the tax framework supports new company creation and investment. In order to be competitive with those of its major trading partners, the U.S. should take measures to address policies that actually disadvantage U.S.-based industry. (Recommendation 3)

Governments at the Federal and state levels should regularly benchmark tax policies and regulatory costs against those of other nations. Where they are found to be serious impediments to corporate investment and innovation, every effort should be made to close gaps or seek ways to reduce the negative impact through compensating incentives. The U.S. should consider reducing corporate taxes and rely increasingly on consumption taxes. Efforts should be made to ensure that changes in taxation and government spending to shrink the federal deficit are made with a full understanding of the potential consequences for future growth. The U.S. should also make current tax credits for research and experimentation permanent, and incentivize commercial credit to innovative manufacturing, particularly the scale-up of an initial production process.

Build a 21st Century Innovation Infrastructure: The U.S. should increase dramatically investment in state-of-the-art broadband networks and other infrastructure required to maintain American leadership in a 21st century global knowledge economy. (Recommendation 4.)

The U.S. should consider the feasibility of a National Infrastructure Bank that can leverage more private investment in highways and railways, renewable-energy systems, water and sewerage and other public works that both meet critical national needs and deploy emerging technologies. The Federal government should increase R&D investments in new materials and sensors for highways, ports, and bridges, as well as technologies to improve energy efficiency in buildings. Incentives to encourage expansion of the high-speed Internet backbone should be strengthened to sharply increase broadband penetration in homes, schools, and businesses.

Capitalize on Globalization of Innovation: The United States should capitalize on the globalization of research and innovation to cooperate with other nations to advance innovations that address shared global challenges in energy, environment, health, and security.(Recommendation 7.)

Just as other nations establish R&D institutions in the U.S. and actively seek to acquire American technology, the United States should recognize the many opportunities presented by the rapid growth in research and innovation activity abroad.

- **Research Collaboration**: The U.S. needs to strengthen and expand research collaborations with growing economies such as China, India, and Brazil; new European Union members such as Poland, the Czech Republic, and Hungary; and historical partners like Sweden, Germany, and Japan to advance research that can lead to innovations in biomedicine, energy, environment, security, and other shared global challenges. To stay abreast of important technological developments abroad, the United States should expand exchanges of researchers, scholars, and students, and support these objectives. (Recommendation 7a.)
- **Network and Engage Globally.** We now operate in global systems of innovation and new knowledge creation. Leading scientists at American universities work in collaborative teams and cohorts that are multinational and dispersed across the globe joined together by strong information technology networks. We need to better leverage these networks and capture value from them. (Recommendation 7b.)

Monitor and Evaluate Investments, Measures, and Innovation Policies of other Nations: In a world where other nations are investing very substantial resources to create, attract and retain the industries of today and tomorrow, the United States needs to increase its understanding of the swiftly evolving global innovation environment and learn from the policy successes and failures of other nations (Recommendation 1.)

The United States needs to understand the swiftly evolving global innovation environment and the implications for America's competitive position and national security. The government should, as a priority, gather current information and assess current implications for the U.S. economy of foreign programs, and at the same time maintain and support regular, on-going efforts to engage with policymakers, business leaders, and academics from around the world. These steps will enable the benchmarking of U.S. policies, programs and measures in light of those of other countries. The U.S. needs to be able to draw upon international best practices aimed at advancing innovation in order to inform its own policies and programs and understand the potential impact of these programs on U.S. industries.

Recognize that Trade and Innovation are Closely Linked:
(Recommendation 6.) It is the responsibility of the U.S. government to provide a rules-based global playing field for its industries. Foreign trade- and investment-distorting measures should be rooted out or offset, especially when U.S. innovation will be stifled. This will require support

---

Box O-1
Four Core Goals

1. **Monitor and learn from what the rest of the world is doing:** The United States needs to increase its understanding of the swiftly evolving global innovation environment and learn from the policy successes and failures of other nations. It is generally recognized that there is much to be learned from the rest of the world in science. This is equally true with regard to innovation policy. See Recommendation 1.
2. **Reinforce U.S. innovation leadership:** It is very important that the United States reinforce the policies, programs, and institutions that provide the foundations for our own knowledge-based growth and high value employment. These include measures to strengthen our research universities and national laboratories, renew our infrastructure, and revive our manufacturing base. See Recommendations 2, 3, and 4.
3. **Capture greater value from its public investments in research:** The United States should improve its ability to capture greater value from its public investments in research. This includes reinforcing cooperative efforts between the private and public sectors that can be grouped under the rubric of public-private partnerships, as well as expanding support for manufacturing. See Recommendations 5 and 6.
4. **Cooperate more actively with other nations:** In an era of rapid growth in new knowledge that is being generated around the world, the United States should cooperate more actively with other nations to advance innovations that address shared global challenges in energy, health, the environment, and security. See Recommendation 7.

from U.S. industry, but ultimately be founded on an independent and well-informed judgment on the part of the U.S. government as to the policy responses that are in the national interest. The United States government should begin to focus attention on the composition of its economy and the extent to which it is being shaped by foreign industrial and trade policies.

Based on intelligence gathered as recommended in this report, and without waiting for the filing by private parties of trade cases, the U.S. government should determine whether the national interest requires that solutions need to be put into place. It needs to vigorously pursue changes in policies of other governments that are harmful to the U.S. industrial base and innovation process and, where policies cannot be changed, offset them with trade measures or financial support for affected domestic industries as necessary.

In addition, every new U.S. international trade or investment agreement should include a comprehensive code of conduct governing the commercial activities of state-owned enterprises, holding their governments accountable for behavior that undermines fair competition and deprives other nations of the economic benefits of their investments in innovation.

CONCLUSION

The U.S. innovation system still enjoys many advantages: the world's largest research infrastructure, a number of the world's greatest universities, the deepest capital markets, and a highly dynamic ecosystem for knowing how to turn inventions into products and businesses. But in a world where other countries are rapidly developing their own innovation capacities, these advantages alone will not guarantee America's future competitive advantage.

Other governments are assertively shaping policies and programs to change the competitive landscape in their favor. U.S. policies and programs are based on a historical position of national leadership and endowment following World War II that has long since been replaced by a broad equilibration of technical and economic capabilities and fundamental changes in the ways in which technologies are developed and implemented. The U.S., while retaining vestiges of its leadership position, should recognize that merely maintaining the current policies and programs will lead to continued erosion of our economic capabilities, especially in high technology industries that are the basis for future prosperity.

The U.S. has every opportunity to secure its economic leadership and national security well into the future. But it will require a fresh policy approach, one that ensures that the United States can compete, cooperate, and prosper in this new world of competitive innovation. The recommendations of this report strongly urge a reformulation of U.S. innovation policies to address this changing competitive environment.

PART I

THE INNOVATION CHALLENGE

Chapter 1

The Innovation Challenge

America has faced many kinds of global competiveness challenges in the post-War era. They ranged from Sputnik-era fears of being technologically eclipsed by the Soviet Bloc to waves of imports from Germany, Japan, and East Asian Tigers that shook one industry after another. Through these challenges, one factor changed little: Thanks to its robust innovation ecosystem and high levels of investments in research, America maintained its leadership in innovation as its entrepreneurs launched new products, companies, and industries, and created high-paying jobs.

While America's innovation system has enabled the nation to weather previous competitive challenges, the nature of global competition has changed in fundamental ways. A number of economies have matured and grown their own innovation systems over the last 15-20 years; many of them actively pursue national policies aimed at rapidly capturing strategic industries and the high-value employment they bring. This means that in today's world, the dynamic of moving to newly created industries to sustain our prosperity is less and less sustainable as a strategic option. Efforts need to be made to retain, grow, and reinforce the industries we have as well as those we wish to develop.[1]

Innovation remains the wellspring of America's economic growth.[2] The challenge for the nation in the new global environment is to

---

[1] For a detailed review of structural changes in the innovation process in 10 service as well as manufacturing industries, see National Research Council, *Innovation in Global Industries: U.S. Firms Competing in a New World*, Jeffrey T. Macher and David C. Mowery, Editors, Washington, DC: The National Academies Press, 2008. While many industries and some firms in nearly all industries retain leading-edge capacity in the United States, the book concludes that this is "no reason for complacency about the future outlook. Innovation deserves more emphasis in firm performance measures and more sustained support in public policy."

[2] Leading economists, including Robert Solow, Trevor Swan, Edwin Mansfield, Zvi Grillichs, and Paul Romer have calculated that technological innovations have made powerful and very substantial contributions to U.S. economic growth. See, for example, Robert M. Solow, "A Contribution to the Theory of Economic Growth," *Quarterly Journal of Economics,* 1956, 70(1):65-94. In a latter article

17

continue to benefit from this innovation while also encouraging regional development and much higher levels of employment.[3]

AMERICA'S INNOVATION CHALLENGES

America is a world leader in innovation capacity, according to several rankings.[4] While not as pre-eminent as in the decades following World War II, the U.S. still leads the world in research spending and patents. U.S. universities and research laboratories continue to produce technological breakthroughs and spin off dynamic start-ups. U.S. companies still create products and business models that transform entire industries.[5] Concern is mounting, however, that America is not capturing enough of value of that innovation in terms of economic growth and employment.[6]

Solow estimated that technological progress accounted for seven-eighths of the increase in real GNP per man-hour from 1909 to 1949 in the United States. "It is possible to argue that about one-eighth of the total increase is traceable to increased capital per man hour, and the remaining seven-eighths to technical change." Robert M. Solow, "Technical Change and the Aggregate Production Function," *The Review of Economics and Statistics*, 1957, 39 (3): 312-320. Often, as Richard Nelson and others point out, this technological progress has been based on a framework of supporting national policies. See Richard Nelson, *Technology, Institutions and Economic Growth*, Cambridge MA: Harvard University Press, 2005. In addition, Harvard's Dale Jorgenson documented that the pervasive use of information technologies, developed through the nation's investments in semiconductor research and early procurement, have actually pushed upwards the nation's long term growth trajectory. See Dale W. Jorgenson et al., *Productivity: Information Technology and the American Growth Resurgence*, Cambridge MA: MIT Press, 2005.

[3] The Honolulu Declaration of the November 2011 APEC meeting affirmed the importance of promoting effective, non-discriminatory, and market-driven innovation policy. The agreement text notes that "Encouraging innovation – the process by which individuals and businesses generate and commercialize new ideas – is critical to the current and future prosperity of APEC economies. Our collective economic growth and competitiveness depend on all our peoples' and economies' capacity to innovate. Open and non-discriminatory trade and investment policies that foster competition, promote access to technology, and encourage the creation of innovations and capacity to innovate necessary for growth are critical aspects of any successful innovation strategy."

[4] The World Economic Forum ranks the United States as fifth in innovation capacity. See Center for Global Competitiveness and Performance, "The Global Competitiveness Report: 2011-2012," World Economic Forum (http://www3.weforum.org/docs/WEF_GCR_Report_2011-12.pdf). Insead's latest global innovation index ranks the United States seventh, down from number one in 2009. Insead, "The Global Innovation Index 2011," (http://www.globalinnovationindex.org/gii/GII%20COMPLETE_PRINTWEB.pdf).

[5] While the U.S. still leads the world in R&D spending, the growth of Chinese R&D spending has shifted the share of global R&D spending over the past ten years with China overtaking Japan in 2010. The U.S. accounted for 32.8 percent of global R&D spending in 2010, compared to 24.8 percent for Europe, 12.0 percent for China and 11.8 percent for Japan. Battelle and R&D Magazine, *2012 Global R&D Funding Forecast*, December 2011.

[6] See Tyler Cowen, *The Great Stagnation: How America Ate All The Low-Hanging Fruit of Modern History, Got Sick, and Will (Eventually) Feel Better.* New York: Dutton, 2011. Cowen argues that on the margin, innovation no longer produces as much additional GDP growth as it used to. In part, this may be an issue of not adequately measuring the contributions of modern information and communications in the national accounts. See National Research Council, *Enhancing Productivity Growth in the Information Age*, D. Jorgenson and C. Wessner, eds., Washington, DC: The National

Capturing the Economic Value of Innovation

This concern is based on the fact that what is innovated in America is increasingly industrialized elsewhere. Even in industries where labor cost is not a deciding factor, the high-paying production and engineering jobs that go with large-scale manufacturing often end up offshore.[7] Increasingly, experts believe that this off-shoring of manufacturing is contributing to the decline in the innovative capacity of the United States.[8] Gary Pisano and Willy Shih have argued, for example, that the "ability to develop very complex, sophisticated manufacturing processes is as much about innovation as dreaming up ideas."[9] And as more and more production moved offshore, other industries in the host countries increasingly benefit from the knowledge, networks and capabilities that are also relocated.

The result has been a loss of opportunity to lead in major emerging industries. The key technologies for rechargeable lithium-ion batteries and liquid-crystal displays were developed in the U.S., for example, yet were commercialized in Japan and now are almost entirely produced in Asia.[10] Other materials and product technologies where the United States was the innovator, but then lost significant market share include oxide ceramics; semiconductor memory devices; semiconductor manufacturing equipment such as steppers; flat panel displays; robotics; solar cells; and advanced lighting.[11]

Academies Press, 2007. Jorgenson, Ho, and Stiroh have documented the step-up in total factor productivity introduced by these semiconductor-based technologies. See Dale W. Jorgenson, Mun S. Ho, and Kevin J. Stiroh, *Productivity, Volume 3, Information Technology and the American Growth Resurgence*, Cambridge MA: MIT Press, 2005.

[7] See Gary P. Pisano and Willy C. Shih, "Restoring American Competitiveness," *Harvard Business Review*, July-August 2009. For an analysis of why the U.S. is losing new high-tech manufacturing industries, also see Pete Engardio, "Can the Future be Made in America?" *BusinessWeek*, Sept. 21, 2009.

[8] See for example, Roger Thompson, Why Manufacturing Matters, Harvard Business School, March 28, 2011. Access at http://hbswk.hbs.edu/item/6664.html. See also Stephen Ezell and Robert D. Atkinson," The Case for a National Manufacturing Strategy." Washington, DC: ITIF, April 2011. To some extent, the off-shoring of manufacturing may be reversing. A recent survey of manufacturing executives found that 85% of them identified low-volume, high-precision, high-mix operations, automated manufacturing and engineered products requiring technology improvements or innovation as the primary forms of manufacturing returning to the U.S. The survey was conducted by Cook Associates Executive Search, which polled nearly 3,000 manufacturing executives primarily in small- to mid-sized U.S. companies from October 13 through November 18, 2011.

[9] Pisano, Gary P., and Willy C. Shih. "Does America Really Need Manufacturing?" *Harvard Business Review* 90(3), March 2012.

[10] See Chapter 6 of this volume for case studies of the advanced battery and flexible display industries. See also Ralph Brodd, "Factors Affecting U.S. Production Decisions: Why are There No Volume Lithium-Ion Battery Manufacturers in the United States?" ATP Working Paper Series Working Paper 05–01, June 2005.

[11] Gregory Tassey, "The Manufacturing Imperative," presentation at NAS Conference on the Manufacturing Extension Partnership, November 14, 2011.

The potential for growing major new U.S. industries that can provide a sizeable return on federal investments in university research is not being realized as manufacturing moves offshore.[12] One barometer of this trend is that America's strong trade surpluses in advanced-technology products in the 1990s have swung to annual deficits and reached $99 billion in 2011.[13] [See Figure 1.1]

At the same time, the United States is not paying sufficient attention to the essential pillars of the innovation ecosystem that have helped make the U.S. a global leader for so long. America's research universities are facing severe financial constraints. The U.S. high-tech manufacturing base is eroding. The U.S. is less welcoming to highly skilled immigrants. Physical infrastructure is crumbling for lack of investment, and data communications networks are slipping below global standards. Severe budget problems are exerting intense pressure on federal and state lawmakers to cut successful programs aimed at commercializing technology and helping small business.

As this report documents, this comes at a time when many other nations are investing aggressively to upgrade their universities, woo top foreign talent, attract investment in advanced manufacturing, build next-generation transportation systems, and connect their entire populations to high-speed broadband networks.

---

[12] Eastman Kodak, which invented OLED technology, recently sold its core technologies to South Korean and Taiwanese interests that are now releasing commercial display products. See the presentation by John Chen, "Taiwan's Flexible Electronics Program," at the National Research Council conference on *Flexible Electronics for Security, Manufacturing, and Growth in the United States*, Washington, DC, September 24, 2010. The U.S. has 9 percent of global manufacturing capacity for solar cells and modules, while Europe has 30 percent, China 27 percent, and Japan 12 percent. See Michael J. Ahearn, "Opportunities and Challenges Facing PV Manufacturing in the United States." *The Future of Photovoltaics Manufacturing in the United States; Summary of Two Symposia,* C. Wessner, ed., Washington, DC: The National Academies Press, 2011. Concerning solar cells, GE recently announced that it would build the largest solar panel factory in the United States in Aurora, Colorado. Kate Linebaugh, "GE to Build Solar-Panel Plant in Colorado, Hire 355 People," *Wall Street Journal*, October 13, 2011.
[13] Advanced technology products defined by the U.S. Census Bureau categorize U.S. international trade into 10 major technology areas: advanced materials, aerospace, biotechnology, electronics, flexible manufacturing, information and communications, life science, optoelectronics, nuclear technology, and weapons. The United States registered trade surpluses in five of the ten categories in 2010 – aerospace, biotechnology, electronics, flexible manufacturing and weapons. But a very large deficit in information and communications offset these surpluses. U.S. Census Bureau, Foreign Trade, Country and Product Trade, Advanced Technology Products. Because the value of trade in the final product is credited to the country where the product was substantially transformed, data for products produced with components from multiple countries are imperfect. To the extent that U.S. imports of advanced technology products contain components manufactured in the United States and previously exported (microprocessors, for example) the import value will overstate the actual foreign value-added.

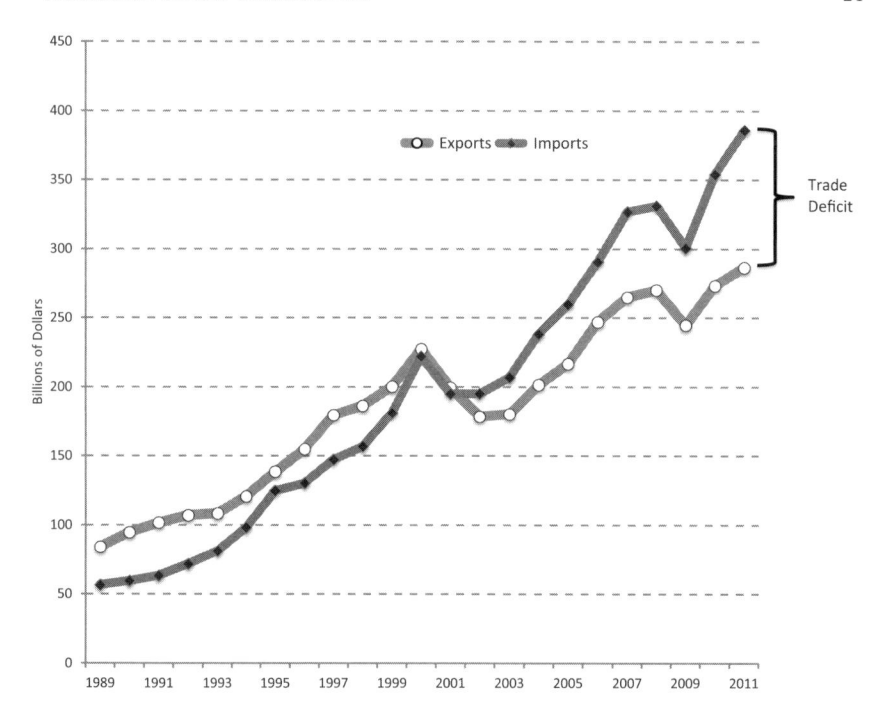

FIGURE 1.1 U.S. exports and imports of advanced technology products. SOURCE: U.S. Census Bureau, Foreign Trade, Trade in Goods with Advanced Technology Products.

Coping with the Growth of New Competitors

The reshaping global environment is affecting U.S. competitiveness.[14] The rest of the world has become smarter, more focused, and more financially committed to developing globally competitive national innovation systems—the networks of public policies and institutions such as businesses, universities, and national laboratories that interact to initiate, develop, modify, and commercialize new technologies.[15]

---

[14] A recent survey of its alumni by the Harvard Business School supports the view that the United States faces a deepening competitiveness challenge. A large majority believed that the United States not keeping pace with other economies, especially emerging economies, as a place to locate business activities and jobs. See Michael E. Porter and Jan W. Rivkin, "Prosperity at Risk," Harvard Business School, January 2012. Access at http://www.hbs.edu/competitiveness/pdf/hbscompsurvey.pdf.

[15] Nelson and Rosenberg popularized the term National Innovation System See Richard R. Nelson and Nathan Rosenberg, "Technical Innovation and National Systems," in *National Innovation Systems: A Companion Analysis*, Richard R. Nelson, ed., Oxford: Oxford University Press, 1993, pg.

As documented in this report, nations in Asia, Europe, and Latin America are boosting investments in both basic research and applied technologies in everything from nano-materials and renewable energies to life sciences. These nations also are encouraging once-cloistered universities and national laboratories to partner with industry, and wooing multinational factories and R&D centers into world-class technology parks with generous tax incentives.

China is making an especially concerted drive to bridge the innovation gap with the U.S.[16] As Yang Xianwu of China's Ministry of Science and Technology explained in a National Academies conference, "The ultimate goal is to make China sufficiently innovative to match the level of countries such as the United States."[17] As this competition intensifies, the United States has tumbled relative to other nations in several global rankings of competitiveness and innovation. For example, the U.S. dropped from No. 1 to No. 5 among 142 nations in the most recent World Economic Forum rankings of "total competitiveness." While ranking No. 5 overall in "innovation," the WEF ranked the U.S. 13th in higher education and training, 16th in infrastructure, 20th in technological readiness, 2nd in "goods market efficiency," 22nd in "financial market development," and 39th in institutions.[18]

---

4. The term "national innovation system" was coined by Christopher Freeman. See Christopher Freeman, " Japan: A New National Innovation System," in G.Dosi, et al, Technology and Economy Theory (London: Pinter, 1988). Charles Wessner initially presented the term "innovation ecosystem," which highlights the complex and non-linear characteristic of innovation processes, to the PCAST. See, for example, Charles W. Wessner, "Entrepreneurship and the Innovation Ecosystem,' in David B. Audretsch, Heike Grimm and Charles W. Wessner, *Local Heroes in the Global Village: Globalization and the New Entrepreneurship Policies*, New York, NY: Springer, 2005. Influential earlier works on global policies to promote innovation include Charles Freeman, *Theory of Innovation and Interactive Learning*, London: Pinter, 1987; Bengt-Åke Lundvall, ed., *National Innovation Systems: Towards a Theory of Innovation and Interactive Learning*, London: Pinter, 1992; and Michael Porter, *The Competitive Advantage of Nations*, New York: The Free Press, 1990. Influential earlier works on global policies to promote innovation include Charles Freeman, *Theory of Innovation and Interactive Learning*, London: Pinter, 1987; Bengt-Åke Lundvall, ed., *National Innovation Systems: Towards a Theory of Innovation and Interactive Learning*, London: Pinter, 1992; and Michael Porter, *The Competitive Advantage of Nations*, New York: The Free Press, 1990.

[16] Chapter 5 of this report provides a detailed case study of China's push to industrialize and develop an innovation-based economy.

[17] See Yang Xianwu, "International Collaboration and Indigenous Innovation," in *Building the 21st Century: U.S. - China Cooperation on Science, Technology, and Innovation*. C. Wessner, ed., Washington, DC: The National Academies Press, 2011.

[18] See World Economic Forum, *The Global Competitiveness Report 2011-2012* (2011), table 5.

NEW TRENDS IN GLOBAL INNOVATION

Strong Policy Focus on Innovation

The twenty-first century is witnessing a rapidly evolving, intensely competitive global landscape. Political and business leaders in both advanced and emerging economies see innovation-led development as central to growth. China, India, Russia, Germany, and Singapore are among the many nations that are formulating comprehensive national strategies for improving their innovation capacity.[19] In many cases, this objective is being pursued with sustained high-level policy attention and substantial funding for applied research and development. Governments also are providing support for innovative small and medium-sized enterprises and are forging innovation partnerships—often based on U.S. models—to bring new products and services to market. They also are investing aggressively to create, attract and retain industries in strategic sectors.

This strong focus on innovation as the basis for economic development is a significant development. Traditional approaches to development followed the prescriptions of Neoclassical Economists who traditionally viewed factors such as capital, labor costs, and business climate as the keys to a nation's growth.[20] Today's focus on knowledge-based growth draws more on the ideas of New Growth economists, including Paul Romer and Robert Lucas, who have put greater emphasis on a nation's innovation capacity.[21]

---

[19] China's 15-year comprehensive innovation strategy is described in the National Medium- and Long-Term Program for Science and Technology Development, 2006-2020, op. cit. An early outline of India's new innovation strategy is found in National Innovation Council, *Towards a More Inclusive and Innovative India*, September 2010. Russia adopted a comprehensive game plan in November 2008 called *The Concept of Long-Term Socio-Economic Development of the Russian Federation for the Period of up to 2020.* Germany's innovation strategy is described in Federal Ministry of Education and Research, *Ideas. Innovation. Prosperity. High-Tech Strategy 2020 for Germany*, Innovation Policy Framework Division, 2010, Canada's national strategy is described in Industry Canada, *Achieving Excellence: Investing in People, Knowledge and Opportunity—Canada's Innovation Strategy*, 2001. An explanation of South Korea's long-term science, technology, and innovation strategy, Vision 2025, can be accessed at http://unpan1.un.org/intradoc/groups/public/documents/APCITY/UNPAN008040.pdf.

[20] See Carl J. Dahlman, *The World Under Pressure: How China and India Are Influencing the Global Economy and Environment*, Palo Alto: Stanford UP, 2011. See also, Carl J. Dahlman, "The Innovation Challenge: Drivers of Growth in China and India," in National Research Council, *Innovation Strategies for the 21st Century: Report of a Symposium*, Charles W. Wessner, editor, Washington, DC: The National Academies Press, 2007.

[21] For a recent review of New Growth Theory, see Daron Acemoglu, "Introduction to Economic Growth," *Journal of Economic Theory*, Volume 147, Issue 2, March 2012, Pages 545-550.

Box 1.1
The Complexity of Innovation

Innovation is the transformation of ideas into new products, services, or improvements in organization or process. Some innovations are incremental; others are disruptive, displacing exiting technologies while creating new markets and value networks.[22] These innovations can lead to new economic opportunities, job growth, and increased competitiveness. A key characteristic of innovation is that it is highly collaborative and often multidisciplinary and multidirectional. To be effective, policies to encourage and accelerate innovation need to recognize this reality.

Innovation is often described in terms of stages: basic research, applied research, followed by development and commercialization. In the real world, this process is often not linear, leading from one stage to the next. Technological breakthroughs (such as in semiconductor research) can precede, rather than stem from, basic research. Often, research can, in parallel, address challenges that are both fundamental and applied.[23] Many products are the result of multiple R&D iterations and draw upon technical sources other than their immediate R&D progenitors; many research projects generate results that are not anticipated – sometimes the unexpected outcomes are extremely important; and innovations often result from the manufacturing process itself.

Ideas that result from the formalized exploration of knowledge do lead, in the long run, to innovations, but to expect this to be the case in the short run is misguided for both firms and governments. While innovation is not a direct consequence of R&D, it is also clear that continuous public investment has been critical in training a large number of people over many years and in creating the necessary environment to foster new technology-based businesses.

This complexity of the innovation process also highlights the role that a variety of intermediating institutions play in fostering collaboration among the many participants—including individual researchers, universities, banks, angel investors, venture capitalists, small and large companies, and governments—across the innovation ecosystem. Connections among these participants are often imperfect. In some cases, for example, a venture capitalist may not realize the true significance of a new idea, meaning that it does not receive the funding needed to develop. In other cases, an individual firm may be reluctant to incur the high costs of research and development for knowledge that will benefit others as much or more than the investor; what economists call "public goods."

---

[22] Clayton M Christensen and Michael Overdorf, "Meeting the Challenge of Disruptive Change" *Harvard Business Review*, March–April 2000.
[23] Donald Stokes, *Pasteur's Quadrant, Basic Science and Technological Innovation*, Washington, DC: Brookings, 1997.

> These are but two common situations where the process of innovation can stall. Intermediating institutions, often with funding from both public and private sources, have often provided the way forward. The U.S. has a rich history of public-private partnerships that have provided a platform for successful cooperation.[24]
>
> What sets the United States apart from most other industrial nations is that there is no overarching national innovation strategy to support, much less coordinate, disparate initiatives to build commercially oriented industries. Instead, as Charles Vest of the National Academy of Engineering has pointed out, the U.S. system consists of multiple centers of activity that are loosely organized but often highly entrepreneurial.[25] Invention and product development are the result of knowledge that flows back and forth among complex, inter-linked, and often ad-hoc sub-ecosystems at universities, corporations, government bodies, and national laboratories. Dr. Vest concludes that the U.S. innovation system "frankly is not really a system. It is not designed or planned very explicitly." Nevertheless, as Dr. Vest notes, it has worked remarkably well at producing commercial products, processes, and services.[26]
>
> Paradoxically, this complexity with its many opportunities for entrepreneurship may be a major strength of the U.S. innovation system. Indeed, Nobel laureate economist Elinor Ostrom has extensively documented the adaptive advantages of open, institutionally diverse systems over linearly designed systems.[27]

Rapid Growth in R&D Spending

The front end of any innovation system is research and development. Since World War II, the United States has enjoyed an overwhelming advantage over the rest of the world in R&D investment. With annual R&D spending for 2012 forecast on the basis of purchasing power parity at $436 billion, the U.S. remains far ahead of the next-largest forecasted R&D investor, China, at $199 billion.[28] Among corporations, 9 of the world's 20 largest investors in R&D are American-based.[29]

---

[24] For a review of best practices among recent U.S. partnership programs, see National Research Council, *Government Industry Partnerships for the Development of New Technologies*, C. Wessner, ed., Washington, DC: The National Academies Press, 2003.
[25] See Charles Vest, "Universities and the U.S. Innovation System," *Building the 21st Century: U.S. - China Cooperation on Science, Technology, and Innovation.* C. Wessner, ed., Washington, DC: The National Academies Press, 2011.
[26] Charles Vest, op. cit.
[27] Elinor Ostrom, *Understanding Institutional Diversity,* Ewing, N. J.: Princeton University Press, 2005.
[28] Battelle and *R&D Magazine, 2012 Global R&D Funding Forecast*, December 2011.
[29] Barry Jaruzelski and Kevin Dehoff, "How the Top Innovators Keep Winning," Booz & Co., 2010 (http://www.booz.com/media/file/sb61_10408-R.pdf).

**Box 1.2
Overcoming the Barriers to Innovation**

As noted in Box 1.1, the process of innovation is itself a complex one involving a variety of participants across the economy. Given the complex and multifaceted nature of innovation, policies to encourage innovation need to reflect this reality.

Support for innovation first requires attention to key framework conditions including adequate investments in R&D, the security of intellectual property, a strong scientific and skills base, and a modern physical, legal, and cyber infrastructure. This includes business regulations that are simple and transparent as possible, consonant with public policy objectives such as health and environmental safety.

Support for innovation also requires our attention to common barriers that can forestall the cooperation needed to bring new ideas to the marketplace. For example, cultural barriers often separate those in industry from academia, where the focus is more on understanding basic phenomenon than on achieving concrete results.[30] These barriers are often reinforced by a legacy of organizational incentives; universities have traditionally emphasized the need to publish rather than commercialize research. Cooperation can also stall when there are information asymmetries—situations where some have better (or worse) information than others in a potential transaction. For example, a venture capitalist may not realize the true significance of a researcher's new idea, with the result that it does not receive the funding needed to develop. Indeed, the economics literature has identified a variety of contexts where the wrong incentives lead to a failure of cooperation.[31]

Pro-innovation policies need to strengthen the framework conditions but also address these barriers to innovation. Successful American innovation policies do just that. The Bayh-Dole Act, for example, encourages innovation by changing the incentives faced by university faculty and administrators.[32]

---

[30] For an illustrative example of barriers to innovation in the food industry, see Sam Saguy, "Paradigm shifts in academia and the food industry required to meet innovation challenges." *Trends in Food Science & Technology*, Volume 22, Issue 9, September 2011, pp. 467-475.

[31] The analysis of incentives in economics can be divided into research on issues related to distorted motivations (including public goods problems, and common pool resource problems) and issues related to incomplete or missing information (including moral hazard and adverse selection problems.) Theoretical work in this area of economics has been richly recognized by the Central Bank of Sweden in awarding Nobel Prizes to George Akerlof, Michael Spence, Joseph E. Stiglitz, Leonid Hurwicz, Eric S. Maskin, Roger B. Myerson and Elinor Ostrom, among others.

[32] For a comparative review of the effectiveness of Swedish and U.S. policies to commercialize university intellectual property see Brent Goldfarb, and Magnus Henrekson, "Bottom-up versus top-down policies towards the commercialization of university intellectual property." *Research Policy* 32 (2003) 639–658. The authors note that Swedish policies "have been largely ineffective due to a

> And the competitive evaluations of the Small Business Innovation Research program (SBIR) create new information for use by market participants about the technological and commercial potential of new ideas. These and other "best practices" in policy are being widely emulated around the world as policymakers in other nations seek to improve the innovative potential of their own economies.

This overwhelming advantage is starting to slip, however. While American R&D spending has risen 3.3 percent a year on average in real terms over the past decade[33], for example, growth in South Korea has averaged 9.2 percent annually and China has averaged 19.4 percent, albeit from a smaller base.[34] As a result, the U.S. share of global R&D spending dropped from 43.1 percent in 1998 to 37.3 percent in 2008.[35] China's share, by contrast, leapt from 3 percent to 11.4 percent over that period, both as a result of increasing R&D intensity and a rapidly industrializing economy.[36]

America's edge in research intensity (R&D as a percent of GDP) also is fading. America once was the most research-intensive nation on earth. America now ranks 8th in the most recent OECD tabulation of R&D intensity by country.[37] This is a disturbing trend. U.S. investment in R&D amounts to around 2.9 percent of GDP, a level that has changed little in three decades. South Korea, by contrast, has boosted R&D spending from less than 2 percent of GDP in the early 1990s to 3.4 percent. Japan's ratio has gone from 2.8 percent to 3.3 percent and China's R&D spending has risen from 0.7 percent of GDP to 1.7 percent. [See Figure 1.2] The Chinese Government has announced plans to boost R&D intensity to 2.5 percent by 2020.[38] Overall, Asia surpassed the U.S. in 2010 in R&D spending and the gap is expected to widen.[39]

---

lack of incentives for academic researchers to become involved in the commercialization of their ideas."

[33] National Center for Science and Engineering Statistics, National Patterns of R&D Resources: 2008 Data Update, Detailed Statistical Tables, NSF 10-314 (March 2010), Table 13, R&D spending from 1998 to 2008.

[34] UNESCO, Institute for Statistics Database, Table 25, gross expenditures on research and development in constant prices from 1998 to 2008.

[35] Ibid.

[36] Ibid.

[37] OECD, *OECD Science, Technology and Industry Scorecard 2011* (September 20, 2011), p. 76.

[38] UNESCO, *UNESCO Science Report 2010* (UNESCO Publishing: Paris, 2010), p. 389.

[39] See NSF Science and Engineering Indicators, 2012. Access at http://www.nsf.gov/statistics/seind12/slides.htm. See also Battelle, op. cit. Battelle estimated U.S. R&D spending at $415.1 billion in 2010 with Asia as a whole at $429.9 billion. The Goldman Sachs Global Markets Institute also estimates that research and development in Asia as a whole will likely overtake U.S. levels in the next five years. Goldman Sachs Global Markets Institute, "The New Geography of Global Innovation," September 2010 http://www.innovationmanagement.se/wp-content/uploads/2010/10/The-new-geography-of-global-innovation.pdf.

TABLE 1.1 Global R&D Spending Forecast

Region	2010 GERD PPP (Billion U.S.)	2010 R&D as Percent of GDP	2011 GERD PPP (Billion U.S.)	2011 R&D as Percent of GDP	2012 GERD PPP (Billion U.S.)	2012 R&D as Percent of GDP
Americas	473.7	2.3	491.8	2.3	505.6	2.3
U.S.	415.1	2.8	427.2	2.8	436.0	2.8
Asia	429.9	1.8	473.5	1.9	514.4	1.9
Japan	148.3	3.4	152.1	3.4	157.6	3.4
China	149.3	1.5	174.9	1.6	198.9	1.6
India	32.5	0.8	38.0	0.8	41.3	0.8
Europe	310.5	1.9	326.7	1.9	338.1	2.0
Rest of World	37.8	1.0	41.4	1.1	44.5	1.1
Total	1,251.9	2.0	1,333.4	2.0	1402.6	2.0

SOURCE: Battelle and R&D Magazine, 2012 Global R&D Funding Forecast, December 2011.
NOTE: GERD: Gross Expenditures on R&D, PPP, Purchasing Power Parity. The Chinese government reports somewhat different estimates of 1.83% for 2011 and 1.76% for 2010. See "Statistical Bulletin on National Science and Technology Expenditures in 2010 and in 2011.

The composition of the U.S. R&D effort has evolved over the years, with the share going to military R&D increasing since the mid-1990s.[40] At $72.6 billion projected for fiscal year 2013, Defense R&D expenditures will make up over half of the federal government's total R&D expenditures of $142.2 billion.[41] Within that component, as Figure 1.3 shows, much greater priority is devoted to later-stage systems development. This is significant in that the aggregate data may be overstating the actual level of basic and early stage applied R&D in the United States. Further, the majority of federal R&D is focused on specific national objectives in defense, health, space, energy and the environment. This has resulted in total federal R&D spending being concentrated in just a few industries. Seventy-five percent of federal R&D allocated to manufacturing goes to aerospace and instruments.[42] Yet these two industries only account for about 10 percent of high technology value-added in

---

[40] Patrick J. Clemins, Presentation of May 25, 2011, "R&D in the Federal Budget." Access at http://www.aaas.org/spp/rd/presentations/aaasrd20110525.pdf.
[41] Matt Hourihan, "R&D in the FY 2013 Budget," AAAS, April 26, 2012.
[42] Gregory Tassey, "The Manufacturing Imperative," presentation at NAS Conference on the Manufacturing Extension Partnership, November 14, 2011. See also Gregory Tassey, *The Economics of R&D Policy*, Westport CT: Greenwood Publishing, 1997.

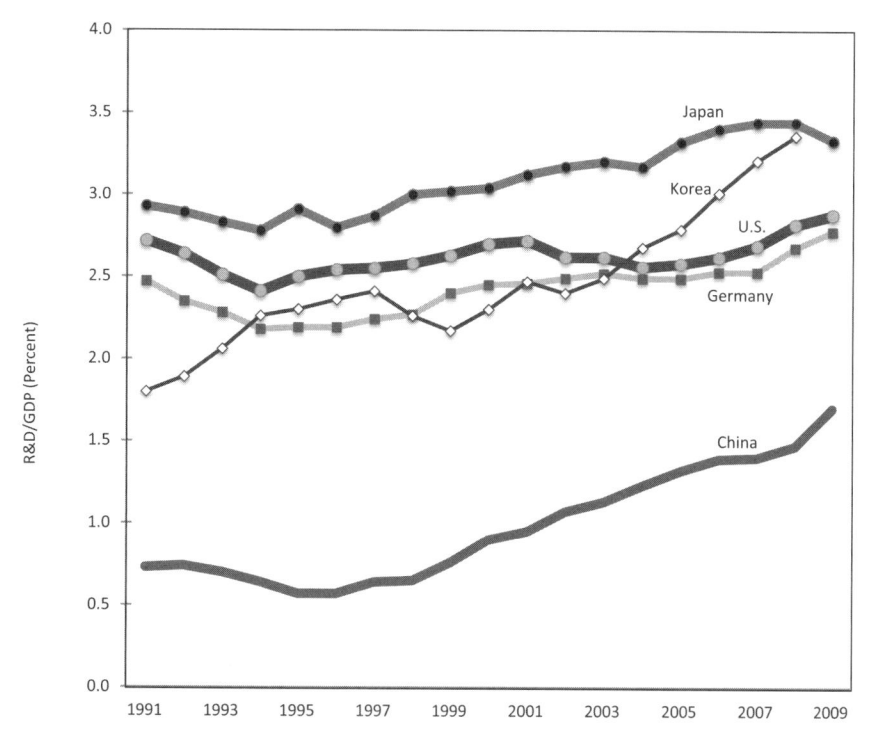

FIGURE 1.2 R&D expenditures as a share of gross domestic product. SOURCE: National Science Foundation, National Center for Science and Engineering Statistics, *Science and Engineering Indicators 2012* (NSB 12-01), January 2012, Appendix Table 4-43.

the economy.[43] This means, in Gregory Tassey's assessment, that federal R&D spending is not optimized for economic growth of the economy as a whole.[44]

On top of these concerns, federal R&D investments—the nation's main source of funding for basic research-- have been declining as a percentage of GDP since the mid 1980s and have been trending downward since the early 1960s.[45] [See Figure 1.4]

---

[43] Id.
[44] Id.
[45] For an analysis of the ratio of public vs. private R&D expenditure from the postwar period to 2006, see Conceição et al. *Knowledge for Inclusive Development*, Westport CT: Praeger, 2002.

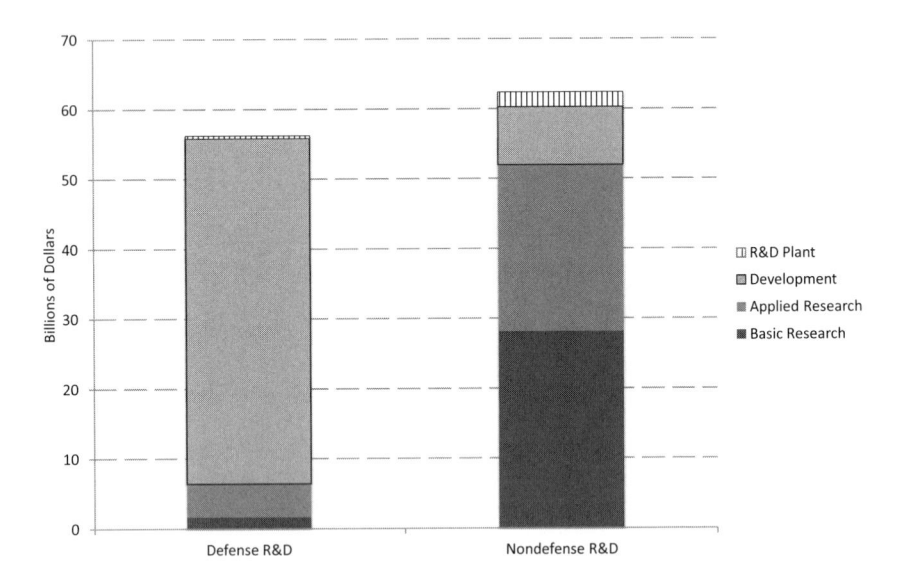

FIGURE 1.3 Federal Obligations for R&D by Character of Work - FY 2010.
SOURCE: National Science Foundation, National Center for Science and Engineering Statistics, Federal Funds for Research and Development: Fiscal Years 2008–10 Detailed Statistical Tables, NSF 12-308 (April 2012), Tables 1, 3 and 7.
NOTE: Eighty-eight percent of Defense related R&D is in development research. FY 2010 data are preliminary.

21st Century Mercantilism

There are growing signs that America's position as the best place to commercialize technology is not as secure as it once was. The reasons for this are multiple and typically revolve around the role that governments around the world play to protect and nurture their domestic industries. The first has to do with markets and market access—with related government subsidies and inducements—for commercializing new technologies. A second major aspect of this relates to favorable access and terms for investment capital. The third factor is infrastructure provision and support, where some high tech industries—notably the semiconductor industry—require billions to set up new plants. A fourth reason relates to taxes and other financial incentives provided by some governments.[46] A final key factor is governmental support for high risk "big

---

[46] Comparatively high corporate taxes and regulatory hurdles and inadequate financing have made America less competitive for capital investment An analysis by economist Jeremy A. Leonard found

THE INNOVATION CHALLENGE 31

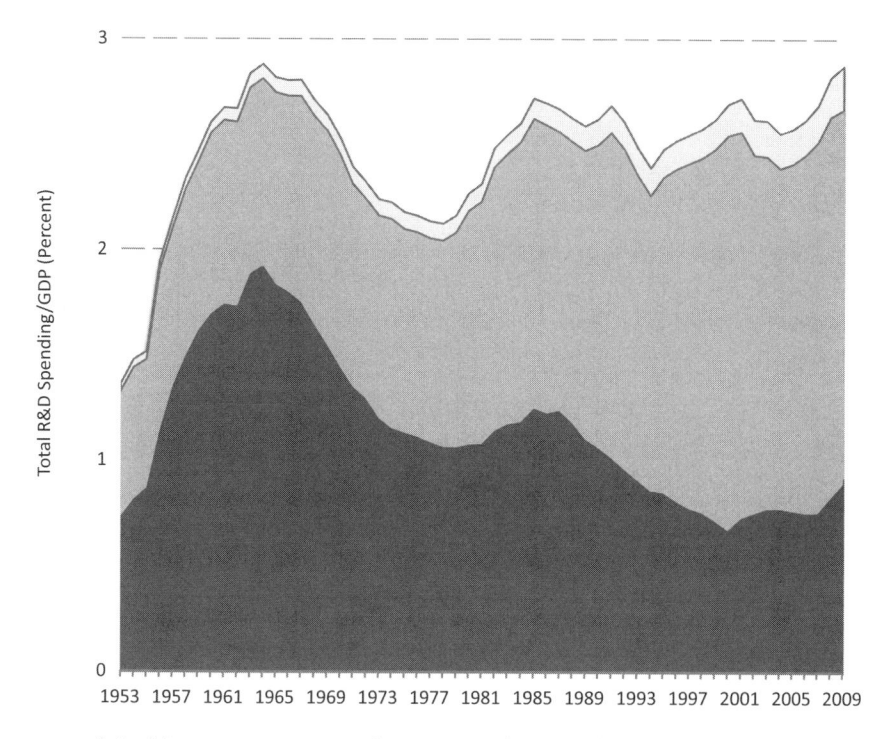

■ Federal Government   ▨ Private Industry   ☐ Other (Non-Profit/University/Other Government)

FIGURE 1.4 Total U.S. R&D spending as a percentage of GDP by funding source.
SOURCE: National Science Foundation, National Center for Science and Engineering Statistics, *Science and Engineering Indicators 2012*, NSB 12-01 (January 2012), Appendix Tables 4-1 and 4-7.

bets" that require all of the above—that is a willingness to foster large-scale endeavors with a long term perspective, not just a quick payoff.

The wave of economic liberalization and free-trade agreements that swept the world in the late 20th century had led some analysts to conclude that

---

that non-production costs such as taxes put U.S. manufacturers at a nearly 18 percent cost disadvantage compared to other nations. See Jeremy A. Leonard, "The Tide Is Turning: An Update on Structural Cost Pressures Facing U.S. Manufacturer," The Manufacturing Institute and Manufacturers Alliance/MAPI, November 2008 (http://www.deloitte.com/assets/Dcom-UnitedStates/Local%20Assets/Documents/us_pip_TideIsTurning_093009.pdf). For another analysis of declining U.S. competitiveness see Aleda V. Roth, et. al, "2010 Global Manufacturing Competitiveness Survey," Deloitte Touche Tohmatsu and U.S. Council on Competitiveness, June 2010.

all major nations were converging toward free-market economic policies[47] and liberal democracy.[48] However, mercantilism is alive and well in the 21st century. One obvious indicator is the persistently large trade surpluses of nations that stress exports and, in some cases, seek to limit imports.

More disconcerting for U.S.-based innovation is the persistence of state capitalism overseas.[49] Government support for homegrown industries, which was instrumental in the ascent of Japan and South Korea in industries such as automobiles, electronics, and steel in the 20th century, plays a heavy role in the economic strategies of nations such as China, Russia, and the Gulf States in the 21st century, notes the National Intelligence Council.[50] The council also noted that state-owned enterprises not only are far from extinction, but actually "are thriving, and in many cases seek to expand beyond their own borders."[51] State enterprises, especially those based in China, often benefit from privileged access to land, labor, capital, government purchases, and industrial subsidies.

Indeed, state enterprises have become a major means of circumventing World Trade Organization rules.[52] Secretary of State Hillary Clinton noted that the world trade system needs institutions to address new challenges from some activities of state-owned enterprises.[53] The OECD also has been seeking to

---

[47] Economist John Williamson in 1989 coined the term "Washington Consensus," referring to the seeming widespread adoption of neoliberal economic policies advocated by the International Monetary Fund, World Bank, and U.S. Treasury. See John Williamson, "What Washington Means by Policy Reform," in John Williamson, editor, *Latin American Readjustment: How Much has Happened*, Washington, DC: Institute for International Economics, 1989.

[48] Francis Fukuyama argued in 1992 that the evolution toward liberal democracy marked "the end of history." See Francis Fukuyama, *The End of History and the Last Man*, New York: The Free Press, 1992.

[49] The term state capitalism has various meanings. Recently, it has been used to describe commercial economic activity undertaken by the state-owned business enterprises that are also supported by the state. For a contemporary review of the scale and scope of modern state capitalism and the challenges it poses, see the *Economist*, "The Rise of State Capitalism." January 21, 2012. The term can also refer to an economic system where the means of production are owned privately but the state plays an active role in the allocation of credit and investment to support the development of major industries. Even in the United States, the state has sometimes played a sustaining role. See, for example, the review of the role of U.S. support for the development of the aircraft industry, in John Birkler et al, "Keeping a Competitive U.S. Military Aircraft Industry aloft." Santa Monica CA: RAND, 2011.

[50] The National Intelligence Council notes that more global wealth is concentrating in emerging economies such as China, Russia, and Gulf States that "are not following the liberal model for self development but are using a different model—'state capitalism.'" The Council describes state capitalism as a loose term used to describe a system of economic management that gives a prominent role to the state." See National Intelligence Council, *Global Trends 2025: A Transformed World*, U.S. Government Printing Office, November 2008. The report can be accessed at http://www.dni.gov/nic/PDF_2025/2025_Global_Trends_Final_Report.pdf.

[51] Ibid.

[52] See Alan Wolff, "America's Trade Policy Agenda and the Future of U.S. Trade Negotiations." Testimony before the House Ways and Means Committee, February 29, 2012.

[53] Hillary Rodham Clinton, "On Principles of Prosperity in the Asia Pacific, speech at Shangri-La Hotel, Hong Kong, July 25, 2011. The address can be accessed at

address this issue through guidelines for the governance of state-owned enterprises.[54]

Even if these government-owned enterprises are not particularly innovative, they have the potential to cause competitive harm to foreign competitors, given their scale, preferential treatment, and access to protected markets.[55] Nations such as Vietnam, Malaysia, and Singapore also have large state enterprises that could evolve into global players and pose challenges to traditional trade agreements.[56] Because many state enterprises are tasked with building state-of-the-art infrastructure, they are gaining experience in deploying the newest technologies for transportation, energy, telecommunications, and clean water.

China is the major source of foreign complaints about policies that distort trade and investment.[57] A recent report by the U.S. International Trade Commission detailed China's lack of enforcement of intellectual property rights, discrimination in government procurement against imported technology products or even those made by multinationals in China, and pressure on multinationals to transfer core technology to domestic Chinese companies.[58]

Due to government policies that favor Chinese producers and compel foreign manufacturers to transfer their technology to sell into the fast-growing domestic market for wind farms, for example, China has become one of the world's biggest producers and exporters of wind turbines and generators. The foreign share of China's annual new purchase of wind power equipment has fallen from 75 percent in 2004 to just 11 percent in 2010.[59] [See Figure 1.5] Rapid expansion of production capacity of photovoltaic modules, fueled by $30 billion in low-cost loans from the China Development Bank, has enabled

---

http://iipdigital.usembassy.gov/st/english/texttrans/2011/07/20110725082343su0.7651876.html#axzz1XIxnKpNm.

[54] OECD, "OECD Guidelines on Corporate Governance of State Owned Enterprises," Paris: OECD, 2005.

[55] Steven Ezell, Fighting Innovation Mercantilism, *Issues in Science and Technology*, Winter 2011.

[56] Bob Davis, "U.S. Targets State Firms in Vietnam, China in Trade Talks," *Wall Street Journal*, October 25, 2011.

[57] For an extensive examination of the implications of Chinese government "indigenous innovation" policies for foreign companies and trade, see Alan Wm. Wolff, "China's Indigenous Innovation Policy," testimony before the U.S. China Economic and Security Review Commission, Washington, DC, May 4, 2011.

[58] U.S. International Trade Commission, *China: Intellectual Property Infringement, Indigenous Innovation Policies, and Frameworks for Measuring the Effects on the U.S. Economy*, Investigation No. 332-514, USITC Publication 4199 (amended), November 2010.

[59] For a review of China's policies to promote its renewable energy equipment sector, see Thomas Howell, William A. Noellert, Gregory Hume, and Alan Wm. Wolff,, *China's Promotion of the Renewable Electric Power Equipment Industry: Hydro, Wind, Solar, Biomass*, Dewey & LeBoeuf LLP prepared for National Foreign Trade Council, March 2010.

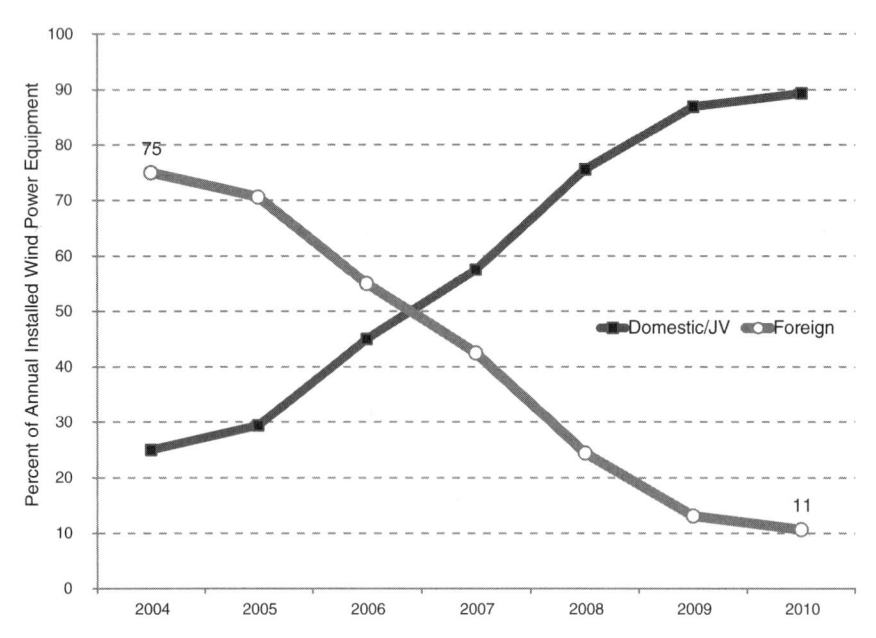

FIGURE 1.5 Foreign share of annual wind power equipment sales within China.
SOURCE: China Wind Energy Association.
NOTE: Foreign share for 2010 for companies other than Vestas, Gamesa, GE, Suzlon and Nordex were estimated based on previous years.

China to dominate the global market.[60] The resulting flood of PV modules has driven down the cost of solar electricity, forcing U.S. manufacturers with alternate but higher priced solar power technologies, such as Solyndra, Evergreen, SpectraWatt, to file for bankruptcy.[61]

Government bodies also essentially require makers of lithium-ion batteries for cars to manufacture in China in order to sell into the growing domestic automobile market.[62] The Chinese government also has refused to allow Chevrolet Volt plug-in hybrid passenger cars to qualify for subsidies totaling $19,300 unless General Motors transfers core technologies to a Chinese

---

[60] Stephen Lacey, "How China Dominates Solar Power: Huge Loans from the Chinese Development Bank are Helping Chinese Solar Companies Push American Solar Firms Out of the Market," Guardian Environment Network, guardian.co.uk, September 12, 2011.
[61] Keith Bradsher, "China Benefits as U.S. Solar Industry Withers," *New York Times*, September 1, 2011.
[62] See Jason M. Forcier, "The Battery Industry Perspective," at the National Research Council conference on *Building the U.S. Battery Industry for Electric-Drive Vehicles: Progress, Challenges, and Opportunities*, Livonia, Michigan, July 26, 2010.

partner.[63] Leveraging its large and growing market for aircraft, China is using technology transferred by U.S. and European aircraft, engine, and avionics suppliers to support its ambitious plans to build a globally competitive commercial aerospace industry.[64] The government also aims to increase to 30 percent the self-sufficiency ratio of integrated circuits used in communications and digital household products and to 70 percent in products relating to national security and defense.[65] China also uses its control over rare-earth metals used in electronics products to its advantage by making it difficult for foreign manufacturers to obtain the critical materials unless they build factories in China.[66]

The Chinese government has adopted a formal policy of favoring products incorporating "indigenous innovation" as a means of cutting dependence on imported technology and building domestic innovation capacity.[67] These goals are embedded in procurement, Chinese technology standards, anti-monopoly law, and tax regulations and laws. "The indigenous innovation 'web of policies' is expected to make it difficult for foreign companies to compete on a level playing field in China," according to the U.S. International Trade Commission (ITC).[68]

The United States lacks an effective policy to prevent the compulsory transfer of cutting-edge technology—much of it developed through federal subsidies—to build new industrial rivals in other nations. The U.S. has various policy tools to fight unfair trade practices. The President, for example, is empowered by Congress to "take all appropriate action" to oppose "unjustified, unreasonable, or discriminatory" polices or practices by foreign governments that restrict U.S. commerce.[69] Although the United States Trade Representative is authorized to initiate retaliatory action by itself, in practice federal agencies react to documented petitions filed by industry. The problem with this procedure is that few U.S.-based multinationals wish to jeopardize their business in China—a critical market—by initiating a trade action.

---

[63] Keith Bradsher, "Hybrid in a Trade Squeeze," *New York Times*, September 5, 2011.
[64] See Roger Cliff, Chad J. R. Ohlandt, and David Yang, *Ready for Takeoff: China's Advancing Aerospace Industry*, RAND National Security Research Division for U.S.-China Economic and Security Review Commission, 2011. See also David Barboza, Christopher Drew and Steve Lohr, "GE to Share Jet Technology with China in New Joint Venture," *New York Times*, January 17, 2011.
[65] Chinese Ministry of Industry and Information Technology, "Outline of the 11th Five-Year Plan and Medium-and-Long-Term Plan for 2020 for Science and Technology Development in the Information Industry," Xin Bu Ke [2006] No. 309, posted on ministry website Aug. 29, 2006. This effort, while well funded, has nonetheless encountered substantial and persistent challenges.
[66] Keith Bradsher, "Chasing Rare Earths, Foreign Companies Expand in China," *New York Times*, August 24, 2011.
[67] See State Council of China, "National Medium- and Long-Term Program for Science and Technology Development, 2006-2020," op. cit.
[68] U.S. International Trade Commission, 2010, op. cit.
[69] Section 301 (a) of the U.S. Trade Act of 1974 (P.L. 93-618).

Techno-nationalism and state-supported enterprises may not, in the end, prove successful at spawning innovation. Yet, these measures do distort investment flows that determine where U.S. inventions are converted into manufacturing industries and thus they limit the economic gains to the U.S. from research and development. If emerging economies such as India and Brazil also rely heavily on state capitalism, the threat to U.S. innovation will grow.

The Search for Talent

America no longer holds an overwhelming advantage in producing skilled talent. In 1975, the U.S. led the world in the proportion of 20- to 24-year-olds who received their first university degrees. The U.S. fell to second place as of 1990. It has since dropped to 14th.[70] America's relative decline has been especially sharp in the proportion of students earning engineering and science degrees.[71] Charles M. Vest, President of the National Academy of Engineering, highlighted in his 2011 President's Address that just 4.5 percent of U.S. college and university students graduate in engineering fields compared to more than 21 percent in Asia and just under 12 percent in Europe.[72]

China and India now award more four-year engineering bachelor's degrees than the U.S., although the quality and nature of these degrees vary.[73] This is perhaps not surprising given their populations and increasing expenditures on education, but it does suggest a long-term shift in engineering capacity.[74]

---

[70] McKinsey & Company, "The Economic Impact of the Achievement Gap in America's Schools." April 2009.

[71] Joan Burrelli and Alan Rapoport, *Reasons for International Changes in the Ratio of Natural Science and Engineering Degrees to the College-Age Population,* InfoBrief National Science Foundation, Directorate for Social, Behavioral and Economic Sciences NAF 09-308, January 2009. See also Anthony P. Carnevale, Nicole Smith and Michelle Melton, *STEM*, Georgetown University Center on Education and the Workforce, October 2011. The authors point to the fact that the United States is relying on foreign-born workers to fill the gap in the STEM (science, technology, engineering and mathematics) workforce. "As a result of STEM talent shortages throughout the U.S. education and workforce pipeline, many technical industries have come to rely on immigrants to fill the gap between supply and demand for skilled scientific and technical workers."

[72] Charles M. Vest, "Engineers: The Next Generation," President's Address, National Academy of Engineering Annual Meeting, October 16, 2011. Vest argues that the United States has a "work force train wreck" coming in engineering. Not only does the nation not graduate enough U.S. engineers but: (1) the fastest growing segment of college graduates have been women, yet women earn less than 20 percent of U.S. engineering degrees; and (2) Asian and African Americans, who represent one-third of college-age people in the country, earn less than 13 percent of U.S. engineering degrees, and their share of the college-age population is projected to steadily increase.

[73] See Vivek Wadhwa, "Chinese and Indian Entrepreneurs Are Eating America's Lunch." *Foreign Policy*, December 28, 2010.

[74] Gary Gereffi, Vivek Wadhwa, Ben Rissing, and Ryan Owen, "Getting the Numbers Right: International Engineering Education in the United States, China, and India," *Journal of Engineering Education,* Vol. 97, No. 1, pp. 13-25, 2008.

The Growth of Foreign Research Centers of U.S. Multinationals

Technology-intensive multinational corporations have established numerous research centers in emerging economies, largely staffed with local talent.[75] The first MNC R&D centers were primarily concerned with development of technology to adapt companies' global products to local needs and conditions[76]. It became apparent that in a number of countries a significant pool of R&D talent existed which was at a far lower cost than comparable workers in developed economies, and that MNCs could dramatically reduce their R&D costs and increase productivity by shifting some research functions to emerging markets[77]. The 9/11 attacks led to a tightening of U.S. immigration policy and a number of MNCs which relied heavily on foreign-born researchers, accelerated the move offshore to retain access to foreign talent[78]. More recently, MNC offshore R&D centers have been the source of some remarkable achievements, demonstrating that they are becoming integral to the R&D strategies of global technology leaders.[79]

---

[75] The seminal work of Sylvia Ostry and Dick Nelson (1995)7, among many others for the last twenty years, has called for our attention of the relationship between the globalism of firms and the nationalism of governments, as well as the related interplay of cooperation and competition that characterizes high technology and knowledge-based environments. See Sylvia Ostry, Richard R. Nelson, *Techno-Nationalism and Techno-Globalism: Conflict and Cooperation*, Washington, DC: Brookings, 1995.

[76] For example, in 2007, DuPont, a major producer of titanium dioxide for use in industrial coatings, opened a technical center in Dzershinsk, Russia to provide support for Russian manufacturers using DuPont's titanium dioxide in their paint, paper, and plastic products. In 2008, DuPont opened an R&D center in Yaroslavl, Russia, to concentrate on the adaptation of DuPont's new coating materials to assembly line conditions at Russia's manufacturer's automobile plants. "DuPont opens Tech Center in Russia," Chemical Week (March 21, 2007); "DuPont opens High-Performance Coatings R&D Center in Russia," Special Chem Coatings and Inks (July 28, 2008).

[77] In 2010, Zinnov Management Consulting released a widely-cited study of MNC R&D centers which concluded that during the preceding three years MNC R&D centers in India alone had helped the parent organizations cut R&D costs by $40 million. "MNC R&D Centers Generate $40bn in savings: Study," The Financial Express (July 18, 2010).

[78] Semiconductor Industry Association, Maintaining America's Competitive Edge: Government Policies Affecting Semiconductor Industry R&D and Manufacturing Activity (March 2009) pp. 29-31. Most engineering PhD graduates from U.S. universities received their bachelor's degrees in other countries. Foreign nationals make up half of the masters' and 71 percent of the PhD candidates graduating from U.S. universities in the engineering fields relevant to the design and manufacture of integrated circuits. National Science Foundation, Division of Resource Statistics, http://www.nsf.gov/statistics/. A number of emerging economies have large pools of highly-educated science and engineering talent, which can staff major research infrastructures. DuPont India indicated in 2011 that it planned to recruit 800 scientists, mostly PhDs, in the next two years. "DuPont India to Recruit 800 Scientists in Two Years," India Business Insight (February 21, 2011).

[79] In 2008, it was announced that for the first time in history, an entire micro[processor had been designed in India at Intel's Design Enterprise Group in Bangalore, where a 7400-series Xeon core x86 processor was created entirely from scratch by an all-Indian design team. Praveen Vishakantaiah, President of Intel India, commented that "within six years of the inception of the India Design Centre, it has rolled out a chip from design to tape out. This is the fastest ramp up in the history of Intel." "India Inside Intel Chips," Financial Express (September 25, 2008). "Intel India

Illustrative of this trend is DuPont's expanding R&D investments in India. The company, which in the early Twentieth Century pioneered the business model of systematic R&D for the purpose of generating a constant stream of new products, remains a global leader in fields such as chemistry, biotechnology, and materials science.[80] At present, Uma Chowdhry, a native of India, supervises all of DuPont's global R&D centers and the company's significant R&D footprint in India underscores the depth and diversity of MNC R&D activity in the country.[81]

- DuPont has established a network of agricultural seeds research centers in India to develop high-yield hybrid crops adapted to local growing conditions.[82]
- In 2008, DuPont opened the DuPont Knowledge Center in Hyderabad, with 300 scientists pursuing research themes in solar energy, biotechnology, and crop science—the only DuPont engineering competence center outside the U.S. and the company's first biotech research center outside the U.S. DuPont is expanding the scope of the centers research to include packaging, safety and protection, biofuels construction and transportation.[83]
- DuPont has established a ballistics facility at the Hyderabad Knowledge Center which develops protective products such as Kevlar to meet "very specific protection needs" applicable to domestic defense procurement, the first such DuPont facility in the Asia-Pacific region.[84]

Team Lofts a Sixer," The Hindu (September 21, 2008). E-Silicon, a fables producer of ASICs, established an R&D center in Bucharest, Romania, and observed that Romanian talent was particularly strong in designing analog and mixed signal devices. An E-Silicon executive commented that "there seems to be a greater skill set of these disciplines in Romania than in other locations". "ESilicon Accelerates Expansion to Europe," Hugin (October 28, 2008); "ESilicon to expand Romanian Chip Design Chip Operation," EE Times Eastern Europe (November 13, 2008).

[80] See Alfred Chandler, Jr., Scale and Scope: The Dynamics of Industrial Capitalism (Cambridge and London: Harvard University Press, 1990) pp. 181-193.

[81] Dr. Chowdry commented that "in India we find the very best talent, entrepreneurship, and skill and language which blends well with the future of the company." "Developing Technology to Meet Market Needs is DuPont's Priority," Business Line (January 19, 2008).

[82] "DuPont Adds Seed Research Centers," India Business Line (September 22, 2009).

[83] "Diane Gulyas, DuPont Group VP, in "DuPont India Growing by Leaps and Bounds Despite Slowdown," The Economic Times (April 5, 2009); "DuPont plans to Double Manpower in India, "India Business Insight (March 28, 2008). In 2010, DuPont disclosed plans to invest $100 million to expand the Knowledge Center in Hyderabad. "DuPont to Invest $100 million to step-up R&D base," The Economic Times (October 5, 2010).

[84] "DuPont Opens World-Class Ballistics Facility in City," The Times of India (April 14, 2012) DuPont reportedly plans to seek collaboration with India's Defense Research and Development Organization to develop new kinds of protective gear such as helmets and vests. "DuPont Bets on Helmet, Vest Maker (Who Use its Products Made Under the Kevlar Brand)," India Business Insight (April 13, 2012)

- In 2011, DuPont established an Innovation Center in Pune, India, to develop materials and technologies with applications to the automotive sector.[85]

China, like India, has experienced a proliferation of MNC R&D centers.[86] Zinnov Management Consulting estimates that as of March 2011, multinational corporations had established over 1300 R&D centers in the country, more than double the number that existed in 2003-04. 400 of the Fortune 500 have R&D centers in China and technology leaders with Chinese R&D centers include IBM, Cisco, Eli Lilly, Microsoft, GE, Panasonic, Motorola, Toshiba, Broadcom, Nortel, DuPont, Fujitsu, Nokia, and British Telecom.[87] Concerns about China's protections of intellectual property, however, have inhibited many multinationals from conducting cutting-edge R&D in China, although they do conduct some R&D, particularly with respect to products aimed at the Chinese market.[88]

As corporations cut or hold flat their R&D operations in the U.S., they are rapidly expanding their offshore design and engineering centers.[89] This enables corporations to draw on strong local talent and adapt to fast-growing markets. As noted above, in some cases, they are responding to foreign government pressure to transfer technology and know-how.

The Rise of Open Innovation

At the same time that new players are rising, the process of innovation itself is undergoing revolutionary change. As Henry Chesbrough has pointed out, the traditional internally focused model for innovation is becoming obsolete. To remain competitive in today's information rich environment, companies need to leverage both "internal and external sources of ideas and take them to market through multiple paths."[90] Indeed, companies such as Apple have prospered in an environment of open innovation, integrating new technologies, components, design expertise, and low-cost Asian manufacturing capabilities into breakthrough products.

---

[85] "India Will Be 3rd Biggest Carmaker: Diane Gulyas," The Economic Times (September 4, 2011).
[86] See also the discussion of MNCs in China in Chapter 5 of this report.
[87] Zinnov Management Consulting, MNC R&D Landscape: A China Perspective.
[88] A 2009 Survey of its members by the U.S. Semiconductor Industry Association indicated that most companies surveyed would not locate their most advanced and critical R&D facilities in China despite encouragement by the government to do so. SIA, Maintaining America's Competitive Edge (2009) op. cit. pg. 31. For a review of the limited nature and scope of research and development by U.S. affiliates in China, see Lee Branstetter and C. Fritz Foley, "Facts and Fallacies about U.S. FDI in China." NBER Working Paper 13470, 2007.
[89] See Steven D. Eppinger and Anil R. Chitkara, "The New Practice of Global Product Management," *MIT Sloan Management Review*, 47(4) Summer 2006.
[90] See Henry Chesbrough, *Open Innovation: The New Imperative for Creating and Profiting from Technology*, Boston: Harvard Business School Press, 2003.

India also has thrived in the new age of globally networked innovation, emerging as a major source of drug-discovery work and semiconductor, software, medical equipment, and auto part design.[91] Companies in India also have excelled at an "inclusive" approach to innovation that addresses the needs of the low-income masses.[92] Indian companies have developed innovative business models selling high-quality but ultra-low-cost goods and services ranging from cellular phone services to simple passenger cars and computers to surgical procedures aimed at the what late management thinker C. K. Prahalad described as the "bottom of the pyramid."[93] As innovation capacity grows abroad, U.S. companies will likely source more new knowledge abroad, just as companies from other countries have done in the U.S.[94]

Growth of Innovative Regions Around the World

Silicon Valley, greater Boston, San Diego, Austin, Seattle and other U.S. innovation zones for decades have been magnets for the world's brightest and most visionary innovators, technology entrepreneurs, and financiers. Now these hubs face greater competition as places to commercialize new technology and launch new companies. Taipei, Shanghai, Helsinki, Tel Aviv, Hyderabad, Singapore, Sydney, and Suwon, South Korea, are among the many cities that now boast high concentrations of technology entrepreneurs and are launching important companies.[95] According to a map of global innovation clusters by the

---

[91] For example, see presentations by Swati Piramal of Nicholas Piramal, Robert Armstrong of Eli Lilly, Kenneth Herd of General Electric, and Ram Sriram of Google in National Research Council, *India's Changing Innovation system: Achievements, Challenges, and Opportunities for Cooperation,* Charles W. Wessner and Sujai J. Shivakumar, editors, Washington, DC: The National Academies Press, 2007. For additional examples of R&D performed for multinationals in India, see National Research Council, *The Dragon and the Elephant: Understanding the Development of Innovation Capacity in China and India—Summary of a Conference,* Stephen Merrill, David Taylor, and Robert Poole, rapporteurs, Washington, DC: The National Academies Press, 2010.
[92] See C. K. Prahalad, *The Fortune at the Bottom of the Pyramid: Eradicating Poverty Through Profits,* Wharton School Publishing, 2005. See also C.K. Prahlad and R.A. Mashelkar 'Innovation's Holy Grail,' *Harvard Business Review,* July 2010.
[93] For example, see the summary of presentations by Kapil Sibal and M. P. Chugh in *India's Changing Innovation System,* op. cit.
[94] Proctor & Gamble, for example, has drawn on research done at India's National Chemical Laboratory to market innovative household products worldwide. Getting fragrance onto clothes had presented a long standing challenge for detergent companies and their suppliers. The key idea of using a unique microencapsulation technology for accomplishing this was revealed in a Ph.D. thesis done at National Chemical Laboratory (NCL) in Pune (India) in the year 1998. Procter & Gamble spotted it, partnered with NCL and developed it further into polymer microcapsules for fiber use. This is a great commercial success today.
[95] Chapter 7 of this report highlights policy instruments being adopted by countries and regions around the world and across the U.S. to rise to the challenges of building innovation clusters.

McKinsey Global Institute and World Economic Forum, some U.S. cities are losing ground to emerging "hot springs" of innovation in Asia and Europe.[96]

Other nations are getting better at replicating the features that once made American innovation hubs unique, such as access to early-stage risk capital, strong R&D linkages between universities and business, modern science parks, and entrepreneurial support networks. In Finland, where annual technology exports leapt five-fold between 1992 and 2008,[97] the government agency Tekes invested €343 million ($494 million) in 2009 directly with enterprises—most of them with fewer than 500 employees--developing technologies in partnerships with universities.[98] Chinese government agencies have mobilized $2.5 billion in venture capital to fund start-ups in the immense Zhangjiang science park outside Shanghai.[99] Singapore, a fast-growing hub for industries such as biotechnology and digital media, is investing $275 million over five years to establish "enterprise boards" at each university, seed money for venture-capital funds, capital for start-ups, and an incubator for "disruptive innovation."[100]

THE PILLARS OF U.S. INNOVATIVE STRENGTH

The U.S. innovation system remains the most dynamic in the world. It is highly decentralized, highly competitive, and highly entrepreneurial. Over the past few decades, the U.S. has been the leading source of game-changing products in fields as diverse as semiconductors, software, medicine, finance, Internet services, and mass entertainment, to name a few. Most recently, a U.S. company, Apple, has launched such revolutionary products such as the iPad and iPod, and Internet leaders such as Google, Facebook and LinkedIn have come into existence in the United States. While the U.S. government has contributed to enabling platform technologies, many of its biggest corporate successes occur

---

[96] A McKinsey & Co. and the World Economic Forum "Innovation Heat Map," which rates on 700 variables such as business environment, human capital, patent applications, economic value added, and industrial diversity, labeled U.S. cities such as Philadelphia, St. Louis, and Indianapolis "silent lakes" or "shrinking pools" while cities such as Shenzhen, Hyderabad, Singapore, and Cheonan, South Korea, are classified as rapidly growing "hot springs." See Juan Alcacer and McKinsey & Co., "Mapping Innovation Clusters," McKinsey Digital, March 19, 2009,
(http://whatmatters.mckinseydigital.com/flash/innovation_clusters/). Also see Andre Andonian, Christoph Loos, and Luiz Pires, "Building an Innovation Nation," McKinsey & Co., March 4, 2009.
[97] Finnish Science and Technology Information Service data. Access at http://www.research.fi/en. This surge would, of course, include the Nokia effect.
[98] Data from Tekes Annual Review 2009,
(http://www.tekes.fi/en/community/Annual%20review/341/Annual%20review/1289).
[99] Data from Zhangjiang High-Tech Park Web site
http://www.zjpark.com/zjpark_en/zjgkjyq.aspx?ID=7.
[100] National Research Foundation, "National Framework for Innovation and Enterprise," Prime Minister's Office, Republic of Singapore, 2008,
(http://www.nrf.gov.sg/nrf/otherProgrammes.aspx?id=1206).

> **Box 1.3**
> **The Postwar Rise of U.S. Pre-eminence in Science and High-technology Industry**
>
> America's pre-eminence in both scale intensive industries and in science based and in high technology industries following the Second World War were the result of an unusual set of circumstances.[101] First, significantly before World War II U.S. industry had taken the lead in a number of industries where economies of scale and scope were significant (like steel, sewing machines, and later automobiles.). The reason was that the U.S. then was by far the world's largest "common market."[102] With the opening of trade after WWII and the significantly lower costs of transport, even firms in small countries could take advantage of large markets and operate at scale. Second, World War II devastated the economies that had been strong competitors for technological leadership prior to the war. Prior to the war, Germany was the leader in many fields and Britain was in a few.[103] The war severely damaged much of the German scientific establishment. The magnitude of U.S. postwar finance of science and new technologies helped the U.S. overtake the British. Third, after the Second World War, the U.S. pioneered in large-scale public finance of university-based scientific research as well as large-scale government support of the development of high tech industries related to defense and space.[104] This is the era in which the United States took the lead in many high technology industries. Political support for these programs in the U.S. depended to a good extent on our sense of being challenged and threatened by the Soviet Union.[105] By the end of the 20th Century two things had changed. One was that other countries were greatly expanding their own finance of university science. The other was that the end of the cold war eroded the political support for programs to support and grow high technology industries in the United States.

---

[101] See Richard R. Nelson and Gavin Wright, "The Rise and Fall of American Technological Leadership: The Postwar Era in Historical Perspective," *Journal of Economic Literature* 30(4), December 1992.

[102] Ibid.

[103] For a review of prewar German leadership in the Chemical Industry, see Ashish Arora, Ralph Landau, and Nathan Rosenberg, "Dynamics of Comparative Advantage in the Chemical Industry," in *Industrial Leadership, Studies of Seven Industries*, David C. Mowery and Richard R. Nelson, eds., Cambridge: Cambridge University Press, 1999.

[104] John Thelin, *A History of American Higher Education,* Baltimore: Johns Hopkins University Press, 2004. See also Hugh Davis Graham, Nancy A. Diamond, *The Rise of American Research Universities: Elites and Challengers in the Postwar Era*, Baltimore: Johns Hopkins University Press, 1997.

[105] See Vernon Ruttan, *Is War Necessary for Economic Growth? Military Procurement and Technological Development*, Oxford: Oxford UP, 2006. See also Stuart W. Leslie, *The Cold War and American Science: The Military-Industrial-Academic Complex*, New York: Columbia University Press, 1993.

with little or no direct government involvement at the point of innovation application.[106]

America's innovation system also is extremely complex. It is characterized by myriad varieties of interactions among government agencies, universities, private industry, financiers, and intermediary organizations.[107] The system is fed by research-and-development spending that still far exceeds that of any other nation. The innovation system is supported by the world's best university system and deepest pools of private angel and venture investment capital.

Strong Protection of Intellectual Property

Strong protection of intellectual property rights, business-friendly bankruptcy laws, a flexible labor force, and an entrepreneurial culture and legal system that favor risk-taking and tolerate failure are among the framework conditions that have kept the U.S. at the forefront of innovation. Another crucial American advantage has been its openness to foreigners. Scientists fleeing European fascism helped develop atomic energy in the U.S. and spurred its post-War ascendance in natural sciences. An influx of top talent from Taiwan, India, South Korea, China and other regions and nations who came to the U.S. to study and then settled were instrumental in U.S. pre-eminence in industries such as semiconductors, computers, software, and biotechnology. Foreign-born talent also has accounted for a disproportionate share of U.S. high-tech start-ups.[108]

---

[106] As Mary Meeker's Kleiner of Perkins has observed, "private investment may have given us Facebook and Garmin, but public sector investment gave us the Internet and GPS." As Roger Noll and Linda Cohen point out, "the foundations of the modern economy" were laid by the long-term public investments in enabling technologies such as nuclear energy, satellites, and computers. See Linda R. Cohen and Roger G. Noll, *The Technology Pork Barrel*, Washington, DC: Brookings Institution. 1991.

[107] A good overview of the U.S. innovation system is provided in Philip Shapira and Jan Youtie, "The Innovation System and Innovation Policy in the United States," Chap. 2 in Rainer Frietsch and Magrot Schüller, editors, *Competing for Global Innovation Leadership: Innovation Systems and Policies in the USA, EU, and Asia*, Fraunhofer IRB Verlag, Stuttgart, 2010.

[108] AnnaLee Saxenian of the University of California at Berkeley estimated that Chinese and Indian engineers were represented on the founding teams of 24 percent of Silicon Valley technology businesses founded between 1980 and 1998. See AnnaLee Saxenian, *Silicon Valley's New Immigrant Entrepreneurs*, San Francisco: Public Policy Institute of California, 1999. A follow-up study found that in one-quarter of all U.S. technology companies founded between 1995 and 2005, one-quarter had chief executive officers or chief technology officers who were foreign-born. See Vivek Wadhwa, Ben Rissing, AnnaLee Saxenian, Gary Gereffi, "Education, Entrepreneurship and Immigration: America's New Immigrant Entrepreneurs, Part II," Duke University Pratt School of Engineering, U.S. Berkeley School of Information, Ewing Marion Kauffman Foundation, June 11, 2007.

Federal Funding of Research

At the front end of America's innovation system is basic research that is largely funded by the federal government and carried out by research universities. In contrast to many other nations, civilian research spending by the federal government is not coordinated by a single agency but instead distributed among a large number of mission agencies and departments.[109] The Department of Defense accounts for a little over half of federal R&D; other funding agencies are the National Institutes of Health; the departments of Defense, Energy, and Agriculture; the National Aeronautics and Space Administration, and the National Science Foundation, which allocates research grants on a peer-review basis.

Research Universities

Research universities are the engines of the U.S. innovation system. Of these, the nearly 200 public research universities conduct more than 60 percent of federally funded basic research.[110] These institutions educate 85 percent of undergraduates and 70 percent of graduate students in U.S. science and technology fields.[111] Since passage of the Bayh Dole Act of 1980,[112] which made it easier for universities to sell and license technology generated from federally funded research, the role of research universities in starting new high-

---

[109] Vannevar Bush, the advisor to President Franklin D. Roosevelt and regarded as the architect of the modern U.S. science and technology policy system, recommended that a National Science Foundation organize and coordinate under "one tent" all of the nation's research activities. See Pascal Zachary, *Endless Frontier: Vannevar Bush, Engineer of the American Century*, Cambridge MA: MIT Press, 1999. President Roosevelt's successor, President Truman disagreed, vetoing Bush's NSF legislation. Meanwhile, as William Bonvillian recounts, "science did not stand still. New agencies proliferated, and by the outbreak of the Korean War, led to the renewal of defense science efforts. By the time NSF was established and funded, its potential coordinating role had been bypassed. It also became a much smaller agency than Bush anticipated, only one among many." William Bonvillian, "The Connected Science Model for Innovation," The DARPA Role," in National Research Council, *21st Century Innovation Systems for Japan and the United States*. Sadao Nagaoka et al, eds., Washington, DC: The National Academies Press, 2009. Indeed, although Congress created the National Science Foundation in 1950, control over research funds was dispersed across different federal agencies.

[110] The Center for Measuring University Performance counts 163 U.S. institutes receiving at least $40 million in federal research expenditure a year as of 2008. See http://mup.asu.edu/research2010.pdf.

[111] Association of Public and Land-grant Universities data. See Peter McPherson, David Shulenburger, Howard Gobstein, and Christine Keller, "Competitiveness of Public Research Universities & the Consequences for the Country: Recommendations for Change," Association of Public and Land-Grant Universities, March 2009, (http://www.aplu.org/NetCommunity/Document.Doc?id=1561).

[112] The University and Small Business Patent Procedure Act (Bayh–Dole Act), Public Law 96-517, permits universities own and license inventions developed through federally funded research.

tech companies and commercializing technology has increased dramatically.[113] Universities also host a range of public-private research centers and consortia that bring together federal agencies, corporations, and national laboratories. The NSF sponsors a network of 55 Industry-University Cooperation Research Centers and a number of Engineering Research Centers at universities around the nation.

National Laboratories

While defense contractors and other private companies receive the lion's share of federal R&D dollars for applied research and development, the U.S. also has 37 federally funded research and development centers, 16 of which are national laboratories sponsored by the Department of Energy (DOE). The Office of Energy Efficiency and Renewable Energy (EERE) within the Department of Energy works with a number of these laboratories on high-risk, high-value research and development in the fields of energy efficiency and renewable energy technologies. A review by the National Academies found that "DOE's RD&D programs in fossil energy and energy efficiency have yielded significant benefits (economic, environmental, and national security-related), important technological options for potential application in a different (but possible) economic, political, and/or environmental setting, and important additions to the stock of engineering and scientific knowledge in a number of fields."[114]

Other research centers are sponsored by the armed forces and agencies such as the Department of Homeland Security, the National Science Foundaton, the Department of Health and Human Services, and the Internal Revenue Service. National laboratories focus on critical national needs such as defense, energy security, and space flight, but have been increasing their roles as partners with private industry. The DOE's four biggest national laboratories—Los Alamos, Lawrence Livermore, Sandia, and Oak Ridge—and NASA's Jet Propulsion Laboratory together account for 55 percent of the $20 billion of U.S. funding for federally funded R&D centers.

In addition to awarding research grants in response to proposals, federal agencies also operate a number of mission-specific programs devoted to accelerating development of high-priority technologies through public-private partnerships with industry and academia. The DOE, for example, awards grants to companies and universities to accelerate development of specific technologies relating to advanced batteries, electric-drive vehicles, and photovoltaic cells.

---

[113] For a review of the diverse channels by which commercialization took place in the 20th century and since the passage of the Bayh-Dole Act, see David Mowery, Richard Nelson, Bhaven Sampat, and Arvids Ziedonis, *Ivory Tower and Industrial Innovation: University-Industry Technology Transfer Before and After the Bayh-Dole Act,* Palo Alto: Stanford Business Books, 2004.
[114] See National Research Council, *Energy Research at DOE: Was It Worth It? Energy Efficiency and Fossil Energy Research 1978 to 2000*, Washington, DC: National Academy Press, 2001.

The Technology Innovation Program, supervised by NIST, was designed to fund high-risk research addressing critical national needs, such as sensors for monitoring civil infrastructure, nano-scale materials, and advanced manufacturing processes for electronics, and genetic engineering. The National Cancer Institute at the NIH funds research into cures and treatments for diseases such as bladder, breast, colon, and kidney cancer. Federal programs aimed at disseminating technology to the private sector include the Hollings Manufacturing Extension Partnership of the Commerce Department.

The Private Sector

The innovation process itself—that of developing marketable products—has traditionally been the realm of the U.S. private sector. Private industry over the past six decades has assumed an ever-greater share of U.S. R&D spending, and now accounts for around two-thirds. Corporate R&D funding has increasingly concentrated on development, as opposed to basic and applied research.[115] More than 70 percent of that private investment is devoted to product development and another 20 percent or so to applied research.[116] The private sector invested $201.8 billion developing new and improved goods, services, and processes in 2008, 84 percent of U.S. spending on developmental research.[117]

This means that private innovation increasingly is carried out on the back of investment by the federal government, which shoulders an estimated 53 percent of funding for basic research that underpins the scientific discoveries and the technologies of tomorrow.[118] [See Figure 1.6] The federal government, in fact, has long played a much bigger role in the U.S. innovation system than many assume.

Public-Private Partnerships

As explained in more detail below, public-private collaborations have been woven into the fabric of the U.S. economic system since the early days of the republic. The armed forces, recognizing that innovation is critical to national defense, have played an instrumental role in funding and procuring platform

---

[115] This development comes even as many large U.S. corporate laboratories, such as Bell Labs and Sarnoff have either closed down or have reduced significantly in scope. Others, such as IBM, GE, and DuPont have maintained their laboratories but have changed their scope of activities to focus more on product development. Other firms, including Intel have systematically supported academic research. At the same time, some foreign based firms, such as Samsung and Novartis, have established advanced technology research centers in the United States, though these facilities also focus on applied research.

[116] National Science Foundation, *Key Science and Engineering Indictors: 2010 Digest*, Arlington, VA, January 2010.

[117] National Science Foundation, op. cit.

[118] Ibid.

technologies such as airframes and engines, satellites, semiconductors, computers, the global positioning systems, nuclear energy, and the Internet. Technology for national defense and economic growth are both part of the same innovation system.[119] In his 2001 book *Technology, Growth and Development: An Induced Innovation Perspective*, the late development economist Vernon Wesley Ruttan concluded that "government has played an important role in the development of almost every general purpose technology in which the United States was internationally competitive."[120] In 2006, Dr. Ruttan took a step further to argue that large-scale, long-term government investment is necessary for the development of general-purpose technologies that spur economic growth. That is because the private sector has little incentive to invest in the scientific research to produce radically new technologies because the gains are too diffuse to be captured by any one corporation.[121]

The military's involvement in U.S. innovation goes far beyond funding R&D projects. For decades, the Defense Advanced Research Projects Agency (DARPA) has helped orchestrate collaborations and social networks among researchers and industry that have identified new technology trends and developed broad technology platforms that cut across industries.[122]

Federal agencies also have been more important to the funding of U.S. technology start-ups than many assume. The Small Business Innovation Research (SBIR) and other federal programs provide up to one-quarter of early-stage technology funding.[123] Importantly, these federal efforts are often

---

[119] "Defense technology cannot be discussed as though it is separate and apart from the technology that drives the expansion of the economy—they are both part of the same technology paradigms." William B. Bonvillian, "The Connected Science Model for Innovation – The DARPA Role," in National Academy of Sciences, Board on Science, Technology, and Economic Policy, *21st Century Innovation Systems for Japan and the United States: Lessons from a Decade of Change*, Washington, DC: The National Academies Press, 2009, pp. 206-237.

[120] Vernon Wesley Ruttan, *Technology, Growth, and Development: An Institutional Design Perspective*, New York: Oxford University Press, 2001. See also Gregory Tassey, *R&D and Long-Term Competitiveness: Manufacturing's Central Role in a Knowledge-Based Economy*, NIST Planning Report 02-2, February 2002, pp. 31-40.

[121] Vernon Wesley Ruttan, *Is War Necessary for Economic Growth? Military Procurement and Technology Development*, New York: Oxford University Press, 2006.

[122] Erica R. H. Fuchs describes DARPA program managers as "embedded agents" in the national innovation ecosystem who maintain constant contact with researchers, track emerging themes, bet on the right people, bring together disconnected researchers, stand up competing technologies against each other, and maintain the systems-level perspective critical to orchestrate these disparate research activities. See Erica R. H. Fuchs, "Rethinking the Role of the State in Technology Development: DARPA and the Case for Embedded Network Governance," *Research Policy* 39 (2010) 1133-1146. Bonvillian and Van Atta emphasize how DARPA has also worked over an extended period of time to change the technological landscape, essentially undertaking "multigenerational technology thrusts." Notable examples include work in information technologies, stealth and stand-off precision strike. William B. Bonvillian and Richard Van Atta, "ARPA-E and DARPA: Applying the DARPA Model to Energy Innovation," *Journal of Technology Transfer* 36 (2011): 469-513.

[123] Lewis M. Branscomb and Philip E. Auerswald, *Between Invention and Innovation: An Analysis of Funding for Early-Stage Technology Development*, NIST GCR 02–841, Gaithersburg, MD: National Institute of Standards and Technology, November 2002.

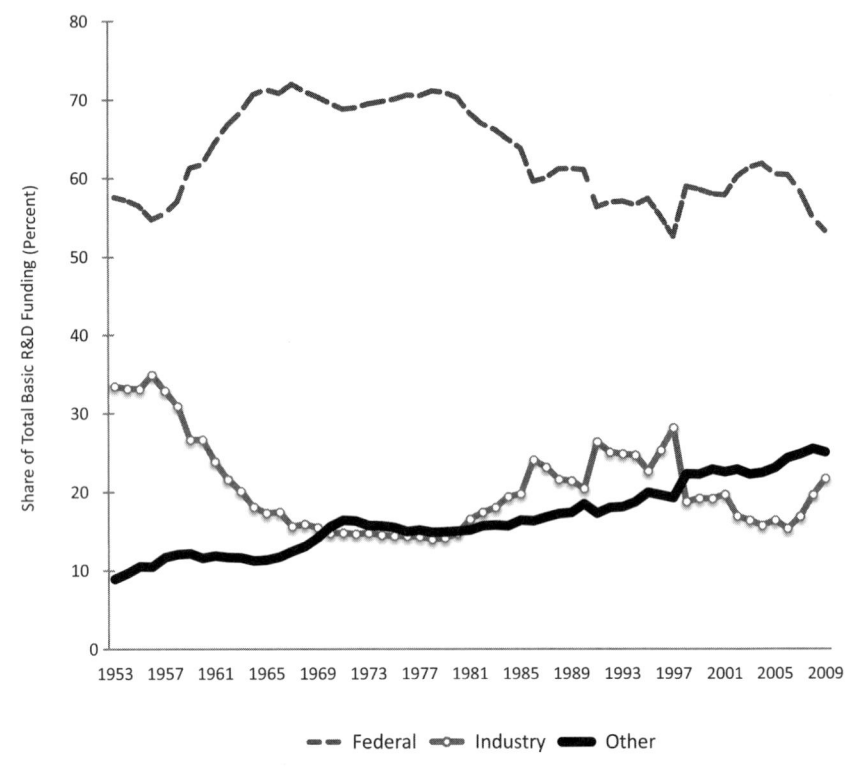

FIGURE 1.6 Funding for basic research in the United States by source of funding.
SOURCE: National Science Foundation, National Center for Science and Engineering Statistics, *Science and Engineering Indicators 2012*, NSB 12-01 (January 2012), Appendix Table 4-8.

successful. Some two-thirds of award-winning inventions honored by *R&D Magazine* stem in part from partnerships between government and business.[124]

---

[124]See Fred Block and Matthew Keller, "Where Do Innovations Come From? Transformations in the U.S. National Innovation System, 1970-2006," The Information Technology and Innovation Forum, July 2008. Accessed at <http://www.itif.org/files/Where_do_innovations_come_from.pdf>. The authors analyze a sample of innovations recognized by R&D Magazine as being among the top 100 innovations of the year over the last four decades. They find that while in the 1970s almost all winners came from corporations acting on their own, more recently over two-thirds of the winners have come from partnerships involving business and government, including federal laboratories and federally-funded university research. In 2006, 77 of the 88 U.S. entities that produced award-winning innovations were beneficiaries of federal funding. Over the past decade, SBIR awards have accounted for 20 to 25 percent of all 'U.S. R&D 100' winners. Block and Keller note that "the R&D 100 Awards carry considerable prestige within the community of R&D professionals, comparable to

> **Box 1.4**
> **Federal Mission Needs Drive Innovation**
>
> Today, the armed forces have a major interest in accelerating development of technologies that conserve energy and reduce dependence on fossil fuel, which they regard not only as important to future weapons systems that can provide strategic advantage in the battlefield but also can reduce America's dependence on distant nations for energy.[125] As the largest energy consumer in the world, the United States Department of Defense (DoD) has realized the value and practicality of energy efficiency, officially codifying it as "a force multiplier" in the 2010 Quadrennial Defense Review.[126] As Admiral Mullen has noted, advances in energy, such as increasing the use of renewable energy supplies and reducing energy demand, simultaneously enhance operational capability in forward deployed combat environments while generating enormous cost savings to U.S. military installations. – all while making U.S. troops and mission critical systems more secure and cutting the risks of climate change.[127]

And while it is true that companies like Apple and Facebook flourish without direct government help, their innovations would not have been possible without previous federal investments in the Internet, computers, and semiconductors, not to mention in the university systems that produced their technology talent.

RESPONDING TO THE INNOVATION CHALLENGE

The growing competition among countries to influence the international location of production of high-technology and high value-added industries requires a fresh approach to science and technology policy. The underlying premise of U.S. policy since World War II has been that big

---

the Oscars for the motion picture industry. Organizations nominate their own innovations. All entries are initially evaluated by outside juries that include representatives of business, government, and universities." (Block and Keller, 2008, page 6).

[125] There is a growing recognition in the military that developing renewable energy sources is important not only for greater energy independence in general but also for specific missions, such as the current military operation in Afghanistan. According to a recent article in the *Wall Street Journal*, "U.S.'s Afghan Headache: $400-a-Gallon Gasoline." (Dec 6, 2011), "The cost and difficulty of fuel deliveries in places like Afghanistan is one major reason the Pentagon is working to overhaul the way the armed forces use energy, from developing aircraft that can run on biofuels to powering remote bases with solar panels or wind turbines."

[126] United States Department of Defense "Quadrennial Defense Review Report" February 2010.

[127] Energy Security Forum Speech as Delivered by Admiral Mike Mullen, chairman of the Joint Chiefs of Staff, Washington, DC Wednesday, October 13, 2010, http://www.jcs.mil/speech.aspx?id=1472.

investments in science and technology will ultimately translate into the growth of new and more productive domestic industries.[128] The assumptions underpinning this premise with regard to market size, first adopters, availability of finance and skilled labor are no longer assured, particularly in a number of promising emerging industries.

For the United States, adjusting to the new challenges of 21st century global competition means not only taking steps to improve its own competitive position but also by recognizing and taking advantage of opportunities that arise with the increasingly dynamic and globally distributed geography of innovation. Often this involves corporate investment in research and production in rapidly growing markets overseas. Yet, there are also opportunities for public policy to enhance the attractiveness of the United States as a place for investments in promising new technologies.

A well-trained workforce is a key component in any national strategy to exploit these emerging opportunities. This is why the 2007 National Academies report *Rising Above the Gathering Storm* documented how underinvestment in R&D, training of engineers, and falling education standards is eroding America's lead in science and technology.[129] Noting that "weakening commitments to S&T puts future U.S. prosperity in jeopardy," the report warns of the risk of an abrupt loss of U.S. leadership in science and technology. The *Gathering Storm* argued that substantial increases in federal and corporate R&D are required to assure America's long-term prosperity.[130] The more recent update of the Gathering Storm Report noted that, due in part to the rising investments in science and innovation by other countries and regions, "the unanimous view of the committee members ... is that our nation's outlook has worsened."[131]

However, public investments in research alone are unlikely to be sufficient. The *Gathering Storm* addresses the challenge of increasing the inputs to innovation. This report addresses the need to renew and broaden America's innovation ecosystem to better capitalize on these inputs to generate commercial products, grow new industries, and, most importantly, create jobs that guarantee high living standards for millions of Americans. In other words, how can

---

[128] For a review of the origins of postwar U.S. science and technology policy, see G. Pascal Zachary, *Endless Frontier: Vannevar Bush, Engineer of the American Century*, New York: The Free Press, 1997. Also, see Harvey Brooks, "The Evolution of U.S. Science Policy," Chap. 2 in Bruce L. R. Smith and Claude E. Barfield, editors, *Technology, R&D, and the Economy*, Washington, DC, The Brookings Institution and American Enterprise Institute, 1996.

[mlb] National Academy of Sciences, National Academy of Engineering, and the Institute of Medicine, *Rising Above the Gathering Storm: Energizing and Employing America for a Brighter Economic future*, Washington, DC: The National Academies Press, 2007. An update of this report was recently published called *Rising Above the Gathering Storm, Revisited: Rapidly Approaching Category 5*, Washington, DC: The National Academies Press, 2010.

[130] Ibid (2007).

[131] Norman Augustine, et al. *Rising Above the Gathering Storm, Revisited, Rapidly Approaching Category 5*. Washington, DC: The National Academies Press, 2011.

America produce more economic value from its tremendous investments in research and development? To sustain public support for current levels of taxpayer funding for research, innovation ultimately needs to pay off in the form of jobs and economic growth.

Policies to Capture the Value of Innovation in Some Leading Countries and Regions

The successes of other nations and regions show that it is possible to benefit from the global flows of goods, technology, capital, talent, and creative ideas in ways that also generate dynamic growth industries at home. Some of these strategies, policies, and programs being undertaken abroad offer valuable lessons deserving study by American policymakers at the federal, state, and local level. The Committee does not endorse these initiatives, though some offer positive lessons on what could be adopted and adapted to the U.S. context. Indeed, the focus of these programs, the instruments they use, and their funding levels may have important lessons for U.S. policy.

- **Germany** is proving that even a high-wage nation can compete globally in manufacturing. The German government invests $2.3 billion a year in industrial production and technology research—*six times more* than the United States.[132] A surge in exports from small and large firms alike of everything from kitchen equipment and industrial machinery to high-speed trains and wind turbines helped power Germany out of the recent recession.[133] German exports to China have soared. One of Germany's secrets: Strong and consistent investment in job training, worker retention, and applied research programs such as the Fraunhofer Institutes that partner with companies to turn advanced technologies into production processes and commercial products, coupled with active export promotion support from the highest level of government.[134]
- **Singapore** has shown that steady investment in S&T higher education and world-class research infrastructure, combined with the right financial incentives and policy climate, can attract substantial investment by multinationals that can turn a region into a global R&D

---

[132] Sridhar Kota, "Stimulating Manufacturing in Ohio" Presentation at National Research Council symposium, "Building the Ohio Innovation Economy," April 25, 2011.
[133] Anthony Faiola, "Germany Seizes on Big Business in China," *Washington Post*, September 18, 2010.
[134] See presentation by Roland Schindler, executive director of Fraunhofer, at the National Research Council conference on, *Meeting Global Challenges: U.S.-German Innovation Policy,* Berlin, May 24-25, 2011. With regard to worker retention, see Klaus Zimmerman, "Germany's Support for Manufacturing and Export Performance." Presentation at the National Academies conference on Meeting Global Challenges: U.S.-German Innovation Policy, November 1, 2010, Washington, DC.

hub.[135] Now Singapore is investing aggressively to build an "innovation-driven economy." Among other things, Singapore is investing some $10 billion in a network of research parks in a 500-acre urban district called One North. They include Biopolis, a 4.5 million-square-foot campus housing 5,000 life science researchers from universities, hospitals, and multinationals such as Eli Lilly and Novartis, and Fusionopolis, a futuristic 24-story tower filled with media, communications, and information-technology companies.[136]

- **China** is leveraging its enormous talent pool, domestic market, foreign investment, and mounting wealth to make significant progress in growing technology-intensive industries. China has doubled its share of global R&D spending from 6 percent in 1999 to 12 percent in 2010.[137] Already a leading exporter of everything from computers and telecom networking equipment to solar modules, China is investing aggressively to become a dominant producer of advanced products like electric vehicles, solid-state lighting devices, and commercial aircraft. Among China's ambitious goals are to sell 1 million electric vehicles a year by 2015, have renewable energies account for 15 percent of energy consumption, and generate 1 million patents a year by 2015. [138] It is determined to become a world-leader in manufacturing everything from automobiles to advanced computers and seems prepared to make the investments and use its market power to do so.[139]

- **Finland**'s success in telecommunications and electronics shows that even a relatively small nation or region can become a global leader in high-tech industries if high levels of government investment in R&D are aligned with skillfully applied research by corporations and universities. Tekes, Finland's funding agency for technology and innovation, invests some €600 million a year in hundreds of research projects in emerging technologies. Much of that funding is direct grants to companies, which match the funds and work in collaborations lasting three to five years with universities and research institutes.[140]

- **Taiwan** has demonstrated that focused investments in applied research and a systematic system for absorbing and disseminating foreign

---

[135] See Yena Lim, "The Singapore S&T Park", National Research Council, *Understanding Research, Science and Technology Parks: Global Best Practices.* C. Wessner, ed., Washington, DC: National Research Council, 2009.

[136] Source: Singapore Economic Development Board. Access at http://www.edb.gov.sg/edb/sg/en_uk/index.html.

[137] Battelle, op. cit.

[138] *The Guardian*, "China plans to make a million electric vehicles a year by 2015," February 18, 2011.

[139] BBC, "China claims supercomputer crown." October 28, 2010.

[140] Heikki Kotilainen, "The TEKES experience and new initiatives," National Research Council, *Innovation Policies for the 21st Century*, op. cit.

technology can produce globally competitive high-tech industries. Thanks largely to public-private partnerships led by the Industrial Technology Research Institute (ITRI), Taiwan has become a world leader in semiconductor manufacturing, digital displays, and notebook computers.[141] Now Taiwan is developing fast-growing innovation clusters in fields such as semiconductor design, flexible displays, and biomedical devices.[142]

- **Canada** has invested heavily over the past decades to upgrade its university research system and draw international talent. Through the Foundation for Innovation, the government has committed more than $5 billion since 1997 to fund 6,300 projects at 130 research institutions. Of the thousands of new faculty and researchers hired by universities through such grants, more than 40 percent were recruited abroad.[143] With the Canada Chairs program, 30 percent of the nearly 2,000 department chairs hired through another program also were recruited outside of Canada.[144]

The Growing U.S. Response: Federal Government

In his 2011 State of the Union address, President Barack Obama declared that "we need to out-innovate, out-educate, and out-build the rest of the world." The President also observed that "none of us can predict with certainty what the next big industry will be or where the new jobs will come from." But "what we can do—what America does better than anyone else—is spark the creativity and imagination of our people."[145] The recognition of the global innovation challenge at the highest levels of the government is as exceptional as it is welcome.

Unlike most other industrial nations, the United States does not have a comprehensive national innovation strategy. The U.S. instead has tended to address specific needs and goals through targeted, short-term legislation and with programs that shift from one Administration to the next. The federal government has paid more attention to innovation and economic

---

[141] Alice H. Amsden, "Taiwan's Innovation System: A Review of Presentations and Related Articles and Books," Memorandum on the National Academies symposium "21st Century Innovation Systems for the U.S. and Taiwan: Lessons from a Decade of Change," January 4-6, 2006, Taipei.
[142] Taiwanese researchers have won a number of recent R&D 100 Awards is these categories. For example, see *R&D Magazine*, "R&D 2010 Winners," July 7, 2010.
[143] Canada Foundation for Innovation, *2009 Report on Results: An Analysis of Investments in Infrastructure*, (http://www.innovation.ca/docs/accountability/2009/2009%20Report%20on%20Results%20FINAL EN.pdf).
[144] Government of Canada website on "Canada Research Chairs." Access at http://www.chairs-chaires.gc.ca/home-accueil-eng.aspx.
[145] See the address by President Obama to the National Academy of Sciences, April 27, 2009. See also the President's 2011 State of the Union Address, White House, January 25, 2011.

competitiveness issues in recent years, driven in part by efforts to recover from the recession and the job crises it engendered. Many states have tended to be more active. States like Ohio and Michigan, which have been hard hit by the nation's manufacturing decline, are making substantial investments in future industries.

The federal government has acted upon some of the *Gathering Storm* recommendations to shore up America's performance in science in technology. Research budgets for the Department of Energy, National Science Foundation, and National Institute of Science and Technology (NIST) and federal funding for K-12 science, technology, and mathematics education have increased substantially, for example.[146] The government established the Advanced Research Projects Agency-Energy (ARPA-E), [147] and has speeded up processing of student visas.[148] Legislation to expand R&D tax credits and make them permanent is being considered by Congress. [149] Overall, however, the 2010 *Gathering Storm, Revisited* report concluded that "our nation's outlook has worsened" over the previous five years relative to the rest of the world.

On January 6, 2011, President Obama signed into law the reauthorization of the America COMPETES Act, a modified version of a law passed in 2007 but one not funded by Congress.[150] Among other things, the law further increases federal research budgets of the NSF, NIST, and the DOE's Office of Science and seeks to better coordinate federal science, technology, and math education programs. The act also provides funding for "high risk, high reward" research and several multi-agency collaborations.[151] Again, however, these provisions have not yet been funded.

President Obama also has unveiled a national innovation strategy that calls for increasing U.S. investments in R&D, higher education, and information-technology and transportation infrastructure. The plan also calls for

---

[146] *Gathering Storm Revisited*, op. cit.
[147] The Advanced Research Projects Agency-Energy (ARPA-E) was established under H. R. 364 in 2007 to conduct cross-disciplinary high-risk, high-reward research on new energy technologies and is modeled after the Defense Advanced Research Projects Agency (DARPA). Its initial budget was included in the American Recovery and Reinvestment Act of 2009.
[148] *Gathering Storm Revisited*, op. cit.
[149] Originally created in 1981, the Research and Experimentation Tax Credit has been renewed 14 times, mostly recently when President Obama signed the Tax Relief, Unemployment Insurance Reauthorization, and Job Creation Act of 2010 in December 2010.
[150] The America Creating Opportunities to Meaningfully Promote Excellence in Technology, Education and Science Reauthorization Act of 2010 (P. L. 111-358) is better known as the America COMPETES Act. The earlier version of this act (P.L. 110-69) was signed into law by President George W. Bush on August 9, 2007.
[151] For an analysis of the America Competes Reauthorization Act of 2010, see Heather B. Gonzalez, John F. Sargent, and Patricia Moloney Figliola, "America COMPETES Reauthorization Act of 2010 (H.R. 5116) and the America COMPETES Act (P. L. 110-69): Selected Policy Issues," Congressional Research Service, July 28, 2010 (http://www.ift.org/public-policy-and-regulations/~/media/Public%20Policy/0728AmericaCompetesAct.pdf).

reforming government regulations and creating new incentives to improve America's competitiveness as a place to do business.[152]

There also is a growing emphasis on coordination among federal agencies around initiatives by state and local governments to support specific regional innovation clusters aimed at meeting national needs. Under White House leadership, the SBA, NIST, EDA, NSF, and EDC, for example, joined an effort by the DOE to establish "energy-innovation hubs"—regional innovation clusters in solar power, energy-efficient buildings, nuclear energy, and advanced batteries. The first $129.7 million project seeks to create an innovation hub devoted to developing technologies, designs, and systems for energy-efficient buildings that will be based at the Philadelphia Navy Yard Clean Energy campus.[153] President Barack Obama's 2009 budget also allocated $50 million administered by the Commerce Department's Economic Development Agency to assist regional cluster initiatives,[154] while the SBA is working with state agencies and the DOD to help launch robotics clusters in Michigan, Virginia, and Hawai'i.[155]

The U.S. government has stepped up financial incentives to support commercialization of technologies. Under the American Recovery and Reinvestment Act (ARRA) of 2009,[156] for example, the DOE extended $6 billion in loan guarantees for renewable-energy and electricity transmission projects, $11 billion in spending and loan guarantees for "smart grid" projects, $117 million to expand the development, deployment and use of solar energy throughout the U.S., and $2.4 billion in grants for manufacturers of advanced batteries and key materials.[157] It is important to note, however, that the ARRA

---

[152] Executive Office of the President, *A Strategy for American Innovation: Driving Towards Sustainable Growth and Quality* Jobs, National Economic Council and Office of Science and Technology, September 2009.
http://www.whitehouse.gov/assets/documents/SEPT_20_Innovation_Whitepaper_FINAL.pdf).
Also see White House, Fact Sheet: Obama's Plan to Win the Future," Office of the Press Secretary, January 25, 2011.
[153] Department of Energy press release, "Penn State to Lead Philadelphia-based team that will pioneer new energy-efficient building designs," August 24, 2010,
(http://www.energy.gov/news/9380.htm). Details of the energy innovation research cluster can be found in the funding opportunity announcement for Fiscal year 2010 on the DOE Web site at http://energy.gov/hubs/documents/eric_FOA.pdf.
[154] President Obama's fiscal 2009 budget provided $50 million in regional planning and matching grants within the Economic Development Administration to "support the creation of regional innovation clusters that leverage regions' existing competitive strengths to boost job creation and economic growth." See Executive Office of the President, *A Strategy for American Innovation*, op. cit.
[155] See Karen Mills, "Luncheon Address," in *Growing Innovation Clusters for American Prosperity, Report of a Symposium*, C. Wessner, ed., Washington, DC: The National Academies Press, 2011.
[156] The American Recovery and Reinvestment Act of 2009, the $787 billion U.S. economic stimulus legislation passed by Congress, includes $59 billion in new spending and tax credits for the development and expansion of energy technology.
[157] SmartGridNews, "$2.4 Billion Going to Accelerate Advanced Battery and EV Manufacturing." August 5, 2009.

(stimulus programs and funding) was a one-time, non-recurring event which has now ended. The DOE loan guarantee program has also been shutdown in the midst of political controversy.[158]

This lack of policy continuity and sustained support for emerging technologies separates the U.S. from many of its global competitors. While fossil and nuclear subsidies have been long term and therefore bankable, subsidies for renewable energy technologies have been subject to short term changes. In wind technology, unstable funding of the production tax credit has resulted in huge drops in investment – with damaging consequences to the development of a robust U.S. wind industry and competitiveness.[159]

Because most of these programs are in the very early stages, it is difficult to measure their impact on the U.S. economy and regional economies. If successful, they can potentially serve as models for additional efforts in other sectors. Their success, however, will depend upon sustained funding over the long term and will benefit from a sustained partnership between federal, state, and local agencies.

The Growing U.S. Response: State and Regional Initiatives

While few regional innovation initiatives in the U.S. can match the financial resources and policy force of those launched by foreign governments, a number of states are starting to achieve impressive results in building innovation-led industries with bold and comprehensive strategies. Promising state and regional initiatives often reflect a holistic understanding of what it takes to build a 21st century innovation ecosystem and compete globally in specific industries. They include public-private partnerships in which corporations, universities, and governments pool resources to establish R&D centers,[160] train workforces, develop supply and support industries, and provide risk capital to starts-ups where angel and venture funding is lacking.[161] To help offset the gap between financial incentives at offshore locations, state governments also are deploying a wider range of policy tools, from tax credits

---

[158] Forbes, "DOE Rescinds Solar Loan Guarantees in Wake of Solyndra Bankruptcy." September 23, 2011.
[159] Institute for Energy Research, "Assessing the Production Tax Credit," April 24, 2012, Access at http://www.instituteforenergyresearch.org/2012/04/24/assessing-the-production-tax-credit/.
[160] The Pew Center on the States, Investing in Innovation, 2007. Access at http://www.pewtrusts.org/uploadedFiles/wwwpewtrustsorg/Reports/State-based_policy/NGA_Report.pdf.
[161] A National Research Council Committee led by Gordon Moore concluded that "Public-private partnerships, involving cooperative research and development activities among industry, government laboratories, and universities, can play an instrumental role in accelerating the development of new technologies to the market." See National Research Council, *Government-Industry Partnerships for the Development of New Technologies*, Charles W. Wessner, ed., Washington, DC: The National Academies Press, 2003, page 23.

and R&D grants to low-cost loans and free or subsidized workforce training.[162] These are a few examples of promising regional strategies—

- **New York:** The Capitol Region in upstate New York, hard hit by a decades-long decline in manufacturing, has become one of the world's premier hubs of semiconductor and nanotechnology R&D. As a result, it is attracting new investment in high-tech manufacturing, including a $4.5 billion silicon wafer fabrication plant by Global Foundries. The catalyst: State investments in public-private research centers, academic programs, and state-of-the-art research laboratories at the State University of New York at Albany that have drawn more than $5 billion in investment by companies such as IBM, AMD, Applied Materials, and Tokyo Electron.[163]
- **Michigan:** In a little over four years, Michigan established itself as one of the world's primary production centers of lithium-ion batteries for future electrified vehicles and power-grid storage—an industry that Asia was poised to dominate. By combining generous manufacturing tax credits with a comprehensive game plan to leverage the state's existing strengths in automotive R&D, engineering, and advanced components manufacturers, Michigan attracted $1.3 billion in one-time Recovery Act (ARRA) funds and $6 billion in private investment in 16 battery-related factories that are expected to create 62,000 jobs in five years.[164]
- **Ohio:** The Northeast Ohio Technology Coalition, an organization funded by foundations and business associations to develop high-tech economy in a 21-county region devastated by the decline of manufacturing, is spearheading programs to build a manufacturing base in flexible electronics and advanced energy with the help of $2.3 billion in state funding for cluster initiatives.[165] State initiatives include the Ohio Third Frontier program, which provides early-stage capital for start-ups and funds applied research, working training, and entrepreneurial assistance. The JumpStart program seeks to enhance the state's entrepreneurial ecosystem through advice from successful

---

[162] See National Research Council, *Growing Innovation Clusters for 21st Century Prosperity*, C Wessner, ed., Washington, DC: The National Academies Press, 2011. Also see Pete Engardio, "State Capitalism," *BusinessWeek*, February 6, 2009.
[163] See Pradeep Haldar, "New York's Nano Initiative," in *Growing Innovation Clusters for 21st Century Prosperity*. Op cit.
[164] Data from Michigan Economic Development Corp. See National Research Council, *Building the U.S. Battery Industry for Electric-Drive Vehicles: Progress, Challenges, and Opportunities*, C. Wessner, ed., Washington, DC: The National Academies Press.
[165] See the presentation by Rebecca Bagley, "The Role of NorTech: Promoting Innovation and Economic Development" at the National Research Council conference on *Building The Ohio Innovation Economy*, Cleveland, April 25, 2011.

entrepreneurs and selective investments in high potential companies.[166] A network of seven Edison Technology Centers help manufacturers commercialize technologies.

- **New Mexico:** Even though research universities and national laboratories based in the state received $6 billion in federal research funding a year, New Mexico had few high-tech start-ups until recently. Clusters are emerging in renewable energy, aerospace, information technology, and digital media. Catalysts include the nation's first science park connected to a national laboratory—located next to the Sandia National Laboratories' Albuquerque campus—and large state investments in early-state capital funds, high-performance computer infrastructure, public-private research partnerships, tax credits for targeted industries, and worker training.[167]
- **West Virginia:** Morgantown, West Virginia, has become the hub of rapidly growing clusters in biometrics and new energy technologies by building alliances between industry, national laboratories, and regional universities such as West Virginia University, Carnegie Mellon, and the University of Pittsburgh. The cluster in biometrics technologies that identify individuals through biological traits, for example, leverages research partnerships with the Federal Bureau of Investigation, a pioneering degree-granting program at West Virginia University, and CITeR, the Center for Identification Technology Research.[168] Morgantown has attracted operations by Booz Allen Hamilton, Northrup Grumman, Lockheed Martin, and other corporations.[169]
- **South Carolina:** The state has been a low-cost base for car assembly for decades. Now Clemson University is helping South Carolina become a hub for advanced systems design and manufacturing. Clemson converted an empty 250-acre site into the Industrial Center for Automotive Research that has "generated more than $220 million in

---

[166] The Jumpstart program was launched in 2003 with founding grants from the Cleveland Foundation, Cleveland Tomorrow, Ohio Department of Development, and the George W. Codrington Foundation. See http://www.jumpstartinc.org/.

[167] See Richard Stulen, "The Sandia Science & Technology Park" in National Research Council, *Understanding Research Science & Technology Parks: Global Best Practice*, C. Wessner, ed., Washington, DC: The National Academies Press, 2009. See also the presentation by Thomas Bowles at the National Academies Symposium, Critical National Needs in New Technologies: Opportunities for the Technology Innovation Program," April 24, 2008. For an analysis of Sandia National Laboratory's science park initiative, see National Research Council, *Industry-Laboratory Partnerships: A Review of the Sandia Science and Technology Park* Initiative, C. Wessner, editor, Washington, DC: National Academy Press, 1999, and presentation by J. Stephen Rottler, "Sandia National Laboratories as a Catalyst for Regional Growth" in the National Academies Symposium on *Clustering for 21st Century Prosperity*, February 25, 2010.

[168] CITeR is an Industry/University Cooperative Research Center funded by the National Science Foundation. The center was founded by West Virginia University and is the I/UCRC's lead site for biometrics research and related identification technologies.

[169] James Clements in *Growing Innovation Clusters for 21st Century Prosperity,* op. cit.

public and private investment and created more than 500 new jobs with an average salary of $72,000."[170] Partners with Clemson include BMW, Timken, Michelin, IBM, Dale Earnhardt, Inc., Sun Microsystems, the Society of Automotive Engineers, and the Richard Petty Driving Experience.[171]
- **Kansas** has developed a thriving cluster in aerospace, and deployed hundreds of millions of dollars of state income-tax withholdings from employees of bioscience-related companies to grow a bioscience cluster focusing on agriculture.[172]

With a few notable exceptions, most state innovation strategies have received little federal support—even though a number of federal agencies have long had economic-development programs seeking to achieve similar aims.[173] "All of this is occurring on an ad-hoc basis without a formal U.S. policy," noted Ginger Lew, then of the White House National Economic Council.[174] In addition, federal programs to support state and regional initiatives are often viewed as being too small in scale or possessing timelines that are too short to provide the confidence needed by businesses to make sizeable investments over the long term.[175]

Looking Ahead

The changing global context raises questions about whether the traditional basis for America's innovation policies is adequate for addressing the competitive challenges of the 21st century. The rapid globalization of innovation has diminished what were once overwhelming American advantages as the prime location for creating, commercializing, and industrializing technology. Basic research and world-class engineering talent now are highly dispersed around the world, especially in important fields such as nanotechnology,

---

[170] See presentation by Clemson University President James Barker in *Understanding Research, Science, and Technology Parks*, op. cit.
[171] Ibid.
[172] Presentation by Richard Bendis," Innovation Infrastructure at the State and Regional Level: Some Success Stories," at the National Academies Symposium on *Building the Arkansas Innovation Economy*, March 8, 2010.
[173] See Karen G. Mills, Elisabeth B. Reynolds, and Andrew Reamer, "Clusters and Competitiveness: A New Federal Role for Stimulating Regional Economies," Metropolitan Policy Program at Brookings, April 2008. Also see Michael E. Porter, "Clusters and Economic Policy: Aligning Public Policy with the New Economics of Competition," Institute for Strategy and Competitiveness White Paper, revised May 18, 2009.
[174] Remarks by Ginger Lew, "The Administration's Cluster Initiative," in at the National Academies Symposium on *Clustering for 21st Century Prosperity; Summary of a Symposium,* February 25, 2010.
[175] Remarks by Sridhar Kota at the National Academies Symposium on *Building the U.S. Battery Industry*, July 26, 2010.

computer science, and renewable energies. How, then, can the U.S. maintain its leadership in innovation? The next chapter addresses this challenge.

Chapter 2

Sustaining Leadership in Innovation

The United States faces new competitive challenges in the 21st century. Globalization is diminishing what once were overwhelming American advantages as the prime location for creating, commercializing, and industrializing technology. Basic research and world-class engineering talent now are highly dispersed around the world, especially in important fields such as nanotechnology, computer science, and renewable energies. How, then, must the U.S. adapt to maintain its leadership in innovation?

IMPROVING FRAMEWORK CONDITIONS

One of America's most fundamental strengths as a place to commercialize innovation has been its overall investment climate. For much of the post-war era, America's boasted some of the world's best transportation, energy, and communication infrastructure.[1] In the 1980s, America's corporate tax rates were among the lowest in the industrialized world.[2] The U.S. also has had one of the world's strongest legal systems for protecting intellectual property rights.[3]

---

[1] Michael Porter observed that American communication, power transportation, and transportation infrastructure was "arguably the best in the world" after World War II, and the fact that infrastructure companies were privately owned "was a stimulus to investment and innovation." See Michael E. Porter, *The Competitive Advantage of Nations*, New York: Simon and Schuster, 1990, p. 297.

[2] The U.S. statutory corporate tax rate dropped from 52 percent to 35 percent in the 1980s, well below the average for OECD nations. See Congressional Budget Office, "Corporate Income Tax Rates: International Comparison," November 2005 (http://www.cbo.gov/ftpdocs/69xx/doc6902/11-28-CorporateTax.pdf). Data from M. P. Devereaux, R. Griffith, and A. Klemm, "Corporate Income Tax Reforms and International Tax Competition," *Economic* Policy, vol. 35 (October 2002).

[3] The United States still has the lowest rate of computer software piracy in the world, followed by Japan and Luxembourg, according to the International Data Corporation (IDC). See Business

Corporate Taxes: There are concerns that America now is at a competitive disadvantage in some of these areas.[4] After the U.S. cut corporate taxes in the 1980s, other industrialized nations cut taxes even further. When state corporate taxes are taken into account, the U.S. corporate statutory rate of 39.3 percent is third highest among OECD nations, which have a median rate of 33 percent.[5] What's more, the tax codes of countries such as Germany, Singapore, Malaysia, and China favor investment in certain industries through such incentives as 10-year tax holidays. While U.S. states offer such tax breaks, the federal government does not. The U.S. is one of the few major trading nations with a tax code that does not treat investment in globally traded industrial activity any differently than non-mobile activity.[6] This means "inefficiency and biases in the corporate tax code fail to promote the productivity and innovative capability of businesses in America, hampering the economy and indirectly affecting all Americans."[7] Business advocacy groups argue that executives find the current tax burden to be an impediment to the competitiveness of their companies operating in the United States."[8]

Infrastructure: Some analysts regard America's aging infrastructure as a competitive disadvantage.[9] The U.S. ranks only No. 27 in terms of infrastructure, according to the World Economic Forum, a major factor in America's falling place in the WEF's overall global competitiveness rankings.[10] That compares to seventh place in 2000, observes the McKinsey Global Institute.[11] The American Society of Civil Engineers asserts that most of America's infrastructure is in poor shape due to delayed maintenance and lack

---

Software Alliance and IDC, *08 Piracy Study*, May 2009, (http://portal.bsa.org/globalpiracy2008/studies/globalpiracy2008.pdf).

[4] It is important to note that the Committee did not conduct a study comparing the U.S. tax system to that of other countries. The Committee did want to draw attention to the growing body of evidence that, in some cases, U.S. tax policy creates a less competitive environment.

[5] Congressional Budget Office, op. cit., citing data from Devereaux, Griffith, and Klemm.

[6] Robert D. Atkinson, "Effective Corporate Tax Reform in the Global Innovation Economy," The Information Technology & Innovation Foundation, July 2009, (http://www.itif.org/files/090723_CorpTax.pdf)

[7] Ibid.

[8] Roth, et al, "2010 Global Manufacturing Competitiveness Survey," Deloitte Touche Tohmatsu and U.S. Council on Competitiveness, June 2010.

[9] For an analysis of the positive link between good infrastructure and innovation and development, see Tony Ridley, Lee Yee-Cheong, Calestous Juma. "Infrastructure, Innovation, and Development," *International Journal of Technology and Globalisation, Volume 2, Number 3-4/2006, Pages 268-278*. For an industry view, see the interview with Eric Spiegel, the president and CEO of Siemens Corporation in *Harvard Business Review*, "Investing in Infrastructure Means Investing in Innovation." March 15, 2012.

[10] World Economic Forum, *Global Competitiveness Report*, op. cit.

[11] James Manyika, et al., *Growth and Renewal in the United States: Retooling America's Economic Engine*, McKinsey Global Institute, February 2011, (http://www.mckinsey.com/mgi/publications/growth_and_renewal_in_the_us/pdfs/MGI_growth_and_renewal_in_the_us_full_report.pdf).

of modernization.[12] The Society reports that an estimated 25 percent of America's bridges need significant repairs, one-third of major roadways are in substandard condition, and that "America's sewer systems spill an estimated 1.26 trillion gallons of untreated sewage every year."[13] More recently the Society called for investments in the nation's transmission, generation, and distribution systems in order to prevent significant costs to businesses and households.[14]

Likewise, a bipartisan study of America's aging transportation infrastructure concluded that it is in "bad shape." The poor condition "compromises our productivity and ability to compete internationally," it added. The study estimated the U.S. needs to spend $134 billion to $262 billion per year more than current plans call for until 2035 to get this infrastructure into proper condition.[15]

Other nations are investing aggressively to build and upgrade their transportation infrastructure. China spent $713 billion--twice as much as the U.S.--just on transportation and water infrastructure over the past five years[16] and is investing an estimated $500 to 700 billion to build the world's biggest high-speed rail network.[17] In 2008, the European Investment Bank lent 58 billion Euros ($81 billion) to finance infrastructure projects, and had a target of $112 billion in 2009.

---

[12] ASCE has assigned a C grade to bridges, C- to rail, D+ for energy, D for aviation, dams, transit, dams, and D- to drinking water. See American Society of Civil Engineers, *2009 Report Card for America's Infrastructure*, March 25, 2009,
(http://www.infrastructurereportcard.org/sites/default/files/RC2009_full_report.pdf).
[13] Data from U.S. federal agencies cited in Eric Kelderman, "Look Out Below! American's Infrastructure is Crumbling," *Stateline.org,* Pew Research Center, January 22, 2008, (http://pewresearch.org/pubs/699/look-out-below).
[14] ASCE, Failure to Act: The Economic Impact of Current Investment Trends in Electricity Infrastructure. April, 2012.
[15] See Miller Center of Public Affairs, *Well Within Reach: America's New Transportation Agenda,* David R. Goode National Transportation Policy Conference. Posted on October 4, 2010 at http://www.infrastructureusa.org/well-within-reach/.
[16] Cathy Yan, "Road-Building Rage to Leave U.S. in Dust," *Wall Street Journal,* January 18 2011.
[17] See Sean Tierney, "High-speed rail, the knowledge economy, and the next growth wave," *Journal of Transport Geography*, Volume 22, May 2012, pages 285-287. Tierney notes that failure to invest in economic development "concedes considerable ground to those countries with whom we are trying to compete. Compare the $8 billion that President Obama set aside in the stimulus bill as a down payment for HSR [High Speed Rail], with the estimated $500 - $700 billion that China plans to invest for its 19,000 km HSR network." For a review of the economic benefits of large scale transportation projects, see T.R. Lakshmanan, "The broader economic consequences of transport infrastructure investments." *Journal of Transport Geography.* Volume 19(1), 2011. For a review of recent China's investments in rail, Will Freeman, "The Big Engine That Can: China's High-Speed Rail Project," *China Insight Economics,* May 28, 2010. Problems have emerged with regard to the rapid construction of China's rail network, its cost, the revenues it is generating, and its relevance to the needs of the general population. Recent train disasters in China have further spotlighted challenges related to the rapid growth of that nation's high-speed rail system. See *Financial Times,* "China's Rail Disaster." July 27, 2011 and Keith B. Richburg, "Are China's High-Speed Trains Heading Off the Rails?" *Washington Post*, April 23, 2011.

To address this competitive disadvantage in infrastructure, some analysts have called for a U.S. infrastructure bank that, like the EIB, could leverage private capital.[18] The purpose of such a National Infrastructure Bank (NIB) would be to invest in merit-based projects of national significance that span both traditional and technological infrastructure by leveraging private capital. Phillips, Tyson and Wolf argue that "the NIB could attract private funds to co-invest in projects that pass rigorous cost-benefit tests, and that generate revenues through user fees or revenue guarantees from state and local governments. Investors could choose which projects meet their investment criteria, and, in return, share in project risks that today fall solely on taxpayers."[19]

Energy Efficiency: Reliable, clean, and relatively inexpensive energy has long been an important competitive advantage for the United States. As a recent UNIDO report notes, "Energy efficiency contributes toward reducing overall company expenses, increases productivity, has effects on competitiveness and the trade balance on an economy-wide level, and, by creating a home market for energy efficient technologies, supports the development of successful technology supply industry in that field."[20] Energy efficiency also represents a major opportunity to increase energy security while also limiting carbon dioxide emissions.

An accelerated deployment of existing and emerging energy-supply and end-use technologies has the potential to yield substantial improvements to energy conservation and efficiency.[21] America's buildings, which alone use more energy than any other entire economy of the world except China, are a key area for conservation efforts.[22] U.S. buildings are generally grossly inefficient; it has been widely documented that energy use in new and existing buildings can be cut by 50% or more cost-effectively.[23] Lowering the cost base for location of

---

[18] Felix Rohatyn, The Case for an Infrastructure Bank, *Wall Street Journal*, September 15, 2010. In the U.S. Senate, legislation, known as the "BUILD Act, was introduced on May 15, 2011 to fund an infrastructure bank.

[19] See Charles Phillips, Laura Tyson, and Robert Wolf, "The U.S. Needs an Infrastructure Bank," *Wall Street Journal*, January 15, 2010.

[20] Wolfgang Eichhammer and Rainer Walz, "Industrial Energy Efficiency and Competitiveness," Vienna: United Nations Industrial Development Organization, 2011.

[21] See National Academy of Sciences, et al., *America's Energy Future, Technology and Transformation*, Washington, DC: The National Academies Press, 2009. The report notes that "The deployment of existing energy efficiency technologies is the nearest-term and lowest-cost option for moderating our nation's demand for energy, especially over the next decade. The committee judges that the potential energy savings available from the accelerated deployment of existing energy-efficiency technologies in the buildings, transportation, and industrial sectors could more than offset the Energy Information Administration's projected increases in U.S. energy consumption through 2030."

[22] U.S. Green Building Council, "Buildings and Climate Change," Accessed on November 3, 2011 at http://www.documents.dgs.ca.gov/dgs/pio/facts/LA%20workshop/climate.pdf.

[23] Greg Kats, *Greening Our Built World, Costs, Benefits, and Strategies*, Washington, DC: Island Press, 2010.

production in the United States can be fostered by improving conservation, and the techniques learned are themselves marketable globally as innovative services.

Broadband: The U.S. is regarded as lagging in broadband infrastructure. In the U.S., 27 of every 100 households subscribe to high-speed Internet service. In Germany, broadband penetration is at 30 percent. The rate is 31 percent in France, 34 percent in South Korea, 38 percent in Denmark, and 41 percent in Sweden.[24] While recognizing that a number of these countries do not have the same geographical spread as the United States, the McKinsey Global Institute nonetheless estimates that the U.S. loses $450 billion in purchasing power annually due to subpar Internet connections.[25]

Intellectual Property: The U.S. still has one of the best legal systems in the world to protect intellectual property rights. This has made America a leader in IP-intensive industries such as pharmaceuticals, software, and entertainment.[26] NDP Consulting estimates that workers in IP-intensive industries generate more than twice the output and sales per employee than do workers in non-IP-based industries. IP-intensive industries also account for around 60 percent of U.S. exports.[27]

Counterfeiting and patent infringement abroad undermine the economic contribution of these industries, however. An estimated 80 percent of software used in China is pirated, IDC estimates. The piracy rate stands at 61 percent in the entire Asia-Pacific region, 65 percent in Latin America, and 66 percent in Central and Eastern Europe, compared to 21 percent in North America.[28] This level of piracy has a substantial effect on U.S. companies' revenues, and therefore their long-term capacity to innovate and compete.

SUBSTANTIALLY INCREASING R&D FUNDING

As mentioned above, the United States still enjoys a clear lead over other nations in total R&D spending. [See Figure 2.1] But as also noted earlier,

---

[24] International Telecommunication Union and Federal Communications Commission data cited in Manyika, op. cit.
[25] Ibid.
[26] In many fields intellectual property protection plays only a small role in enabling firms to reap returns from their innovations. And in some fields it would appear that for the industry as a whole aggressive patenting is a negative sum game. For a survey of the economic literature, both theoretical and empirical, on the choice of intellectual property protection by firms, see Bronwyn H. Hall, Christian Helmers, Mark Rogers, and Vania Sena, "The Choice between Formal and Informal Intellectual Property: A Literature Review," NBER Working Paper No. 17983, April 2012.
[27] See Nam d. Pham, "The Impact of Innovation and the Role of Intellectual Property Rights on U.S. Productivity, Competitiveness, Jobs, Wages, and Exports," NDP Consulting, April 2010 (http://www.theglobalipcenter.com/sites/default/files/reports/documents/IP_Jobs_Study_Exec_Summary.pdf).
[28] Business Software Alliance and IDC, *08 Piracy Study*, May 2009, (http://portal.bsa.org/globalpiracy2008/studies/globalpiracy2008.pdf).

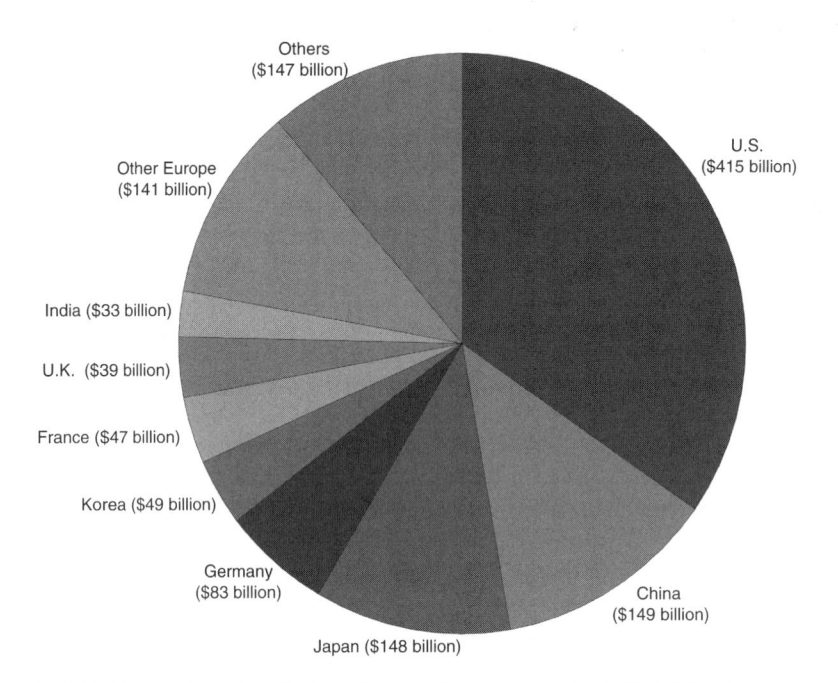

FIGURE 2.1 Total global R&D spending reached $1,252 billion in 2010.
SOURCE: Battelle and R&D Magazine, *2012 Global R&D Funding Forecast*, December 2011.

this lead is eroding as other nations dramatically increase their investments in research—both in real terms and as a percentage of GDP.

The most dramatic gains are being made by China. R&D spending as a percentage of GDP rose from only 0.6 percent in 1996 to 1.7 percent in 2009—a period during which China's economy grew by an astounding 12 percent a year.[29] Between 2002 and 2007, the percentage of the world's researchers living in China rose from 13.9 percent to 19.7 percent.[30] Since then, China has continued to increase R&D investment by around 10 percent a year, even during the global recession. China's long-term plans call for boosting R&D to 2.5 percent of GDP by 2020.[31] The government also has set an ambitious target of

---

[29] National Science Foundation *Science and Engineering Indicators: 2010* and Ministry of Science and Technology of the People's Republic of China, *China S&T Statistics Data Book 2010*, Figure 1-1.
[30] UNESCO Science Report 2010, Paris: United Nations Educational, Scientific and Cultural Organization. Access at http://unesdoc.unesco.org/images/0018/001899/189958e.pdf.
[31] China State Council, "National Medium- and Long-Term Program for Science and Technology," op. cit.

> **Box 2.1**
> **The European Union's Growing Investments in Research and Innovation**
>
> Complementing the rising R&D expenditures of its member states, the European Union is dramatically increasing its investments in research and innovation. The new Horizon 2020 program, which succeeds the Seventh Framework Program, will invest 80 billion Euros over seven years, beginning in 2013, an increase of some 45 percent. This includes a dedicated budget of € 25 billion to strengthen the EU's position in science; € 18 billion to strengthen Europe's industrial leadership in innovation including greater access to capital and support for SMEs; and € 32 billion to help address global challenges such as climate change, renewable energy, and health care.[32]
>
> According to the European Commissioner for Research, Innovation, and Science Máire Geoghegan-Quinn, the goal of the Horizon 2020 program is designed to transform Europe's "world-class science base into a world-beating one."[33]

producing 2 million patents of inventions, utility models, and designs annually by 2015.[34]

Investment in R&D has risen sharply in other nations as well. Japanese spending on research and development surged from 2.9 percent of GDP in 1995 to 3.6 percent in 2009.[35] India doubled national R&D spending between 2002 and 2008, to Rupees 378 billion ($8.7 billion) annually[36], and plans another 220 percent increase by 2012.[37] South Korea has boosted R&D spending by an average of 10 percent annually from 1996 to 2007,[38] and reportedly plans to increase the R&D-to-GDP ratio from an already-high 3.2 percent to 5 percent by 2012.[39] Brazil nearly tripled R&D expenditure between 2000 and 2008, to $24.4 billion.[40] Finland has boosted R&D spending from 2 percent of GDP in 1991 to

---

[32] Access at http://ec.europa.eu/research/horizon2020/index_en.cfm?pg=h2020.
[33] Neil McDonald, "Euro Commissioner visits US," *Federal Technology Watch*, 10(4) January 23, 2012.
[34] China State Intellectual Property Office, "National Patent Development Strategy (2011-2020)."
[35] Japanese Ministry of Internal Affairs and Communications, Statistics Bureau, accessed at http://www.stat.go.jp/english/data/kagaku/index.htm. Data refer to fiscal years.
[36] UNESCO, *UNESCO Science Report 2010*, p. 371.
[37] Government of India Planning Commission, "Report of the Steering Committee on Science and Technology for Eleventh Five-Year Plan (2007-2012)," December 2006.
[38] Battelle, op. cit.
[39] Kim Tong-hyung, "5% of GDP Set Aside for Science Research," *Korea Times*, December 12, 2009.
[40] Brazil Innovation Secretary Francelino Grando, "Brazil's New Innovation System," National Academies symposium, *Clustering for 21st Century Prosperity,* Washington, DC, February 25, 2010.

3.9 percent in 2010, one of the highest levels in the world.[41] In 2006, the Singapore government tripled its five-year R&D budget and set a target of pushing national spending to 3.5 percent of GDP by 2015.[42]

In the United States the growth in pubic R&D funding has been more uneven. Public research spending received an $18.7 billion temporary boost under the 2009 American Recovery and Re-investment Act of 2009. Congress approved significant long-term increases to non-defense R&D investment when it passed the America COMPETES Act, which pledges to double the research budget of the NSF, the DOE's Office of Science, and NIST over seven years. However, the COMPETES Act has not yet been funded by Congress and its prospects are uncertain in the current budgetary environment.

Federal commitments to higher research spending have been flat or falling. Overall federal funding for R&D in the United States has not increased significantly since 2004,[43] and the full-year continuing resolution passed by Congress for fiscal year 2011 cut R&D spending by 3.5 percent to $144.4 billion. Under the resolution, the NIH budget was reduced by 1.1 percent, the DOE's energy programs by 14.6 percent, the Office of Science by 1.6 percent, the NSF by 1.3 percent, and NIST by 2.5 percent.[44] The Obama Administration proposed a substantial 7.3 percent increase in non-defense R&D spending for fiscal year 2011-2012. Federal support for basic and applied research, in fact, would reach its highest level in history under the proposed budget. Under the President's plan, the NSF, NIST, and DOE would see especially large percentage increases.[45] However, fiscal challenges, precipitated by concerns about the rapid growth in the federal debt, leave the prospect of rising budgets for research and development uncertain.

These developments come at a time when federal spending on R&D as a share of GDP has been in long-term decline.[46] This decline has been masked by rising private-sector R&D spending, which has maintained total U.S. R&D spending as a percentage of GDP at a roughly constant level over the past few decades. [See Figure 2.2] The increased business R&D intensity has enabled

---

[41] Statistics Finland, Science and Technology Statistics accessed at http://www.research.fi/en/resources/R_D_expenditure/R_D_expenditure_table and Statistics Finland, "R&D Expenditure in the Higher Education Sector Up by 11 Per Cent," October 27, 2011.
[42] See Ministry of Trade and Industry, *Sustaining Innovation-Driven Growth, Science and Technology*, Government of Singapore, February 2006.
[43] Patrick J. Clemens, "Historical Trends in Federal R&D," in *AAAS Report XXXVI: Research and Development FY 2012*, Intersociety Working Group, American Association for the Advancement of Science, May 2011.
[44] See analysis by American Association for the Advancement of Sciences, "R&D in the FY 2011 year-Long Continuing Resolution," May 2, 2011.
[45] *AAAS Report XXXVI*, op. cit.
[46] Ben Bernanke, "Promoting Research and Development: The Government's Role." *Issues in S&T*, Volume XXVII (4) Summer 2011.

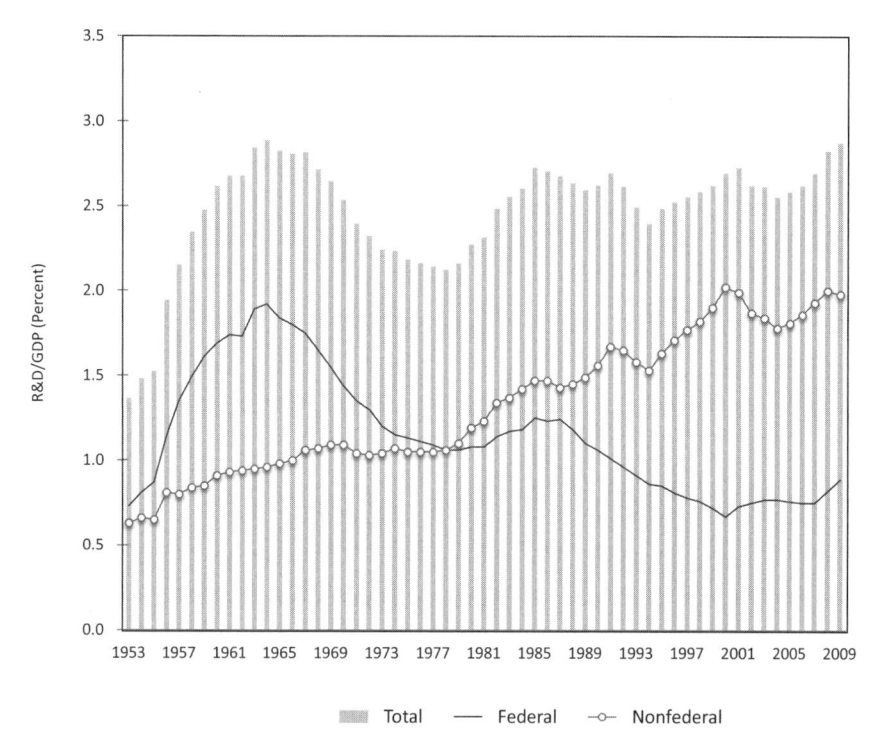

FIGURE 2.2 Federal funding for R&D as a share of GDP has been in long-term decline.
SOURCE: National Center for Science and Engineering Statistics, *U.S. R&D Spending Suffered a Rare Decline in 2009 but Outpaced the Overall Economy*, NSF 12-310 (March 2012), Figure 4.

total U.S. R&D spending to grow by 3.1 percent in constant dollars over the past 20 years.[47]

The private sector, however, spends nearly three-fourths of its R&D budget on applied R&D activities. [See Figure 2.3] The federal share, with its greater focus on basic R&D, has fallen steadily since the mid 1980s and now is about 0.7 percent of GDP —its lowest level since World War II.[48]

---

[47] National Science Foundation, *Science and Engineering Indicators: 2010*, Chapter 4.
[48] National Science Foundation Science and Engineering Indicators, 2010.

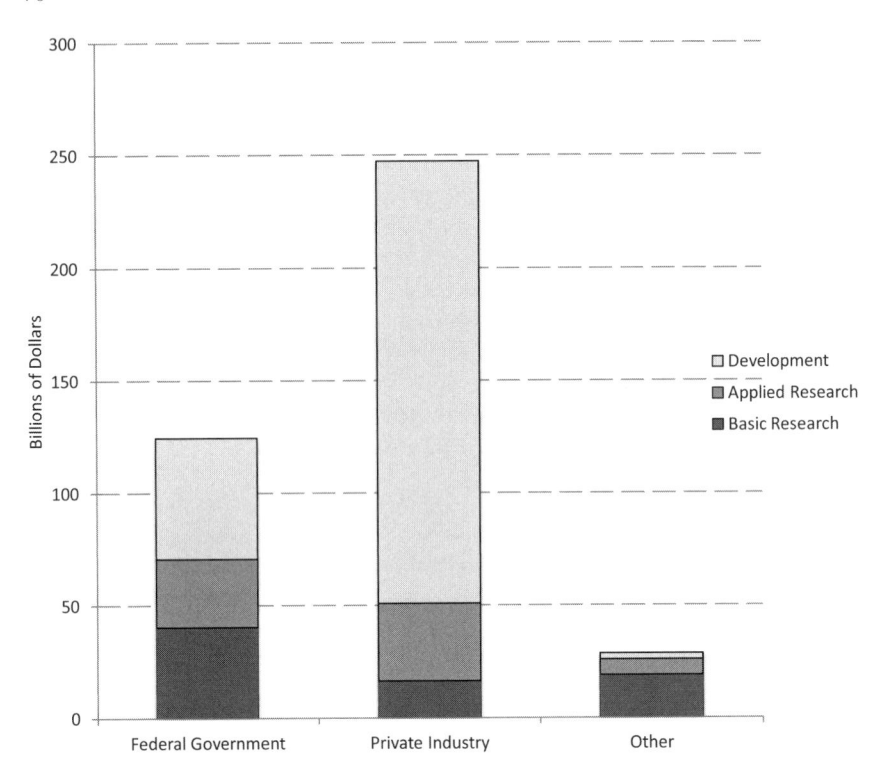

FIGURE 2.3 U.S. R&D spending by source of funding and character of expenditure, 2009.
SOURCE: National Science Foundation, National Center for Science and Engineering Statistics, *Science and Engineering Indicators 2012*, NSB 12-01 (January 2012), Appendix Tables 4-8, 4-9 and 4-10.

While the overall growth in total absolute R&D spending is good news, the downward trend in federal spending as a percent of GDP is less propitious for it is investments in basic research that generate the discoveries that lie behind future innovation. The burden of funding basic research is increasingly falling upon the federal government as U.S. corporations focus more of their R&D dollars on later-stage development.

The share of federal R&D that is targeted to basic research has also declined. The Department of Defense—which accounted for more than 52 percent of the federal research budget in 2011—invests around 90 percent of its R&D funds on weapons systems development, rather than on basic or applied research. [See Figure 1.4]

This does not mean the federal government can cut back on applied research. It does mean that the United States is spending a great deal less on early stage research than the official figures might suggest. It also means that much of the U.S. R&D effort is for later-stage military purposes with limited civil applications. The R&D spending of U.S. competitors tends to be the reverse, with heavier emphasis on later-stage R&D for commercial applications. As explained below, a greater emphasis on civilian applied research will be needed in order to compete with other nations that invest more to turn new technology into products and industry, keeping in mind that many of these products eventually have military applications.

These trends in R&D spending are not, of course, entirely uniform. Not all nations are meeting their research investment targets. In 2000, for example, the European Union set a target of 3 percent of GDP by 2010 for its members. But collectively the EU remains at 1.9 percent.[49] (There are notable exceptions: Germany and France are both significantly increasing their R&D budgets.[50]) In addition to the recent recession and financial crises, Battelle attributes the shortfall in part to high labor costs, which equal 70 percent of total R&D spending in Europe compared to 45 percent in the U.S. and 30 percent in non-Japan Asia.[51] Despite strong growth since 2002, R&D spending in Brazil remains below 1 percent of GDP, although this is counterbalanced by a substantial investment in FINEP, the Brazilian Technology Agency. FINEP has a $2.5 billion budget and focuses on applied research.[52]

While governments have increased research funding, some are having a difficult time getting the private sector to do the same. Chinese industry accounts for just 21 percent of the nation's R&D spending, and the vast majority of enterprises do not conduct continuous R&D.[53] In Canada, business spending on R&D has remained at only around 1 percent of GDP—compared to 1.6 percent for average OECD countries[54]--and fell in 2010 for the third year.[55] Singapore also has struggled to increase spending on innovation by private

---

[49] Börje Johansson, Charlie Karlsson, Mikaela Backman and Pia Juusola, "The Lisbon Agenda from 2000 to 2010," CESIS Working Paper No., 106, December 2007.

[50] Chancellor Merkel's government in Germany has proposed increasing R&D expenditures to 3 percent of GDP, up from 2.5 percent. See also remarks regarding European R&D targets by the European Commissioner for Research, Innovation, and Science Máire Geoghegan-Quinn, "Innovation for stronger regions: opportunities in FP7 Committee of the Regions" Brussels, July 14, 2011.

[51] Battelle and R&D Magazine, *2011 Global R&D Funding Forecast*, December 2010.

[52] Xinhua, "Financing agency boosts Brazil's innovation, productivity," March 6, 2011.

[53] See Chunlin Zhang, Douglas Zhihua Zeng, William Peter Mako, and James Seward, *Promoting Enterprise-Led Innovation in China*, Washington, DC: The International Bank for Reconstruction and Development/The World Bank, 2009.

[54] Science, Technology, and Innovation Council, *State of the Nation 2008*. Ottawa: CSTI Secretariat, 2008.

[55] The Daily, "Spending on Research and Development," Statistics Canada, December 24, 2010. Access at: http://www.statcan.gc.ca/daily-quotidien/101224/dq101224a-eng.htm.

domestic companies.[56] In the United States, by contrast, industry's share of R&D funding has risen steadily and is expected to reach 64 percent in 2012.[57] Industrial spending on R&D is forecast to account for all of the increase in U.S. R&D spending from 2011 to 2012.[58]

INSTITUTIONAL SUPPORT FOR APPLIED RESEARCH

One feature of several successful exporting nations and regions is strong public support for programs that help industries convert new technologies into manufacturing processes and products. In the United States, such collaboration on applied research typically occurs at universities that receive part of their funding from industry. Several other countries and regions have large national institutions employing thousands of scientists and engineers devoted to applied research. In such nations and regions, big public-private research institutes play a vital role in developing globally competitive industries: These institutions can effectively disseminate new technologies to a variety of domestic manufacturers. Small companies can often benefit from the lower cost through shared use of R&D personnel and equipment required to develop proofs-of-concept and to hone the manufacturing processes required for scale production.

As we see below, leading examples of institutions that support applied research include Germany's Fraunhofer, Taiwan's ITRI, and South Korea's ETRI.

Germany's Fraunhofer

Germany's Fraunhofer Gesellschaft is a network of institutes that offer some of the world's most successful applied-research programs.[59] Fraunhofer employs 4,000 Ph.D. and master's students and has a $2.2 billion annual budget. It essentially is a contract research organization, but Germany's federal government supplies a third of its budget. Another third is funded by the Länder, or state, governments. Private companies account for the final third. Fraunhofer operates 59 well-staffed Institutes of Applied Research across the country working closely with German manufacturers in 16 different innovation clusters. Fraunhofer Executive Director Roland Schindler described the organization as a "technology bridge," helping industry partners develop production processes,

---

[56] For example, see Richard W. Carney and Loh Yi Zheng, "Institutional (Dis)Incentives to Innovate: An Explanation for Singapore's Innovation Gap," *Journal of East Asia Studies* 9 (2): 291-319.
[57] Battelle, op. cit.
[58] Ibid.
[59] For a case study of the Fraunhofer Gesellschaft, see the annex to Chapter 5 of this volume.

materials, and product designs. Fraunhofer also contributes global market research and helps promote German products abroad.[60]

Taiwan's ITRI

Taiwan's government-owned Industrial Technology Research Institute (ITRI) is one of the foremost institutes of applied industrial research in the world. Half of its $600 million annual operating budget is provided by the government and half is derived from the private sector in the form of licensing fees and payments for contract R&D. It has a staff of 5,728 personnel, of which 1,163 hold PhD's and 3,152 Master's degrees. ITRI functions as a technology intermediary between the domestic and international research community, on the one hand, and Taiwanese Industry, on the other hand. It is "arguably the most capable institution of its kind in the world in scanning the global technological horizon for developments of interest in Taiwanese industry, and executing the steps required to import the technology—either under license or joint development...and then absorbing and adopting the technology for Taiwanese firms to use"[61]. Technology is transferred to Taiwanese industry through licensing arrangements, demonstration of process technologies on internal pilot manufacturing lines, incubation of start-ups spun off from ITRI labs, and the migration of ITRI personnel to Taiwanese companies. ITRI spinoffs were the genesis of Taiwan's semiconductor industry, a process which has been repeated in personal computers, lighting, displays, and photovoltaics[62]. ITRI fosters not only the start-up of companies to manufacture new products, but of complete industry chains, including design, materials, process technology development, equipment, packaging, testing, and applications.[63]

South Korea's ETRI

In South Korea, the government-funded Electronics and Telecommunications Research Institute (ETRI) plays a similar role. With

---

[60] Presentation by Roland Schindler at the National Academies Symposium on "Meeting Global Challenges: US-German Innovation Policy" November 11, 2010.
[61] John A. Matthews and Dong-Sung Cho, *Tiger Technology: The Creation of a Semiconductor Industry in East Asia,* Cambridge: Cambridge University Press, 2000.
[62] Sridhar Kota, "Technology Development and Manufacturing Competitiveness," Presentation to NIST, Extreme Manufacturing workshop, January 11, 2011. Chun-yen Chang, who founded Taiwan's first semiconductor research center at National Chiao Tung University, observed in a 2011 oral history interview that "[Y]ou can see that all the Taiwan high tech industry was originally from...the success of the semiconductor industries in Taiwan. We spun off [from the semiconductors] to LCD displays and then to the computer business. "Interview with Chun-yen Chang, *Taiwanese IT Pioneers: Chun-yen Chang,"* recorded February 16, 2011 (Computer History Museum, 2011), p. 11.
[63] Presentation by ITRI Display Technology Center Director John Chen, Hsinchu, Taiwan (February 14, 2012).

roughly 1,700 researchers with doctoral and master's degrees, ETRI is South Korea's largest research institute. ETRI was central to the development of the Korean semiconductor industry, participating in the industry-government research consortia that developed Korea's 256 megabit and 1 gigabit dynamic random access memories[64]. ETRI currently is number one in the world among public research organizations in terms of patents generated, with second place going to the University of California and third to MIT.[65] ETRI laboratories now specialize in fields such as information technology convergence, new materials, next generation semiconductors, and new broadcast and telecom technologies.[66] In the emerging field of flexible electronics, in which Korea is becoming a major player, ETRI is developing flexible memristor memory technology, utilizing graphenes, which are highly-conductive carbon nanoparticles seen as having a vast range of potential applications in electronics.[67]

U.S. Applied Engineering Programs

Federal applied R&D is fragmented among many agencies. A 2010 survey by MIT found that direct manufacturing R&D spending by the federal government, totaling over $700 million, is spread across four agencies. This number has risen significantly with new DARPA and DOE programs in 2011.[68]

The Manufacturing Extension Program of the U.S. Commerce Department, which helps small businesses apply new techniques and technologies, has a modest $125 million annual budget spread among 66 centers across the country, supported on a matching basis by the states as well as through fees.[69]

The National Science Foundation supports a network of more than 60 Industry/University Cooperative Research Centers specializing in fields such as advanced electronics, materials, and manufacturing, including a photovoltaic consortium involving four universities, several national laboratories, and 15 industry partners.[70] NIST supports programs such as the National

---

[64] "Taedok to Become Mecca for Venture Firms," *Chonja Sinmun* (April 10, 1998).
[65] "Korea's ETRI: World Top Agency in Patents," Korea Times (April 4, 2012).
[66] Electronics and Telecommunications Research Institute, accessed at http://www.etri.se.kr/eng/.
[67] "Flexible Graphene Memristors," Printed Electronics World (December 9, 2010).
[68] MIT Washington Office, Survey of Federal Manufacturing Efforts, September 2010. Access at http://web.mit.edu/dc/policy/MIT%20Survey%20of%20Federal%20Manufacturing%20Efforts.pdf.
[69] For a comparative assessment of the MEP partnership, see Philip Shapira, Jan Youtie, and Luciano Kay. "Building Capabilities for Innovation in SMEs: A Cross-Country Comparison of Technology Extension Policies and Programs" *International Journal of Innovation and Regional Development*, 3-4 (2011): 254-272. See also Philip Shapira, "US manufacturing extension partnerships: technology policy reinvented?" *Research Policy*, Volume 30, Issue 6, June 2001, Pages 977–992.
[70] Thomas Peterson, "The NSF Model: The Silicon Solar Consortium." In National Research Council, *The Future of Photovoltaic Manufacturing in the United States*, C. Wessner, ed., Washington, DC: The National Academies Press, 2011.

Nanoelectronics Initiative[71] with a set of four research centers around the country[72] in which 35 universities, companies such as IBM and Texas Instruments, and government agencies are striving to develop semiconductor technologies that eventually will replace CMOS as the core technology in most integrated circuits.[73]

National laboratories also are playing a growing role in helping industry turn technology into products. The National Renewable Energy Laboratory in Boulder, Colo., is one of the few national laboratories where commercializing technology is a top mission. Since it was founded in the 1970s, NREL has helped a number of U.S. businesses pioneer new technologies in solar power, wind energy, and bio-fuels, although its budget has fluctuated widely. Some of America's largest applied technology programs are run by the military. The U.S. Army Tank Automotive Research, Development and Engineering Center (TARDEC), for example, collaborates extensively with private industry to apply advanced technologies in vehicles it develops.[74] TARDEC's mission, however, is to apply technologies for military needs, not commercial industries.

The U.S. government has recently launched several initiatives to boost federal support for programs aimed at translating new technology into commercial products. The DOE's Advanced Technology Vehicle Manufacturing program, for example, provides $25 billion in direct loans to automobile and component manufacturers to fund projects aimed at improving fuel-efficiency and reducing dependence on petroleum,[75] $2.4 billion of which is being used to develop advanced batteries and electrified vehicles. The Obama Administration's 2013 Fiscal Year budget request called for $500

---

[71] For the latest assessment of this initiative, see the President's Council of Advisors on Science and Technology, "Report to the President And Congress on the Fourth Assessment of the National Nanotechnology Initiative," Washington, DC: The White House, April 2012. See also Semiconductor Industry Association, "Nanoelectronics Research Initiative: A Model Government-Industry Partnership Promoting Basic Research." Access at http://www.sia-online.org/clientuploads/One%20Pagers/Nanoelectronics_SRC_FINAL.pdf.

[72] The four institutes are the South West Academy of Nanoelectronics (SWAN), headquartered at the Microelectronics Research Center at The University of Texas at Austin; The Western Institute of Nanoelectronics (WIN) in California, headquartered at the UCLA Henry Samueli School of Engineering and Applied Science; The Institute for Nanolectronics Discovery and Exploration (INDEX) in Albany, NY, headquartered at the College of Nanoscale Science and Engineering of the University at Albany; and The Midwest Institute for Nanoelectronics Discovery (MIND), led by the University of Notre Dame and includes Pennsylvania State University, Purdue University, and University of Texas-Dallas.

[73] CMOS, patented by Frank Wanlass in 1967, stands for complementary metal-oxide semiconductor. CMOS is a technology for constructing integrated circuits that is used in devices such as microprocessors, static random-access memories, and image sensors.

[74] See presentations by Grace Bochenek and Sonya Zanardelli of the U.S. Army Tank and Automotive Research, Development, and Engineering Center at the National Research Council conference on *Building the U.S. Battery Industry for Electric-Drive Vehicles: Progress, Challenges, and Opportunities,* Livonia, Michigan, July 26, 2010.

[75] The Advanced Technology Vehicles Technology Loan Program was authorized under section 136 of the energy Independence and Security Act of 2007 (P. L. 110-140).

million for the DOE to aid advanced manufacturing in flexible electronics and lightweight vehicles, $200 million to DARPA for advanced manufacturing research, and increases for NSF programs relating to cyber physical systems, robotics, and advanced manufacturing.[76]

Another new U.S. government initiative is aimed at boosting federal assistance to development of commercial drugs. The National Institutes of Health announced Dec. 7, 2010, it would create the National Center for Advancing Translational Sciences (NCATS) by reallocating $700 million from other programs. The aim is to accelerate the pace of new drug development being brought to market by the pharmaceutical industry.[77] However, this reallocation has not taken place; instead other programs, such as Therapeutics for Rare and Neglected Diseases (TRND), have been merged and now continue under the NCATS title. Should this trend continue, it would mean that a lower program level will be available for new translational drug R&D than initially announced.

State Programs

Several state governments have begun to invest in public-private applied research institutes aimed at stimulating local manufacturing industries. One of the biggest is the Albany NanoTech Complex at SUNY Albany. The complex was launched by the state government in cooperation with corporations such as IBM, Applied Materials, and Tokyo Electron. It includes one of the world's most advanced 300 mm research fabrication plants devoted to developing prototypes of semiconductors. The complex has generated $5 billion in private investment, has 250 corporate partners, and houses 2,500 researchers, students, faculty, and staff.[78] SUNY Albany's College of Nanoscale Science and Engineering also runs a $50 million prototyping facility for micro-electromechanical systems (MEMs) and optoelectronics devices in Canandaigua, N. Y. The goal is to accelerate development of commercial devices that will be manufactured in the region.[79] Other public-private programs for assisting manufacturing at the state level include the Florida Center for Advanced Aero-Propulsion and the Laboratory for Surface Science and Technology at the University of Maine and the Ohio's Edison Technology

---

[76] Sridhar Kota, "Opening Remarks" at the National Research Council conference on *Building the U.S. Battery Industry for Electric-Drive Vehicles: Progress, Challenges, and Opportunities,* Livonia, Michigan, July 26, 2010.
[77] See Gardiner Harris, "Federal Research Center Will Help Develop Medicines," *New York Times*, January 22, 2011.
[78] Source: College of Nanoscale Science and Engineering at the University of the University of New York at Albany (SUNY-Albany). Also Pradeep Haldar "New York's Nano Initiative," in National Research Council, *Growing Innovation Clusters for American Prosperity,* C. Wessner, ed., Washington, DC: The National Academies Press, 2011.
[79] College of Nanoscale Science & Engineering press release, October 23, 2010.

Centers, which includes the Northeast Ohio Manufacturing Advocacy and Growth Network (MAGNET).[80]

Reflecting what they see as an institutional gap in the U.S. innovation system, Germany's Fraunhofer institutes are helping fill what they see as a gap in the U.S. innovation system by opening a number of U.S. applied technology institutes, often in collaboration with U.S. industries. Fraunhofer USA opened a non-profit state-of-the-art center to develop prototypes for laser components and systems in Plymouth, Mich., for example, and a center in Brookline, Mass., for manufacturing innovation. Other Fraunhofer centers in the U.S. focus on products such as advanced coatings, clean-energy devices, software, and molecular biotechnology applications.[81]

Lessons and Calls for New U.S. Institutions

The decades-old experience of organizations such as Fraunhofer, ITRI, and ETRI suggest that applied research programs run most effectively with significant, reliable, and steady financial commitment from both the government and the private sector to develop new technological options and sustain new or existing industries. Such programs also require the flexibility to adjust to new technology trends and to capture new commercial opportunities. At the same time, much of the focus of these institutions is on incremental improvements to existing industries and firms to enable them to remain globally competitive.

Some experts recommend that the federal government support new public-private intermediary institutions to accelerate industrialization of new technologies. Sridhar Kota, formerly assistant director for advanced manufacturing at the White House Office of Science and Technology Policy, has called for the U.S. to establish "Edison Institutes" modeled after those of Fraunhofer to help make maturing technologies ready for manufacturing. "We need strategic and coordinated investments to transition home-grown discoveries into home-grown products," Dr. Kota contends.[82]

---

[80] The NSF Science and Engineering Indicators for 2012 (Chapter 4) reports that $28.6 billion in 2009, or about 7% of all funding in the US. comes from sources that include academia's own institutional funds (which support academic institution's own R&D), other nonprofits (the majority of which fund their own R&D, but also contribute to academic research), and state and local governments (primarily for academic research).
[81] Presentation by Roland Schindler at the National Academies Symposium on "Meeting Global Challenges: US-German Innovation Policy" November 11, 2010.
[82] Sridhar Kota, "Opening Remarks" at the National Research Council conference on *Building the U.S. Battery Industry for Electric-Drive Vehicles: Progress, Challenges, and Opportunities,* Livonia, Michigan, July 26, 2010.

Recent Initiatives

In its most recent report to the President on Advanced Manufacturing, the PCAST characterizes U.S. private sector's under-investment in important emerging technologies and in the infrastructure to support advanced manufacturing as a market failure. The report notes that individual companies cannot justify such investments because they cannot capture all the benefits for themselves. Instead, the benefits would spill over to many competitors. As a result, PCAST argues, the public sector has an important role in ensuring that new technologies are not only developed but also produced in the U.S.[83]

A number of government policy proposals have been offered to bolster U.S. manufacturing through support for applied research. The most recent PCAST report, for example, called for an Advanced Manufacturing Initiative spearheaded by the departments of Commerce, Defense, and Energy and coordinated by the Office of Science and Technology Policy, the National Economic Council, or the Office of the Assistant to the President for Manufacturing. Among other things, PCAST calls for federal investment of $1 billion annually for four years to support applied-research programs in potential transformational technologies, public-private partnerships to facility development of broadly applicable technologies, dissemination of new design methodologies, and shared technology infrastructure that would help U.S. manufacturers. PCAST also calls for reforms in corporate income taxes and measures to expand the skilled workforce.[84] So far, however, no legislation establishing these programs has been introduced into Congress. Spence and Hlatshwayo advocate co-investment with the private sector to better align private incentives with social objectives. "It is probably a good idea to explicitly target some of the public-sector investment at technologies with the potential to expand the scope of the tradable sector and employment."[85]

This call has been followed up with the recently announced National Network for Manufacturing Innovation (NNMI)— an association of precompetitive public-private consortia to conduct applied research on new technologies and design methodologies.[86] According the Federal Register

---

[83] PCAST, *Report to the President on Ensuring American Leadership in Advanced Manufacturing,* op. cit.
[84] Ibid.
[85] Michael Spence and Sandile Hlatshwayo, "The Evolving Structure of the American Economy and the Employment Challenge," *Council on Foreign Relations Working Paper*, March 2011.
[86] NNMI appears to be modeled in concept on Germany's Fraunhofer-Gesellschaft, NNMI. See Chapter 5 of this report for a description of the Fraunhofer Gesellschaft. See also the presentation by Roland Schindler, Executive Director of Fraunhofer CSE, at the National Academies Symposium on Meeting Global Challenges: U.S.-German Innovation Policy, Washington, DC, November 1, 2010. Germany's Fraunhofer system has established seven research institutes based at U.S. universities, including Michigan State University, Boston University, Massachusetts Institute of Technology, the University of Maryland, the University of Michigan, Johns Hopkins University, and the University

notice, "The proposed Network will be composed of up to fifteen Institutes for Manufacturing Innovation (IMIs or Institutes) around the country, each serving as a hub of manufacturing excellence that will help to make United States (U.S.) manufacturing facilities and enterprises more competitive and encourage investment in the U.S. ... The NNMI program will be managed collaboratively by the Department of Defense, Department of Energy, Department of Commerce's NIST, the National Science Foundation, and other agencies. Industry, state, academic and other organizations will co-invest in the Institutes along with the NNMI program."[87]

STRENGTHENING MANUFACTURING

The innovation challenge the United States faces in the 21st century was brought about by the transformation of the global economy in the last decades of the 20th century. Dramatic changes in the location of international production and in the direction of international trade flows resulted from the integration of the emerging economies into world commerce. Foreign direct investment into emerging markets transferred capital and know-how. World trade expanded more rapidly than world output, and trade in high-technology products expanded more rapidly than trade in general. This was due in large part to an increase in the growth of knowledge- and technology-intensive industries worldwide, but especially in emerging economies as they liberalized markets, increased spending on R&D and education, and adopted policies to encourage high-technology manufacturing production and exports.[88] The development of global supply chains initially increased specialization as lower value-added production was moved to lower cost locations.

Emerging economies increasingly have moved up the value-added supply chain so that they are now competing in the same product and technology space as the United States. One measure of this increased competition is the deterioration in the U.S. trade balance in advanced-technology products that began in the late 1990s. [See Figure 2.4] The trade deficit in advanced technology products, based on data through August, will set an all-time high in 2011.

The policy objective of other nations, including emerging economies like China, and India is to move up the manufacturing value-added chain by driving innovation in their economies and increasing the technology intensity of their manufactured exports. As they do so, the United States faces increased

---

of Delaware. These institutes provide research and development services to help translate the fruits of research at U.S. academic institutions into products for the marketplace.

[87] Federal Register Notice, May 4, 2012. The President's FY 2013 budget requests $1 billion for the NNMI program.

[88] National Science Foundation, *Science and Engineering Indicators 2010*, chapter 6.

competition in the tradable goods manufacturing sector and increased pressure on domestic manufacturing production and employment.

To be sure, other countries are pursuing these innovation-led policies not out of any desire to cause economic disadvantage to the United States, but because it offers them the best prospects for economic growth and a high standards of living for their citizens. A recent IMF study summarized it as follows: "Technology intensive export structures generally offer better prospects for future economic growth. Trade in high-technology products tends to grow faster than average, and has larger spillover effects on skills and knowledge-intensive activities. The process of technological absorption is not passive but rather 'capability' driven and depends more on the national ability to harness and adapt technologies rather than on factor endowments."[89]

These changes in technology and trade are massive and are occurring with great rapidity from a historical perspective. In little over a decade, for example, China has increased its share of world high-technology manufactured exports from 6 percent to 22 percent and is now the world's largest exporter of these products. Over the same period, the U.S. share of high-technology manufactured exports fell from 21 percent to 15 percent.[90] [See Figure 2.4]

China's increase in its share of high-technology exports is reflected in statistics published by the U.S. Census Bureau on trade in advanced-technology products.[91] As shown in Table 2.1, the U.S. trade deficit in advanced-technology products in 2011 was concentrated in China. But this is more a reflection of U.S. loss of competitiveness with the Pacific Rim area in general because China primarily is an assembler of high-technology components made in nations and regions such as Japan, Taiwan, Korea, and the United States.[92] China and other emerging economies, however, are continuing to move

---

[89] The traditional factor endowments are labor and capital. See International Monetary Fund, "Changing Patterns of Global Trade," June 15, 2011, pp. 8-9. Paul Romer much earlier stated the same idea differently. "But our knowledge of economic history, of what production looked like 100 years ago, and of current events convinces us beyond any doubt that discovery, invention, and innovation are of overwhelming importance in economic growth and that the economic goods that come from these activities are different in a fundamental way from ordinary objects." Paul Romer, "Idea Gaps and Object Gaps in Economic Development," *Journal of Monetary Economics* 32 (1993): 562.

[90] National Science Foundation, *Science and Engineering Indicators 2010*, chapter 6. Data published by the World Bank show similar, but somewhat different results, with China's share at 20.4 percent in 2008 and the U.S. share at 12.4 percent. World Bank, World Development Indicators at http://data.worldbank.org/indicator/TX.VAL.TECH.CD.

[91] The data for advanced technology products put together by the Census Bureau is constructed from more highly disaggregated product definitions allowing for a more precise measure of U.S. trade in technology intensive products than the high technology industry-based OECD classification used in Figure 1.11. National Science Foundation, *Science and Engineering Indicators 2010*, pp. 6-34.

[92] Robert Koopman, William Powers, Zhi Wang and Shang-Jin Wei, "Give Credit Where Credit Is Due: Tracing Value Added in Global Production Chains," *NBER Working Paper No. 16426*, September 2010. See also Robert Koopman, Zhi Wang and Shang-Jin Wei, "A World Factory in Global Production Chains: Estimating Imported Value Added in Chinese Exports," *Centre for Economic Policy Research Discussion Paper No. 7430*, September 2009.

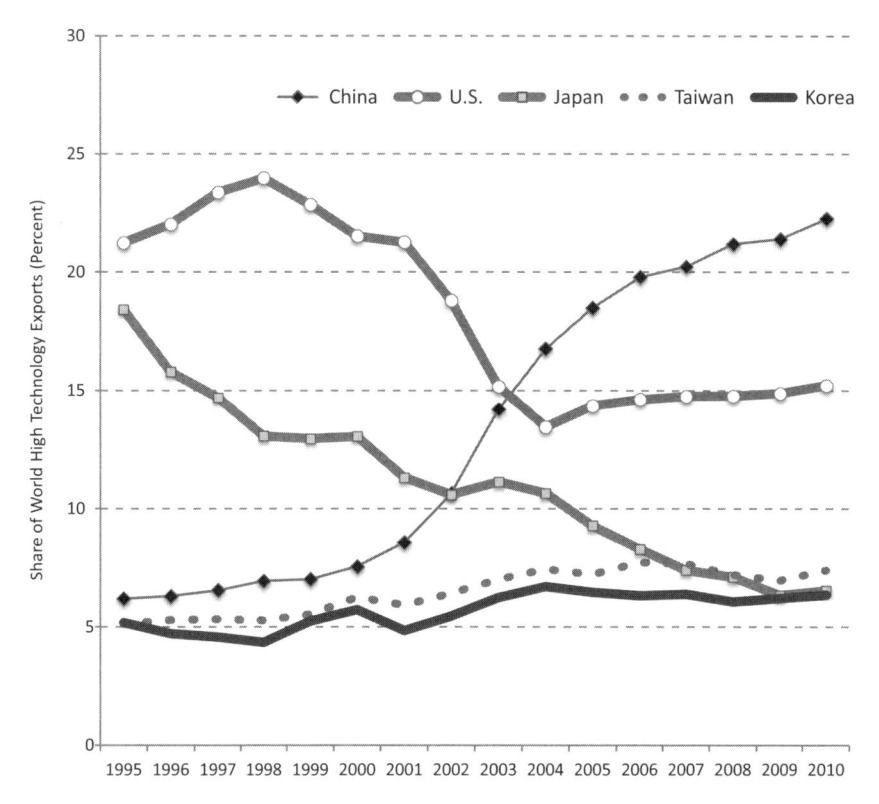

FIGURE 2.4 World export shares of high-technology goods.
SOURCE: National Science Foundation, National Center for Science and Engineering Statistics, *Science and Engineering Indicators 2012*, NSB 12-01 (January 2012), Appendix Table 6-24.

upstream in the global supply chain, increasing competition for U.S. based manufacturing.[93]

By shifting and reorganizing global supply chains, the globalization of the world economy has also affected the price of products, employment patterns and wages in advanced and emerging economies alike. One of the most significant changes for the United States, as documented in a recent study by Spence and Hlatshwayo, is that from 1990 to 2008, almost all incremental employment growth came from the non-tradable sector of the U.S. economy,

---

[93] George Tassey, "Rationales and Mechanisms for Revitalizing US Manufacturing R&D Strategies," *Journal of Technology Transfer* (2010) 35, pp. 283–333 and International Monetary Fund, "Changing Patterns of Global Trade," June 15, 2011, pp. 27-29.

primarily government and health care jobs.[94] There were job gains in the tradable sector in high-end services (management and consulting, computer systems design, finance and insurance) but these were offset by losses in most areas of manufacturing.[95] The authors state that the manufacturing job losses were due to lower value-added positions moving offshore while higher value-added positions remained in the United States. Looking ahead, with budget constraints at all levels of government and growing pressures to rein in the rate of growth in health care costs, major gains in future employment are unlikely to come from the non-tradable sector. The authors believe the answer lies in expanding the U.S. export sector in both high-end manufacturing and services. "To create jobs, contain inequality, and reduce the U.S. current-account deficit, the scope of the export sector will need to expand. That will mean restoring and creating U.S. competitiveness in an expanded set of activities via heightened investment in human capital, technology, and hard and soft infrastructure. The challenge is how to do it most effectively."[96]

Because of the interrelationships between manufacturing and services, expanding the scope of the U.S. export sector will also necessarily expand high value-added services. As manufacturing has become more technology-intensive, the scope and nature of manufacturing has changed, increasing the demand for service occupations and service inputs at the expense of machine operators and assembly-line workers.[97] "Data on occupations show that in the last decade there has been a steady increase in the share of employees in the manufacturing sector who are employed in occupations that can be considered as services-related" while at the same time in countries like the United States manufacturing has become more service intensive.[98] For example, industrial products increasingly are comprised of a combination of mechanical, electrical and software components that make them more innovative, more capable and more easily updated and enhanced.[99] Thus as Gregory Tassey has stated, "the fast-growing high-tech services sector must have close ties to its manufacturing base."[100]

---

[94] Michael Spence and Sandile Hlatshwayo, "The Evolving Structure of the American Economy and the Employment Challenge," *Council on Foreign Relations Working Paper*, March 2011.

[95] The authors state that the manufacturing job losses were due to the lower value added positions moving offshore while higher value added positions remained. *Id.* at 31.

[96] *Id.* at 5.

[97] OECD, *OECD Science, Technology and Industry Scoreboard 2011*, Paris: OECD, September 20, 2011, p. 168 and Dirk Pilat and Anita Wölfl, "Measuring the Interaction Between Manufacturing and Services," OECD STI Working Paper, DSTI/DOC(2005)5, May 31, 2005.

[98] OECD, *OECD Science, Technology and Industry Scoreboard 2011*, id. The OECD estimated that in 2008 services-related occupations in manufacturing in the United States were just over 50 percent of all employees in manufacturing.

[99] Jim Brown, "Issue in Focus: Systems and Software Driven Innovation," Tech-Clarity, 2011. As Janos Sztipanovits, director of Vanderbilt University's Institute for Software Integrated Systems, stated "More and more industrial products internal complexity is concentrating in software." Kate Linebaugh, "GE Makes Big Bet on Software Development," *The Wall Street Journal*, November 17, 2011.

TABLE 2.1 U.S. Trade in Advanced Technology Products by Country and Region in 2011

By Country and Region (Billions of Dollars)

Country	Exports	Imports	Balance
China	20.1	129.5	-109.4
Ireland	2.5	21.6	-19.1
Mexico	31.9	47.8	-15.9
Taiwan	8.8	18.7	-9.9
Japan	15.7	25.5	-9.8
Korea	11.3	17.5	-6.2
Malaysia	8.0	14.1	-6.1
Thailand	2.9	7.9	-5.0
France	10.5	11.2	-0.7
Germany	13.4	12.7	0.7
Singapore	10.3	8.1	2.1
U.K.	14.0	10.1	3.9
Brazil	11.7	1.0	10.6
Canada	30.3	13.5	16.8

By Region (Billions of Dollars)

Region	Exports	Imports	Balance
Pacific Rim	95.7	219.9	-124.2
EU	67.0	75.3	-8.3
Other	124.0	90.8	33.2
World	286.7	386.0	-99.3

SOURCE: U.S. Census Bureau, Foreign Trade, Trade in Goods with Advanced Technology Products.

Seen in this context, the innovation challenge that the United States faces is at the same time a trade competitiveness challenge and a high-tech manufacturing and services challenge. Therefore, a fundamental objective of capturing the economic value of innovation has to be increasing the output of manufacturing in the United States for high-technology, high valued-added products to grow U.S. exports and employment.[101]

---

[100] Gregory Tassey, "The Manufacturing Imperative," presentation at NAS Conference on the Manufacturing Extension Partnership, November 14, 2011. Tassey also points out that the manufacturing sector accounts for 67 percent of R&D performed by industry and 57 percent of scientists and engineers in industry are employed by manufacturing.
[101] Tassey argues that "Once the premise is accepted that the only way to achieve long-term growth in jobs for a high-income economy such as the United States is through investment in technology,

84 *RISING TO THE CHALLENGE*

The Link between Manufacturing and Innovation

Manufacturing is integral to new product development. Production lines are links in an iterative innovation chain that includes pre-competitive R&D, prototyping, product refinement, early production, and full-scale production.[102] U.S. corporations still dominate a number of industries, such as personal computers and certain semiconductors, even though end products are produced offshore.[103] America's logic chip-design industry, which includes companies like Qualcomm, Nvidia, and Broadcom, relies almost entirely on silicon wafers fabricated in Asian foundries, while Apple iPods, iPhones, and iPads are assembled in China by the Taiwanese firm Hon Hai Precision Industry. In such products, the greatest economic value is in software, microprocessors, and proprietary designs, while the hardware is generally comprised of standardized parts and assembled with standard production processes.

In many high technology industries, however, design is not so easily separated from manufacturing. Production processes for advanced solar cells, lithium-ion vehicle batteries, and next-generation solid-state lighting devices are highly proprietary to the producing company and often constitute a competitive advantage. If new U.S. companies lack the domestic capability to scale up, Intel founder Andy Grove warns, "we don't just lose jobs -- we lose our hold on new technologies. Losing the ability to scale will ultimately damage our capacity to innovate."[104]

---

innovation, and subsequent productivity increases, the key policy issue becomes how to promote desired long-term investment in a domestic economy that must save more and consume less, while reducing budget deficits through decreased spending and increased taxes." George Tassey, "Rationales and Mechanisms for Revitalizing US Manufacturing R&D Strategies," *Journal of Technology Transfer* (2010) 35, pp. 303-304.

[102] See President's Council of Advisors on Science and Technology, "Sustaining the Nation's Innovation Ecosystems: Information Technology Manufacturing and Competitiveness," January 2004. (http://www.choosetocompete.org/downloads/PCAST_2004.pdf). See also President's Council of Advisors on Science and Technology, "Report to the President on Ensuring American Leadership in Advanced Manufacturing," June 2011.

[103] A recent National Research Council study of a range of technology-intensive industries found that in many cases U.S. companies dominated market share, profits, and innovation despite a considerable shift of manufacturing and R&D work offshore. See National Research Council, *Innovation in Global Industries: U.S. Firms Competing in the World*, Jeffrey T. Macher and David C. Mowery, editors, Washing The National Academies Press, 2008.

[104] Andy Grove, "How to Make an American Job Before it is Too Late," *Bloomberg BusinessWeek*, July 1, 2010.

> **Box 2.2**
> **The Case of the Display Industry**
>
> A clear example of how loss of one manufacturing industry prevents development of others is computer and TV displays. Asian producers assumed dominance of liquid-crystal displays in the 1990s as U.S. producers abandoned the industry.[105]
>
> The development by U.S. companies of key technologies and materials for displays on flexible, rather than glass, substrates would seem to present a fresh opportunity for America to re-enter the potentially huge display industry. According to Ross Bringans of the Palo Alto Research Center, "flexible electronics is a very exciting direction, and there will be a lot of new technologies. We are certain that interesting business opportunities will flow out of that." According to Dr. Bringans, these opportunities are beginning to open, particularly in Europe and East Asia.[106]
>
> Two major barriers stand in the way of developing a robust U.S. based flexible electronics industry. The first is the commercial challenge of launching the industry. Bob Street of the Palo Alto Research Center has observed that Asian manufacturers such as Samsung will likely dominate this industry because the entry barriers are too high for U.S. production of displays: The ecosystem of production capacity, expertise in volume production, local equipment manufacturers, materials suppliers and technology developers reside in Asia.[107] The second challenge concerns the role of the government support. In this regard, a recent study commissioned by the National Science Foundation and the Office of Naval Research of European programs to support the development and commercialization of flexible electronics technologies found that "…the relatively low prevalence of actual manufacturing and advanced systems research and development in the United States has led to an incomplete hybrid flexible electronics R&D scenario for this country…."[108]

---

[105] Jeffrey Hart, "Flat Panel Displays," in National Research Council, *Innovation in Global Industries: U.S. Firms Competing in a New World,* Jeffrey T. Macher and David C. Mowery, Editors, Washington, DC: The National Academies Press, 2008. For a history of the flat panel display industry, see Thomas P. Murtha, Stefanie Ann Lenway, and Jeffrey A. Hart, *Managing New Industry Creation: Global Knowledge Formation and Entrepreneurship in High Technology*, Palo Alto: Stanford Business Books, 2002.

[106] Ross Bringans, "Challenges and Opportunities for the Flexible Electronics Industry," Presentation at the National Academies conference on "Flexible Electronics for Security, Manufacturing, and Growth In the United States." September 24, 2010.

[107] See Bob Street, "Next Generation: The Flex Display Opportunity" in *The Future of Photovoltaic Manufacturing in the United States,* C. Wessner, ed., Washington, DC: The National Academies Press, 2011.

[108] Ananth Dodabalpur et al., "European Research and Development in Hybrid Flexible Electronics." Baltimore MD: WTEC, 2010.

Support for Manufacturing Overseas

Some nations aggressively support manufacturing in favored industries with a range of policy tools. They include—

- **Financial Incentives:** China, Singapore, Malaysia, and other nations offer 10-year tax holidays to foreign companies building factories in desired industries. The use of tax credits that eventually refund a portion of a company's investment in plants or laboratories also is quite common. In Canada, for example, federal, provincial, and local governments offer some of the world's most generous tax incentives for aerospace manufacturing, including investment rebates and high depreciation allowances for machinery and equipment. Non-discretionary tax incentives for aerospace manufacturing equal $1,569 per job in Montreal and $2,617 in Winnipeg, compared to $624 in Seattle and $1,240 in Wichita.[109] Canada has become a major global manufacturer of civil helicopters, flight simulators, landing gear, and gas-turbine engines.[110]
- **Workforce Training:** Some nations design the curricula of universities and polytechnics to meet the projected needs for skilled workers in desired industries. They also cover the costs of worker training for foreign investors. For example, the mission of Singapore's Workforce Development Agency (WDA) is to "enhance the employability and competitiveness of everyone in the workforce, from the young to old workers, from the rank-and-file to professionals, managers and executives." It realizes this mission through training and education programs as well as workshops to upgrade worker skills.[111]
- **Leveraging Domestic Markets:** A number of countries use the buying power of the government and consumer subsidies to build local demand for domestic industries. Germany's feed-in tariffs, which are high enough to guarantee a financial return for both utilities and manufactures, largely explain why that nation has emerged as a global manufacturing leader of photovoltaic systems, for example.[112] Indeed,

---

[109] Invest in Canada Bureau, "Canada—A Strategic Choice: Canada as an Investment Destination for Aerospace" (undated).
[110] Ibid.
[111] Website of Singapore's Workforce Development Agency. Access at http://app2.wda.gov.sg/web/Common/homepage.aspx.
[112] A feed-in tariff is an incentive structure that sets by law a fixed guaranteed price at which power producers can sell renewable power into the electric power network. The tariff obligates regional or national electricity utilities to buy renewable electricity, such as electricity generated from solar photovoltaic panels, at above-market rates. See presentation by Bernhard Milow of the German Aerospace Center at the National Academies symposium on *Meeting Global Challenges: U.S.-German Innovation* Policy, November 1, 2010. Also see Michael J. Ahearn. "Opportunities and Challenges Facing PV Manufacturing in the United States." *The Future of Photovoltaics*

Germany's renewable-energy sector now employs 340,000, more than the auto industry.[113] To help meet its goal of having 2 million electric vehicles on the roads by 2020, the French government awards up to €5,000 to buyers of electric vehicles and plans to have state-owned companies and government agencies order 50,000 such vehicles for their fleets.[114] China offers a $9,036 subsidy to buyers of electric cars and subsidizes fleet operations in 25 cities as part of its target of selling 1 million electric vehicles *per year* by 2020.[115] To promote domestic manufacturers of solid-state lighting, which the government hopes will be a $30 billion export industry by 2015, China is rolling out a program to help 21 major cities install 1 million street lamps using light-emitting diodes.[116]

- **Trade Policy:** Although trade barriers have fallen dramatically around the world in recent decades, some nations continue to use a variety of official and unofficial policy tools to support domestic manufacturing. It is common for countries to require foreign defense and aerospace contractors, as well as vendors of big-ticket items such as power plants and rail stock, to source some parts or to perform final assembly domestically, for example. Of major trading nations, China has the most aggressive such "import substitution" policies. The government, which has not signed World Trade Organization protocols on government procurement, essentially compels foreign makers of everything from wind turbines to high-speed trains to manufacture in China and transfer technology to domestic companies.[117] Already a big exporter of solar panels, China requires at least 80 percent of equipment for its own solar power plants to be domestically produced.[118] A particularly controversial policy directs state agencies to

---

Manufacturing in the United States; Summary of Two Symposia, C. Wessner, ed., Washington, DC: The National Academies Press, 2011.
[113] Solar Progress, December 2010 Issue. Access at http://www.auses.org.au/wp-content/uploads/2010/12/SP_DEC10.pdf.
[114] David Pearson, "France Backs Battery-Charging Network for Cars," *Wall Street Journal*, Oct. 1, 2009.
[115] *People's Daily*, "China to Sell 1 Million New-Energy Cars Annually by 2015," Nov. 223, 2010. English translation viewable at http://english.peopledaily.com.cn/90001/90778/90860/7207607.html.
[116] China Research and Intelligence, "Brief of the LED Lighting Program of 10,000 Lights in 10 Cities in China," July 23, 2009. This article can be accessed at http://www.articlesbase.com/press-releases-articles/brief-of-the-led-lighting-program-of-10000-lights-in-10-cities-in-china-1061573.html.
[117] See Jason M. Forcier, "The Battery Industry Perspective," presented at the National Research Council conference on *Building the U.S. Battery Industry for Electric-Drive Vehicles: Progress, Challenges, and Opportunities*, Livonia, Michigan, July 26, 2010.
[118] Keith Bradsher, "China Builds High Wall to Guard Energy Industry." *International Herald Tribune*, July 13, 2009.

buy high-technology products that incorporate "indigenous innovation."[119]

U.S. Support for Manufacturing

The explicit national support for domestic manufacturing in Asia and European nations such as Germany has been in sharp contrast to the United States, where support for industry has tended to be limited to defense-related manufacturing and enforcing free-trade rules. A recent report by the President's Council of Advisors for Science and Technology (PCAST) warned that the U.S. is losing leadership in manufacturing, not only in low-tech industries that depend on low-cost foreign labor but also in high-tech products that result from U.S. innovation, inventions, and manufacturing-associated research and development.[120]

America's advanced manufacturing base faces formidable competitive challenges. In some cases, according to an analysis by Erica Fuchs and Randolph Kirchain, the cost gaps between manufacturing in the U.S. and Asia are so large that they discourage innovation. It makes more economic sense for companies to import products made with mature technologies than to domestically produce advanced, better-performing products made with new technologies.[121]

Offshore cost advantages in high-technology products often have little to do with labor rates because manufacturing is highly automated. According to an analysis by the Manufacturing Institute, non-production expenses such as high U.S. corporate taxes, employee benefits, torts, and pollution control put American-based manufacturing at an 18 percent structural cost disadvantage compared to major trading partners and more than a 50 percent disadvantage compared to China, although rising costs elsewhere and a weaker dollar have help narrow these gaps substantially since 2006.[122] Manufacturing executives

---

[119] The State Council, People's Republic of China, "National Medium- and Long-Term Program for Science and Technology Development, 2006-2020," (undated).
[120] President's Council of Advisors on Science and Technology, *Report to the President on Ensuring American Leadership in Advanced Manufacturing,* Executive Office of the President, June 2011.
[121] Fuchs and Kirchain demonstrated the "dilemma" of manufacturing products with prevailing designs offshore in order to reduce as opposed to manufacturing new-technology products in the U.S. by analyzing the optoelectronic device industry. See Erica R. H. Fuchs and Rondolph Kirchain, "Design for Location? The Impact of Manufacturing Off-Shore on Technology Competitiveness in the Optoelectronics Industry," *Management* Science, 56(12), pp. 2323-2349, 2010. In an analysis of optoelectronics devices, Fuchs found that U.S. manufacturing yields would have to increase.
[122] Jeremy A. Leonard, "The Tide Is Turning: An Update on Structural Cost Pressures Facing U.S. Manufacturer," The Manufacturing Institute and Manufacturers Alliance/MAPI, November 2008 (http://www.deloitte.com/assets/Dcom-UnitedStates/Local%20Assets/Documents/us_pip_TideIsTurning_093009.pdf). A recent Boston Consulting Group report predicts that, with respect to China, some manufacturing operations will return to the United States as wages increase in China and the U.S. dollar weakens. Harold L.

addressing NRC symposia also cited availability of workers, the lack of a domestic supply base, and inadequate access to capital for new plants or expansion as serious obstacles to keeping production in the United States.

State and federal policies and programs can help industry ameliorate these competitive gaps. Strategies for addressing these challenges include—

- **Financial Incentives:** The U.S. federal and state governments have increased incentives for domestic manufacturing. Michigan's $1.02 billion Advanced Battery Tax Credits program was instrumental in the state's success in drawing private investment in lithium-ion battery plants, for example.[123] New York and New Mexico are among the other states that have used aggressive tax credits to lure advanced manufacturing in desired industries. The federal government also has introduced a number of such incentives, especially over the past three years. The 2009 Recovery Act (ARRA) provided a one-time boost of $2.4 billion in grants earmarked for 48 advanced-battery manufacturing projects, for example. The Advanced Energy Manufacturing Tax Credit program provides $2.3 million to companies to cover 30 percent of investments in new, expanded, or refurbished manufacturing plants producing renewable-energy equipment.[124]

    The Department of Energy says the credits, which were matched by $5 billion in private investment, funded 183 projects in 43 states and created tens of thousands of jobs.[125] The federal government also has encouraged domestic manufacturing with tax deductions for consumer purchases of electrified vehicles, loan guarantees for green-technology projects, and greater access to export financing.

Sirkin, Michael Zinser and Douglas Hohner, *Made in America, Again: Why Manufacturing Will Return to the U.S.*, The Boston Consulting Group, August 2011.

[123] Michigan's Advanced Battery Tax Credits initiative was created through an amendment to the Michigan Business Tax Act, Public Act 36 of 2007, to allow the Michigan Economic Development Authority to tax credits for battery pack engineering and assembly, vehicle engineering, advanced battery technology development, and battery cell manufacturing. Under the scheme, Michigan refunds up to $100 million of a company's capital investment. Battery pack manufacturers receive a credit for each pack they assemble in Michigan. See presentation by Eric Shreffler, "Michigan Investments in Batteries and Electric Vehicles," at the National Academies Symposium on *Building the U.S. Battery Industry for Electric Drive Vehicles*, Livonia, Michigan, July 26, 2010.

[124] The Advanced Energy Manufacturing Tax Credit was authorized in Section 1302 of the American Recovery and Reinvestment Act and also is known as Section 48C of the Internal Revenue Code. It authorizes the Department of Treasury to award $2.3 billion in tax credits to cover 30 percent of investments in advanced energy projects, to support new, expanded, or re-equipped domestic manufacturing facilities.

[125] Carol Browner, "White House Blog: 183 projects, 43 states, Tens of Thousands of High Quality Clean Energy Jobs." January 8, 2010. Access at http://www.whitehouse.gov/blog/2010/01/08/183-projects-43-states-tens-tthousands-high-quality-clean-energy-jobs.

Incentive packages in the U.S. are still unable to match many of those offered by foreign governments, however. These broad-based packages of incentives, which range from tax holidays to free infrastructure, to cheap capital, to lax environmental and labor regulations, offer a coordinated program to create a non-market advantage using state resources. Such practices by China, Korea, and Taiwan (among others) have introduced a fundamental shift in cost and revenue that essentially changes the economic game.

While U.S. states have wide latitude to waive corporate taxes, for example, manufacturing plants still are required to pay federal corporate taxes, which are among the highest in the industrialized world. The Milken Institute argues that reducing the U.S. corporate tax rate to match the OECD average would create 350,000 new manufacturing jobs by 2019, while increasing the R&D tax credit by 25 percent and making it permanent would create 270,000 manufacturing jobs.[126] Financial analyst Steve O'Rourke of Deutsche Bank explained in a National Academies conference on photovoltaic manufacturing that "manufacturing migrates to where companies are most profitable, and the single biggest issue in this analysis is taxes."[127] Federal incentives have closed some of those cost gaps. In the case of photovoltaic manufacturing, Department of Energy official John Lushetsky estimated that the combination of U.S. and state incentives have closed about two-thirds of the cost advantage of operating a factory in China that is attributable to that country's incentives.[128]

A major concern voiced in STEP symposia about current federal incentives is that they are too short-term and unpredictable for long-term investments, with funding requiring frequent renewal by Congress. The controversy over the bankruptcy of Solyndra (a manufacturer of novel cylindrical solar panels) after receiving $535 million in federal loan guarantees, moreover, has raised concerns over

---

[126] Ross DeVol and Perry Wong, "Jobs for America: Investments and Policies for Economic Growth and Competitiveness," Milken Institute, January 26, 2010. Also see John Neuffer, "China: Intellectual Property Infringement, Indigenous Innovation Policies, and Frameworks for Measuring the Effects on the U.S. Economy," written testimony to the United States International Trade Commission Investigation No. 332-514 Hearing on behalf of the Information Technology Industry Council, June 15, 2010. (http://www.itic.org/clientuploads/ITI%20Testimony%20to%20USITC%20Hearing%20on%20China%20%28June%2015,%202010%29.pdf).
[127] Steve O'Rourke, "Financing Photovoltaics in the United States," in *The Future of Photovoltaics Manufacturing in the United States*, C. Wessner, ed., Washington, DC: The National Academies Press, 2011.
[128] From presentation by John Lushetsky of the Department of Energy at National Academies symposium "Meeting Global Challenges" in Washington, DC, November 1, 2010.

how such programs are administered—and highlight the political risks of supporting emerging technologies in the face of fierce import competition.[129]

- **Workforce Availability and Location:** Availability of engineers and workers with the right skills is another oft-cited reason for America's declining competitiveness in advanced manufacturing.[130] In addition, there is a growing concern that U.S. business school programs are not turning out enough graduates who can run manufacturing operations.[131] Availability of talent is the most important factor in a company's decision where to locate production, according to a recent survey of 400 global manufacturing executives. That report suggested the hollowing out of U.S. manufacturing is taking a toll on America's skill base. Once this "high degree of accumulated tacit knowledge" is lost, it warned, it "is difficult, if not impossible, to recover."[132] Some 60 percent of the science and engineering workforce will be eligible for retirement in the next five years, a prospect that former Under Secretary of Energy Kristina Johnson described as "a real national crisis."[133] In the field of electrical power engineering, an essential skill for the advanced-storage industry, an analysis by the Institute of Electrical and Electronics Engineers' Power & Energy Society concludes that U.S. graduation rates do not meet the nation's current and future needs.[134]

---

[129] See Eric Lipton and John M. Broder, "In Rush to Assist Solar Company, U.S. Missed Signs," *New York* Times, Sept. 22, 2011, and Melissa C. Lott, "Solyndra—Illuminating Energy Funding Flaws?", *Scientific* American, September 27, 2011.

[130] While the overall number of scientists and engineering graduates has grown over the past 3 decades to about 4.3 percent of all U.S. jobs (NSF S&E Indicators 2012, Chapter 3), industry surveys show a shortage of workers with the necessary level and mix of skills needed on the factory floor. See Deloitte and U.S. Council on Competitiveness, "2010 Global Manufacturing Competitiveness Index." Demand for industrial engineers has remained high even in the recent recession. For a review of the rapidly changing nature of factory employment, see also *The Economist*, "Factories and Jobs: Back to Making Stuff," April 12, 2012.

[131] See Jack R. Meridith, "Hopes for the future of operations management," *Journal of Operations Management* 19 (2001) 397–402. The author notes that "Operations Management has a much longer history than our sister functions in business: finance, marketing, accounting, etc. Yet, we still struggle with fewer majoring students, fewer and newer journals, less academic respect, greater student fear, and fewer professors."

[132] Aleda V. Roth, et. al, "2010 Global Manufacturing Competitiveness Survey," Deloitte Touche Tohmatsu and U.S. Council on Competitiveness, June 2010.

[133] Kristina Johnson, "Advancing Solar Technologies: The U.S. Department of Energy's Perspective," in *The Future of Photovoltaics Manufacturing in the United States*, C. Wessner, ed., Washington, DC: The National Academies Press, 2011.

[134] Amy Fischbach, "Engineering Shortage Puts Green Economy and Smart Grid at Risk," Transmission and Distribution World, April 21, 2009, (http://blog.tdworld.com/briefingroom/2009/04/21/engineer-shortage-puts-green-economy-and-smart-grid-at-risk/).

A number of innovative partnerships between industry, schools, and state government agencies are underway to address this skills gap.[135] Indiana has launched a new kind of community college called Ivy Tech with 23 campuses and 130,000 students. One of its strengths is working with industry to train "middle-skill workers," those with two years of college but who did not earn a bachelor's degree in engineering. Fifty-six percent of demand for all workers in Indiana is classified as middle skill, while only 45 percent of Indiana's workforce has such training.[136] The state of Michigan has a number of programs to train workers and engineers for emerging industries such as advanced batteries, electric vehicles, and solar power with financial support from the DOE and the U.S. Army's TARDEC.[137]

The federal government also provides training for trade-affected workers who have lost their jobs as a result of increased imports or shifts in production out of the United States through the Trade Adjustment Assistance Program (TAA).[138] Administered by the Department of Labor, this $1 billion a year program includes assistance for displaced workers to find and relocate to new jobs, and training for workers to develop skills demanded in existing labor markets. This includes classroom and on-the-job training, as well as customized training to meet the needs of a specific employer. In some instances, the program also provides income support to workers who are participating in full-time training.

- **Promoting Markets:** U.S. competitors in Asia and Europe recognize that emerging technologies, such as solar photovoltaics or lithium-ion batteries for vehicles, generally do not have existing market structures and in fact have almost always been established by some sort of non-market support.[139] This is especially true of the first instantiations of

---

[135] For a revealing comparative case studies of the importance of social networks in the divergent trajectories of post-industrial regions, see Sean Stafford, *Why the Garden Club Couldn't Save Youngstown: The Transformation of the Rust Belt,* Cambridge MA: Harvard UP, 2009.
[136] Data from Indiana Department of Workforce Development and U.S. Census Bureau.
[137] Presentation by Simon Ng, "Technical Training and Workforce Development," at the National Academies Symposium on *Building the U.S. Battery Industry for Electric Drive Vehicles.* July 26, 2010.
[138] See Harold F. Rosen, "Strengthening Trade Adjustment Assistance," Policy Brief 08-02, Peterson Institute for International Economics, 2008. Access at http://www.iie.com/publications/pb/pb08-2.pdf. For a review of issues relating to training programs and global competitiveness focusing on the TAA program, see the transcript of the Hearing before the House Ways and Means Committee, "Promoting U.S. Worker Competitiveness in a Globalized Economy." June 14, 2007. Serial No. 110-47, Washington, DC: USGPO, 2008. Access at http://www.gpo.gov/fdsys/pkg/CHRG-110hhrg43113/pdf/CHRG-110hhrg43113.pdf.
[139] The instrumental role of procurement in the development of leading U.S. industries is exemplified by support by the Department of Defense for integrated circuits and advanced

new technologies as contrasted with derivative technologies and products within a technology area (such as the tablet computer as a melding of the laptop computer and cell phone). While other nations pursue active commercial market development strategies through subsidies and other preferential treatment, the debate on this issue continues in the U.S., with some contending that, for example, alternative energy technologies ought not to be subsidized, even though they cannot compete on a cost per kilowatt basis with entrenched incumbent technologies.

In the absence of initial markets of sufficient scale puts the U.S. at a competitive disadvantage in several promising emerging technology industries. Because the largest markets for solar panels have been in Europe and Asia, the U.S. accounts for just 9 percent of global manufacturing capacity of photovoltaic cells and modules, even though American companies are at the forefront of new technologies and production of key materials. European companies control 30 percent of the market.[140] Pike Research predicts Asia will account for 53 percent of global demand for lithium-ion batteries for vehicles in 2015, thanks in large part to the supportive policies by governments such as those of China, Japan, South Korea, and Taiwan.[141] Roland Berger Strategy Consultants, on the other hand, forecasts that Asia will only account for 26 percent of the global automotive lithium-ion battery market of $8.9 billion in 2015, increasing to 38 percent by 2020 when the market is forecast to reach $15.7 billion.[142]

If the U.S. does not have a sufficient domestic market in emerging technologies, domestic manufacturers may well lack the scale needed to compete and survive. America's fledgling advanced battery industry illustrates this paradox. Some 48 factories funded by private investors and government incentives are being established, but industry analysts project serious overcapacity for at least five years before the hybrid and

---

computing, including the internet. An extended list also includes jet engines, satellite communications, and the cell phone.

[140] Michael J. Ahearn "Opportunities and Challenges Facing PV Manufacturing in the United States." *The Future of Photovoltaics Manufacturing in the United States,* C. Wessner, ed., Washington, DC: The National Academies Press, 2011. See also the summary of remarks by Ken Zweibel of the George Washington University Solar Institute, Subhendu Guha of UniSolar, and Dick Swanson of SunPower in the same volume.

[141] Pike Research, "Asian Manufacturers Will Lead the $8 Billion Market for Electric Vehicle Batteries," June 1, 2010 (ttp://www.pikeresearch.com/newsroom/asian-manufacturers-will-lead-the-8-billion-market-for-electric-vehicle-batteries).

[142] RolandBerger Strategy Consultants, *Global Vehicle LiB Market Study*, Detroit/Munich, August 2011.

electric-vehicle markets are big enough to absorb the output.[143] "We can create the best battery in the world, but without vehicles to put them in this industry will go back overseas and we will have stimulated another country's industries," said A123 Systems executive Les Alexander.[144]

To spur demand, the U.S. General Services Administration has announced a goal to buy more than 40,000 alternative-fuel and fuel-efficient vehicles to replace aging, less-efficient sedans, trucks, tankers, and wreckers in the fleets of federal agencies.[145] The federal government also is creating a market for advanced batteries through its programs to promote solar and wind projects. Currently, however, there is no requirement that such batteries be purchased from domestic suppliers. This means that national subsidies to foreign manufacturers will have the desired effect by lowering their immediate costs and allowing them to capture overseas markets from less well-subsidized competitors.

- **Supporting Exports:** Global exports of U.S. manufactured goods and services are important to our balance of payments and economic growth. A key task of the U.S. Commercial Service is to support firms in identifying and exploiting new market opportunities abroad. In 2010, the U.S. Commercial Service directly helped generate $34.8 billion in US exports, assisting over 18,000 business clients. However, while the rest of the world, especially China, India, and Germany, has been augmenting their export assistance, the U.S. has reduced the size of its Commercial Service from over 1,275 employees in 2000 in the international field to barely 900 in 2011.[146] By comparison, Germany fields a staff of 100 in Shanghai alone.[147] To address the need and opportunity to increase U.S. exports, US Commerce Secretary John Bryson has called for growing and restructuring the Foreign Commercial Service in order to intensify its focus on identifying

---

[143] See Boston Consulting Group, "Batteries for Electric Cars: Challenges, Opportunities, and the Outlook to 2020," accessible at http://www.bcg.com/documents/file36615.pdf. Also see presentation by Mohamed Alamgir of Compact Power in *Building the U.S. Battery Industry for Electric Drive Vehicles*.
[144] From presentation by Les Alexander at the National Academies Symposium on *Building the U.S. Battery Industry for Electric Drive Vehicles*. July 26, 2010.
[145] Department of Energy press release, January 26, 2011, (http://www.energy.gov/10034.htm).
[146] See testimony of Keith Curtis of the American Foreign Service Association before House Committee on Appropriations. March 22, 2012.
[147] American Chamber of Commerce, Shanghai, "US Export Competitiveness in China, Winning in the World's Fastest-Growing Market," September 2010. Access at http://www.amcham-shanghai.org/ftpuploadfiles/publications/viewpoint/us_export.pdf.

markets where U.S. exports have the best potential for continued growth, including China, Brazil, India, Saudi Arabia and Turkey.[148]

Export finance, often on concessional terms, is also a major source of support to foreign manufacturers. Many U.S. trade competitors invest significantly more in export credit assistance as both a share of GDP and exports than the United States does.[149] The U.S. Export-Import Bank, which provides financing and insurance for export transactions, plays an important role in supporting these manufacturers by expanding the financing of sales of U.S. exports to international buyers.

Why Manufacturing Matters for the U.S.

Concern over America's declining manufacturing base was a recurring theme of STEP board symposia. Leading executives, industry analysts, and military officials warned that the U.S. is losing competitiveness as a location for new investment in advanced manufacturing capacity, even in industries where the U.S. is at the technological forefront. PCAST also warns that continued erosion of America's high-tech manufacturing base threatens to undermine U.S. leadership in next-generation technologies.[150] Manufacturing matters to the health of the U.S. economy and its innovation ecosystem. The reasons include—

- **Jobs:** U.S. manufacturing shed 5.5 million jobs between 2000 and 2010. At a time when unemployment remains around 9 percent, the loss of manufacturing jobs takes on greater significance. The Milken Institute estimates that every computer-manufacturing job, for example, creates an additional 15 jobs elsewhere in the economy.[151] It also notes that the average manufacturing job in California paid $66,200 a year, roughly 50 percent more than jobs in health care, the state's fastest-

---

[148] See Department of Commerce Press Release, "Commerce Secretary John Bryson Lays Out Vision for Department of Commerce." December 15, 2011.

[149] Export-Import Bank of the United States, Report to the US Congress on Export Credit Competition and the Export-Import Bank of the United States, June 2010. The U.S. Chamber of Commerce is partnering with the Export-Import Bank of the United States on its Global Access for Small Business initiative to help more than 5,000 small companies export goods and services produced by U.S. workers. For a concise review of the role and performance of the Export-Import Bank in promoting U.S. exports, see Stephen Ezell, "Understanding the Importance of Export Credit Financing to U.S. Competitiveness." Washington, DC: ITIF, June 2011.

[150] President's Council of Advisors on Science and Technology, *Report to the President on Ensuring American Leadership in Advanced Manufacturing*, op. cit.

[151] Ross C. DeVol, et. al., "Manufacturing 2.0: A More Prosperous California," Milken Institute, June 2009.

growing industry. Overall, manufacturing contributes $1.6 trillion to GDP and employs 11 million workers.[152]

- **Innovation:** A strong manufacturing base is an integral, though often under-appreciated, part of America's innovation ecosystem. Manufacturing companies account for nearly 70 percent of U.S. industrial research and development[153] and employed 63.4% of all domestic scientists and engineers in 2007.[154] Domestic manufacturing is a critical element in the creation of new technologies. NIST economist Gregory Tassey notes that most modern technologies are actually "systems" that evolve from an interdependent network of "industries that contribute advanced materials, various components, subsystems, manufacturing systems, and eventually service systems based on sets of manufactured hardware and software."[155]

    A 2003 report by to President George W. Bush by the President's Council of Advisors on S&T, which included past and present CEOs of Dell, Intel, Lockheed Martin, and Autodesk, underscored the link between innovation and manufacturing. The study concluded that nations that manufacture commoditized products increasingly are able to develop the capacities to compete directly with the U.S. on "innovating new products and new industries." The PCAST report stated "with manufacturing leaving the country, the United States runs the risk of losing the strength of its innovation infrastructure of design, research and development and the creation of new products and industries."[156]

- **National Security:** Large-scale domestic industries also are vital to national defense. Not only does the military need secure supplies of critical components such as semiconductors and sensors. Scale production also is necessary for controlling costs of materiel. Consider the military's growing requirements for fuel-efficient vehicles. The U.S. Army has committed to cutting fuel consumption by 20 percent in the next 10 to 15 years. At the same time, new weapons and

---

[152] Gregory Tassey, "The Manufacturing Imperative," presentation at NAS Conference on the Manufacturing Extension Partnership, November 14, 2011.
[153] The Manufacturing Institute, "The Facts About Modern Manufacturing-8th Edition," Gaithersburg MD: NIST, 2009. Access at http://www.nist.gov/mep/upload/FINAL_NAM_REPORT_PAGES.pdf.
[154] Wolfe, 2009, cited by George Tassey, "Rationales and mechanisms for revitalizing US manufacturing R&D strategies," *Journal of Technology Transfer*, DOI 10.1007/s10961-009-9150-2, 2010.
[155] Tassey, ibid.
[156] President's Council of Advisors on Science and Technology, "Sustaining the Nation's Innovation Ecosystems: Information Technology Manufacturing and Competitiveness," January 2004, (http://www.choosetocompete.org/downloads/PCAST_2004.pdf).

communications systems are boosting the need for power in combat and non-combat vehicles.[157] Converting much of the Army's 400,000-vehicle fleet to hybrids would reduce fuel costs, ease dependence on imported petroleum, and provide important logistical advantages in the battlefield. Greater fuel efficiency enabled by light-weight, high-density lithium-ion batteries would mean fewer dangerous truck convoys through deserts, tanks that can travel and fight longer without refueling while operating next-generation weapons.[158] High U.S. production volumes of such batteries will make wide deployment of such equipment more feasible, explained John Pellegrino of the U.S. Army Research Laboratory. "We don't want each of those vehicles to cost $1 billion," Dr. Pellegrino said.

PROVIDING EARLY-STAGE FINANCE

The ecosystem for providing risk capital to promising new technology companies not only has been one of the greatest advantages of America's innovation system—but also one of the most difficult to replicate by other nations. The U.S. still has the world's biggest pool of private angel, venture capital, and private-equity funds. It also has the strongest equity markets for taking successful start-ups public.

However, the availability of angel and venture funding has shrunk dramatically in the U.S. over the past decade.[159] What's more, investors have become far more averse to risk, and therefore devote more of their capital to later-stage companies that already have established a position in the market. As a result, many promising start-ups—especially in capital-intensive sectors, such as bio-medical, struggle to raise the funds needed to survive the perilous period of transition when a developing technology is deemed promising, but too new to validate its commercial potential and thereby attract the capital necessary for its continued development.

---

[157] See presentations by Grace Bochenek and Sonya Zanardelli at the National Academies Symposium on *Building the U.S. Battery Industry for Electric Drive Vehicles*, July 26, 2010.

[158] See presentations of John Pellegrino of the Army Research Laboratory and Grace Bochenek in of TARDEC at the National Academies Symposium on *Building the U.S. Battery Industry for Electric Drive Vehicles*, July 26, 2010.

[159] For an analysis of the effect of the recent financial crisis on the venture capital market, see Joern Blockab and Philipp Sandnerc, . "What is the effect of the financial crisis on venture capital financing? Empirical evidence from US Internet start-ups." *Venture Capital: An International Journal of Entrepreneurial Finance,* Volume 11, Issue 4, 2009. For a review of the impact of the financial crisis across industries and countries, see Block, Joern Hendrich, De Vries, Geertjan and Sandner, Philipp G., "Venture Capital and the Financial Crisis: An Empirical Study Across Industries and Countries" (January 24, 2010). HANDBOOK OF VENTURE CAPITAL, Oxford University Press, Forthcoming. The authors' research suggests that the financial crisis has led to a severe 'funding gap' in the financing of technological development and innovation around the world.

> **Box 2.3**
> **The Economic Debate on Manufacturing**
>
> Economists hotly debate the degree and significance of America's decline in manufacturing. Although manufacturing employment dropped sharply over the past decade—after remaining stable at around 17 million for 35 years--some economists contend that decline is explained by higher productivity by U.S. manufacturers.
>
> The National Association of Manufacturers notes that U.S. manufacturing generates $1.6 trillion in value each year, accounts for the lion's share of exports, directly employs 10 percent of the American workforce, and overall supports one is six private-sector jobs. U.S.-based manufacturers also conduct half of private R&D.[160] Economists also have estimated that manufacturers pay 30 to 40 percent of all corporate taxes collected by the federal, state, and local governments and that each $1 of final manufacturing output creates another $1.43 in economic output when services such as finance, construction, and transportation are included.[161]
>
> The Heritage Foundation notes that while U.S. manufacturing employment dropped by one-third since 1987, output rose by 46 percent, thanks to a 114 percent increase in productivity.[162] Such data show there is no empirical evidence of a U.S. manufacturing crisis, the RAND Corp. concluded in 2004.[163]
>
> Recent economic analysis shows, however, that U.S. statistics may be overstating the gains in manufacturing productivity because they fail to adequately reflect the value of imported inputs in manufactured products and because they do not adequately account for the growing use of temporary factory workers. They also note that gains in manufacturing productivity are unevenly distributed, with the significantly higher productivity in computer and electronics manufacturing masking the trends in other sectors.[164] Foreign

---

[160] National Association of Manufacturers data.
[161] *National Review*, "Why Manufacturing Matters," December 3, 2008.
[162] James Sherk, "Technology Explains Drop in Manufacturing Jobs," Backgrounder #2476, Heritage Foundation, October 12, 2010.
[163] Charles Kelley, et al. "High-Technology Manufacturing and U.S. Competitiveness," TR-136-OSTP, prepared for the Office of Science and Technology Policy, RAND Corp., March 2004.
[164] See Susan Houseman and others. "Offshoring Bias in U.S. Manufacturing," *Journal of Economic Perspectives* 25: 111-132. 2011. Susan Helper, Timothy Krueger, and Howard Wial, "Why Does Manufacturing Matter? Which Manufacturing Matters? A Policy Framework." Washington, DC: Brookings, February 2012. See also Robert D. Atkinson and others, "Worse than the Great Depression: What Experts Are Missing about American Manufacturing Decline." Washington, DC: ITIF, March 2012; and Robert D. Atkinson, "Commentary on Gregory Tassey's 'Rationales and Mechanisms for Revitalizing U.S. Manufacturing R&D Strategies,'" *Journal of Technology Transfer* DOI 10.1007/s10961-010-9164-9, 2010 (http://www.itif.org/files/2010-Atkinson-JTT.pdf). See also IDA, "Global Trends in Advanced Manufacturing," March 2012. Access at https://www.ida.org/upload/stpi/pdfs/p-4603_final2a.pdf .

> manufacturing and trade practices, particularly those of China in the first decade of this century, have also negatively impacted U.S. manufacturing employment. Measured in terms of value-added, U.S. production of computer and electronic products—a high-performing sector from 1985 to 2000—dropped by 21 percent in the past decade.[165]

Public-private partnership programs such as the Small Business Innovation Research program (SBIR) and NIST's Advanced Technology Program (ATP) have proved successful—and, in the case of SBIR, increasingly important—sources of early-stage capital for new, innovative U.S. companies.

The Technology Innovation Program (TIP) is the successor to NIST's highly regarded Advanced Technology Program (ATP). Independent evaluations by the National Research Council found ATP to be "an effective federal partnership program." [166] One of its strengths was to bring together small and large companies (and universities) in partnership to develop new high technology products, such as amorphous silicon detectors that digitally enhance MRI images for improved breast cancer detection. As ATP's successor, TIP sought to accelerate "innovation in the United States through high-risk, high-reward research in areas of critical national need" through "targeted investments in transformational R&D that will ensure our nation's future through sustained technological leadership."[167] Despite its broad mandate, funding for TIP was modest and no funds were appropriated for this program in the FY 2012.

The Small Business Innovation Research Program (SBIR) provides more than $2.5 billion annually in competitively awarded R&D grants and contracts to qualified small businesses. In comparison, the private venture capital industry invested $919 million at the seed stage in 2011.[168] Eleven federal agencies are required by law to provide these funds by setting aside 2.5% of their annual extra-mural R&D budgets for small businesses innovation. In a recent assessment, the National Academies found the program to be "sound in concept and effective in practice." It highlighted the program's important role as a source of start-up and seed capital for small businesses to develop new innovative product concepts for the market as well as develop products and

---

[165] Data cited by Tassey, op. cit.
[166] Independent evaluations by the National Research Council found ATP to be "an effective federal partnership program." See National Research Council, The Advanced Technology Program, Assessing Outcomes, Washington, DC: National Academy Press, 2001. The successor to the Advanced Technology Program, the Technology Innovation Program (TIP) at the National Institute for Standards and Technology "supports, promotes, and accelerates innovation in the United States through high-risk, high-reward research in areas of critical national need" through "targeted investments in transformational R&D that will ensure our nation's future through sustained technological leadership." See http://www.nist.gov/tip/.
[167] See http://www.nist.gov/tip/.
[168] Data from PWC-MoneyTree, January 2012.

carry out contract research for specific agency mission needs. It is valued both as start-up funding and as a "low cost technological probe."[169]

A 2008 study by Block and Keller, based on a sample of top inventions recognized by R&D Magazine over 40 years, found that over the past two decades, about two-thirds of the top innovations have roots in government-industry partnerships.[170] This contrasts with the awards in the previous period that were predominantly funded by either private or federal sources. Their study also found that 20 to 25 percent of the R&D 100 inventions awards over the past decade benefitted from SBIR awards. [See Figure 2.5]

SBIR is increasingly seen as "Best Practice" around the world. As we see below, a growing list of countries have adapted the SBIR program within their own innovation systems.[171]

- **Brazil**: The Brazilian Innovation Agency, better known by its acronym FINEP, operates the PIPE and Pappe programs that provide grants to hundreds of small companies that are commercializing technologies.[172]
- **Japan:** Japan is expanding the scope and scale of the Small Research Innovation Research program, which was established in 2003 and is directly modeled after the U.S. SBIR. The Small and Medium Enterprise Agency manages the program, and funds are contributed by various ministries involved in areas such as energy, information and communications technologies, and bio and medical sciences. Plans call for increased lending to small- and medium-sized enterprises, more hands-on support for start-ups, and making the application process more flexible.[173]

---

[169] National Research Council, *An Assessment of the SBIR Program*, C. Wessner, ed., Washington, DC: The National Academies Press, 2008. Also from the National Research Council, see *An Assessment of the SBIR Program at the Department of Defense* (2009) and *An Assessment of the SBIR Program at the National Institutes of Health* (2009).

[170] Block and Keller, op. cit. The authors analyze a sample of innovations recognized by *R&D Magazine* as being among the top 100 innovations of the year over the last four decades. They find that while in the 1970s almost all winners came from corporations acting on their own, more recently over two-thirds of the winners have come from partnerships involving business and government, including federal labs and federally-funded university research. The authors note that "the R&D 100 Awards carry considerable prestige within the community of R&D professionals, comparable to the Oscars for the motion picture industry. Organizations nominate their own innovations. All entries are initially evaluated by outside juries that include representatives of business, government, and universities." (Block and Keller, 2008, page 6).

[171] See OECD Innovation Policy Platform, "Public Procurement Programmes for Small Firms—SBIR-type programs," which can be accessed at http://www.oecd.org/dataoecd/33/37/48136807.pdf. This publication describes the evaluation programs for SBIR in the United Kingdom and the Netherlands.

[172] An explanation of these activities can be found in Odilon Antonio Marcuzzo do Canto, "Incentives to Support Innovation in the Private Sector: The Brazilian Experience," Brazil Innovation Agency. Access at http://idbdocs.iadb.org/wsdocs/getdocument.aspx?docnum=976023.

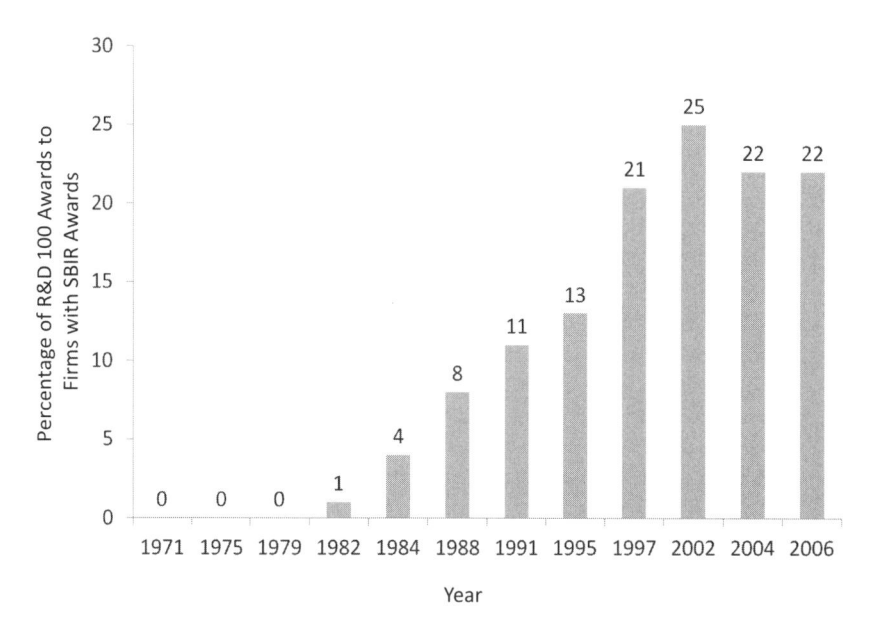

FIGURE 2.5 Percentage of R&D 100 Awards to firms with SBIR Awards. SOURCE: Fred Block and Matthew R. Keller, "Where Do Innovations Come From? Transformations in the U.S. National Innovation System, 1970-2006," Washington, DC: The Information Technology and Innovation Foundation, July 2008.

- **India**: The Small Business Innovation Research Initiative (SBIRI), launched in 2005, supports high-risk R&D projects by Indian biotech start-ups in sectors such as health care, agriculture, industrial technology, and environment. The program has phases for early-stage research and for later development and commercialization.[174]
- **Netherlands**: The Dutch government launched the Small Business Innovation Research program in 2004. The program fully funds the first phases of pre-commercial R&D up to €50,000 and up to €450,000 for a second two-year phase.[175]
- **Finland:** Of the €343 million that the government invested directly in enterprises in 2009, 61 percent went to small and midsized companies.

---

[173] See Ministry of Economy, Trade and Industry, "!00 Actions to Launch Japan's New Growth Strategy," Action 73, pg. 23, August 2010. Access at http://www.meti.go.jp/english/aboutmeti/policy/2011policies.pdf.
[174] See Indian government Web site http://india.gov.in/sectors/science/sbiri.php.
[175] See SCI-Network, "Case Study: Small business Innovation Research (SBIR) in the Netherlands," March 2011. Access at http://www.sci-network.eu/fileadmin/templates/sci-network/files/Resource_Centre/Case_Studies/Case_Study_-_Dutch_SBIR_-_Final.pdf.

Eighty-seven percent of those companies had fewer than 500 employees. Companies use the funds to develop technologies in partnership with universities.[176]
- **United Kingdom**. The Small Business Research Initiative, run by the Technology Strategy Board, was established in 2000 to earmark a share of the government's procurement budget to contracts with small- and midsized enterprises. The program was revised and expanded in 2009.
- **Germany**: Among several government programs to help start-ups commercialize technology are the Central Innovation Programme SME (ZIM), which has an annual budget of €300 million and received another €900 million in 2009.[177] Another program, EXIST, awards grants to technology start-ups and stipends to cover costs of equipment, materials, coaching, and childcare for scientists.[178]

Despite its validation though the National Academies assessment as well through its emulation abroad, the U.S., the U.S. Congress delayed reauthorization of SBIR for several years, creating uncertainty about the future of the program. The SBIR/STTR Reauthorization Act of 2011 has extended the programs through September 2017.. NIST's Technology Innovation Program, which replaced the Advanced Technology Program, is now unfunded, despite a proven track record of success, at least for the ATP antecedent. A major advantage of ATP was its focus on linking small and large companies, along with universities to develop new high technology products.

DEVELOPING TWENTY-FIRST CENTURY UNIVERSITIES

A major asset for America's innovation system is the strength and independence of its research universities. U.S. research universities serve as the funnel point for the entry into the U.S. of foreign talent in Science, Technology, Engineering, and Mathematics (STEM). Twenty-first century universities are also playing a growing role in supporting regional innovation ecosystems by transferring new technologies to the private sector. Universities such as Stanford, the Massachusetts Institute of Technology, Georgia Tech, and the University of Texas-Austin have acted as powerful engines of innovation, often

---

[176] Tekes Annual Review 2009, (http://www.tekes.fi/en/community/Annual%20review/341/Annual%20review/1289).
[177] Federal Ministry of Economics and Technology, *Central Innovation Programme (ZIM)*, January 2011 (http://www.zim-bmwi.de/download/infomaterial/informationsbroschuere-zim-englisch.pdf). For an analysis of ZIM, see European Commission, *ZIM, the Central Programme for SMEs (Zentrales Innovationsprogramm Mittelstand)*, PRO INNO Europe, INNO-Partnering Forum, Document ID: IPF 11-005, 2010.
[178] Details of the EXIST program can be found on the German federal government Web site http://www.hightech-strategie.de/en/879.php.

generating promising new companies and industries.[179] Passage of the Bayh Dole Act of 1980, which incentivized universities to sell and license technology generated from federally funded research, is widely associated with contributing to a boom in new technology companies.[180]

America's 198 public research universities have long been the backbone of this system, conducting 62 percent of federally funded research.[181] These institutions educate 85 percent of undergraduates and 70 percent of graduate students enrolled in all research universities. They account for 60 percent to 80 percent of doctorates degrees in computer and information sciences, engineering, math and statistics, physical sciences, and security—and from 78 percent to 95 percent of bachelor's degrees in all of these areas of national need.[182]

U.S. Universities Face Financial Challenges

Yet this invaluable national asset is in financial trouble.[183] Charles M. Vest recently summed up the problem: "In the last decade, the real state appropriation to public colleges and universities per student has dropped by 20 percent overall. But the total cost to students and their families of attending a state university has increased by 52 percent during this same decade. Such declining state support and the resultant tuition increases are not a sustainable

---

[179] See Edward P. Roberts and Charles Eesley, Entrepreneurial Impact; the Role of MIT, Kauffman Foundation Report (2009). Access at http://web.mit.edu/dc/policy/MIT-impact-full-report.pdf.
[180] According to Arundeep S. Pradhan of the Association of University Technology Managers, "since 1980, American universities have spun off more than 5,000 companies, which have been responsible for the Introduction of 1.25 products per day into the marketplace and have contributed to the creation of more than 260,000 jobs. The result has been a contribution of over $40 billion dollars annually to the American economy." Testimony before the House Committee on Science and Technology, July 17, 2007. For an academic assessment of the impact of the Bayh-Dole Act of 1980, see Rosa Grimaldi, Martin Kinney, Donald S. Siegel, and Mike Wright, "30 years after Bayh–Dole: Reassessing academic entrepreneurship" *Research Policy*, Volume 40, Issue 8, October 2011, Pages 1045–1057. See also the analysis of Mowery and Sampath who note that the catalytic effects of the Bayh-Dole Act may be overrated. Mowery, David C. and Sampat, Bhaven N., The Bayh-Dole Act of 1980 and University Industry Technology Transfer: A Model for Other OECD Governments? (2005). *The Journal of Technology Transfer*, Vol. 30, Issue 1-2, p. 115-127 2005. Finally see Roberts and Eesley, Entrepreneurial Impact, Kauffman Foundation Report (2009), http://web.mit.edu/dc/policy/MIT-impact-full-report.pdf
[181] Association of Public Land-Grant Universities, "Ensuring Public Research Universities Remain Vital," November 2010. Access at https://www.aplu.org/NetCommunity/Document.Doc?id=2819.
[182] Association of Public and Land-grant Universities data. See Peter McPherson, David Shulenburger, Howard Gobstein, and Christine Keller, "Competitiveness of Public Research Universities & the Consequences for the Country: Recommendations for Change," Association of Public and Land-Grant Universities, March 2009, (http://www.aplu.org/NetCommunity/Document.Doc?id=1561).
[183] The National Academies of Sciences is undertaking a competitiveness study focusing on the health of U.S. research universities at the request of Senators Barbara Mikulski (D-MD) and Lamar Alexander (R-TN) and Representatives Bart Gordon (D-Tenn.) and Ralph Hall (R-Texas).

situation."[184] An article in *The Chronicle of Higher Education* noted that state funding for public universities has been declining for two decades on a per-student basis and is reaching levels that are "threatening to cripple many leading public universities and erode their world-class quality."[185]

- **State Budget Cutbacks:** Cutbacks have been especially harsh in the past few years, as state-government budget deficits widened dramatically as a result of the recession. In 2009, the University of California's budget was cut by 20 percent, or $813 million. The university expects a further 16.4 percent cut in Fiscal Year 2011, reducing state funding to 1999 levels even though there now are 73,000 more students.[186] At the University of Georgia, state funds per student have dropped from $8,191 in FY 2009 to $6,242 in FY 2011 and also now are at 1999 levels not adjusted for inflation, despite 4,000 more students, additional buildings, and higher teacher salaries.[187] Arizona State University's budget was slashed by $88 million in 2009, and a further cut of 20 percent for four-year colleges in universities in the state has been proposed for 2011.[188]

    In all, 32 U.S. states cut their support for higher education in 2010 by between 0.3 percent and 13.5 percent, with double-digit declines in Missouri, Delaware, Iowa, Minnesota, Arizona, and Oregon.[189] "Given the national reliance on public universities for majority contributions to the nation's need to advance knowledge and prepare new scientists and engineers," warns the Association of Public and Land-Grant Universities, "a serious decline in the capacity of public research universities critically risks the attainment of these national goals."[190]

- **Limited use of Dedicated Taxes:** Funding from states for universities is especially vulnerable to state budget cuts because it often comes

---

[184] Charles M. Vest, "Chancellor's Colloquium," University of California, Davis, November 30, 2011, p. 8.
[185] See Paul Courant, James Duderstadt, and Edie Goldenberg, "Needed: A National Strategy to Preserve Public Universities," *The Chronicle of Higher Education*, January 3, 2010, (http://milproj.dc.umich.edu/pdfs/2010/2010-Chronicle-Commentary.pdf).
[186] Carolyn McMillan, "Regents Scrutinize Fiscal Crisis," *UC Newsroom*, March 16, 2011 (http://www.universityofcalifornia.edu/news/article/25150).
[187] University of Georgia President Michael F. Adams Budget Update, May 19, 2010 (http://www.uga.edu/Budgetslides_05-19-2010.pdf).
[188] *The Arizona Republic*, " Arizona Board Approves Steep Tuition Hikes," April 8, 2011.
[189] Center for the Study of Education Policy data cited in *Inside Higher Ed*, "The Sinking States," January 24, 2011, (http://www.insidehighered.com/news/2011/01/24/states_make_more_cuts_in_spending_on_higher_education).
[190] Peter McPherson, Howard J. Gobstein, and David E. Schulenburger, "Forging a Foundation for the Future: Keeping Public Research Universities Strong," Association of Public and Land-Grant Universities, 2010 (http://www.aplu.org/NetCommunity/Document.Doc?id=2263).

from general tax revenues. While programs such as highway maintenance and construction of sports stadiums and convention centers often are funded from recurring revenue streams such as lotteries, casino proceeds, and gasoline and alcohol taxes, only a handful of states dedicate such funds to universities.[191]

The advantage of these dedicated taxes is that these sources provide recurring revenue streams for important programs, even during times of economic downturn, and are not subject to government budgetary restraints. Notably, the State of Texas uses a Permanent University Fund (PUF), established in its 1876 Constitution, to fund higher education. Currently, PUF land assets deliver proceeds through oil, gas, sulfur, and water royalties, rentals on mineral leases, and gains on fiduciary investments.

- **Declines in private funding**: Reversing a three-decades-long trend of increasingly strong ties between industry and universities, the absolute value of industrial R&D dollars to academic institutions—funds provided directly to academic institutions for the conduct of research—began to decline beginning in 2002 after reaching a high of $2.2 billion in 2001. Also, industrial R&D support to academia has historically been concentrated in relatively few institutions.[192] Leading university and industry leaders have pointed out that U.S. companies increasingly choose to work with foreign rather than U.S. universities, encouraged by the more favorable IP rights that foreign universities offer and the strong incentives for joint industry-university research that foreign governments provide.[193]

Growing Investments in Universities Abroad

As U.S. universities struggle, other nations are increasing investment and overhauling their higher-education systems to turn universities into engines of innovation-led growth. Strengthening university commercialization programs, breaking down barriers between academia and industry, and freeing university researchers to start or join companies are standard features of many national innovation strategies. Governments in emerging economies also are aiming to

---

[191] Alene Russell, "Dedicated Funding for Higher Education: Alternatives for Tough Economic Times," American Association of State Colleges and Universities, Higher Education Policy Brief, December 2008. Access at http://www.aascu.org/uploadedFiles/AASCU/Content/Root/PolicyAndAdvocacy/PolicyPublications/08.decpm(2).pdf.
[192] See National Science Foundation, "Where has the Money Gone? Declining Industrial Support of Academic R&D," InfoBrief, NSF 06-328, September 2006.
[193] GUIRR, "Re-Engineering the Partnership: Summit of the University-Industry Congress," Meeting of 25 April 2006, Washington, DC.

upgrade universities to world standards and establish innovative new ones more attuned to needs of the global economy. For example—

- **Flanders:** The Flemish government launched a large €232 million program in 2006 for strategic basic research at universities of benefit to industry, the non-profit sector, and government policy objectives. The biggest investments have gone to large university-based R&D centers for microelectronics, biotechnology, and broadband technology. Each Flemish university has been instructed to keep a portfolio of industry-oriented projects and operate a technology-transfer office.[194] At the Katholieke University Leuven, which has launched 100 companies, each of the 50 research divisions can reinvest proceeds from industrial involvement into equipment and infrastructure.[195]
- **India:** India's 358 universities and famed Indian Institutes of Technology (IIT)traditionally have played little role in commercializing technology.[196] The government is starting to overhaul the entire system of science and engineering education to promote collaboration with industry and allow faculty to work with industry.[197] A committee studying reforms of IITs is expected to call for granting them greater management and financial autonomy from the government and to encourage research partnerships with private companies.[198] India's Five-Year plan, for example, calls for establishing a network of globally competitive "centers of excellence" in certain technologies based at universities.[199]
- **Canada:** As part of Canada's efforts to promote commercialization by universities, the Foundation for Innovation since 1997 has allocated $5.2 billion to research projects, new laboratories, industry collaborations, and recruitment of foreign faculty. The government also has increased the number of Centers of Excellence based at universities

---

[194] Fientje Moerman "Keynote Address," National Research Council, *Innovative Flanders, Innovation Policies for the 21st Century*, C. Wessner, ed., Washington, DC: The National Academies Press, 2008.
[195] Koenraad Debackere, "Leuven as a Hotspot for Regional Innovation," in *Innovative Flanders*, op. cit.
[196] Martin Gruber, and Tim Studt. "2011 Global R&D Funding Forecast: The Globalization of R&D," *R&D Magazine*, December 15, 2010.
[197] Ramesh Mashelkar I, "Renewing the National Laboratories," in *India's Changing Innovation System*, op. cit. Also see P. V. Indiresan, "National and State Investments in Science and Engineering Education," in *India's Changing Innovation System*, op. cit.
[198] *Hindustan Times*, "More Autonomy, New Programmes for IITs," January 16, 2011.
[199] Government of India Planning Commission, "Report of the Steering Committee on Science and Technology for Eleventh Five Year Plan (2007-2012)," December 2006, (http://planningcommission.gov.in/plans/planrel/fiveyr/11th/11_v1/11th_vol1.pdf). Also see *Hindustan Times*, "More Autonomy, New Programmes for IITs," January 16, 2011. India's Eleventh Five-Year plan also sets high targets for expanding university enrollment.
http://planningcommission.gov.in/plans/planrel/fiveyr/11th/11_v1/11th_vol1.pdf.

devoted to research collaborations with industry. The centers, in fields ranging from optics to brain research, are credited with spinning off more than 100 companies.[200]
- **Singapore:** Singapore is seeking to upgrade schools such as the National University of Singapore and Nanyang Technological University, which are strong in science and technology, to become world-class research institutions and fonts of entrepreneurialism. The government established a high-level Enterprise Board at each university and an innovation fund to supplement each school's own resources to finance entrepreneurship education, technology incubators, commercialization programs, and entrepreneurs-in-residence programs. Polytechnics are receiving grants to help universities bring research to the market.[201]
- **Japan:** In 1999, Japan enacted a law similar to the Bayh-Dole Act of 1980 allowing universities and research institutes to patent investments derived from publicly funded research.[202] The government also boosted funding for joint university-industry research programs and helped universities set up 45 centers for commercializing research. In 2004, universities were given more autonomy to allocate resources, collaborate with industry, and set their own research priorities.[203] The reforms led to a sharp rise in university spin-offs and industry research collaborations.[204]
- **Brazil:** Between 2000 and 2008, the number of master's degrees and doctorates awarded by Brazilian universities annually has both doubled, to more than 36,000 and 10,000, respectively. From 2002 to 2010, the government invested $550 million to build 226 new technology schools.[205]

---

[200] Networks of Centers of Excellence Web site. Also see Peter J. Nicholson, "Converting Research into Innovation," in *Innovation Policies for the 21st Century*, op. cit.
[201] National Research Foundation, "National Framework for Innovation and Enterprise," Prime Minister's Office, Republic of Singapore, 2008,
(http://www.nrf.gov.sg/nrf/otherProgrammes.aspx?id=1206).
[202] See Sadao Nagaoka and Kenneth Flamm, "The Chrysanthemum Meets the Eagle— The Co-evolution of Innovation Policies in Japan and the United States," in National Research Council, *21st Century Innovation Systems for Japan and the United States: Lessons from a Decade of Change*, Masayuki Kondo, Kenneth Flamm, and Charles Wessner, Editors, Washington, DC: The National Academies Press, 2009.
[203] A concise analysis of Japan's shift in innovation policy is found in National Research Council, *S&T Strategies of Six Countries: Implications for the United States*, Committee on Global Science and Technology Strategies and Their Effect on U.S. National Security, Washington, DC: The National Academies Press, 2010.
[204] Masayuki Kondo, "Kyutenkaishihajimeta Nippon no Daigakuhatsubencha no Genjou to Kadai" ("The Current State and Issues of Rapidly Increasing University Spin-offs in Japan"), *Venture Review*, No. 3, 101-107. 2002.
[205] Francelino Grando, "Brazil's New Innovation System," National Academies symposium, *Clustering for 21st Century Prosperity,* Washington, DC, February 25, 2010.

- **China:** China has an ambitious $2.8 billion plan called Project 211 to create 100 higher education facilities that are on par with the best in the world.[206]
- The United States is one of the few industrial nations without a national strategy for sustaining the quality of its research universities.[207] Some higher-education experts contend that public research universities' reliance on state funding is a flaw in the U.S. innovation system because state lawmakers do not recognize a direct payoff from such investments. The authors of the *Chronicle of Higher Education* article note that "many of the benefits from graduate training—like the benefits of research—are public goods that provide only limited returns to the states in which they are located. The bulk of the benefits are realized beyond state boundaries." Several higher-education advocates contend the U.S. federal government needs to assume more responsibility for funding public research universities.[208] To provide more stable funding for higher education, the American Association of State Colleges and Universities has called upon states to earmark more revenue from recurring sources such as excise taxes, gaming, and land-use rights.[209]

New Models of 21st Century Universities

While it's often the case that other nations are adapting best practices from the United States, new schools are being established around the world based on innovative models designed to meet the needs of the 21st century global economy. Several of these new institutions deserve study by American educators. Finland's Aalto University, for example, merges three existing universities that specialized in technology, economics, and art and design to integrate students and faculty in all of these disciplines into a single community.[210] Since the 1990s, Sweden's Chalmers University of Technology has transformed itself into one of Europe's most entrepreneurial universities.[211] The new Singapore University of Technology and Design, a collaboration with MIT and China's Zhejiang University, will have a multi-disciplinary curriculum

---

[206] See China Education and Research Network, "Project 211: A Brief Introduction," (http://www.edu.cn/20010101/21852.shtml).
[207] Courant, Duderstadt, and Goldenberg, op. cit.
[208] The chancellor and vice-chancellor of the University of California at Berkeley, for example, have called for the federal government to provide basic funding for a limited number of top public research universities. See Robert J. Birgeneau and Frank D. Yeary, "Rescuing Our Public Universities," *Washington Post*, September 27, 2009.
[209] American Association of State Colleges and Universities, op. cit.
[210] Aalto University Web site: http://www.aalto.fi/en/.
[211] See Merle Jacob, Mats Lundqvist, and Hans Hellsmark, "Entrepreneurial Transformations in the Swedish University System: The Case of Chalmers University of Technology," *Research Policy*, Volume 32, Issue 9, October 2003, pp. 1555-1568.

and research focus that will strive to teach students to be creative and solve problems.[212] The university will house an International Design Center modeled after a smaller facility at MIT and intends to "become the world's premier hub for technologically intensive designs." MIT will help design programs to encourage innovation and entrepreneurship.

New Opportunities for Global Collaboration

The globalization of universities is helping to foster a 21st century learning environment for American students by proving them greater opportunities to work with partners and in teams that are cross cultural as well as cross functional. Technological advances are also allowing for students and faculty to work across borders while avoiding the time and costs of travel. These potentials can be further developed through encouraging U.S. faculty and students to collaborate more extensively with their peers abroad, as demonstrated by leading U.S. universities like MIT and Carnegie Mellon University.

Indeed, a number of U.S. academic institutions are now operating internationally, addressing not only potential students individually (per the traditional paradigm), but increasingly addressing foreign universities, foreign local authorities and governments, in order to develop new types of institutional arrangements. These include helping creating, monitoring or evaluating emerging institutions in other countries, transferring organizational skills, operating training programs for teachers and researchers, contributing to higher education and research capacity building abroad and to the marketing of its benefits for economic and social progress in other societies. Such new arrangements may also include the coaching and steering of research programs in emerging countries, their early inclusion in international networks, and the affiliation of private companies to academic and research programs.

On the other hand, many emerging nations are now facing the need and the opportunity of large investments in science, technology and higher education (public and private), aiming at responding to the explosive social demand for higher education and to the vast social and political transformations already induced by new waves of educated youth. These investments not only seek new skills and but also the certification of quality that may be expected from working along together with well established academic and scientific institutions from the United States. For these institutions, including the American universities, such institutional arrangements provide new forms of expansion, as they tend to help securing new financial or human

---

[212] Brochure of Singapore University of Technology and Design, 2010. Access at http://www.sutd.edu.sg/cmsresource/brochure/undergraduate_brochure.pdf.

resources, and to challenge their own traditional competences and agendas. Above all, they provide unique access to new pulls of talent worldwide, benefiting above all leading American universities. [213]

Collaboration with Industry

The culture of academic collaboration with industry is well established in the United States. Notable among these is the Semiconductor Research Corporation Focus Center Research Program, a multi-million dollar, 30-university research collaboration to address long-term technology issues of relevance to the semiconductor industry.[214]

University-industry collaboration, particularly with regard to technology-transfer programs, offers a mixed picture.[215] Over all, the number of start-ups spun out of elite research universities in the United States has risen from 200 in 1994 to 651 in 2010. Successful patent applications and new technology licenses, had remained flat for a decade, but were up in the latest survey by the Association of University Technology Managers (AUTM). According to AUTM's 2010 survey, the number of startups formed increased 10.6 percent and the number of licenses/options executed to startups increased 14 percent. At the same time, "the total number of licenses and options executed remained essentially flat, increasing only 0.6 percent. The number of licenses executed decreased two percent, whereas the number of options increased 13 percent. However, there was a strong 15 percent increase in the total number of active licenses and options through the close of 2010."[216] Of 20,309 invention disclosures by universities in 2010, about 22 percent resulted in issued U.S. patents.

The performance of university technology-transfer offices varies. Fifty-two percent of the 130 technology-transfer programs studied do not have revenues to cover their costs. Some 16.2 percent of U.S. institutions surveyed reported that their programs are financially self-sustaining, meaning they do not

---

[213] D. Bruce Johnstone et al., eds., *Higher Education in a Global Society,* New York: Northampton MA: Edward Elgar Publishing, 2010.

[214] The Focus Center Research program is aimed at solving the long-range (normally 5 years or more), difficult challenges outlined in the International Technology Roadmap for Semiconductors, which is a forward looking assessment that is sponsored by several industry groups.

[215] For a review of the challenges universities face in technology transfer, see DiGregorio, D., and Shane, S. "Why do some universities generate more start-ups than others?" *Research Policy*, 32(2), 209-227, 2003. (Reprinted in D. Siegel (ed.) *Technological Entrepreneurship: Institutions and Agents Involved in University Technology Transfer*, Aldershot, UK: Edward Elgar) and Siegel, D. et al, 2003. "Assessing the impact of organizational practices on the relative productivity of university technology transfer offices." *Research Policy*, 32(1):27-48. See also Thursby, J. and Thursby M. "Who is Selling the Ivory Tower? Sources of Growth in University Licensing," *Management Science*, 48:1, January 2002, 90-104.

[216] See AUTM U.S. Licensing Survey Highlights, 2010.

depend on the university's operating budget.[217] Many technology transfer offices are not only underfunded but also labor under federal rules that make it difficult for principal investigators to commercialize federally funded research.[218] In addition, the system for allocating federal R&D funds and for rewarding faculty focuses overwhelmingly on scientific discovery, rather than on applied research or development of prototypes.[219]

A recent National Research Council study affirmed that the primary mission of university technology transfer activities is the dissemination of technologies for the public good and recommended that the current system of technology transfer be improved.[220] To this end, the study on *Managing Intellectual Property in the Public Interest* noted that university leadership should more clearly articulate the mission of technology transfer activities and adopt organizational changes to make them more effective.[221]

INVESTING IN MODERN S&T PARKS

The United States pioneered the use science and technology parks—typically with research universities at their core--as platforms for launching new companies and creating regional innovation clusters. Now, research parks are proliferating across the world. While key aspects are borrowed from successful U.S. science and technology parks, many new parks overseas have a greater scope and scale, and in many cases benefit from substantial government funding. Here are some examples.[222]

- **Singapore:** Singapore is building a network of science parks in a 500-acre urban district called One North, located close to the National University of Singapore, National University Hospital, and Singapore Polytechnic. The $10 billion master plan includes Biopololis, a 4.5 million-square-foot campus that aspires to be Asia's biomedical hub. The complex houses 5,000 life science researchers from universities, hospitals, and multinationals such as Eli Lilly and Novartis in disciplines ranging from X-ray crystallography to DNA sequencing. One North also includes Fusionopolis, a futuristic 24-story tower intended as a one-stop R&D shop mixing companies in energy

---

[217] Presentation by Ashley Stevens at the National Academies Symposium on *Clustering for 21st Century Prosperity*, February 25, 2010.
[218] National Research Council, *Managing University Intellectual Property in the Public Interest*, Stephen A. Merrill and Ann-Marie Mazza, editors, Washington, DC: National Academy Press, 2010.
[219] Darmody, "University Based Clusters," op. cit.
[220] National Research Council, *Managing University Intellectual Property in the Public Interest*, op. cit.
[221] Ibid.
[222] See National Research Council, *Understanding Research, Science, and Technology Parks*, op. cit. Also, see National Research Council, *Innovative Flanders: Innovation Policies for the 21st Century*, Charles W. Wessner, ed., Washington, DC: The National Academies Press, 2008.

technologies, aerospace, nanotechnology, sensors, cognitive science, and devices for wired homes.[223]

- **Russia:** The government is investing $3 billion over three years in an attempt to develop a 400-hectare Skolkovo district in Moscow into an innovation hub for multinationals and Russian start-ups. Siemens, GE, and Nokia-Siemens have all pledged to build R&D centers, and Dow, Intel, and Cisco have signed memorandums of understanding to do so. Skolkovo will include a new university being developed in a partnership with MIT that is to open in 2014. The central government also has earmarked $172 million to be given to 130 start-ups.[224]

- **China**: China has a number of mega-parks larger in size than North Carolina's Research Triangle and that typically feature a diversity of industries and a high concentration of R&D facilities by universities, corporations, and government research institutes. The Chinese government invested $1.4 billion in Suzhou Industrial Park, for example, home to operations of 113 of the Fortune 500 companies. The more established Zhongguancun Science Park in Beijing hosts more than 20,000 enterprises and 950,000 employees, and has produced $110 billion worth of income as of 2009.[225] The Zhangjiang High-Tech Park in Shanghai's Pudong district, which was farmland in 1992, has more than 4,000 companies and 100,000 workers and covers 17 square kilometers. Zhangjiang includes more than 30 government research institutes and 91 R&D centers by multinational corporations in such industries as life sciences, information technology, semiconductors, and multimedia gaming. It also has a $2.5 billion venture capital fund for start-ups and nearly 100 multinational corporate R&D centers, including major expansions by Novartis, General Electric, Pfizer, Novartis, and AstraZeneca.[226]

- **Mexico:** The new Research and Technology Innovation Park (PIIT) on the outskirts of Monterrey, Mexico, has strong ties to Tecnologico de Monterrey, the nation's premier engineering school. Spread over 172 acres near the airport, PIIT will the first in Mexico to integrate the laboratories in an array of technologies by leading universities, foreign and domestic corporations, small-business incubators, and national laboratories at a single site. PIIT's first $145 million phase includes major laboratories by companies as diverse as Motorola, PepsiCo, and India's Infosys. It also is building public R&D centers for electronics,

---

[223] See Yena Lim, "The Singapore S&T Park," in National Research Council, *Understanding Research, Science and Technology Parks*, op. cit.
[224] Courtney Weaver, "Welcome to Russia's Silicon Valley," *Financial Times*, August 21, 2011.
[225] See Zhu Shen, "China: Navigating the Frontier of Life Sciences Silk Road," in National Research Council, *Understanding Research, Science, and Technology Parks, op. cit.*
[226] Data from Zhangjiang High-Tech Park Web site, http://www.zjpark.com/zjpark_en/zjgkjyq.aspx?ID=7.

biotechnology, mechatronics, advanced materials, the food industry, product design, IT, and water research. The University of Texas at Austin will run an IC2 business incubator.[227]
- **France:** Minatec, a campus of 3,000 students and researchers in Grenoble, has emerged as one of Europe's premier hubs for nano-technology and micro-system research. The French government has invested €3.2 billion and regional and local governments have provided another €150 million for the 20-hectare campus, which in the lynchpin of a €4 billion government initiative to make Grenoble a world center for development of next-generation chips. Minatec has 200 industrial partners, including Mitsubishi, Philips, Bic, and Total, and has launched startups in fields such as optronics, biotechnology, circuit design, and sensing.[228]

Measuring the performance of science and research parks is difficult, and empirical literature on the topic has been described as "embryonic."[229] Several experts note that better metrics are required to evaluate research parks in order to justify the substantial public investment.[230] In their seminal study of research parks, Michael I. Luger and Harvey A. Goldstein observed that one reason measuring performance is difficult is that "there is no consensus about the definition of success." Goals cited by developers, universities, and public officials include economic development, technology transfer, land development, and enhancement of research capabilities at affiliated universities.[231]

GROWING INNOVATION CLUSTERS

Companies in similar industries have long tended to locate close to each other for centuries.[232] In the United States, innovation clusters in regions such as Silicon Valley and greater Boston have tended to flourish close to major research universities without government coordination. In the past two decades,

---

[227] Jaime Parada, "Monterrey-International City of Knowledge Program," National Research Council, *Understanding Research, Science and Technology Parks,* op. cit.
[228] David Holden, "Initiatives in France," National Research Council, *Understanding Research, Science, and Technology Parks.* op. cit.
[229] For a review of the empirical literature on research parks, see Albert Link, "Research, Science, and Technology Parks: An Overview of the Academic Literature," in National Research Council, *Understanding Research, Science, and Technology Parks.* op. cit.
[230] In his presentation at the March 13, 2008, National Academies symposium "Understanding Research, Science, and Technology Parks," William Kittredge of the U.S. Department of Commerce described effective performance-measurement metrics for research parks and for economic development in general remains a "work in progress."
[231] Michael I. Luger and Harvey A. Goldstein, *Technology in the Garden,* Chapel Hill: University of North Carolina Press, 1991, p. 34.
[232] Alfred Marshall was one of the first economists to develop a theory about regional agglomerations of industries. See *Principles of Economics*, London: Macmillan, 1920. The first edition of Marshall's classic textbook appeared in 1890.

however, regional innovation clusters have become a matter of more focused public policy in the U.S. and around the world.[233] Of 260 cluster initiatives studied in 2003, government supported two-thirds. In 52 percent, government was the primary funder.[234]

International Cluster Initiatives

Regional cluster initiatives linking universities, industry, government economic-development agencies, and investor groups now are found across Asia-Pacific, Europe, and Latin America. The European Union even operates a European Cluster Observatory that maps clusters across the continent.[235] In some cases, clusters receive significant government financial assistance and are integral components of comprehensive national or regional innovation strategies.

These examples offer a flavor of the public-private strategies being deployed around the world—

- **Germany:** The German government is investing €500 million and private industry €2.6 billion in "innovation alliances" that aim to develop nine innovation clusters.[236] An initiative for a cluster in molecular imaging for medical engineering, for example, includes Bayer Schering Pharma, Goehringer Ingelheim Pharma, and Siemens. Other innovation alliances include photovoltaic cells, lithium-ion batteries for energy storage, and automotive electronics.[237] Germany's Fraunhofer has established pilot production centers in a program to accelerate development of cluster in organic electronics in Heidelberg involving a coalition of universities and companies.[238]
- **Brazil:** Brazil's Minas Gerais state is supporting emerging clusters in microelectronics, bio-fuels, and software. The state also has identified

---

[233] See Örjan Solvell, Göran Lindqvist, and Christian Ketels, "The Cluster Initiative Greenbook" (Stockholm: The Competitiveness Institute, 2003). Of 260 cluster initiatives around the world studied for this report, 72 percent had been established in 1999 or later.
[234] Ibid.
[235] Presentation by Andrew Reamer, "Stimulating Regional Economies: The Federal Role," in the National Academies Symposium, *Growing Innovation Clusters for American Prosperity*, June 3, 2009.
[236] German Federal Ministry for Education and Research, *Ideas. Innovation. Prosperity: High Tech Strategy 2020 for Germany*, Berlin: BMBF, 2010 (http://www.bmbf.de/pub/hts_2020_en.pdf). Details of the "European Cluster Alliance" can be found at http://www.proinno-europe.eu/index.cfm?fuseaction=page.display&topicID=223&parentID=50.
[237] Information on Germany's Innovation Alliances is found on the Research in Germany Web site, http://www.research-in-germany.de/coremedia/generator/research-landscape/rpo/networks-and-clusters/41832/10-3-innovation-alliances.html.
[238] Presentation by Christian May, "German Policy Initiatives," in the National Academies Symposium on Flexible Electronics for Security, Manufacturing, and Growth in the United States. September 24, 2010.

several hundred "poles of excellence" in traditional industries that it is seeking to consolidate into hubs based in one location so that they can achieve bigger scale and support larger concentrations of public and private investment. To advance these clusters, the new agency Sistema Mineiro de Inovação, or SIMI, is supporting science parks, incubators, and training programs and helping establish linkages between government programs, researchers, and investors across the state.[239]

- **Taiwan:** The Taiwanese government was instrumental in launching the island's semiconductor, notebook computer, and liquid-crystal display industrial clusters, among others in the 1980s and 1990s.[240] Now, the Industrial Technology Research Institute is coordinating public-private to establish Taiwan as a global leader in industrials such as flexible displays, solid-state lighting devices, and solar modules.[241] The government has invested more than $50 million to help Taiwan develop a comprehensive supply chain for flexible electronics, for example, and has helped acquire key U.S. technologies.[242]

- **Hong Kong:** The Hong Kong government began a concerted cluster-development program following the 1997 Asian financial crisis. It began by targeting areas like green technology, precision engineering, communications technologies, and biotechnology. The goal is to leverage Hong Kong's strategic location on the border of mainland China. Hong Kong is promoting such new clusters as thin-film photovoltaic panels, chips wireless telecom devices, and smart cards. The government has invested $1.5 billion in a science park that is the focal point of these clusters.[243]

- **Singapore:** Singapore's Ministry of Trade and Industry announced $10 billion in R&D spending over five years to accelerate development of clusters such as life sciences, environmental and water technologies, interactive and digital media. The government wants Singapore to become a "global talent hub" in these industries and expects they will employ 80,000 by 2015 and that their value-added will triple to

---

[239] Alberto Duque Portugal, "An Integrated Approach: Brazil's Minas Gerais Strategy," in the National Academies Symposium on Clustering for 21st Century Prosperity, February 25, 2010.
[240] See Alice H. Amsden, "Taiwan's Innovation System: A Review of Presentations and Related Articles and Books," Memorandum on the National Academies Symposium, "21st Century Innovation Systems for the U.S. and Taiwan: Lessons From a Decade of Change," Taipei, January 4-6, 2006.
[241] Chu Hsin-Sen, "The Taiwanese Model: Cooperation and Growth" in National Research Council, *Innovation Policies for the 21st Century.* Op. cit.
[242] See presentation by John Chen, "Taiwan's Flexible Electronics Program," at the National Academies Symposium on Flexible Electronics for Security, Manufacturing, and Growth in the United States." September 24, 2010.
[243] Presentation by Nicholas Brooke "Optimizing Synergies: The Hong Kong Science Park" at the National Academies Symposium on Clustering for 21st Century Prosperity, February 25, 2010.

$27 billion.[244] Government support for these clusters includes new incubators and funding for early-state capital programs.[245]

U.S. Regional Cluster Initiatives

As previously mentioned, many promising regional innovation cluster initiatives are underway across the U.S. Many of cluster-building strategies at the state level reflect a holistic understanding of what it takes to build a 21st century innovation ecosystem and compete globally in specific industries.[246] Promising state and regional initiatives often involve public-private partnerships in which corporations, universities, and governments pool resources to establish R&D centers, train workforces, develop supply and support industries, and provide risk capital to starts-ups where angel and venture funding is lacking.[247]

State governments are deploying a wider range of policy tools, from tax credits and R&D grants to low-cost loans to free workforce training, in the attempt to close the gap with financial incentives offered by offshore locations in the intense competition for investment.[248] Few of these initiatives, however, can match the financial resources and policy support of those in other nations.[249]

The U.S. Federal Role

In remarks at a STEP Board symposium, then Commerce Secretary Gary Locke declared that "regional innovation clusters have a proven track record of getting good ideas more quickly into the marketplace. The burning question becomes, 'How do we create more of them?'"[250]

---

[244] Singapore Ministry of Trade and Industry, *Sustaining Innovation-Driven Growth, Science, and Technology*, Government of Singapore, February 2006,
(http://app.mti.gov.sg/data/pages/885/doc/S&T%20Plan%202010%20Report%20(Final%20as%20of%202010%20Mar%2006).pdf).

[245] Singapore National Research Foundation, "National Framework for Innovation and Enterprise," Prime Minister's Office, Republic of Singapore, 2008,
(http://www.nrf.gov.sg/nrf/otherProgrammes.aspx?id=1206.

[246] For review of cluster growth in the U.S. states, see Mary Jo Waits, "The Added Value of the Industry Cluster Approach to Economic Analysis, Strategy Development, and Service Delivery." *Economic Development Quarterly*, 14(1):35-50, February 2000.

[247] A National Research Council Committee led by Gordon Moore concluded that "Public-private partnerships, involving cooperative research and development activities among industry, government laboratories, and universities, can play an instrumental role in accelerating the development of new technologies to the market." See National Research Council, *Government-Industry Partnerships for the Development of New Technologies*, C. Wessner, ed., Washington, DC: The National Academies Press, 2003, page 23.

[248] See National Research Council, *Growing Innovation Clusters for American Prosperity*, Charles W. Wessner, Rapporteur, Washington, DC: The National Academies Press, 2011.

[249] For a review of scope, as well as advantages and disadvantages of state capitalism, See *The Economist*, The Rise of State Capitalism, January 21, 2012.

[250] Keynote address by then Commerce Secretary Gary Locke at the National Academies Symposium on *Clustering for 21st Century Prosperity*, Washington, DC, February 25, 2010.

A number of analysts, policy institutes, and non-government organizations have published studies in recent years urging the federal government to make regional initiatives a core element in economic development.[251] Rather than calling for massive new funding, several of these same studies call on federal agencies to make more effective and efficient use of scattered resources they already deploy. Michael Porter, for instance, has criticized existing federal programs as "often fragmented, duplicative, and inefficient."[252]

One new federal approach is for several agencies to pool efforts with state and local governments and universities to support specific regional clusters aimed at meeting national needs. Under White House leadership, the SBA, NIST, EDA, NSF, and EDC, for example, are joining an effort by the DOE to establish "energy-innovation hubs," regional innovation clusters in solar power, energy-efficient buildings, nuclear energy, and advanced batteries. The first $129.7 million project seeks to create an innovation hub devoted to developing technologies, designs, and systems for energy-efficient buildings that will be based at the Philadelphia Navy Yard Clean Energy.[253] President Barack Obama's 2009 budget also allocated $50 million in funds administered by the Commerce Department's Economic Development Agency to assist regional cluster initiatives,[254] while the SBA is working with state agencies and the DOD to help launch robotics clusters in Michigan, Virginia, and Hawai'i.[255]

---

[251] For example, see Karen G. Mills, Elisabeth B. Reynolds, and Andrew Reamer, "Clusters and Competitiveness: A New Federal Role for Stimulating Regional Economies," Metropolitan Policy Program at Brookings, April 2008. Also see Michael E. Porter, "Clusters and Economic Policy: Aligning Public Policy with the New Economics of Competition," Institute for Strategy and Competitiveness White Paper, revised May 18, 2009. Mark Muro and Bruce Katz, "The New Cluster Moment: How Regional Innovation Clusters Can Foster the Next Economy," Washington, DC: Brookings Institution, September 2010,
http://www.brookings.edu/papers/2010/0921_clusters_muro_katz.aspx.
[252] Porter, op. cit.
[253] Department of Energy press release, "Penn State to Lead Philadelphia-Based Team that will Pioneer New Energy-Efficient Building designs," August 24, 2010,
(http://www.energy.gov/news/9380.htm). Details on the energy innovation research cluster can be found in the funding opportunity announcement for FY 2010 on the DOE Web site. See http://www.energy.gov/hubs/documents/eric_foa.pdf.
[254] President Obama's fiscal 2009 budget provided $50 million in regional planning and matching grants within the Economic Development Administration to "support the creation of regional innovation clusters that leverage regions' existing competitive strengths to boost job creation and economic growth." See Executive Office of the President, "A Strategy for American Innovation: Driving Towards Sustainable Growth and Quality Jobs," National Economic Council Office of Science and Technology Policy, September 2009.
[255] Presentation by Karen Mills, "Building Regional Innovation Clusters" at the National Academies Symposium on *Clustering for 21st Century Prosperity*, February 25, 2010.

HUNTING FOR GLOBAL TALENT

One of the keys to America's post-war dominance of high-technology industries has been its ability to attract the world's best and brightest scientific, technological, and entrepreneurial talent. European immigrants such as Alexander Graham Bell helped fuel America's industrial takeoff, and the U.S. assumed world leadership in physical sciences with the help of an influx of physicists who fled European fascism, including such Albert Einstein and Enrico Fermi.[256] Since the 1970s, immigrant engineers and scientists from India, Taiwan, South Korea, and then China have been instrumental to the success of the U.S. semiconductor, computer, software industries, and biotechnology industries and have founded an inordinate share of U.S. technology companies.[257]

America is as dependent as ever on imported brainpower as a pipeline for future innovation: Foreign students earned 40 percent of U.S. science and engineering doctorate degrees in 2005, compared to 16 percent in 1980. In engineering, the share was 61 percent.[258] One telling sign of this foreign dominance is to look at where recipients of U.S. engineering Ph.D. have earned their bachelor's degrees. Of the 10 schools with the highest representation of alumni in 2008, six are from China.[259] The Massachusetts Institute of

---

[256] These scientists and engineers were highly esteemed by society though public perceptions may have changed. Recent research suggests that public perceptions of science are highly contextual, with people making judgments about the relative trust to be placed in traditional scientific expertise (which often is generated by government institutions) and in local knowledge based in the local context. See, Lewenstein, Bruce V. 1992. "The Meaning of 'Public Understanding of Science' in the United States After World War II." *Public Understanding of Science* 1 (1):45-68. Recent research also reveals that that social support contributes directly to men's and women's ability to envision themselves in a future science career, which, in turn, predicted their interest in and motivation for a science career. See Sarah K. Buday, Jayne E. Stake and Zoë D. Peterson, "Gender and the Choice of a Science Career: The Impact of Social Support and Possible Selves." *Sex Roles-Journal of Research*, 66(3-4):197-209, 2012.

[257] AnnaLee Saxenian of the University of California at Berkeley estimated that Chinese and Indian engineers were represented on the founding teams of 24 percent of Silicon Valley technology businesses founded between 1980 and 1998. See AnnaLee Saxenian, *Silicon Valley's New Immigrant Entrepreneurs*, San Francisco: Public Policy Institute of California, 1999. A follow-up study found that in one-quarter of all U.S. technology companies founded between 1995 and 2005, one-quarter had chief executive officers or chief technology officers who were foreign-born. See Vivek Wadhwa, Ben Rissing, AnnaLee Saxenian, Gary Gereffi, "Education, Entrepreneurship and Immigration: America's New Immigrant Entrepreneurs, Part II," Duke University Pratt School of Engineering, U.S. Berkeley School of Information, Ewing Marion Kauffman Foundation, June 11, 2007.

[258] Robert V. Hamilton presentation at Brookings Institution conference on "Immigration Policy: Highly Skilled Workers and U.S. Competitiveness and Innovation," Washington, February 7, 2011.

[259] Semiconductor Industry Association, Maintaining America's Competitive Edge: Government Policies Affecting Semiconductor R&D and Manufacturing Activity, prepared by Dewey & LeBoeuf, March 2009, (http://www.sia-online.org/galleries/default-file/Competitiveness_White_Paper.pdf).

Technology ranks No. 10. Chinese students alone accounted for 30 percent of all U.S. doctorate degrees granted in natural sciences.[260]

Now the competition for non-native talent is becoming global as more countries take an activist approach to recruiting talent.[261] To address skill shortages exacerbated by an aging population, the European Union has promulgated a "blue card" that allows highly skilled migrants from non-EU nations to live and work on a temporary base, and also allows them to move freely among most member countries.[262] The EU also is simplifying procedures for obtaining legal resident status for foreign workers to by setting up a "one-stop-shop" system for applicants.[263] Canada has made recruiting foreign talent a top priority in its national innovation strategy.[264] Forty percent of the 8,053 new faculty members and 44 percent of the 1,806 new researches recruited by Canadian universities and the Foundation for Innovation as of the fall of 2009 came from other nations, for example.[265] Thirty percent of the nearly 2,000 department chairs hired the Canada Research Chairs program also were recruited outside of Canada.[266] Singapore's innovation strategy puts a heavy emphasis on "drawing creative and talent people from all corners of the world to live and work in Singapore."[267] Among its prize recruits are eminent scientists from the National Cancer Institute, MIT, and the University of California at San Diego.[268]

While other nations step up recruiting, it has been getting more difficult for highly skilled foreigners to live and work in the U.S. The backlog for permanent resident visas grew so long amid tightened scrutiny after the Sept. 11,

---

[260] Robert V. Hamilton, "Foreign Natural Sciences Doctoral Attainment at U.S. Universities, 1980 to 2005, George Mason University, prepared for Brookings Institution conference on "Immigration Policy: Highly Skilled Workers and U.S. Competitiveness and Innovation, " Washington, February 7, 2011.
[261] See Devesh Kapur and John McHale, *Give us Your Best and Brightest*, Washington, DC: Center for Global Development, 2005.
[262] The Blue European Labour Card is an approved EU-wide work permit (Council Directive 2009/50/EC) allowing high-skilled non-EU citizens to work and live in any country within the European Union, with the exception of UK, Denmark, and Ireland.
[263] *Europa*, "Making Europe More Attractive to Highly Skilled Immigrants and Increasing the Protection of Lawfully Residing and Working Migrants," Brussels, October 23, 2007, (http://europa.eu/rapid/pressReleasesAction.do?reference=IP/07/1575).
[264] Industry Canada, Achieving Excellence: Investing in People, Knowledge and Opportunity—Canada's Innovation Strategy, 2001. (http://dsp-psd.pwgsc.gc.ca/Collection/C2-596-2001E.pdf).
[265] Canada Foundation for Innovation, *2009 Report on Result*, op. cit.
[266] Canada Research Chairs data http://www.chairs-chaires.gc.ca/home-accueil-eng.aspx.
[267] Ministry of Trade and Industry, *Sustaining Innovation-Driven Growth, Science, and Technology*, Government of Singapore, February 2006, (http://app.mti.gov.sg/data/pages/885/doc/S&T%20Plan%202010%20Report%20(Final%20as%20of%2010%20Mar%2006).pdf).
[268] Lim Chuan Poh, "Singapore Betting on Biomedical Science," *Issues in Science and Technology*, Spring 2010.

2001, terrorist attacks that an estimated 1 million people were waiting for 120,120 visas issued a year as of 2006—a backlog of nine years.[269]

The tougher immigration climate comes despite forecasts of looming skill shortages due to demographic changes and declining interest by U.S. students in science and engineering. The McKinsey Global Institute, for instance, projects a possible shortfall of nearly 2 million technical and analytical workers in the U.S. over the next 10 years.[270] The National Association of Manufacturers and Deloitte & Touche reported that higher immigration will be necessary to meet a projected need for new skilled workers in manufacturing by 2020. The alternative could be "a significant decrease in manufacturing's competitiveness."[271] The Brookings Institution concludes that the "the U.S. immigration priorities and outmoded visa system discourage skilled immigrants and hobble the technology-intensive employers who would hire them." As a result, these policies "work against urgent national priorities."[272]

Not all analysts agree that dramatic increases in immigration are required to meet future skill needs. Research by Lindsey Lowell and Harold Salzman, for example, concluded that the U.S. actually graduates more STEM students than are hired each year, and that many graduates find work in other fields for economic reasons.[273] Nor is there yet firm evidence that Chinese, Indian, and other foreign nations are returning home in significant numbers after receiving advanced U.S. science and technology degrees.[274] Other studies, however, suggest a significant risk of a "brain drain" as highly skilled Chinese and Indians leave to take advantage of greater career opportunities in their home countries.[275] Continued inaction and complacency threatens over time to undermine an essential pillar of U.S. competitiveness.

Several proposals seek to reform U.S. immigration rules that tilt heavily toward granting citizenship to relatives of current citizens, regardless of

---

[269] See Vivek Wadwha, Guillermina Jasso, et. al, "Intellectual Property, the Immigration Backlog, and a Reverse Brain-Drain," Ewing Marion Kauffman Foundation, August 2007, (http://www.kauffman.org/uploadedFiles/reverse_brain_drain_101807.pdf).
[270] James Manyika, et. al, *Growth Renewal in the United States: Retooling America's Economic Engine*, McKinsey Global Institute, February 2001.
[271] The National Association of Manufacturers, the Manufacturing Institute, and Deloitte & Touche, "Keeping America Competitive: How a Talent Shortage Threatens U.S. Manufacturing," April 21, 2003.
[272] Darrell M. West, "Creating a 'Brain Gain' for U.S. Employers: The Role of Immigration," Brookings Policy Brief Series #178, Brookings Institution, January 2011.
[273] B. Lindsay Lowell, Hal Salzman, Hamutal Bernstein, and Everett Henderson, "Steady as She Goes? Three Generations of Students Through the Science and Engineering Pipeline," paper presented at annual meets of the Association for Public Policy Analysis and management, Washington, DC, October 2009.
[274] See Patrick Gaule, "Return Migration: Evidence From Academic Statistics," National Bureau of Economic Research fellow, draft paper, November 17, 2010.
[275] Vivek Wadhwa, AnnaLee Saxenian, Richard Freeman, and Alex Salkever, "Losing the World's Best and Brightest: America's New Immigrant Entrepreneurs," Ewing Marion Kauffman Foundation, March 2009.

skills. Only 6.5 percent of U.S. immigrant visas are for skilled workers, compared to 36 percent in Canada. And of those holding H-1B visas, only 7 percent are able to change to permanent resident status, notes Darrell West of Brookings.[276] Common reform proposals include easing limits on temporary work visas, streamlining visa procedures, and giving priority for green cards to foreigners with advanced science and technology degrees and needed skills.[277] The McKinsey Global Institute observes that nations such as Australia, the United Kingdom, and Canada have moved to a point-based system for allocating residency based heavily on skill levels. It suggests the U.S. do the same.[278]

Proposed changes in U.S. immigration policy, however, have aroused intense political passions that make it difficult for Congress to consider reform of rules that would attract and retain highly skilled immigrants to the Unites States.[279] In this context, the recent initiatives by the Department of Homeland Security and the Bureau of Citizenship and Immigration Services are welcome. Announced in August 2011, these initiatives now make it possible for foreign entrepreneurs to obtain an EB-2 immigrant visa if they can demonstrate that their business endeavors will be in the national interest of the United States. Also, H-1B beneficiaries who are sole owners of the petitioning company may petition for H-1B non-immigrant visas to employ foreign workers in specialty occupations that require theoretical or technical knowledge.[280]

THE WAY FORWARD

The world of innovation is changing rapidly. Old assumptions about how investments in research result in commercial products and domestic industries are becoming less valuable as frameworks for U.S. science and technology policy.

A New Approach: A new policy approach is required, one based on a richer understanding of the complexity and global dimensions of innovation. While greater investments in research and development are needed to keep the United States at the technology forefront, that alone will not guarantee globally competitive U.S. industries and a prosperous U.S. economy. Intermediating

---

[276] Darrell M. West, "Creating a 'Brain Gain' for U.S. Employers: The Role of Immigration," Brookings Policy Brief Series #178, Brookings Institution, January 2011.
[277] Ibid. Some analysts have emphasized the need to strengthen the U.S. pipeline of scientists and engineers and to create a more competitive immigration policy that admit the "best and brightest" from around the world. See the statement of B. Lindsay Lowell before the House Judiciary Committee "Immigration and the Science & Engineering Workforce: Failing Pipelines, Restrictive Visas, and the 'Best and Brightest'"October 5, 2011.
[278] James Manyika, et. al., *Growth and Renewal in the United States: Retooling America's Economic Engine*, McKinsey Global Institute, February 2011
[279] For a review of potential reforms concerning the H-1B visa, which enables U.S. employers to hire temporary, foreign workers in specialty occupations, see GAO, "Reforms Are Needed to Minimize the Risks and Costs of Current Program." GAO-11-26.
[280] *Wall Street Journal*, "U.S. to Assist Immigrant Job Creators." August 3, 2011.

institutions and new initiatives, both at the state and federal levels, as well as by private foundations, are needed for the United States to capture the benefits of its public investments in research and development.

Indeed, the way forward for the United States is to build on its strengths: open competition, deep private capital markets, leadership in academic research, a flexible labor force, intellectual property protections, and an environment that allows entrepreneurs to quickly respond to new market and investment opportunities. Importantly, these strengths need to be renewed and reinforced, as they have in the past, with federal programs to nurture and grow new technologies and new industries of the future.

The Role of Partnerships: Public-private partnerships have long been a key element of successful U.S. innovation policy.[281] Public-private partnerships can provide incentives for closer collaboration among government industry, higher education, the military, private investment groups, and other institutions to foster an environment in which the United States can thrive in an era of open and global innovation.[282] Well designed public-private partnerships not only can help insure that the U.S. remains a world leader in creating knowledge, but they also can enable America to capture more of the economic value of innovation by making U.S. regions more competitive places to translate inventions into products, companies, industries, and jobs.

This report documents several examples of successful U.S. collaboration between government, industry, and academia. They include federal programs such as the SBIR and the NIST Advanced Technology Program, research consortia such as Sematech, and newer institutions such as the Flexible Display Center at Arizona State University.[283] This report also highlights a number of promising and innovative state and regional public-private initiatives to bolster competitiveness.[284] Such initiatives include regional innovation clusters, new kinds of science parks, workforce-training programs, and efforts to help entrepreneurs obtain access to the facilities, technical assistance, and early-stage capital they need to convert U.S. innovation into a new wave of U.S. industries. Federal agencies can play a valuable support role in aiding these regional initiatives.

What are others doing? American policymakers also need to learn from the experiences of other nations and discern which best practices can be

---

[281] National Research Council, *Government-Industry Partnerships for the Development of New Technologies, Summary Report*, C. Wessner, ed., Washington, DC: National Academy Press, 2001.
[282] National Research Council, *Government-Industry Partnerships for the Development of New Technologies: Summary Report*, C. Wessner, ed., Washington, DC: National Academy Press, 2001.
[283] See Chapter 6 for an illustrative review of national policies and programs to support emerging industries abroad.
[284] See Chapter 7 for an illustrative review of national and regional policies to develop innovation clusters around the world.

adapted to the American context.[285] Well-designed public-private partnerships can address many of the challenges facing the myriad actors of the U.S. innovation ecosystem and can help ensure that more of the fruits of America's tremendous investments in research flow into the American economy.

The bold and innovative strategies being deployed abroad offer valuable lessons for policymakers in the U.S. This report details a great variety of actions governments are taking around the world to both increase their nations' innovation capacity and global competitiveness in emerging technology-intensive industries. In some cases, governments are adapting the most successful features of the U.S. innovation ecosystem—such as university-industry collaboration, public provision and support for early-stage risk capital, strong protection of intellectual property rights, and well-funded, scalable research parks. In other cases, nations in Asia and Europe are pioneering new models of public-private partnerships that far exceed the scale and scope of comparable U.S. programs. This is especially true when it comes to applied technology and support for large-scale manufacturing.

This unprecedented focus around the world on innovation means that American science and technology policies can no longer be based on the outdated assumption that the United States is naturally destined to remain the global center of innovation activity. Nor can it be based on the assumption that bolstering American industrial competitiveness is merely a matter of increasing R&D spending. As innovation becomes more globalized, absorbing and capitalizing on product and process innovations from abroad will become increasingly important for U.S. competitiveness.

Importance of Collaboration: policies also need to take into account the increasingly global and open nature of the innovation process, much of which takes place within collaborative international networks of researchers in universities, companies, and other institutions. As nations around the world increase their innovation capacity and R&D workforces, leveraging technology and brainpower abroad will become increasingly important for the U.S. to achieve its own science and technology goals.

Collaboration in research and development can greatly accelerate discoveries of cures for chronic disease, the development of renewable energies, and technologies to curb the negative impacts of climate change. Open cross-border innovation networks, meanwhile, can help corporations turn new technologies into innovative products faster, at greater variety and at lower cost. It is important, therefore, to insure that the United States can compete, cooperate, and prosper in this new world of innovation. That will require a fresh approach to innovation policy.

---

[285] See Chapter 5 for case study reviews of programs and policies of leading nations and regions, including China, India, and Germany.

Box 2.4
A History of Public Private Partnerships

Public-public-private collaborations have been woven into the fabric of the U.S. economic system from the beginning of the Republic. What became known as the American System of Manufacturing, in which goods from muskets to clocks were made of interchangeable parts, was pioneered in the early 1800s through War Department contracts.[286] Congress funded Samuel Morse's demonstration of the first telegraph with a substantial grant in 1842. America's aircraft industry was nurtured by the 1925 U.S. Air Mail Act.[287] RCA was founded in 1919 at the initiative of the Navy Department, which also held equity and a board seat, so that the U.S. could have a radio communication industry to compete with Britain's Marconi Co.[288] The U.S. Signal Corps funded most of the initial research for transistors and semiconductors, and the military funded the first production lines of Western Electric, General Electric, Raytheon, and Sylvania. It also bought most of the output for weapons and communications systems.[289] Admiral Hyman Rickover and his naval reactor group oversaw the design and construction of America's first civilian light-water nuclear power plant at Shippingport, Penn., in the 1950s.[290] Military research and weapons contracts also have been instrumental in establishing America's aerospace and computer industries and the forerunner of the Internet.[291] Federal programs have been instrumental as well to the U.S. pharmaceutical industry. A recent study found that public-sector research institutions made important contributions to

---

[286] See David A. Hounshell, *From the American System to Mass Production, 1800-1932: The Development of Manufacturing Technology in the United States*, Baltimore, Maryland, USA: Johns Hopkins University Press, 1984.

[287] A stated purpose of the U.S. Air Mail Act of 1925 (also known as the Kelly Act), which authorized the U.S. Postal Service to contract with private aviation companies, was "to encourage commercial aviation." The federal role in their early airline industry is explained in Roger E. Bilstein, *Flight in America: From the Wrights to the Astronauts*, Baltimore: Johns Hopkins University Press, 1984, and in Tim Brady, editor, *The American Aviation Experience: A History*, Southern Illinois University Press, 2001.

[288] An early account of the U.S. Navy's role in establishing RCA and the U.S. radio communication system is found in *The World's Work*, "The March of Events," Volume XLIV, May 1922.

[289] A concise history of U.S. government involvement in establishment of America's electronics industry is found in Kenneth Flamm, *Mismanaged Trade?: Strategic Policy and the Semiconductor Industry*, Washington, DC, Brookings Institution, 1996. pp. 27-38.

[290] Richard Hewlett and Francis Duncan, *The Nuclear Navy*, Chicago: University of Chicago, 1974.

[291] See National Research Council, *Funding a Revolution, Government Support for Computing Research*, Washington, DC: National Academy Press, 1999. The extensive NRC review documents the seminal role o federal funding for the information and communications industries of today. See also the presentation by Kenneth Flamm of the University of Texas at Austin in National Research Council, *Innovation Policies for the 21st Century*, op. cit.

> the discovery of up to 21.2 percent al all new FDA-approved drugs from 1990 through 2007.[292]

Capturing the value of U.S. investments in R&D: The assumption that the output of the U.S. innovation process will be captured by U.S.-based industry has been rendered obsolete by globalization and the rise of corporate open innovation practices. In today's world, knowledge created through federally funded research at universities and national laboratories can be commercialized and industrialized virtually anywhere. The key is to take measures to provide the funding, support services, and to anchor new and existing companies in clusters of competency here in the United States.

This report highlights the features of a more comprehensive innovation policy. It calls for a better understanding by government of the real factors behind corporate decisions on where to develop new technologies, commercialize products, and locate production and help close competitive gaps with other nations to the degree possible. Some of these gaps can be closed with more enlightened tax policy, in others through incentives such as research grants, loans, and credits for U.S.-based manufacturing.

The committee's formal findings and recommendations on how to sustain a strong American innovation system for the 21st century are found in the next two chapters.

---

[292] Ashley J. Stevens, Jonathan J. Jensen, Katrine Wyller, Patrick C. Kilgore, Sabarni Chatterjee, and Mark L. Rohrbaugh, "The Role of Public-Sector Research in the Discovery of Drugs and Vaccines," *The New England Journal of Medicine*, February 9, 2011, (http://healthpolicyandreform.nejm.org/?p=13730&query=home).

Chapter 3

Findings

A fundamental challenge in making recommendations to improve the U.S. innovation system is that it arguably remains the best in the world. The U.S. is home to the vast majority of the world's leading research universities. It has wide and deep capital markets, receptivity to innovative products, a culture and legal system that encourage entrepreneurship, and substantial public and private investments in research and development. The country also makes substantial investments in national security that can generate new products and develop new platform technologies.

The challenge for the United States is that the global environment is changing substantially and rapidly. Some of these changes, although they may require adjustments, are nonetheless quite positive, involving the production of more and better research and more and better-trained students. Globally, these trends represent a potential improvement in human welfare. On the other hand, changes in the competitive environment and, in particular, other countries' focus on the application, commercialization, and local production of new technologies and new products pose challenges to the long-term health of the U.S. innovation system. A global system in which the U.S. does the research and other countries capitalize on the results to enhance the competitiveness and competency of their own economies is not in the U.S. national interest, nor is it sustainable.

Moreover, the security dimension of a robust U.S. innovation ecosystem cannot be ignored. U.S. leadership in innovation has been the source of U.S. economic and military power throughout the post-war era. The United States must continue to lead as an innovator and manufacturer of leading edge technologies and products, especially in the current environment where other nations are pursuing active innovation policies to enhance their world role.

Current financial constraints should not dictate U.S. policy in this crucial arena because the failure to preserve American technological leadership imperils both our long-term prosperity and, very directly, our national security. Although the U.S. must exercise fiscal prudence as it wrestles with its debt and deficits, the Committee believes that the investments advocated below will repay the expenditures in the aggregate, paving the way for the economic growth necessary to help solve our fiscal problems in the long-term.[1]

While it is neither desirable nor possible to freeze the global allocation of production, it is essential that the U.S. recognizes that other countries are pursuing vigorous policies and programs, at increasing funding levels, to nurture and grow the industries of the future as well as revitalize those of today. Some of these policies are mercantilist in nature and include measures that distort the international location of productive activity through national regulation of investment and trade, forced technology transfer, and toleration if not promotion of intellectual property violations that undercut the basis for a rules-based trading system.

Success in promoting innovation – from invention through commercialization – is necessary not only for reasons of national security but to preserve and enhance the economic well-being of the American people. It is the key to maintaining the promise that the opportunities for each future generation will be better than those enjoyed by the preceding one.

This chapter presents the Committee findings. There are seven major findings, which are further elaborated in sub-findings. The organization of these findings and sub-findings is presented in an outline, below, as a guide to the reader.

---

[1] Although the Committee did not do a cost-benefit analysis of the policies and investments recommended in this report, the economics literature strongly suggests that investments in research, education, and infrastructure contribute to U.S. economic growth. See for example, Robert M. Solow, "A Contribution to the Theory of Economic Growth," *Quarterly Journal of Economics,* 1956, 70(1):65-94. Robert M. Solow, "Technical Change and the Aggregate Production Function," *The Review of Economics and Statistics*, 1957, 39 (3): 312-320. Richard Nelson, *Technology, Institutions and Economic Growth,* Cambridge MA: Harvard University Press, 2005. Dale W. Jorgenson et al., *Productivity: Information Technology and the American Growth Resurgence*, Cambridge MA: MIT Press, 2005.

OUTLINE OF FINDINGS

1. The future economic prosperity and security depends on sustaining the nation's capacity to innovate—that is, translate our investments in research into new products for the market and new solutions for national missions.

   a. The global environment is changing rapidly
   b. A vibrant national innovation ecosystem is an essential component of U.S. security
   c. The importance of innovation for jobs and technological leadership

2. Pillars of the U.S. Innovation System

   a. The role of research universities
   b. Research and development by the private sector
   c. Federal support for emerging technologies
   d. Public-private partnerships for the development of new technologies
   e. Small business entrepreneurship
   f. Talented immigrants

3. Advantages and Challenges in the U.S. Innovation System

   a. U.S. advantages

      i. An open innovation system
      ii. Strong intellectual property rights
      iii. Bankruptcy laws that permit risk sharing and recovery
      iv. Worker mobility

   b. Challenges for the U.S.

      i. Fiscal constraints
      ii. Declining federal R&D intensity
      iii. Decline in university funding amid new challenges
      iv. High non-production costs
      v. Infrastructure and broadband enablers

4. Governments around the world have made the development of a globally competitive, innovation-led economy a top strategic priority.

   a. Developing national strategies
   b. Increasing commitments to R&D

c. Emulating global best practices
   d. Pursuing mercantilist policies
   e. Expanding universities
   f. Providing early-stage finance
   g. Attracting global talent
   h. Focusing on building innovation clusters and science parks

5. U.S. leadership in innovation is eroding

   a. The emergence of major global competitors
   b. Growth of innovative regions around the world
   c. Growth of offshore research centers

6. Capturing the Benefits of Investments in R&D

   a. Research is a global public good
   b. The need for a strategic approach
   c. An institutional focus on translational research and applications
   d. A focus on manufacturing
   e. Trade and innovation are closely linked

7. Opportunities for Cooperation:

   a. New opportunities and common challenges
   b. Greater outreach
   c. The internet and cross-border data flows
   d. Greater awareness

FINDINGS IN DETAIL

1. ***The future economic prosperity and security depends on sustaining the nation's capacity to innovate—that is, translate our investments in research into new products for the market and new solutions for national missions.*** Other nations are focused on developing greater capacity to translate research into marketable products. Although the U.S. innovation system remains the world's most dynamic and productive, America's continued standing as the premier location for producing new technologies and new high-technology products and services is no longer assured.

    a. **The Global Environment is Changing Rapidly[2]:** As identified in earlier Academy reports, there are disturbing trends, notably between what the United States is doing and what it needs to do, compared with what the rest of the world is doing in terms of investments in education, infrastructure, research, new technologies, and measures to bring new technologies to the market.[3] The U.S. international position as a location for the production of new processes and products is declining relatively as other nations, especially emerging economies, have accelerated their efforts to catch-up technologically.[4]

    b. **A Vibrant National Innovation Ecosystem is an Essential Component of U.S. Security**: Leadership in innovation has been the source of U.S. economic and military power throughout the

---

[2] For an overview of new trends in global innovation, see Chapter 1 of this report.

[3] As a recent National Academies report has noted, "Although many people assume that the United States will always be a world leader in science and technology, this may not continue to be the case inasmuch as great minds and ideas exist throughout the *world*. We fear the abruptness with which a lead in science and technology can be lost—and the difficulty of recovering a lead once lost, if indeed it can be regained at all." See National Academy of Sciences, *Rising Above the Gathering Storm; Energizing and Employing America for a Brighter Economic Future,* Washington, DC: The National Academies Press, 2007, p. 3.

[4] The recent National Academies report *S&T Strategies of Six Countries* concludes "globalization has facilitated the success of formal S&T plans in many developing countries, where traditional limitations can now be overcome through the accumulation and global trade of a wide variety of goods, skills, and knowledge. As a result, centers for technological research and development (R&D) are now globally dispersed, setting the stage for greater uncertainty in the political, economic, and security arenas." National Research Council, *S&T Strategies of Six Countries: Implications for the United States*, Washington, DC: The National Academies Press, 2010. Some analysts see the focus and investments of others as a challenge and an example of what needs to be done in the United States. For example, Ernst argues that "China's innovation policy and its considerable achievements should serve as a wake-up call for America to mobilize the combined forces of private industry and government to upgrade its own innovation system." Dieter Ernst, "China's Innovation Policy is a Wake-Up Call for America," *Asia Pacific Issues*, No. 100 (May 2011).

post-war era.[5] Nations pursue active innovation policies not just for economic growth and jobs but also to enhance their world role.[6] The United States will not be able to meet its defense needs without a robust economy that is able to and in fact does produce leading edge technologies and products.[7] The composition of the American economy matters. This will require America building on its historical strength of melding of private ingenuity and public support.

c. **The Importance of Innovation for Jobs and Technological Leadership:** An assessment of a nation's economic health must go beyond simple aggregate measures such as gross domestic product and include the ability to innovate and manufacture new products

---

[5] As the "Six Countries" report cited above notes, the globalization of innovation "will have a potentially enormous impact for U.S. national security policy, which for the past half century has been premised on U.S. economic and technological dominance." National Research Council, *S&T Strategies of Six Countries: Implications for the United States.* Washington, DC: The National Academies Press, 2010. Bonvillian argues that "defense technology cannot be discussed as though it is separate and apart from the technology that drives the expansion of the economy—they are both part of the same technology paradigms." William B. Bonvillian, "The Connected Science Model for Innovation – The DARPA Role," in National Academy of Sciences, Board on Science, Technology, and Economic Policy, *21st Century Innovation Systems for Japan and the United States: Lessons from a Decade of Change*, Washington, DC: The National Academies Press, 2009, pp. 206-237. See also David C. Mowery, "National Security and National Innovation Systems," *Journal of Technology Transfer* (2009) 34:455–473. In addition to the security mission, military and defense related research, development and procurement have been major sources of technology development across a broad spectrum of industries that account for an important share of United States industrial production. See Vernon W. Ruttan, *Is War Necessary for Economic Growth.* Oxford: Oxford University Press, 2006.

[6] For example, as Chinese President Hu Jintao noted in his Report to the 17th National Congress of the Communist Party of China, "Innovation is the core of our national development strategy and a crucial link in enhancing the overall national strength."

[7] Jacques Gansler argues that a strong and affordable national security posture must be built on a healthy economy: "a nation that devotes too many of its resources to the military rather than to the growth of its economy is likely to weaken its national power." He further notes that the defense industry must remake itself through innovation to become responsive and relevant to the needs of twenty-first-century security. See Jacques S. Gansler, *Democracy's Arsenal, Creating a 21st Century Defense Industry,* Cambridge MA: MIT Press, 2011. Leadership in enabling technologies such as semiconductors is critical to the U.S. military's strategy of maintaining technological superiority, for example. See U.S. Department of Defense, *Report on Semiconductor Dependency*, Office of the Undersecretary of Defense for Acquisition, prepared by the Defense Science Board Task Force, Washington, DC, February 1987. Acceleration of innovation in clean-energy technologies is vital to the U.S. Army's new advanced weapons programs and development of hybrid and electric-drive combat vehicles, which can provide important tactical advantages in the battlefield. See presentations by Grace Bochenek and Sonya Zanardelli of the U.S. Army Tank and Automotive Research, Development, and Engineering Center at the National Academies conference on Building *the U.S. Battery Industry for Electric-Drive Vehicles: Progress, Challenges, and Opportunities,* Livonia, Michigan, July 26-27, 2010.

FINDINGS

for the market, and the ability to create and sustain high skilled, high pay manufacturing jobs.[8]

2. **Pillars of the U.S. Innovation System**[9]: *The U.S. Innovation system is built on the foundations of its robust research universities, substantial federal and private support for research and development, vibrant entrepreneurship including that of immigrants, and the often catalytic role of public-private partnerships in bringing new technologies to the marketplace.*[10] *These pillars of the U.S. innovation system need to be preserved and reinforced.*[11]

   a. **The Role of Research Universities:** Research universities are engines of the American innovation system and have been a distinct U.S. competitive advantage in the post-War era.[12] Federally funded university research has enabled some of the most important innovations of the modern economy, including computing, the laser, the fundamentals of global positioning systems, numerically controlled machines, the organization and deployment of the World Wide Web, the revolution in genetics, and much of modern medicine.[13]

   b. **Research and Development by the Private Sector:** Private firms have conducted two-thirds of R&D in the United States over the past decade. [See Figure 3.1] Since the late 1980s, nearly all of the growth in R&D spending in the United States has come from the

---

[8] Nelson argues that "technological advance is the key driving force behind economic growth" and highlights the importance of history, culture, and institutions in the development of new technologies. Richard R. Nelson, *The Sources of Economic Growth*, Cambridge MA: Harvard University Press, 2000.

[9] For an overview of key pillars of U.S. innovation, see Chapter 1 of this report.

[10] For example, the June 2011 launch of the Advanced Manufacturing Partnership cited cooperation between industry, universities, and the federal government as a critical component of the effort to enhance U.S. manufacturing and innovation. (http://www.whitehouse.gov/the-press-office/2011/06/24/president-obama-launches-advanced-manufacturing-partnership).

[11] The recommendations to strengthen the pillars of the U.S. innovation system amplify key recommendations of the National Academies report *Rising Above the Gathering Storm*, op. cit.

[12] See David C. Mowery and Bhaven N. Sampat, "Universities in national innovation systems," *Oxford Handbook of Innovation*, 2005. See also, John Aubrey Douglass, "Universities, the US High Tech Advantage, and the Process of Globalization," Berkeley Research Paper CSHE.8, 2008.

[13] As Robert Birgeneau, Chancellor of UC Berkeley has noted, "To suggest that, somehow, universities are not and should not be engines of economic growth is missing the central point of how our economy grows and how we create jobs." Quoted on NPR Morning Edition Date: 08-09-04. See also Kent Hughes and Lynn Sha, eds., *Funding the Foundation: Basic Science at the Crossroads*, Washington, DC: Woodrow Wilson Center, 2006. See Peter McPherson, David Shulenburger, Howard Gobstein, and Christine Keller, "Competitiveness of Public Research Universities & the Consequences for the Country: Recommendations for Change," Association of Public and Land-Grant Universities, March 2009, (http://www.aplu.org/NetCommunity/Document.Doc?id=1561).

private sector.[14] This investment, which is focused more on the application and development of knowledge, has yielded numerous innovations, contributing to U.S. competitiveness and economic productivity.[15] For example, the applied science of drug development and clinical refinement of compounds carried out by the private sector is closely linked to new scientific discoveries that have been translated into new medicines.[16] These major innovations by American private companies are typically built on platforms developed through long-term substantial U.S. public investments in basic research.[17] It is important to understand that these public and private research efforts are complementary, with neither sufficient on its own, and thus the stagnant government R&D spending is a matter of concern.

---

[14] Industry R&D spending (in constant dollars) has increased over two and a half times during the past 20 years while federal R&D spending as a percentage of GDP has remained roughly constant. National Science Foundation, National Center for Science and Engineering Statistics, *Science and Engineering Indicators 2012*, NSB 12-01 (January 2012), Appendix Tables 4-1 and 4-7.

[15] Congressional Budget Office, R&D and Productivity Growth, June 2005, http://www.cbo.gov/ftpdocs/64xx/doc6482/06-17-R-D.pdf.

[16] Benjamin Zycher, Joseph A DiMasi and Christopher-Paul Milne, "The Truth About Drug Innovation: Thirty-Five Summary Case Histories on Private Sector Contributions to Pharmaceutical Science," Medical Progress Report 6, June 2008.

[17] As Zycher et al. (op cit) note, "Both NIH-sponsored and private-sector research are crucial for the advance of pharmaceutical science and the development of new and improved medicines. Research conducted at universities and government laboratories, often funded by the NIH or other government agencies, has been an indispensable component of the advance of pharmaceutical science and the development of new medicines." As the Venture Capitalist Mary Meeker has remarked more generally, "Remember: private investment maybe have given us Facebook and Garmin, but public sector investment gave us the Internet and GPS."

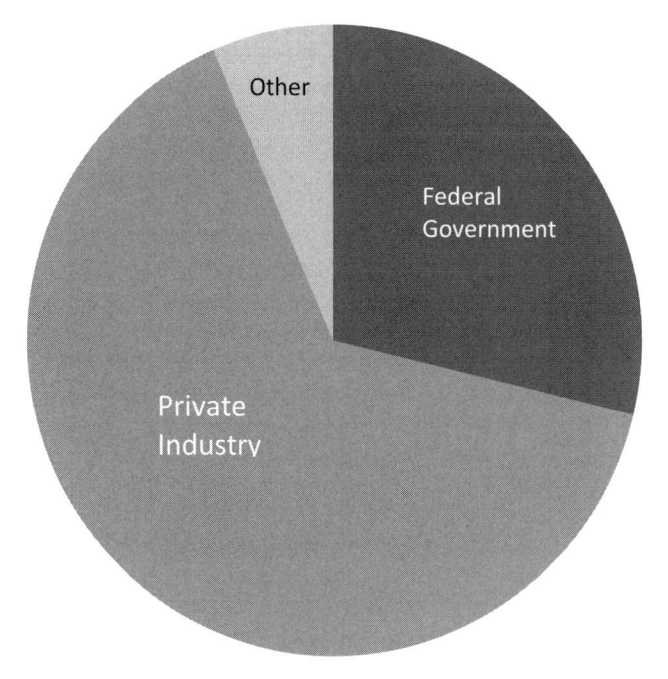

FIGURE 3.1 Private industry has funded almost two-thirds of R&D in the United States over the past ten years.
SOURCE: National Science Foundation, National Center for Science and Engineering Statistics, *Science and Engineering Indicators 2012*, NSB 12-01 (January 2012), Appendix Table 4-7.

    c. **Federal Support for Emerging Technologies:**[18] The United States Government has a long history of supporting the development and domestic production of emerging technologies. Federal support for new technologies played crucial roles in developing industries as diverse as the telegraph, radio, airframes, engines, space, nuclear power, computers, and of course the internet.[19] These pervasive technologies have exerted a significant

---

[18] For a review of support for selected emerging technologies by the U.S. and leading European and Asian nations, see Chapter 6 of this report.

[19] As Vernon Ruttan has observed, "government has played an important role in the development of almost every general purpose technology in which the United States was internationally competitive." Vernon W. Ruttan, *Technology, Growth and Development: An Induced Innovation Perspective*, Oxford: Oxford University Press, 2001. See also Linda Cohen and Roger Noll, *The Technology Pork Barrel*, Washington, DC: Brookings, 1991. Cohen and Noll observe that there although there are failures, there are frequent major successes among federal R&D programs. They count among the successes telegraphy, hybrid seeds, aircraft, radio, radar, computers,

impact on U.S. productivity growth.[20] The prospect that Federal funding for R&D that develops these innovations will diminish due to budget pressures is therefore a cause for major concern.

d. **Public-Private Partnerships for the Development of New Technologies:** Public-private partnerships have often played a powerful role in accelerating the conversion of new technologies into commercial products and in preserving the competitiveness of existing U.S. industries.[21] American research consortia such as SEMATECH[22] and the Department of Energy's recent Sunshot Initiative , long-term investments over many decades such as the Department of Energy's funding for research and development for renewables, fossil fuels, and nuclear technologies, and competitive innovation awards such as the Small Business Innovation Research Program and the Technology Innovation Program[23] are all

---

semiconductors, and communications satellites. In short, much of the foundation for the modern economy. At the same time, Cohen and Noll stress that political capture by distributive congressional politics and industrial interests are one of the principal risks for government-supported commercialization projects.

[20] See National Research Council, *Funding a Revolution: Government Support for Computing Research*, Washington, DC: National Academy Press, 1999. For a review of the positive impact of computers, communications technologies, and software on U.S. total factor productivity, see Dale W. Jorgenson, Mun S. Ho, and Kevin J. Stiroh, Productivity, Volume 3: Information Technology and the American Growth Resurgence, Cambridge MA: MIT Press, 2005. For a review of the positive impact of U.S. investments in energy technologies, see National Research Council, *Energy Research at DoE: Was It Worth It? Energy Efficiency and Fossil Energy Research 1978 to 2000*, Washington, DC: National Academy Press, 2001.

[21] A National Research Council Committee led by Gordon Moore concluded that "public-private partnerships, involving cooperative research and development activities among industry, government laboratories, and universities, can play an instrumental role in accelerating the development of new technologies to the market." See National Research Council, *Government-Industry Partnerships for the Development of New Technologies*, C. Wessner, ed., Washington, DC: The National Academies Press, 2003, page 23. For a brief summary of the role of public-private partnerships through U.S. history, see Box 2.4 in Chapter 2 of this report. According to Kent H. Hughes, public-private collaboration played a key role in the recovery of the U.S. economy from its last period of economic malaise. He argues that similar collaboration is needed to address the competitive challenges of the 21st Century. Kent Hughes, *Building the Next American Century: The Past and Future of American Economic Competitiveness*, Washington, DC: Woodrow Wilson Center Press, 2005.

[22] See Kenneth Flamm and Qifei Wang, "Sematech Revisited: Assessing Consortium Impacts on Semiconductor Industry R&D," in National Research Council, *Securing the Future, Regional and National Programs to Support the Semiconductor Industry*, C. Wessner, ed., Washington, DC: The National Academies Press, 2003. See also Thomas R. Howell, Brent L. Bartlett, and Warren Davis, *Creating Advantage: Semiconductors and Government Industrial Policy in the 1990s*, Semiconductor Industry Association and Dewey Ballentine, 1992.

[23] For a review of these programs and the challenges they address, see National Research Council, *An Assessment of the Small Business Innovation Research Program*, C. Wessner, ed., Washington, DC: The National Academies Press, 2008. See also National Research Council, The Advanced Technology Program, Assessing Outcomes, C. Wessner, ed., Washington, DC: National Academy Press, 2001. Also Lewis M. Branscomb and Philip E. Auerswald, *Between Invention and*

examples of public-private collaboration among researchers, private companies, entrepreneurs, and government agencies.

e. **Small Business Entrepreneurship:** "Equity-financed small firms are a key feature of the U.S. innovation system, serving as an effective mechanism for capitalizing on new ideas and bringing them to the market."[24] In the United States, small firms are also a leading source of employment growth, generating a very high percentage of *net* new jobs in recent years.[25] These small businesses also employ nearly forty percent of the United States' science and engineering workforce.[26] Small businesses renew the U.S. economy by introducing new products and new lower cost ways of doing things, often with substantial economic benefits. They play a key role in introducing technologies to the market, often responding quickly to new market opportunities.[27]

f. **Talented Immigrants:** America's ability to attract the world's best and brightest technological and entrepreneurial talent is an important element of its economic success and global leadership.

Innovation: An Analysis of Funding for Early-Stage Technology Development, NIST GCR 02–841, Gaithersburg, MD: National Institute of Standards and Technology, November 2002.

[24] See National Research Council, An Assessment of the SBIR Program, op. cit., See also Zoltan J. Acs and David B. Audretsch, *Innovation and Small Firms,* Cambridge, MA: MIT Press, 1990. See also Zoltan J. Acs and David B. Audretsch, "Entrepreneurship, Innovation and Technological Change," *Foundations and Trends in Entrepreneurship* 1, no. 5 (2005): 1-65 and Boyan Jovanovic, "New Technology and the Small Firm," *Small Business Economics*, 16(1) (2001): 53-55. The Small Business Administration's Office of Advocacy defines a small business as an independent business having fewer than 500 employees. Access at http://web.sba.gov/faqs/faqIndexAll.cfm?areaid=24.

[25] According to Robert Litan of the Kauffman Foundation," Between 1980 and 2005, virtually all net new jobs created in the U.S. were created by firms that were 5 years old or less." See also Small Business Administration, Office of Advocacy, "Small Business by the Numbers," 2006. This net gain depends on the interval examined since small firms exhibit a much higher frequency of entries and exits than large firms. For a discussion of the challenges of measuring small business job creation, see John Haltiwanger and C. J. Krizan, "Small Businesses and Job Creation in the United States: The Role of New and Young Businesses." In *Are Small Firms Important? Their Role and Impact,* Zoltan J. Acs, ed. Dordrecht: Kluwer, 1999. For a recent robust finding that small businesses do create more jobs, see David Neumark, Brandon Wall, and Junfu Zhang, "Do Small Businesses Create More Jobs? New Evidence for the United States from the National Establishment Time Series," *The Review of Economics and Statistics*, February 2011, Vol. 93, No. 1, Pages 16-29.

[26] Specifically, from 1993 through 2009:Q2, small firms (firms with fewer than 500 employees) accounted for 65 percent of net new jobs. Brian Headd, *An Analysis of Small Business and Jobs*, U.S. Small Business Administration, Office of Advocacy, March 2010. The report also noted that using a different data source and time period (1993-2006), small business accounted for 88 percent of net new jobs. Research commissioned by the Small Business Administration has also found that scientists and engineers working in small businesses produce fourteen times more patents than their counterparts in large patenting firms in the United States—and these patents tend to be of higher quality and are twice as likely to be cited.

[27] For an extended discussion of the empirical evidence supporting the finding of high innovation performance of small firms, see Zoltan J. Acs and David B. Audretsch, Innovation in Large and Small Firms, An Empirical Analysis, *The American Economic Review* Vol. 78, No. 4, 1988, pp. 678-690.

Immigrants have often played a major role in the growth of innovative U.S. firms and are the source of a significant proportion of the startups in places like Silicon Valley.[28] Some analysts find that foreign-born engineers were represented on the founding teams of 24 percent of Silicon Valley technology businesses founded between 1980 and 1998.[29]

3. ***Advantages and Challenges in the U.S. Innovation System:*** *The United States has some of the world's best framework conditions that create a pro-innovation environment. These include an open and flexible innovation system, strong intellectual property-rights protection, constructive bankruptcy laws, well-developed capital markets, and extensive worker mobility. But the U.S. also faces significant challenges including high debt levels, inadequate federal support of R&D, declining university funding, and under-funded, sub-par infrastructure.*

   a. **U.S. Advantages**[30]

      i. **An Open Innovation System:** The U.S. economic system is relatively open to new entrants and that, along with a premium placed by society on entrepreneurship and risk-taking, makes it among the best in the world in terms of encouraging firm formation and growth. The United States consistently ranks high in the World Bank's *Ease of Doing Business* rankings, placing 5th in the 2011 report.[31]

      ii. **Strong Intellectual Property Rights:** Secure rights to intellectual property encourage companies to develop and commercialize new technologies. The new legislation to modernize U.S. patent, trademark and copyright laws along with efficient systems to assign ownership are intended to

---

[28] See the related discussion on "Hunting for Global Talent" in Chapter 2 of this report.
[29] See AnnaLee Saxenian, *Silicon Valley's New Immigrant Entrepreneurs*, San Francisco: Public Policy Institute of California, 1999. A follow-up study found that of all U.S. technology companies founded between 1995 and 2005, one-quarter had chief executive officers or chief technology officers who were foreign-born. See Vivek Wadhwa, Ben Rissing, AnnaLee Saxenian, Gary Gereffi, "Education, Entrepreneurship and Immigration: America's New Immigrant Entrepreneurs, Part II," Duke University Pratt School of Engineering, U.S. Berkeley School of Information, Ewing Marion Kauffman Foundation, June 11, 2007.
[30] See the related discussion in Chapter 1 of this report, which describes the "Pillars of U.S. Innovative Strength."
[31] The World Bank and International Finance Corporation, *Doing Business 2011* (2010), Table 1.2. The U.S. is compared with 182 other countries.

encourage the formation and location of knowledge-intensive industries in the United States.[32]

iii. **Bankruptcy Laws that Permit Risk Sharing and Recovery:** Bankruptcy laws that balance creditor and borrower rights are essential for a well-functioning innovation system. They provide incentives for lenders to select and monitor their investments more carefully, and by permitting recovery, they also allow for borrowers to share some of their risk.[33]

iv. **Worker Mobility:** Employee mobility increases dissemination of knowledge, in turn feeding innovation and economic growth.[34] Significant labor mobility gives the United States advantages vis-à-vis other countries that seek to ensure an unusually high level of protection for workers from dismissal. Strong employment protection is often a disincentive for enterprises seeking to hire new workers and, in aggregate, leads to lower productivity growth.[35]

b. **Challenges for the U.S.**

i. **Fiscal Constraints:**[36] America's high budget deficits and debt burden are exerting extraordinary pressure on lawmakers to cut spending on the very investments needed to keep the U.S. ecosystem competitive and to drive growth: in universities, applied-research programs, incentives for small business, new energy technologies, and improved transportation and

---

[32] The Leahy-Smith America Invents Act was signed into law by President Barack Obama on September 16, 2011. This law, which represents the most significant change to the U.S. patent system since 1952, drew on the recommendations of a National Academies panel. See National Research Council, *Patents in the Knowledge-based Economy*, W. Cohen and S. Merrill eds., Washington, DC: The National Academies Press, 2003.

[33] Joseph Stiglitz, "Bankruptcy Law; Basic Economic Principles," in Stijn Claessens et al. eds., *The Resolution of Financial Distress, An International Perspective on the Design of Bankruptcy Laws*, Washington, DC: The World Bank, 2001. The United States Constitution (Article 1, Section 8, Clause 4) authorizes Congress to enact "uniform Laws on the subject of Bankruptcies throughout the United States." The current U.S. Bankruptcy Code was enacted in 1978: The *Bankruptcy Reform Act of 1978* (Pub.L. 95-598, 92 Stat. 2549, November 6, 1978). Code has since been amended, most recently in 2005.

[34] Tracy R. Lewis and Dennis Yao, "Innovation, Knowledge Flow, and Worker Mobility," Wharton School Working Paper Series, 2001.

[35] In a 2008 review of labor laws in Indian states, for example, the World Bank noted that "States that amended the legislation in the direction of reinforcing security rights of workers and other pro-labor measures had lower output and productivity growth in manufacturing sector than those who did not change it or made it more flexible." World Bank, India Country Overview, 2008.

[36] See the related discussion on the "Rapid Growth of R&D Spending" overseas and the composition of U.S. R&D expenditures in Chapter 1 of this report.

information-technology infrastructure.[37] Failure to sustain adequate investments in these areas will inflict long-term damage on America's innovation ecosystem, economic growth, and the welfare and security of its citizens.

ii. **Declining Federal R&D Intensity:** Federal funding for R&D as a percent of GDP is in a long-term decline.[38] [See Figure 3.2] Total U.S. R&D spending has risen over the past 20 years, driven by a more than two and a half times increase in industry R&D spending. But it is important to note that the private sector spends nearly three-fourths of its R&D budget on applied research and development activities. Given the particular importance of federal R&D expenditures for basic research, the long-run implication of stagnant federal investment is "slower technological progress and hence slower growth."[39]

---

[37] National Research Council, *Choosing the Nation's Fiscal Future*, Washington, DC: The National Academies Press, 2010.

[38] The European Union has adopted a 3% target but with limited success. However, both France and Germany have significantly increased their R&D spending to 2.1% and 2.5% respectively. The Merkel government has committed to 10% of GDP for research (3%) and education (7%). President Obama announced a goal to devote more than three percent of GDP to R&D. "Remarks by the President at the National Academy of Sciences Annual Meeting," The White House, Office of the Press Secretary (April 27, 2009).

[39] For a detailed affirmation of the importance of national investments in R&D for economic growth, see Ben S. Bernanke, "Promoting Research and Development: The Government's Role" Speech presented at the Conference on New Building Blocks for Jobs and Economic Growth, Washington, DC: May 16, 2011. For a review of postwar R&D trends, see Linda Cohen and Roger Noll, "Is U.S. Science Policy at Risk? *Trends in Federal Support for R&D*, Washington, DC: Brookings, 2001.

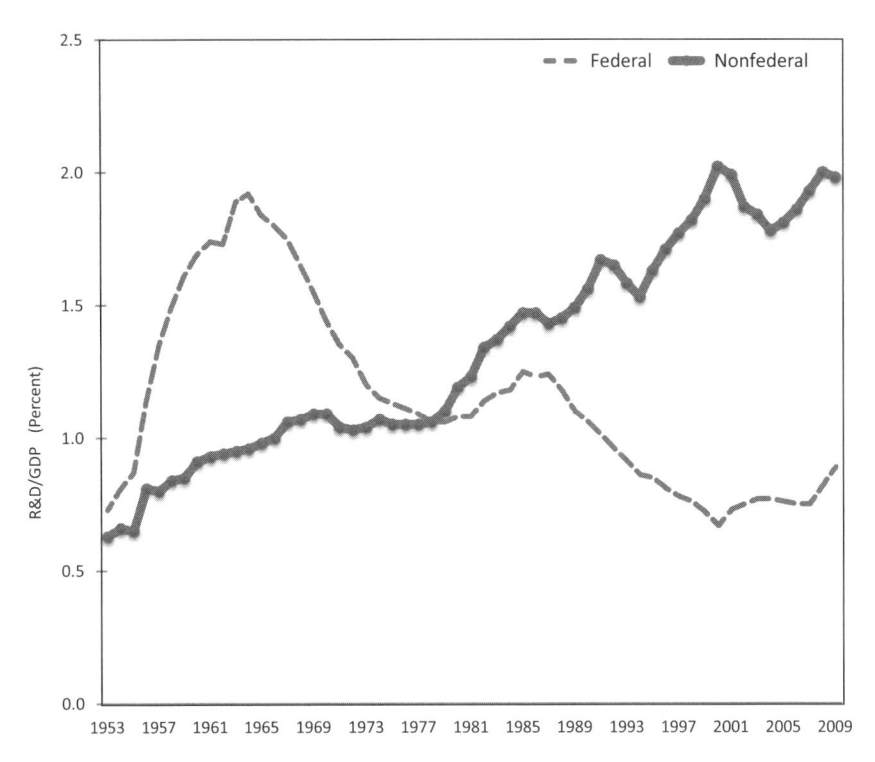

FIGURE 3.2 Federal funding for R&D as a percent of GDP is in long-term decline.
SOURCE: National Science Foundation, National Center for Science and Engineering Statistics, *Science and Engineering Indicators 2012*, NSB 12-01 (January 2012), Figure 4-2.

    iii. **Decline in University Funding Amid New Challenges:** The quality and reputation of U.S. research universities has been built on a foundation of sustained and substantial federal and state funding.[40] Even as countries around the world reform their higher education system, and create new technical institutes and research universities, and increase support for university research, we are underfunding institutions that have

---

[40] Pavitt notes that key features of U.S. innovation policy have been "massive and pluralistic government funding, high academic quality, and the ability to invest in the long-term development of new (often multidisciplinary) fields." See K. Pavitt, (2001) Public Policies to Support Basic Research: What Can the Rest of the World Learn from US Theory and Practice? (And What They Should Not Learn)." *Industrial and Corporate Change* Volume 10, Issue 3 Pp. 761-779.

proven to be enormously successful in sustaining U.S. leadership in science and technology, with their benefits for growth, employment and security.[41]

Although U.S. research universities have long been recognized as the engines of the American innovation system, today's universities face a host of unprecedented challenges:

- **Rapid Expansion in Knowledge:** These include an exponentially expanding knowledge base made possible by new information and communications technologies and the changing needs of a knowledge-driven society.[42]
- **Growth in Regulations:** The growth in federal regulations and reporting requirements, in combination with other factors, is straining university resources and is diverting faculty time from its missions in research, education, and innovation.[43]
- **Increased Competition for Resources:** Universities face the need to be more responsive to competition: for students who demand more value from high tuition bills, for leading professors actively sought by other U.S. (and increasingly overseas) institutions, and for grants and contracts from government agencies, foundations, and private firms.[44]
- **New Mission to Innovate:** Going beyond their traditional missions to educate and conduct research, universities are also increasingly "going to market"—seeking to commercialize their research to raise revenues to sustain academic quality and ensure financial stability.[45] This new mission also addresses the call by states and regions for research universities to serve as sources of

---

[41] See Keld Laursen and Ammon Salter, "The fruits of intellectual production: economic and scientific specialisation among OECD countries," *Cambridge Journal of Economics* Volume 29, Issue 2, 2005, Pp. 289-308. Reviewing data across the OECD, the authors conclude that "it is important to have high levels of relevant to-the-industry scientific strength *per capita* in order to be specialised in science-based industries."

[42] See Robert Zemsky and James J. Duderstadt, "Reinventing the Research University; An American Perspective," in *Reinventing the Research University*, Luc E. Weber and James J. Duderstadt, eds., London: *Economica*, 2004.

[43] See Tobin L. Smith, Josh Trapani, Anthony Decrappeo and David Kennedy, "Reforming Regulation of Research Universities," in *Issues in Science and Technology*, Summer 2011.

[44] See Robert Zemsky and James J. Duderstadt, op. cit.

[45] Ibid.

entrepreneurship and regional growth.[46] While this is in general a positive development, universities need to adapt their organizational culture to support this new mission.

iv. **High Non-production Costs:** Non-production costs, including corporate taxes and health care costs, put the U.S. at a disadvantage as a place to invest.[47] Nations such as Japan, Canada, the Netherlands, and South Korea have sharply lowered their corporate tax rates since the 1990s, leaving the U.S. with one of the highest nominal corporate tax rates among OECD nations, although effective tax rates are considerably less.[48] U.S. businesses are also less competitive globally because they bear the expense of surging U.S. healthcare costs.[49]

v. **Infrastructure and Broadband Enablers:**[50] Other nations are investing heavily in state-of-the-art broadband networks, mass-transit systems, clean power plants, and modern airports while much of the physical infrastructure in the United States

---

[46] See presentation by University of Maryland President Dan Mote, "Universities as Drivers of Growth in the United States," in National Research Council, *Building the 21ˢᵗ Century: U.S. – China Cooperation for Science, Technology, and Innovation*, C. Wessner, rapporteur, Washington, DC: The National Academies Press, 2011. See also the presentation by University of Hawaii President M.R.C. Greenwood, "Presentation of the Hawai'i Innovation Council Report" at the National Academies Conference, *E Kamakani Noi`i—Fostering Knowledge-based Growth in Hawaii*, January 13-14, 2011.

[47] The Manufacturing Institute estimates that non-production expenses such as high U.S. corporate taxes, torts, and pollution control put American-based manufacturing at an 18 percent structural cost disadvantage compared to major trading partners and more than a 50 percent disadvantage to China. Jeremy A. Leonard, "The Tide Is Turning: An Update on Structural Cost Pressures Facing U.S. Manufacturer," The Manufacturing Institute and Manufacturers Alliance/MAPI, November 2008.

[48] Chen, Duanjie, and Jack Mintz, 2010. "U.S. Effective Corporate Tax Rate on New Investments: Highest in the OECD." Tax & Budget Bulletin No. 62. Cato Institute, Washington, DC. However, according to the Government Accountability Office (GAO), "Statutory tax rates do not provide a complete measure of the burden that a tax system imposes on business income because many other aspects of the system, such as exemptions, deferrals, tax credits, and other forms of incentives, also determine the amount of tax a business ultimately pays on its income." The GAO estimated that "[t]he average U.S. effective tax rate on the domestic income of large corporations with positive domestic income in 2004 was an estimated 25.2 percent." GAO, "US Multinational Corporations; Effective Tax Rates are Correlated with where Income is Reported." GAO-08-950 Report to the Senate Committee on Finance, August 2008. Unlike the United States, other countries rely on indirect taxes (such as the VAT) which imposes a portion of the country's social costs on imports and relieves them on its exports. Direct taxes (such as the corporate income tax) are not border-adjustable.

[49] Toni Johnson, "Health Care Costs and U.S. Competitiveness," Washington, DC: Council of Foreign Relations, March 2010. Access at http://www.cfr.org/health-science-and-technology/healthcare-costs-us-competitiveness/p13325. The article lays out divergent views on the competitive impact of health care costs, importantly noting the disparate impacts of these costs on different industries and types of companies.

[50] See the related discussion on "Improving Framework Conditions," in Chapter 2 of this report.

is becoming outmoded and in disrepair due to under-investment.[51]

4. **Governments around the world have made the development of a globally competitive, innovation-led economy a top strategic priority.** To this end, many countries are developing national strategies and adopting, adapting, and strengthening what they see as successful elements of other innovation systems, in particular those of the U.S. system.

   a. **Developing National Strategies:**[52] Both advanced and emerging nations such as China, India, Russia, Germany, South Korea, and Finland, have formulated - or are seeking to formulate - comprehensive national strategies for improving their innovation capacity and are backing them with substantial public investments, broad policy support, and attention at the highest levels of government.[53]

   b. **Increasing Commitments to R&D:**[54] Investments around the world in education, research, and new products are rising. This is an overall a positive development, with benefits for people all over the world— for example, in solving global health problems—as

---

[51] The World Economic Forum now ranks U.S. 16th in infrastructure. World Economic Forum, *The Global Competitiveness Report 2011-2012* (2011), table 5. Also see American Society of Civil Engineers, *2009 Report Card for America's Infrastructure*, March 25, 2009.

[51] Data from U.S. federal agencies cited in Eric Kelderman, "Look Out Below! American's Infrastructure is Crumbling," *Stateline.org,* Pew Research Center, Jan. 22, 2008 (http://pewresearch.org/pubs/699/look-out-below).

[52] For an extended review of the "New Global Competitive Environment," including the accomplishments and ambitions of China, India, Germany, and others, see Chapter 5 of this report.

[53] China's 15-year comprehensive innovation strategy is described in State Council of China, "National Medium- and Long-Term Program for Science and Technology Development, 2006-2020." Germany's innovation strategy is described in Federal Ministry of Education and Research, *Ideas. Innovation. Prosperity. High-Tech Strategy 2020 for Germany*, Innovation Policy Framework Division, 2010. Canada's national strategy is described by Industry Canada, *Mobilizing Science and Technology to Canada's Advantage — 2007.* (Access at http://www.ic.gc.ca/eic/site/ic1.nsf/vwapj/SandTstrategy.pdf/$file/SandTstrategy.pdf) India's National Innovation Council has published a new innovation strategy on March 2011: *Towards a More Inclusive and Innovative India, (*Access at http://www.innovationcouncil.gov.in/images/stories/report/Innovation_Strategy.pdf. For an explanation of South Korea's strategy, see *Vision 2025,* Korea's *Long Term Plan for Science, Technology, and Development.* For a review of Finland's most recent innovation strategy, see "Tekes Strategy: Growth and wellbeing from renewal," Tekes 2011. Access at www.tekes.fi/en/document/49702/tekes_strategy_engl_2011_pdf. A National Academies report on innovation policies in six countries concluded that some countries such as China and Singapore are most likely to achieve their five-year S&T goals while others such as Brazil and India will likely have more limited success. National Academy of Sciences, *S&T Strategies of Six Countries*, op. cit.

[54] See the related discussion, "Substantially Increasing R&D Funding," in Chapter 2 of this report.

FINDINGS

countries seek greater returns on their R&D investments.[55] And although the United States still leads the world in R&D spending, emerging economies are increasing resources to R&D at a much faster rate [See Figure3.3] while real U.S. federal R&D spending has remained roughly constant for the past two decades.[56]

c. **Emulating Global Best Practices:** At a time when U.S. public investments are threatened with major reductions, other nations are devoting ever-greater government funds to develop their innovation systems.[57] In many cases they are actively seeking to replicate what they see as successful U.S. policies and programs. These include policies to strengthen R&D partnerships linking research universities and industry, and programs to provide risk capital and training for technology entrepreneurs.[58]

---

[55] The European Union established in 2000 a three percent of GDP target for R&D spending by 2010 for European nations, but only limited progress toward this goal has been achieved. Recently, both France and Germany have significantly increased their R&D spending to 2.1% and 2.5% respectively. The government of Chancellor Merkel has committed to 10% of GDP for research (3%) and education (7%) by 2015. Federal Ministry of Education and Research, *Federal Report on Research and Innovation 2010*, Innovation Policy Framework Department 2010.

[56] While the U.S. federal government spent approximately $148 billion (FY 2010) on R&D, defense R&D made up over half of this amount. Further, about ninety percent of defense R&D is for defense related technology development (including weapons testing). See AAAS Report XXXVI FY 2012. Thus the effective U.S. expenditures on basic and applied research is much smaller than the overall figure suggests.

[57] South Korea has boosted R&D spending from 2.27 percent of GDP in 1995 to 3.37 percent in 2008, for example. China's R&D spending has risen from 0.57 percent of GDP to 1.54 percent, and China's plans call for it to reach 2.5 percent by 2020 while it's GDP has expanded at a remarkable average rate above 10% per annum since 1990. Japan's ratio has gone from 2.92 percent in 1995 to 3.42 percent in 2008 and Finland's from 2.26 percent to 3.73 percent. OECD, *OECD Main Science and Technology Indicators*, Volume 2010/1, May 2010, Table 2. Under its current five-plan, Singapore tripled R&D investment, to $10 billion.

[58] For example, Japan, Canada, and China are among the countries that have implemented reforms modeled after Bayh-Dole to incentivize universities to commercialize research and encourage universities and national labs to collaborate with industry. Innovation Programs such as the SBIR and the Sematech Consortium have been widely emulated. Countries as diverse as Sweden, the Netherlands, India, South Korea, and Russia have adopted SBIR-type programs. Based on what they saw as the success of the Sematech consortium, Japan established a series of consortia to advance their domestic semiconductor industry in the 1990s. See National Research Council. 2009. *21st Century Innovation Systems for Japan and the United States: Lessons from a Decade of Change*. S. Nagaoka, M. Kondo, K. Flamm, and C. Wessner, eds. Washington, DC: The National Academies Press. For a discussion of the origins and achievements of Sematech, and subsequent emulation, see National Research Council, *Securing the Future, Regional and National Programs to Support the Semiconductor Industry*, op. cit. For an evaluation of SEMATECH, see Kenneth Flamm, "The Impact of SEMATECH on Semiconductor R&D," in that volume.

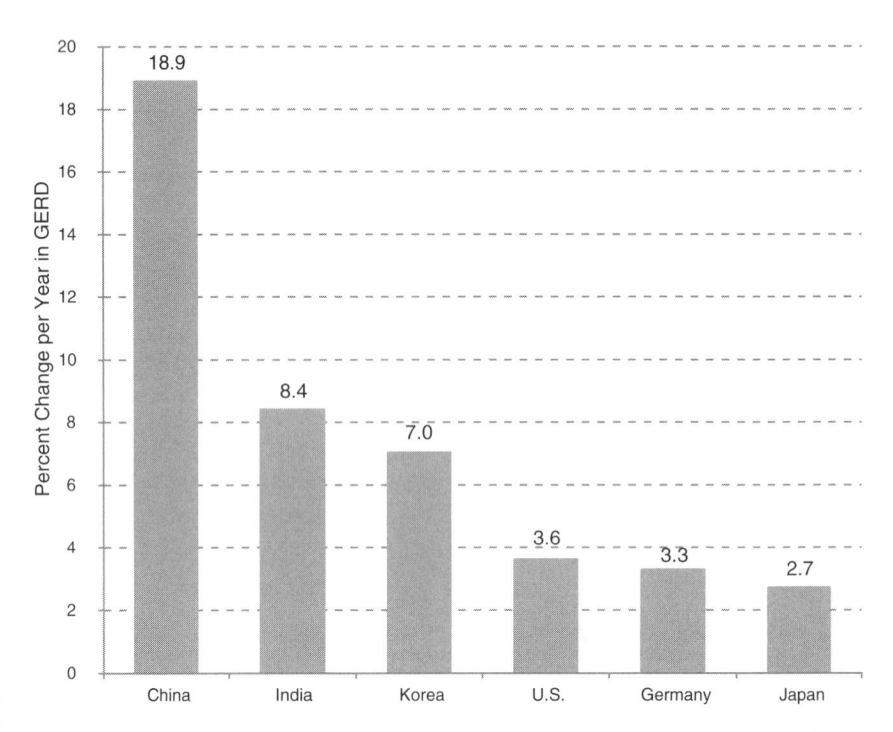

FIGURE 3.3 R&D spending growth by emerging economies is significantly faster than in developed countries.
SOURCE: UNESCO, UNESCO Institute for Statistics, Science and Technology, Table 25.
NOTES: GERD refers to gross domestic expenditure on R&D. Percent refers to average annual growth rate for 1996 to 2008. India growth rate is based on 1996-2007. R&D growth in China, India, and Korea has expanded rapidly, though starting from a small base. Some countries have not been able to maintain their R&D growth targets.[59]

d. **Pursuing Mercantilist Policies:**[60] Government enterprises (state-owned, state-invested or state-supported) engaged in commercial

---

[59] For instance, in his 2000 Presidential address to Indian Science Congress, then Indian Prime Minister Atal Behari Vajpayee promised to raise R&D spending to 2 % of GDP. However, R&D spending in India has yet to cross 1 % of GDP. In 2012, Prime Minister, Dr. Manmohan Singh again pledged the same target.
[60] See related discussion of the character and impact of 21st Century Mercantilism in Chapter 1 of this report.

activities remain a powerful force in the global economy.[61] Their effects on innovation are felt in a variety of ways. The intention of many governments, for example, China, is to develop these enterprises as centers of innovation. The policies have not yet proved themselves sound in terms of creating nodes of innovation, but they do affect the capture of the economic value of global innovation in multiple ways:

   i. Their impact on international trade and investment.
  ii. Lack of enforcement of intellectual property laws is another means affecting the capture of the economic value of innovation that costs foreign private sector competitors tens of billions of dollars of lost revenues.[62]
 iii. Forced or induced technology transfer as a condition of investment further dilutes the value of innovation to the innovator.
  iv. Denial of market access.

Successive U.S. Administrations have made some progress in addressing these problems but progress has been limited.

e. **Expanding Universities:**[63] More positively, other nations and regions are dramatically increasing funding to upgrade, expand, and open new research universities and science-and-technology teaching programs.[64] This comes at a time when U.S. research universities face budget cuts due to state fiscal problems, new

---

[61] The list of policies of other countries that have an impact on U.S. competitiveness includes "currency manipulation and dollar overvaluation, value added taxes and their rebates on exports, mercantilism, "buy national" policies and practices, anti-trust and competition policies, enforcement of global trade rules, financial subsidies aimed at luring the outsourcing of production and technology development abroad, and indigenous technology preferences." See Clyde Prestowitz, "Competitiveness Council wide of its mark," *Foreign Policy*, December 16, 2011. For a compilation of foreign trade barriers by country and quantitative estimates of the impact of these foreign practices on the value of U.S. exports, see National Trade Estimates, Office of the U.S. Trade Representative, Access at http://www.ustr.gov/about-us/press-office/reports-and-publications/2010.
[62] See Matthew J. Slaughter, *How Piracy in China Costs U.S. Jobs*, Tuck School of Business at Dartmouth and NBER, September 2010.
[63] See the related discussion on "Developing 21st Century Universities" in Chapter 2 of this report.
[64] Taiwan plans to invest $1.7 billion over five years to develop world-class universities. India's current five-year plan calls for 1,500 new universities, three new Indian Institutes of Science Education and Research, and seven new Indian Institutes of Technology. The Flemish government launched a €232 million program in 2006 to boost basic research at universities. China's $4.5 billion 985 programs seeks to make 39 universities among the best in the world. Canada has invested $5.2 billion since 1997 in 130 research institutions, while its $300 million Canada Research Chairs program has established 2,000 chairs headed by top-flight academics.

challenges, and new missions in regional development and technology commercialization.[65]

f. **Providing Early Stage Finance:**[66] Other nations are adopting programs often modeled after U.S. programs in order to help promising technology companies survive the gap in funding that frequently occurs between inventing a product and bringing it to market.[67] The U.S. has only recently launched new efforts in this area such as Start-up America to address the need of start-ups for capital and expertise.[68] Proven U.S. programs have faced challenges: SBIR has just emerged from a long and difficult reauthorization. Despite its considerable accomplishment, NIST's Technology Innovation Program is currently without funding.[69] And notwithstanding the recent the efforts to address the early stage funding in biomedicine, funding for translational research at NIH remains a challenge.[70]

g. **Attracting Global Talent:**[71] While strong U.S. investments in research and universities have traditionally enabled it to draw and

---

[65] See Paul Courant, James Duderstadt, and Edie Goldenberg, "Needed: A National Strategy to Preserve Public Universities," *The Chronicle of Higher Education*, Jan. 3, 2010 (http://milproj.dc.umich.edu/pdfs/2010/2010-Chronicle-Commentary.pdf). See also National Research Council, *Breaking Through: Ten Strategic Actions to Leverage Our Research Universities for the Future of America*, Washington, DC: The National Academies Press, 2012.

[66] See the related discussion on "Providing Early Stage Finance" in Chapter 2 of this report.

[67] For example, Japan established a Small Business Innovation Research program modeled after that of the U.S. India's Small Business Innovation Research Initiative, launched in 2007, supports high-risk R&D projects by biotech start-ups. The Netherlands introduced its SBIR program in 2004. The United Kingdom established the Small Business Research Initiative in 2001. Finland's Tekes invested €343 million ($494 million) directly in enterprises, most of them with fewer than 500 employees, to develop technologies in partnership with universities.

[68] Under the Start-up American initiative, the Small Business Administration will commit to a $1 Billion Impact Investment Fund that invests growth capital in companies located in underserved communities. It will also commit to a $1 Billion Early-Stage Innovation Fund that provides a 1:1 match to private capital raised by seed and early stage funds. See http://www.sba.gov/startupamerica.

[69] The Technology Innovation Program (TIP) at the National Institute for Standards and Technology "supports, promotes, and accelerates innovation in the United States through high-risk, high-reward research in areas of critical national need" through "targeted investments in transformational R&D that will ensure our nation's future through sustained technological leadership." See http://www.nist.gov/tip/ . TIP succeeds the Advanced Technology Program, which was assessed by a committee of the National Academies to be "an effective federal partnership program." See National Research Council, *The Advanced Technology Program, Assessing Outcomes*, C. Wessner, ed., Washington, DC: National Academy Press, 2001.

[70] There have been new initiatives to address the need for translational research. The NIH leadership has proposed a new National Center for Advancing Translational Sciences, currently funded at $575 million, against an overall NIH budget of $32 billion. See NIH News, "NIH establishes National Center for Advancing Translational Sciences," December 23, 2011. Access at http://www.nih.gov/news/health/dec2011/od-23.htm.

[71] See the related discussion on "Hunting for Global Talent" in Chapter 2 of this report.

retain top global talent, other governments are intensifying efforts to attract accomplished science and technology talent back home and to recruit star scientists from around the world.[72] The relative loss of global talent is reinforced as foreign-born U.S. graduates and foreign-born entrepreneurs face greater difficulty obtaining U.S. work visas, residency, and citizenship.[73]

h. **Focusing on Building Innovation Clusters and Science Parks:**[74] Governments around the world have recognized the powerful competitive advantages of strong regional innovation clusters and are investing aggressively in developing science parks[75] as part of comprehensive strategies to foster innovative clusters.[76] In the United States, until recently, there has tended to be little alignment between federal economic-development programs and state and

---

[72] Since launching an aggressive campaign to lure top foreign talent a decade ago, Canada has recruited more than 3,000 foreign researchers and more than 600 university department chairs. Among the elite international scientists recruited by Singapore's A*Star agency are senior researchers from The National Cancer Institute, MIT, and the University of Texas at Austin. Under China's Thousand Talents Program, launched in 2008, top Chinese scientists working abroad are offered grants of 1 million Yuan, world-class salaries, and generous lab funding if they return to China.

[73] The Obama Administration has taken a number of new initiatives in this area. On Aug. 2, 2011, for example, the Administration announced that foreign entrepreneurs may obtain EB-2 employment visas set aside for immigrants with advanced degrees and skills and qualify for H-1B visas as self-employed entrepreneurs. Procedures for obtaining EB-5 visas for immigrant investors were streamlined. Department of Homeland Security and U.S. Citizenship and Immigration Service press release, Aug. 2, 2011. These are very positive steps designed to attract and retain foreign talent.

[74] For an expanded discussion of initiatives around the world to develop clusters and science parks, see Chapter 7 of the report.

[75] The level of Chinese central and regional government investment, and the number of parks and their scale, are most impressive. The vast majority of U.S. parks are on a much smaller scale and benefit from much smaller levels of public investment. Only Research Triangle Park approximates the scale of the Chinese efforts. See Rick Weddle, "Research Triangle Park: Past Success and the Global Challenge," in National Research Council, *Understanding Research, Science and Technology Parks*, C. Wessner, ed., Washington, DC: The National Academies Press, 2009, p. 26. It is important to keep in mind that the parks can have substantially different objectives; some are focused on research, some on industrial development, but many combine technology development and industrial applications and can also support national missions, as does the Sandia Science and Technology Park. See http://www.sstp.org/index.html.

[76] Examples include the French *Pôle de Croissance* program, the Chinese drive to build large research parks, and the new Russian Skolkovo innovation hubs. For a review of national strategies in France, China, and elsewhere to develop research parks, see National Research Council, *Understanding Research, Science and Technology Parks: Global Best Practices*, op. cit. For a review of the role of public policy in fostering innovation clusters, see National Research Council, *Growing Innovation Clusters for American Prosperity*, C. Wessner, rapporteur, Washington, DC: The National Academies Press, 2011. With regard to Russian efforts, the *Financial Times* reports "the Kremlin is working hard to position Skolkovo as a hallmark of its modernization program and a key part of its strategy to diversify away from oil and gas." The innovation hub has been promised $3 billion in government funding over the next three years. In addition, the project is seeking an equal amount from private groups. See *Financial Times*, "Welcome to Russia's Silicon Valley," August 21, 2011.

local innovation cluster initiatives. Recent initiatives by the current administration are steps in the right direction; the question is whether the number of clusters receiving support and their funding levels are sufficient.[77]

5. **U.S. leadership in innovation is eroding.**[78] *First, the preeminence of the United States is diminishing in terms of research inputs, from the number of science and technology personnel, to federal research funding, and the number of patents filed and scientific papers produced.*[79] *Second, America's position as the world's pre-eminent ecosystem for turning new technologies into commercial products is also declining relative to both new entrants and established competitors. In part, the U.S. position is less secure as the result of the growing commitments by the rest of the world not only to education and research but also to the commercialization of new technologies. Finally, as emerging nations increase their support for R&D and innovation and insist on commitments to their innovation systems, U.S. companies are performing more of their R&D in those countries.*[80]

---

[77] The U.S. Departments of Energy, Commerce, Defense, Agriculture, Labor and Education now all have cluster-development programs and coordinate activities on specific regional initiatives. See presentation by Ginger Lew, then of the White House National Economic Council at the National Academies conference on *Clustering for Prosperity*, Washington, DC, February 23, 2010.

[78] See the related discussion of "New Trends in Global Innovation" in Chapter 1 of this report.

[79] National Academy of Sciences, National Academy of Engineering, and the Institute of Medicine, *Rising Above the Gathering Storm: Energizing and Employing America for a Brighter Economic future*, Washington, DC: The National Academies Press, 2007. Also see National Academy of Sciences, National Academy of Engineering, and Institute of Medicine, *Rising Above the Gathering Storm, Revisited: Rapidly Approach Category 5,* Washington, DC: The National Academies Press, 2010.

[80] According to the NSB, "the geographic distribution of R&D by overseas affiliates of U.S. MNCs is gradually reflecting the role of emerging markets in global R&D." The share of major developed economies or regions (Canada, Europe, and Japan) "accounted for a decreasing share of the overseas R&D investments of U.S. MNCs, declining from 90% in 1994 to 80% in 2006." At the same time, R&D performed by U.S.-owned affiliates in China and India "increased from less than $10 million in each country in 1994 to $804 million in China and $310 million in India in 2006." National Science Board, *Science and Engineering Indications 2010*, Chapter 4. Accessed at http://www.nsf.gov/statistics/seind10/c4/c4s6.htm#s2. For a survey of factors driving multinational R&D location, see Jerry Thursby and Marie Thursby, *Here or There? -- Report to the Government-University-Industry Research Roundtable*, Washington, DC: The National Academies Press, 2006. For a recent analysis of the evolution of R&D by multinationals in China, see Robert Pearce, ed., *China and the Multinationals, International Business and the Entry of China into the Global Economy*, Aldershot: Edward Elgar Publishing, 2011. The book documents how leading multinationals have drawn their operations in China into their established operations and suggests that the operations of multinationals are "increasingly embedded in the growth and sustainability of the Chinese economy itself, rather than merely serving as a supply base for their global markets."

FINDINGS   151

> **a. The Emergence of Major Global Competitors[81]**: The U.S. is facing major competition from the policies and markets of rising global powers. By tapping into global knowledge and integrating themselves more into to the global economy, emerging nations, like China, India, and South Korea have rapidly become major global players, albeit in different ways.[82] These fast growing economies have critical masses of highly educated people and of scientists and engineers, now matched by rapidly growing expenditures on R&D [See Figure 3.4], as well as large, and in some cases, largely protected domestic markets. They are seeking to perform R&D for multinational companies with the learning that this entails,[83] to deploy this potential to meet their own needs, and to expand their production for export markets.

---

[81] Chapter 5 of this report highlights policies and programs to spur innovation based competitiveness undertaken by leading nations, including China, India, and Germany.
[82] Carl Dahlman, *The World Under Pressure: How China and India Are Influencing the Global Economy and Environment*, Palo Alto: Stanford Economics and Finance, 2011. See also Carl Dahlman, "China and India: Emerging Technological Powers." in *Issues in Science and Technology*, Spring 2007. See also Alice Amsden, *Asia's Next Giant: South Korea and Late Industrialization*, Oxford: Oxford University Press, 1989, Alice Amsden, *Beyond Late Development: Taiwan's Upgrading Policies*, Cambridge: MIT Press, 2003, and AnnaLee Saxenian, *The New Argonauts: Regional Advantage in a Global Economy*, Cambridge: Harvard University Press, 2006.
[83] For a review of India's accomplishments as well as challenges in innovation, See National Research Council, *India's Changing Innovation System*, C. Wessner and S. Shivakumar, eds., Washington, DC: The National Academies Press, 2007. See also World Bank, "Unleashing India's Innovation" Mark A. Dutz, ed., Washington, DC: World Bank, 2007. For a review of recent product and business innovation by Indian pioneers, see R.A. Mashelkar and C. K. Prahalad, "Innovation's Holy Grail," *Harvard Business Review*, July 2010. See Dan Breznitz, Michael Murphree, *Run of the Red Queen: Government, Innovation, Globalization, and Economic Growth in China*, New Haven: Yale University Press, 2011. The authors examine the strengths and weaknesses of the Chinese innovation system, noting that China's sustained economic vitality does not appear to depend on generating cutting edge innovation. See also National Research Council, *Building the 21st Century, U.S. China Cooperation in Science, Technology, and Innovation*, C. Wessner, rapporteur, Washington, DC: The National Academies Press, 2011. Finally, Mu Rongping of the Chinese Academy of Sciences provides a summary of his nation's innovation accomplishments and challenges in the 2010 UNESCO Science Report. See United Nations Educational, Scientific and Cultural Organization, *UNESCO Science Report 2010*, Paris: UNESCO Publishing, 2010, Chapters 17, 18 and 20.

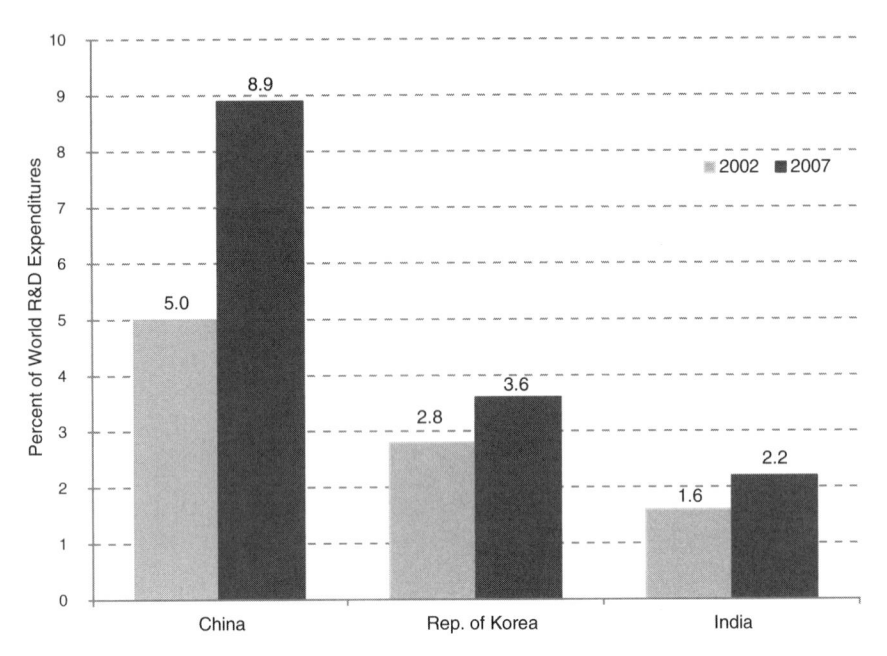

FIGURE 3.4 China, Korea and India increased their share of world spending on R&D from 9.4% to 14.7% from 2002 to 2007.
SOURCE: UNESCO, *UNESCO Science Report 2010* (UNESCO Publishing, Paris, 2010).

b. **Growth of Innovative Regions around the World[84]:** Innovation hubs like Silicon Valley, greater Boston, San Diego, Austin, and Seattle have for decades been magnets for the world's brightest and most visionary innovators, technology entrepreneurs, and financiers. Now these hubs face greater competition as places to commercialize new technology and launch new companies. Taipei, Shanghai, Helsinki, Tel Aviv, Bangalore, Hyderabad, Singapore, Sydney, and Suwon,[85] are among the many cities that now boast

---

[84] Chapter 7 highlights national and regional programs to develop innovation clusters around the world.
[85] Home to a large Samsung Electronics factory, Suwon, South Korea is a major educational center that is home to 14 university campuses. For a review of the impact of Korean innovation clusters, including Suwon, see Doohee Lee, "Regional Innovation Activity: The Role of Regional Innovation Systems in Korea." KIET Occasional Paper No. 78, February 2010.

high concentrations of technology entrepreneurs and are increasingly able to launch innovative companies.[86]

c. **Growth of Offshore Research Centers**:[87] American multinational corporations in sectors ranging from pharmaceuticals to software have, in recent years, set up advanced R&D centers in countries such as India, China, and Russia.[88] This trend was made possible by the liberalization of state controls in these countries, and driven at least initially by the availability of skilled graduates, and lower costs and the need to deploy and adapt products suited to these large, rapidly growing markets.[89] While these R&D centers develop and adapt technologies to domestic markets of the countries where they are located, they also plan to develop products for the global market. Increasingly, these centers are a part of the integrated innovation system of global enterprises including GE, IBM, Intel, 3M, and Microsoft that connects company research across borders.[90]

---

[86] According to a map of global innovation clusters by the McKinsey Global Institute and World Economic Forum, some U.S. cities are losing ground to these and other emerging "hot springs" of innovation in Asia and Europe. See Juan Alcacer and McKinsey & Co., "Mapping Innovation Clusters," McKinsey Digital, March 19, 2009, (http://whatmatters.mckinseydigital.com/flash/innovation_clusters/). Also see Andre Andonian, Christoph Loos, and Luiz Pires, "Building an Innovation Nation," McKinsey & Co., March 4, 2009.

[87] See also the discussion in Chapter 1 on the "Growth of Foreign Research Centers of U.S. Multinationals."

[88] For a review of the drivers and impacts of the growth of advanced R&D centers in emerging economies, see OECD, Science, Technology and Industry Outlook, Chapter 4 "The internationalisation of R&D", Paris: OECD, 2006. See also Pete Engardio, Aaron Bernstein, and Manjeet Kripalani, "The New Global Job Shift" *BusinessWeek,* February 3, 2003 and UNCTAD, *Globalization of R&D and Developing Countries,* New York: United Nations, 2005.

[89] Ashok Deo Bardhan, and Dwight M. Jaffee, "Innovation, R&D and Off-shoring," University of California at Berkeley: Fisher Center Research Reports, 2005.

[90] For example, GE has recently moved its X-ray business headquarters from Wisconsin to China. *Wall Street Journal,* "GE Bases X-Ray Unit in China," July 26, 2011. For a perspective from IBM on the globalization of its research and development operations, see Mark Dean, "ICT development in U.S. and Chinese Contexts", in National Research Council, *Building the 21st Century, U.S. China Cooperation in Science, Technology, and Innovation, op cit.* See also Gert Bruche, "A new geography of innovation – China and India rising," in Karl P. Sauvant et al. (eds.*) FDI Perspectives: Issues in International Investment,* New York: Vale Columbia Center on Sustainable International Investment (January 2011). Bruche notes that while" the dominant share of MNE R&D in China and India comprises routine activities adapting existing designs or processes, or providing modular contributions transformed into innovative products and processes in the triad's higher order R&D centers … scattered evidence points to fast learning and upgrading processes resulting in ever more centers and CROs taking on selective regional or global roles as centers of excellence within MNEs global innovation networks." According to Roland Berger, for example, 3M corporation has R&D locations in 30 countries supported by a central research center at corporate headquarters in St. Paul. Robert Ohmayer, "Globalization of R&D: Drivers and Success Factors," Roland Berger Strategy Consultants, April 19, 2007.

6. ***Capturing the Benefits of Investments in R&D:***[91] *A key challenge for the United States is to capture an important part of the economic benefits of its substantial investments in basic research in an era when other countries are adopting policies and programs focused on translating nationally and globally sourced research into domestic production of new products for the market.*[92]

   a. **Research as a Global Public Good:** Other nations have intensified their efforts to capture the economic value of the world's research efforts, including those financed by U.S. taxpayers. Although the U. S. federal government remains the world's largest sponsor of basic research, and total federal R&D spending reached $148 billion in FY 2010, traditional trading partners and emerging nations alike are more focused than the U.S. in seeking to capture the economic value of these tremendous public investments by channeling their efforts on translating new technology into commercial applications and job-generating industries.[93] Research, especially basic research, is widely recognized as a public good. The full economic value of basic research is unlikely to accrue to private investors, hence the rationale for government support for research.[94] In the new world order of rapid, open global knowledge flows, the gap between federally funded research and U.S. based commercialization means that it is possible for foreign enterprises (often with state support) to capitalize on U.S. investments in basic research. Many countries have focused on commercializing innovations within their national borders, with the goal of creating large-scale industries and high value employment.[95] This is an important paradigm shift. Whereas the commercialization of research funded by the U.S. in

---

[91] See the related discussion, "Capturing the Economic Value of Innovation" in Chapter 1 of this report.

[92] Gary P. Pisano and Willy C. Shih in "Restoring American competitiveness," *Harvard Business Review* 87, Nos. 7-8, (July-August 2009). Some in the U.S. believe that it is inappropriate for government to support and/or encourage downstream development of commercial products. Whatever the merits of this view, most big U.S. trading partners do not share it.

[93] New growth theory models show that R&D spillovers are a major source of endogenous growth. See Zvi Griliches, "The Search for R&D Spillovers," *The Scandinavian Journal of Economics*, Vol. 94, 1992 Supplement, pp. 29-47. Coe and Helpman add that the tendency of research to spillover means that R&D investments by other countries can have substantial beneficial effects on domestic factor productivity. David T. Coe and Elhanan Helpman, "International R&D spillovers," *European Economic Review*, Volume 39, Issue 5, May 1995, pp. 859-887.

[94] See Ben S. Bernanke, "Promoting Research and Development: The Government's Role" *Issues in Science and Technology*, Volume XXVII, Number 4, Summer 2011.

[95] Carl Dahlman, *The World Under Pressure: How China and India Are Influencing the Global Economy and Environment*, Palo Alto: Stanford Economics and Finance, 2011.

FINDINGS

  the postwar era took place mostly in the United States, the globalization of innovative capacity in the 21st Century means that ideas developed in the United States can now be more easily developed and commercialized overseas.[96]

 b. **The Need for a Strategic Approach:**[97] Most of America's major trading partners do not leave the development of their economies solely to the workings of the market. They take a more strategic approach and many are expanding programs and policies aimed at advancing promising technologies and large-scale domestic manufacturing in areas such as electric-drive vehicles, renewable energy equipment, and solid-state lighting in order to secure global competitive advantage, gain or maintain national competency in production, and to keep or create high-quality jobs.[98] Not all of these programs succeed; sometimes they fail or need readjustment. This willingness to readjust and reinvest is fundamental. The United States takes the same approach with U.S. defense or space efforts, where failure elicits renewed efforts. The United States is one of the few industrial nations that have, until recently, tended not to adopt a strategic approach regarding the composition of its economy, although particular sectors with political influence receive substantial support.[99] To some extent, this has not mattered until now due to the momentum gained from past public and

---

[96] Joseph Stiglitz, "Knowledge as a Global Public Good," in I. Kaul, I. Grunberg and M. Stern, eds. *International Cooperation in the 21st Century*, New York: UNDP, 1999. See also Charlotte Hess and Elinor Ostrom eds., *Understanding Knowledge as a Commons*, Cambridge: MIT Press, 2007.

[97] For a review of how some leading economies are addressing their innovation and growth challenges, see Chapter 5 of this report. For a review of national support for emerging industries, see Chapter 6 of this report.

[98] China's most recent Five-Year plan calls for major government investments in seven strategic industries, including biotechnology, alternative energy, and next-generation information technology. For details on Germany's long-term plans to advance transportation-related industries, see *German Federal Government's National Electromobility Development Plan*, August 2009, and for its information and communications technology strategy, see Federal Ministry of Education and Research, *ICT Strategy of the German Federal Government: Digital Germany 2015*, November 2010. Among South Korea's initiatives targeting specific industries are its plan to invest $12.5 billion over 10 years to become the world's dominant producer of advanced batteries. See *Yonhap News Agency*, "S. Korea Aims to Become Dominant Producer of Rechargeable Batteries in 2020," July 11, 2010.

[99] To some extent, these initiatives are now being emulated in the U.S. To ensure that the U.S. has a domestic manufacturing base for advanced batteries, the federal government in 2009 awarded $2.4 billion in grants under the American Recovery and Reinvestment Act to manufacturers of lithium-ion cells, battery packs, and materials. These grants complemented the $25 billion in debt capital made available by the federal government to encourage automakers to produce more energy-efficient cars under the Advanced Technology Vehicles Manufacturing (ATVM) Loan Program. The state of Michigan has also made significant investments to develop an electrified-vehicle industrial cluster. The state offered more than $1 billion in grants and tax credits to manufacturers of lithium-ion battery cells, packs, and components. See chapter 6 on National Support for Emerging Industries in this volume.

private investments.[100] But as the emerging economies have become richer and more advanced economically, they have moved up the value-added chain, increasingly producing "the kind of high-value-added components that 30 years ago were the exclusive purview of advanced economies."[101] This has created economic pressures in the developed economies to more rapidly move into technology-intensive manufacturing industries and knowledge intensive service industries.

c. **An Institutionalized Focus on Translational Research and Applications:**[102] Taiwan, Germany, Finland, China, South Korea and other regions and nations have major institutions focused on applied and translational research aimed at enabling domestic companies to develop manufacturing processes and marketable products. Large, well-funded public-private partnerships such as Germany's Fraunhofer-Gesellschaft, Taiwan's Industrial Technology Research Institute, Korea's Electronics and Telecommunications Research Institute, and Finland's Tekes have proven remarkably successful at helping domestic manufacturers translate new technologies into products and production processes and remain globally competitive despite high or rising labor costs.[103] The U.S. has no equivalent to these large applied research institutions that collaborate with industry to capitalize on national investments in research to develop technology and commercial products that are produced domestically at large-scale.

d. **A Focus on Manufacturing:**[104] Major U.S. trading partners understand that a domestic industrial base that can produce advanced products in high volumes, and the high skilled jobs that this productive activity generates, is integral to maintaining global competitiveness in innovation and increases chances of leading in

---

[100] "Cheaper information technology has given greater importance to more productive forms of capital. The rising contribution of investments in information technology since 1995 has been a key contributor to the U.S. growth resurgence and has boosted growth by close to a percentage point." See National Research Council, *Enhancing Productivity Growth in the Information Age*, D. Jorgenson and C. Wessner, eds., Washington, DC: The National Academies Press, 2007, page 21.
[101] Michael Spence, "Globalization and Unemployment: The Downside of Integrating Markets," *Foreign Affairs* (July/August 2011).
[102] See the related discussion on "Institutional Support for Applied Research" in Chapter 2 of this report. See also a summary description of the Fraunhofer-Gesellschaft in Chapter 5 of the report.
[103] Germany's Fraunhofer-Gesellschaft has more than 80 research units, including 60 Fraunhofer Institutes, with a $2.2 billion annual budget to help Germany manufacturers launch new products and manufacturing processes in 16 industrial clusters. Taiwan's Industrial Technology Research Institute has 6,000 staff that collaborates with manufacturers in emerging industries such as flexible displays, sold-state lighting, photovoltaic cells, and MEMs devices. South Korea's Electronics and Telecommunications Research Institute has 1,700 researchers with doctoral and master's degrees helping industries such as semiconductors, digital mobile communications, and fuel cells.
[104] See the related discussion on "Strengthening Manufacturing," in Chapter 2 of this report.

next-generation technologies.[105] Therefore, many nations and regions support their manufacturing sectors with tax holidays, grants, and credit.[106] They also support domestic manufacturing through trade policy measures and government procurement[107] and programs designed to stimulate large domestic demand in key industries,[108] as well as well-financed institutes to facilitate adoption and importation of new technologies for large and small firms alike.[109]

In the past, the U.S. has successfully driven technology down the cost curve and up the learning curve with defense procurement.[110]

---

[105] Suzanne Berger, "Why Manufacturing Matters," *MIT Technology Review*, July 1, 2011. Access at http://www.technologyreview.com/business/37932/. The Indian Government's recently announced policies for ICTE industries highlights the requirement for a "concerted effort to boost manufacturing activity ... as robust economic growth in the country is leading to extraordinarily high demand for electronic products in general and telecom products in particular." Government of India, "A Triad of Policies to Drive a National Agenda for ICTE," (October 10, 2011). Accessed at http://www.dot.gov.in/NTP-2011/final-10.10.2011.pdf.

[106] For example, China, Malaysia, Singapore, and other nations offer 10-year tax holidays to foreign companies building factories or R&D centers in targeted industries. To convince AMD to build a silicon wafer plant in Germany in 2004, federal and state governments provided $798 million in cash and allowances, guaranteed 80 percent of the value of bank loans, and covered the total product cost of the plant. The Israeli government offered more than $1 billion in aid, including a $525 million to grant, for Intel's 300 mm plant in Kiryat Gat and $660 million in tax benefits to upgrade another plant. Many U.S. states have similar policies, as with Michigan's focus on electric cars and New York's nano initiative in Albany, but often they lack scope, consistency, and/or an overall strategy. The State Science and Technology Institute (SSTI) lists the leading technology based economic development programs of U.S. states and regions.

[107] Perhaps the most explicit use of government policy to support domestic manufacturers are China's "indigenous innovation" regulations, which mandate that purchases of high-tech goods using government funds favor Chinese-owned companies that own the intellectual property rights to the products. see James McGregor, "China's Drive for 'Indigenous Innovation: A Web of Industrial Policies, U.S. Chamber of Commerce, Global Intellectual Property Center, APCO Worldwide (http://www.uschamber.com/sites/default/files/reports/100728chinareport_0.pdf). Also see U.S. International Trade Commission, *China: Intellectual Property Infringement, Indigenous Innovation Policies, and Frameworks for Measuring the Effects on the U.S. Economy,* Investigation No. 332-514, USITC Publication 4199 (amended), November 2010, (http://www.usitc.gov/publications/332/pub4199.pdf) and Alan Wm. Wolff, "China's Indigenous Innovation Policy," testimony before the U.S. China Economic and Security Review Commission, Washington, DC, May 4, 2011.

[108] Germany, Spain, and other nations encouraged large domestic industries in photovoltaic cells and modules, for example, through feed-in tariff systems that compel utilities to purchase solar power at high rates. See Thilo Grau, Molin Huo, and Karsten Neuhoff, *Survey of Photovoltaic Industry and Policy in Germany and China*, Climate Policy Initiative Report, DIW Berlin and Tsinghua University, March 2011. France and China are using government purchases as one way of promoting large-scale production of hybrid and electric-drive vehicles.

[109] These would include, for example, the Fraunhofer-Gesellschaft in Germany, the Industrial Technology Research Institute in Taiwan, and the Korea Institute of Industrial Technology in South Korea., and on a smaller scale, the Industrial Research Assistance Program in Canada.

[110] To cite one example, military purchases of integrated circuits were critical to establishment of America's semiconductor industry in the 1960s and 1970s. See Kenneth Flamm, *Mismanaged*

More recently, the federal government has tended to leave this competition for manufacturing capacity to the states. Some of these efforts have recorded significant success.[111] In other cases, federal initiatives have reinforced state-based programs, such as in Michigan, where the federally funded battery initiative has helped re-shore U.S. production of advanced batteries.[112] Nonetheless, the recent deterioration in the U.S. trade balance in advanced technology products is a troubling indication that the U.S. high-technology manufacturing base is losing ground relative to other global competitors.[113] [See 3.5] And there is growing and authoritative concern that the continued erosion of America's high-tech manufacturing base threatens to undermine U.S. leadership in next-generation technologies[114], while at the same time failing to produce the high value-added employment gains that would follow expanded U.S. high technology exports. Moreover, some analysts argue that with respect to maintaining manufacturing competitiveness and the associated skilled labor and technical institutions, activity that is lost is difficult to recover. They therefore argue that it is important for policy makers to be

---

Trade? Strategic Policy and the Semiconductor Industry, Washington, DC, Brookings Institution, 1996. pp. 27-38. See also William B. Bonvillian and Richard Van Atta, "ARPA-E and DARPA: Applying the DARPA Model to Energy Innovation," *Journal of Technology Transfer* 36 (2011): 469-513. At the state level, California has imposed mandates for fuel economy (leading to increased demand for hybrids) and reduced the use of incandescent bulbs with various regulations.

[111] As noted, New York State's initiative to support semiconductor manufacturing and other nano-scale industries has achieved significant impact in terms of jobs, growth, and competency. See chapter 7 on Regional Innovation Clusters in this volume and Everett M. Ehrlich, *A Study of the Economic Impact of GLOBALFOUNDRIES*, June 2011.

[112] Michigan has succeeded in developing one of the world's largest clusters of advanced battery-related manufacturers. See Chapter 7 on Regional Innovation Clusters in this volume. Whether the demand will be adequate to support these investments remains to be seen.

[113] Advanced technology products defined by the U.S. Census Bureau categorizes U.S. international trade into 10 major technology areas: advanced materials, aerospace, biotechnology, electronics, flexible manufacturing, information and communications, life science, optoelectronics, nuclear technology, and weapons. U.S. Census Bureau, Foreign Trade, Country and Product Trade, Advanced Technology Products. Because the value of trade in the final product is credited to the country where the product was substantially transformed, data for products produced with components from multiple countries are imperfect. To the extent that U.S. imports of advanced technology products contain components manufactured in the United States and previously exported (microprocessors, for example) the import value will overstate the actual foreign value-added.

[114] This concern has been shared by the PCAST in both the Bush and Obama Administrations. See President's Council of Advisors on Science and Technology, *Report to the President on Ensuring American Leadership in Advanced Manufacturing,* Executive Office of the President, June 2011. Also see President's Council of Advisors on Science and Technology, "Sustaining the Nation's Innovation Ecosystems: Information Technology Manufacturing and Competitiveness," January 2004. In addition see Gregory Tassey, "Rationales and mechanisms for revitalizing US manufacturing R&D strategies," *Journal of Technology Transfer*, DOI 10.1007/s10961-009-9150-2, 2010. (http://www.choosetocompete.org/downloads/PCAST_2004.pdf).

concerned with the composition of the economy. "One implication is that long-term policy frameworks should include an evolving assessment of competitive strength and employment potential across sectors and at all levels of the human capital and education spectrum, and a goal of steering or nudging market outcomes to achieve the social objectives. The structural evolution of the economy matters and can be influenced in relatively efficient ways."[115]

e. **Trade and Innovation are Closely Linked:** Trade and investment measures cannot be ignored when examining the location of innovation – from invention to commercialization. Providing a market induces not only original research, but the ability to achieve scale. Open markets foster innovation, although there is a strong school of thought in a number of countries abroad that protection is a more promising tool. For this reason, the "indigenous innovation" policies of China often have taken the form of local content requirement placed on foreign investors and purchasers of goods in China.[116] Open markets, the U.S. policy, can be detrimental to an import-competing industry if another country's industrial policies have created distortions in trade and investment patterns, which can lead to subsidized production and "dumping" of products in foreign markets.[117]

---

[115] Michael Spence and Sandile Hlatshwayo, "The Evolving Structure of the American Economy and the Employment Challenge," *Council on Foreign Relations*, Working Paper, March 2011, p. 37.
[116] For additional discussion of mercantilist policies, see the section on "21st Century Mercantilism" in Chapter 1. Chapter 5 provides a further description of China's trade and innovation policies.
[117] HIER, KEIL and NRC, *Conflict and Cooperation in National Competition for High Technology Industry*, Washington, DC: National Academy Press, 1996. For an illustrative study, see Thomas Howell, *Steel and the State; Government Intervention and Steel's Structural Crisis*, New York: Westview Press, 1988.

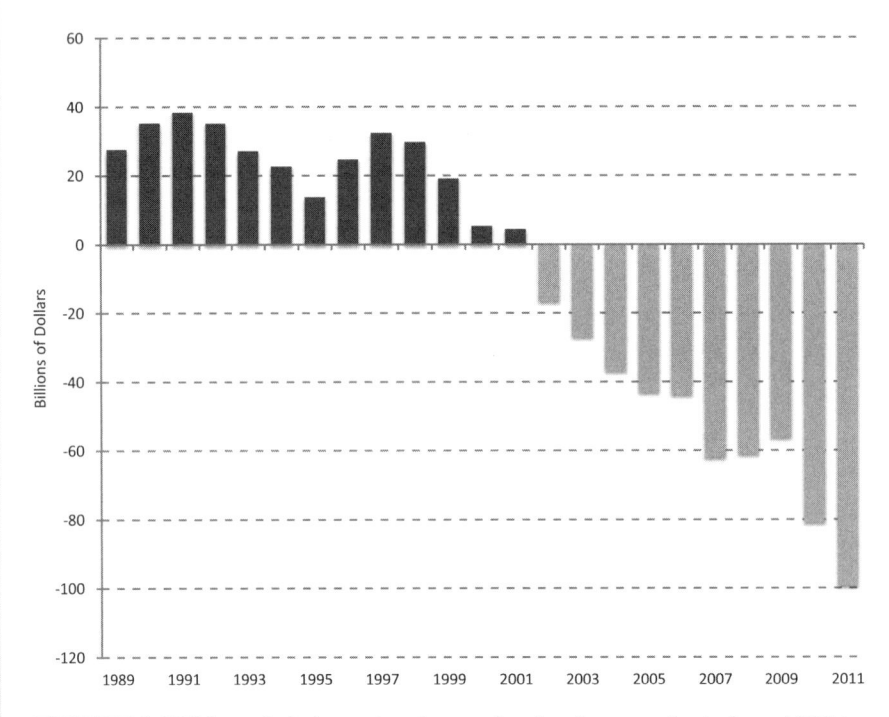

FIGURE 3.5 U.S. trade balance in advanced technology products from 1989 to 2011.
SOURCE: U.S. Census Bureau, Foreign Trade, Trade in Goods with Advanced Technology Products.

7. ***Opportunities for Cooperation:*** [118] *The focus and investments of other nations to accelerate innovation activity opens genuine opportunities for enhancing cooperation on today's global challenges concerning the environment, energy, and health. The globalization of research and innovation presents valuable opportunities for U.S. firms and federally funded research institutes to capitalize on offshore R&D initiatives and growing pools of science and technology talent.*[119] Yet the United States

---

[118] See the related discussion on "The Way Forward," including the need to monitor developments and cooperate globally, in Chapter 2 of this report.
[119] "The 20th-century national S&T innovation environment that has been a hallmark of the United States since World War II, and the model for the world, is evolving into a new 21st-century global S&T innovation environment in which R&D talent, financial resources, and manufacturing facilitated by global communications are geographically dispersed and globally sourced." National Academy of Sciences, *S&T Strategies of Six Countries*, op. cit., p. 93.

currently invests little to stay abreast of foreign science and innovation policies, and the opportunities they present for cooperation.

a. **New Opportunities and Common Challenges:** The rapid growth of R&D activities and research workforces in emerging powerhouses such as India, China, and Brazil—as well as improvements in Internet infrastructure—present greater opportunities for the U.S. to accelerate development of technologies and address common challenges through global partnerships.[120] The innovation strategies of major trading partners place a high priority on expanding international cooperation to accelerate development of technologies and to meet common global needs such as clean energy and cures for disease.[121] This is because our partners recognize that we face common challenges and because they hope to benefit from pooling assets. At the same time, potentially beneficial international cooperation can be challenging. Matching resources and objectives, while equitably sharing the results, is often difficult.[122]

b. **Greater Outreach:** It is also true that many recognize that the United States has committed substantial resources to develop technologies to the point where they can be—with substantial additional resources—developed into marketable products. Research organizations of other nations and regions have established an extensive R&D presence in the U.S. universities to keep abreast of new technologies[123] and U.S. corporations have established extensive offshore innovation networks.[124] U.S.

---

[120] For a review of opportunities as well as challenges for closer U.S. – China cooperation on research and innovation, see National Research Council, *Building the 21st Century, U.S. China Cooperation in Science, Technology, and Innovation,* op. cit. Wagner, Cote and Archambault suggest that the new global innovation environment can benefit the United States if it take advantage of "the distributed knowledge base emerging in science and technology." Caroline S. Wagner, Gregoire Cote and Eric Archambault, "The Shifting Landscape of Global Science: A Challenge for United States Policy," pre-publication version available at http://www.carolinewagner.org/index.php?option=com_content&view=article&id=107.

[121] The national innovation strategies of China, Germany, and India, among others, all call for greater international research collaboration.

[122] Hamburg Institute for Economic Research, Kiel Institute for World Economics, and National Research Council, *Conflict and Cooperation in National Competition for High Technology Industry,* Washington, DC: National Academy Press, 1996.

[123] Taiwan's ITRI, for example, has joint research programs with MIT, the University of California at Berkeley, Carnegie Mellon University, and Stanford Research Institute. Germany's Fraunhofer has seven research institutes based at U.S. universities, including Michigan State University, Boston University, Massachusetts Institute of Technology, the University of Maryland, the University of Michigan, Johns Hopkins University, and the University of Delaware.

[124] Some 249 of America's top 500 corporations have overseas R&D facilities, with China and India the most numerous destinations. Jadeep C. Prabju, Andreas B. Eisengerich, Rajesh K. Chandy, and Gerard J. Tellis, " Patterns in the Global Location of R&D Centres by the World's Largest Firms:

government agencies and national laboratories, however, have a relatively small offshore presence that limits their ability to learn from other nations, but could do so.[125]

c. **The Internet and Cross-Border Data Flows**: The Internet and information and communications technologies (ICT) are at the forefront of developments changing the way business is done internationally. The Internet and ICT have also transformed the way R&D activities are performed by "enabling distributed research, grid and cloud computing, simulation, or virtual worlds."[126] The Internet, because of its global nature, is accelerating the "pace and scope of research and innovation, and encouraging new kinds of entrepreneurial activity."[127] Networked information systems and data flows have become a core component of 21st century innovation. It is important that international consensus be achieved on maintaining the free and open flow of legitimate data and knowledge across borders so that the benefits of the Internet and ICT on world growth and innovation can be preserved and expanded.

d. **Greater Awareness:** The massive investments in innovation capacity and ambitious policy initiatives underway around the world will have an impact on the United States in ways that can scarcely be imagined today. It can be certain, however, that the impact will be immense. Not all of these strategies will work, yet some are likely to transform 21st century global competition, with profound implications for America's well-being and national security. Yet the United States currently invests little to track foreign technology investments, industrial policies, and pro-innovation policies, much less project their implications into the future.

The Role of India and China," paper presented at Druid Summer Conference 2010, Imperial College London Business School, June 16-18, 2010. IBM, Microsoft, General Electric, Pfizer, and other U.S. corporations all perform R&D in India and China for products sold around the world. See Chapter 5 analyses of MNC innovation in India and China in this volume.

[125] The U.S. military has a limited number of science and technology representatives overseas. For example, the Office of Naval Research operates regional offices in places such as Singapore, Prague, Santiago, and London. In addition, and the staff of many U.S. embassies include officers whose portfolios cover science, but they often have many additional responsibilities such as health and the environment, and few may focus on innovative technologies.

[126] OECD, "The Future of the Internet Economy," Policy Brief, June 2008, p. 4. The Internet has also increased R&D efficiency. Marlo I. Kafouros, "The Impact of the Internet on R&D Efficiency: Theory and Evidence," *Technovation*, Volume 26, Issue 7, July 2006.

[127] OECD, "The Future of the Internet Economy," *Ibid.*

Chapter 4

Recommendations

Many of the specific policy measures suggested below have deep historical roots, building on the steps taken by previous administrations and the Congress to nurture and grow the U.S. economy.[1] Taken together, and with adequate and sustained resources, these measures can significantly enhance prospects for the United States to remain a leading center of innovation in the 21st century.[2] Recognizing the fiscal constraints facing the country, our recommendations are limited to policies fostering investments that will, in our Committee's view, repay the expenditures needed many times over.

FOUR CORE GOALS

1. **Monitor and learn from what the rest of the world is doing:** The United States needs to increase its understanding of the swiftly evolving global innovation environment and learn from the policy successes and failures of other nations. It is generally recognized that there is much to be learned from the rest of the world in science. This

---

[1] For a review of the national response to the competitive challenge from Japan in the 1970s and 1980s and a call to develop responses to today's complex challenges, see James Turner, "The Next Innovation Revolution, Laying the Groundwork for the United States," *Innovations*, Spring, 2006. Turner notes that the 1979 President's Industrial Innovation Initiatives, the result of an 18-month Domestic Policy Review, "reflected a strong belief in the free enterprise system and an equally strong belief in the federal government's responsibility to nurture an environment in which industry, universities, and government can function smoothly together." Key bi-partisan legislation of that era includes the Bayh-Dole Act, the expansion of the SBIR program, and the clarification of anti-trust policies to encourage collaborative pre-competitive research by the semiconductor industry. Turner notes the importance on building on previous successes but also the need to articulate a new vision around which policymakers can coalesce.

[2] The Committee does not specify which agency should act on the particular recommendations made in this chapter; one or several agencies could take appropriate actions, depending on the sector, the policies, and the funding available.

is equally true with regard to innovation policy. See Recommendation 1.
2. **Reinforce support for U.S. innovation leadership:** It is very important that the United States reinforce the policies, programs, and institutions that provide the foundations for our own knowledge-based growth and high value employment. These include measures to strengthen our research universities and national laboratories, renew our infrastructure, and revive our manufacturing base. See Recommendations 2, 3, and 4.
3. **Capture greater value from its public investments in research:** The United States should improve its ability to capture greater value from its public investments in research. This includes reinforcing cooperative efforts between the private and public sectors that can be grouped under the rubric of public-private partnerships, as well as expanding support for manufacturing. See Recommendations 5 and 6.
4. **Cooperate more actively with other nations:** In an era of rapid growth in new knowledge that is being generated around the world, the United States should cooperate more actively with other nations to advance innovations that address shared global challenges in energy, health, the environment, and security. See Recommendation 7.

This chapter presents the Committee recommendations. There are seven major recommendations, which are further elaborated in sub-recommendations. The organization of these recommendations and sub-recommendations is presented in an outline, below, as a guide to the reader.

OUTLINE OF RECOMMENDATIONS

1. Monitor and Evaluate Investments, Measures, and Innovation Policies of other Nations

    a. Benchmark best practices
    b. Engage and cooperate abroad
    c. Respond and adapt at home

2. Reinforce the traditional pillars of U.S. economic strength and innovation capacity.

    a. Raise federal support for R&D
    b. Sustain support for university research

        i. Stabilize university funding
        ii. Use dedicated taxes and sources of revenue
        iii. Incentivize private donations
        iv. Increase funding of tuition
        v. Reduce and streamline regulations

    c. Support innovative small businesses

        i. Reauthorize and expand proven innovation programs
        ii. Experiment with and evaluate new initiatives
        iii. Provide policy support for innovation capital

    d. Strengthen the skilled workforce

        i. Support community colleges
        ii. Encourage worker training
        iii. Increase funding and opportunities for dislocated workers
        iv. Create incentives to induce retirees and potential retirees to remain active in contributing to the American economy
        v. Encourage immigration of scientific and entrepreneurial talent

3. Provide a Competitive Tax Framework

    a. Benchmark tax and regulatory policy
    b. Examine the tax code
    c. Pursue prudent deficit reduction
    d. Make the Research and Experimentation tax credit permanent

4. Build a 21st Century Innovation Infrastructure

   a. Build world-class infrastructure
   b. Expand broadband penetration
   c. Secure cross-border data flows
   d. Encourage energy conservation

      i. Smart grid
      ii. Innovative financing

5. Adopt specific policy measures to capture greater economic value from America's public investments in research

   a. Strengthen university links to the market

      i. Provide matching seed funds
      ii. Develop university incubators
      iii. Expand SBIR support for commercialization of university research
      iv. Develop additional Centers of Excellence
      v. Use of innovation prizes
      vi. Encourage private foundations to take equity positions in start-ups by amending SEC rules

   b. Strengthen National Laboratories' links to the market

      i. Expand use of research parks
      ii. Expand SBIR to the National Laboratories

   c. Develop public private partnerships

      i. New initiatives in early-stage finance
      ii. Support for industry consortia

   d. Expand support for manufacturing

      i. Provide incentives for manufacturing
      ii. Expand manufacturing support programs

   e. Sustain federal programs to jump-start new industries
   f. Create new institutions for applied research
   g. Open foreign markets to business services
   h. Expand support for U.S. manufactured exports
   i. Foster cluster development

i. Assess foreign clusters
ii. Support the development of science and research parks

j. Leverage government procurement to establish early markets

i. Leverage defense procurement
ii. Encourage procurement from small businesses

6. Recognize that trade and innovation are closely linked

    a. Provide a rules-based playing field
    b. Develop an enforceable international code of conduct

7. Capitalize on the globalization of research and innovation

    a. Strengthen international cooperation
    b. Expand exchanges of scholars and students

RECOMMENDATIONS

1. ***Monitor and Evaluate Investments, Measures, and Innovation Policies of other Nations:*** *In a world where other nations are investing very substantial resources to create, attract and retain the industries of today and tomorrow, the United States needs to increase its understanding of the swiftly evolving global innovation environment and learn from the policy successes and failures of other nations.*[3]

    a. **Benchmark Best Practices:** The federal government should support a systematic, ongoing process to monitor and evaluate investments, measures, and policies of other nations aimed at improving their capacity to innovate and compete in the industries of tomorrow. This should include ensuring that U.S. science counselors and research agencies support the collection and analysis of relevant information.[4] Foreign innovation programs should be benchmarked against those of the United States. This will require a very substantial investment of dedicated resources across a variety of public and private institutions.[5]

    b. **Engage and Cooperate Abroad:** The governmental institutions of the United States should increase their cooperation and engagement with policymakers, research institutions, academics, and investors from around the world to both gain from their investments and better understand the rationale and objectives of programs to promote innovation, product commercialization, and development of emerging industries and learn best practices that can be applied to programs in the U.S.

    c. **Respond and Adapt at Home:** Knowledge gained from this benchmarking process should be used to inform U.S. policymakers and legislators, and help to shape U.S. innovation programs, R&D investments, and incentives and other policy responses, including, importantly, incentives to encourage investments by industry.

---

[3] See related Finding 7 in Chapter 3.
[4] There are initiatives to capture a broader view of foreign government innovation policies and the opportunities they present for cooperation. For example, the Office of Naval Research, in cooperation with its Global component, have launched a series of outreach activities designed to explore best practices in innovation policy and identify cooperative projects. The National Academies Board on Science, Technology, and Economic Policy is launching an Innovation Forum, with the support of ONR, to provide an on-going institutional mechanism to benchmark national innovation policies and to provide a mechanism for regular policy discussions and learning.
[5] This recommendation complements Recommendation 10-1 of the National Academy report *S&T Strategies of Six Countries*. National Academy of Sciences, S&T Strategies of Six Countries, op. cit., p. 95. Recommendation 10-1 calls more generally for "monitoring the transformation from a national to a global S&T innovation environment."

2. **In a dramatically more competitive world, the United States needs to reinforce the traditional pillars of its economic strength and innovation capacity.**[6]

   a. **Raise Federal Support for R&D**: Federal support for R&D should be raised in line with the goal set by President Obama in a 2009 speech before members of the National Academy of Sciences, to increase the combined public and private investment in R&D in the United States to more than 3 percent of the U.S. Gross Domestic Product.[7]

   b. **Sustain Support for University Research:** Basic research carried out at U.S. research universities and national laboratories is valuable in its own right; it is also the source of the knowledge and insights that drive U.S. innovation and growth. Universities should be provided with the necessary resources to maintain and grow their facilities, attract and retain outstanding faculty and students from around the world, and provide the educational experience necessary to maintain and enhance the innovative capacity that assures America's position in the world.

      i. **Stabilize University Funding:** The federal and state governments should reverse the cyclicality and negative trends in university financing. Steady, sustainable, predictable increases over the long term are needed for universities to plan their own investments in research, and would make federal and state research expenditures more effective and efficient.[8]

---

[6] See related Finding 2 in Chapter 3.

[7] For a transcript of address by President Obama at the annual meeting of the National Academy of Sciences on April 28, 2009, see http://www.issues.org/25.4/obama.html. According to the Congressional Research Service, based on 2008 figures, reaching President Obama's 3% goal would require an 8.4% real increase in national R&D funding. See CRS, "Federal R&D Funding FY 2012" June 21, 2011. Returns on federal R&D are considered to be very substantial. See Robert Solow, "Technical Change and the Aggregate Production Function," in *The Review of Economics and Statistics*, August 1957, 39(3). For a review of the econometric evidence between R&D and productivity, see Zvi Grilliches, "R&D and Productivity," NBER Monograph, 1998. More recently, the Chairman of the Federal Reserve Board has drawn a close link between government support for R&D and economic growth. See Ben S. Bernanke, "Promoting Research and Development: The Government's Role" Speech presented at the Conference on New Building Blocks for Jobs and Economic Growth, Washington, DC: May 16, 2011, page 38. For a review of how the impact of federal investments in R&D can be measured, see, National Research Council, *Measuring the Impacts of Federal Investments in Research*, S. Olson and S. Merrill, rapporteurs, Washington, DC: The National Academies Press, 2011.

[8] See National Research Council, *Breaking Through: Ten Strategic Actions to Leverage Our Research Universities for the Future of America,* Washington, DC: The National Academies Press, 2012. The report calls for "stable, strong, and effective Federal funding for university-performed

ii. **Use Dedicated Taxes and Sources of Revenue:**[9] Dedicated state funds and taxes are potential sources of reliable revenue for research universities.[10] Shifting more university research funding from the general state budget to dedicated revenue sources will provide a more reliable funding stream for vital investments in the state's economic future.[11]

iii. **Incentivize Private Donations:**[12] Consider expanding federal tax credits for companies that fund university research in order to stimulate additional funding for universities.[13] In addition,

---

R&D so that the nation will have a stream of new knowledge and educated people to power our future, helping us meet national goals and ensure prosperity and security."

[9] See discussion in Chapter 2 section "U.S. Universities Face Financial Challenges."

[10] For a review of the use of dedicated taxes as a means of public finance, see Alan J. Auerbach, "Public Finance in Practice and Theory," paper prepared as the Richard Musgrave Lecture, CESifo, Munich, May 25, 2009. A growing number of states earmark all or part of taxes on hotels, cigarettes, and alcohol for specific programs, such as road maintenance, schools, and construction of convention centers and sports stadiums. The attraction is that such sources provide recurring revenue streams for important programs, even during times of economic downturn, and are not subject to government budgetary restraints. Other sources of steady, non-tax state income include proceeds from lotteries, casinos, sales of public land, and oil and mineral rights. Notably, the State of Texas uses a Permanent University Fund (PUF), established in its 1876 Constitution, to fund higher education. Currently, PUF land assets deliver proceeds through oil, gas, sulfur, and water royalties, rentals on mineral leases, and gains on fiduciary investments.

[11] A handful of states use non-tax revenue to fund activities relating to innovation. A New Mexico, for example, has devoted revenues from oil, gas, and land rights in a private-equity fund that has invested nearly $300 million into local companies in fields ranging from solar power to molecular diagnostics. (Details of New Mexico's private-equity investments can be found on the State Investment Council Web site, http://www.sic.state.nm.us/investments.htm. Descriptions of specific investments are provided in Sun Mountain Capital, "New Mexico Private Equity Investment Program: Overview and 2010 Review," June 2011.) Arkansas used a portion of increased cigarette taxes to help fund a campus of the University of Arkansas for Medical Sciences. See John Lyon, "Beebe Signs Tobacco Tax Hike Into Law," Arkansas News, Feb. 17, 2009. Nebraska invested $106 million received from a 2002 court settlement with tobacco companies to fund medical research at state universities, a move that has generated more than $800 million investment and created nearly 1,800 jobs. See Steve Jordon, "Tobacco Money Gives Nebraska an Economic, Research Lifeline," Omaha World-Herald, Feb. 3, 2011. Currently, only 20 states earmark some dollars for higher education, and the sums are quite small, according to the American Association of State Colleges and Universities. See Alene Russell, "Dedicated Funding for Higher Education: Alternatives for Tough Economic Times," American Association of State Colleges and Universities, Higher Education Policy Brief, December 2008.

[12] See discussion in Chapter 2 section "U.S. Universities Face Financial Challenges."

[13] Reversing a three-decades-long trend of increasingly strong ties between industry and universities, the absolute value of industrial R&D dollars to academic institutions—funds provided directly to academic institutions for the conduct of research—began to decline beginning in 2002 after reaching a high of $2.2 billion in 2001. Also, industrial R&D support to academia has historically been concentrated in relatively few institutions. See National Science Foundation, "Where has the Money Gone? Declining Industrial Support of Academic R&D," *InfoBrief*, NSF 06-328 September 2006. Leading university and industry leaders have pointed out that U.S. companies increasingly choose to work with foreign rather than U.S. universities, encouraged by the more favorable IP rights that foreign universities offer and the strong incentives for joint industry-

assess the potential for state matches for certain private endowment donations as a way to provide incentives for private donors or foundations to increase their support for research universities.[14]

iv. **Increase Funding of Tuition.** Students are borrowing more money to pay for college than ever before, with student debt in this country exceeding the level of credit card debt.[15] Tuitions keep rising.[16] No other major economy with which the United States competes places as heavy a financial burden on its students.[17] Although co-investment by students plays an important role in motivating students to capitalize on their education, the necessary growth of enrollment and matriculation as a percent of the U.S. population will be choked off in the absence of increased federal, state and private support for tuitions.

v. **Reduce and Streamline Regulations:** The expanding costs of compliance with federal regulations are making it increasingly expensive for universities to conduct research.[18]

---

university research that foreign governments provide. GUIRR, "Re-Engineering the Partnership: Summit of the University-Industry Congress," Meeting of 25 April 2006, Washington, DC.

[14] See National Research Council, *Breaking Through: Ten Strategic Actions to Leverage Our Research Universities for the Future of America,* Washington, DC: The National Academies Press, 2012. The report calls for the creation of a "R&D tax credit that incentivizes business to develop partnerships with universities (and others as warranted) for research that results in new U.S.-located economic activities." For an analysis of the impact of financial shocks to a university's resource base, see Jeffrey R. Brown, et al., "Why I Lost My Secretary: The Effect of Endowment Shocks on University Operations."NBER, May 29, 2010.

[15] Institute for College Access and Success, "Student Debt and the Class of 2010" November 2011. The report notes that two-thirds of college seniors graduated with loans in 2010, and they carried an average of $25,250 in debt.

[16] Published tuition has barely increased at two-year colleges (by only $68 over the course of nine years), but has increased substantially at four-year colleges (by $3,004 over the same nine year period). From the 1999-2000 academic year to the 2008-09 academic year, Net Student Tuition actually fell by $849 at two-year colleges, representing a fairly dramatic decrease in net tuition at the two-year level, given that the national average for net tuition was never higher than $900 any single year. In contrast, Net Student Tuition has increased by $1,067 at four-year colleges over the same time span. While this absolute growth in net tuition at four-year institutions may not seem particularly high, keep in mind that per capita income in the U.S. declined by $1,325 from 2000 to 2009. See Andrew Gillen, et al., "Net Tuition and Net Price Trends in the United States (2000-2009), Washington, DC: Center for College Affordability, November 2011.

[17] For a comparative review of "who participates in education, how much is spent on it and how education systems operate," see OECD: Education at a Glance 2011: OECD Indicators, Paris: OECD, 2012.

[18] Tobin L. Smith, Josh Trapani, Antony Decrappeo, and David Kennedy, "Reforming Regulations of Research Universities," *Issues in Science and Technology*, Summer 2011. See also the January 21, 2011 filing by the AAU, APLU, and COGR on "Regulatory and Financial Reform of Federal Research Policy." The document notes that "Rationalizing the Federal regulatory infrastructure is essential to the health of the university-government research partnership and to the efficient and

Federal and state policymakers and regulators should review the costs and benefits of federal and state regulations, eliminating those that are redundant, ineffective, inappropriately applied to the higher education sector, or impose costs that outweigh the benefits to society.[19]

c. **Support Innovative Small Businesses:** The availability of early stage funding for small entrepreneurial firms and start-ups is crucial for the vitality of the innovation process. There are three elements to address in this regard: public innovation programs, the policy framework and incentives for angel funding, and other measures and incentives to encourage entrepreneurship.

　　i. **Reauthorize and Expand Proven Innovation Programs:** The U.S. should expand successful innovation programs, as it recently has with the SBIR program, restore funding for NIST's revamped Technology Innovation Program with its current focus on manufacturing,[20] and consider new programs such as the recently announced Start-up America.[21] These early-stage funding programs support the development of new products and help promising small technology companies bring new ideas and products to the market, in part by creating

---

productive use of federal research funding." Access at www.cogr.edu/viewDoc.cfm?DocID=151794.

[19] [19] See National Research Council, *Breaking Through: Ten Strategic Actions to Leverage Our Research Universities for the Future of America,* Washington, DC: The National Academies Press, 2012. The report calls for "a balanced regulatory environment in order to increase the cost-effectiveness of our research universities."

[20] See Chapter 2 of this volume for a discussion of the Advanced Technology Program (ATP) and its successor, the Technology Innovation Program. A National Academies assessment of ATP found it to be "an effective federal partnership program." National Research Council, *The Advanced Technology Program, Assessing Outcomes*, C. Wessner, ed., Washington, DC: The National Academies Press, 2001. See also the discussion of SBIR in Chapter 2. A National Academies assessment of SBIR found it to be "sound in principle and effective in practice." National Research Council, *An Assessment of the SBIR Program*, Washington, DC: The National Academies Press, 2008.

[21] Start Up America, a White House led initiative, is focused on increasing innovation and commercialization and accelerating support for U.S. entrepreneurs though a variety of policies and programs. These efforts are being deployed through federal agencies like the Departments of Energy, Labor, the Small Business Administration and Commerce including the Economic Development Administration. One example of this is the Jobs and Innovation Accelerator Challenge, which according to the EDA is "a multi-agency competition launched in May to support the advancement of 20 high-growth, regional industry clusters. Investments from three federal agencies and technical assistance from 13 additional agencies will promote development in areas such as advanced manufacturing, information technology, aerospace and clean technology, in rural and urban regions in 21 states." See http://www.eda.gov/InvestmentsGrants/jobsandinnovationchallenge.

new information that investors need.[22] In addition, programs that fund collaborations between small businesses, and universities, such as ARPA-E, DARPA, and Energy Efficiency and Renewable Energy (EERE), should be expanded where appropriate and provide sustained support.[23] When new ideas and promising technologies are not funded here in the United States, investors overseas may well fill the gap, thus capitalizing on U.S. investments in R&D.[24]

ii. **Experiment with and Evaluate New Initiatives**: Recent public-private initiatives such as the Administration's Start-Up America program should be given clear metrics and carefully evaluated in conjunction with programs at state and regional levels.

iii. **Provide Policy Support for Innovation Capital:** Market inefficiencies and a long term shift away from seed stage investments have created a substantial gap between the demand by entrepreneurs for seed and early-stage funding and the supply in the risk capital market. Bridging this gap is essential for sustaining the flow of innovation from U.S. R&D investments and the growth and employment they generate.[25]

---

[22] See discussion of SBIR in Chapter 1, section on "Public-Private Partnerships" and Chapter 2, section on "Providing Early-Stage Finance." See the discussion of ATP/TIP in Chapter 2, section on "Providing Early-Stage Finance." See the discussion of StartUp in Chapter 3, Section 4f. Finally see discussion of EERE in Chapter 1, section on "National Laboratories."

[23] The National Academies 2006 report, *Rising Above the Gathering Storm*, (op cit.) recommended the establishment of an Advanced Research Projects Agency—Energy (ARPA-E) within the Department of Energy (DOE). The 2007 America COMPETES Act, which implemented many of the recommendations in the National Academies' report authorized. ARPA-E, but without an initial budget. The new program received $400 million of funding in the 2009 American Recovery and Reinvestment Act. The America COMPETES Reauthorization Act of 2010 made additional changes to ARPA-E's structure. ARPA-E is modeled after the successful Defense Advanced Research Projects Agency (DARPA). For a review of the history and the distinguishing features of DARPA, see William B. Bonvillian, "The Connected Science Model for Innovation: The DARPA Model," in National Research Council, *21st Century Innovation Systems for Japan and the United States, Lessons from a Decade of Change, Report of a Symposium*, Sadao Nagaoka et al., eds., Washington, DC: The National Academies Press, 2009. For a review of the Energy Efficiency and Renewable Energy program, see National Research Council, *Energy Research at DOE: Was It Worth It? Energy Efficiency and Fossil Energy Research 1978 to 2000*, Washington, DC: The National Academies Press, 2001.

[24] For example, Rusnano, the Russian state technology firm, is seeking to make large investments in U.S. life sciences and technology companies whose products are to be manufactured in Russia. See Megan Davis, "Rusnano, US fund to invest $760 mln in pharma venture," Reuters, March 6, 2012.

[25] A study by Gittell, Sohl, and Tebaldi finds that technology-based entrepreneurship, particularly by small businesses, is a more powerful job creator than entrepreneurship in general. See Gittell, Ross; Sohl, Jeffrey; and Tebaldi, Edinaldo (2010) "Is there a Sweet Spot for U.S. Metropolitan Areas? Exploring the Growth in Employment and Wages in U.S. Entrepreneurship and Technology Centers in Metropolitan Areas over the last Business Cycle, 1991 To 2007, *Frontiers of Entrepreneurship Research*: Vol. 30: Issue 15, Article 13.

- **Increase the understanding of the role and evolution of Innovation Capital in the United States**: Capital is essential for innovation, but the market for innovation capital is often poorly understood by policy makers, with myths prevalent about the perfect operation of markets.[26] Potential policy support for innovation capital needs to begin with a better understanding of the current trends, challenges and opportunities inherent in early stage financing of innovation.[27] A careful study of the role that public policies can play to support the formation of innovation capital is needed.
- **Develop complementary funds that Co-invest with angel investors:** Although angel investors play an important role in funding innovation at the seed and early stages of a technology's development, the size and reach of these investments remains limited.[28] Recent initiatives that provide capital as part of co-investments with angel investors on a matching basis, often with the angel leading the deal, can increase the amount of innovation funding available and should be assessed and expanded where models prove successful.[29] These complementary

---

[26] In economics, a perfect market is defined by several simplifying conditions, including perfect market information. The real world is characterized by pervasive information asymmetries. In 2001, the Nobel Prize in Economics was awarded to George Akerlof, Michael Spence, and Joseph E. Stiglitz "for their analyses of markets with asymmetric information."

[27] For example, the 5 to 7 year timeframes of venture capital funding, set by IT sector precedents, do not fit the developmental timeframes of many other sectors, including biomedicine and energy. To address this challenge, some analysts (e.g., Andrew Lo of MIT) have suggested alternative private sector approaches, such as pooling small investments from the public into a multi-portfolio risk pool. Relatedly, the Senate in March 2012 passed a "crowd-funding" bill that allows entrepreneurs to raise up to $1 million per year through approved crowd-funding portals. This legislation was signed into law in April 2012 by the President and became the JOBS Act.

[28] Angel investors are affluent individuals who provide capital for a business start-up, usually in exchange for convertible debt or ownership equity. According to the Center for Venture Research, total investments in 2010 were $20.1 billion, with a total of 61,900 entrepreneurial ventures receiving angel funding in 2010. The number of active investors in 2010 was 265,400 individuals. Access at http://www.unh.edu/news/docs/2010angelanalysis.pdf.

[29] For a review of the potential of the Archimedes fund, see Jeff E. Sohl, "The Organization of the Informal Venture Capital Market," in *Handbook of Research on Venture Capital,* Hans Landström, editor. Northampton MA: Edward Elgar, 2007. In a recent initiative in the Netherlands and Belgium, angels receive a match of up to 50 per cent of the size of the investment, thus limiting the risk exposure of the angel and reducing the price of the deal, both of which can be expected to encourage angels to increase their investment activity. Also being experimented in Europe are hybrid funds that supplement private funds with public money. The funds are targeted to early stage firms with high growth potential in emerging technological sectors. For a review of seed and early-stage financing for high-growth companies in OECD and non-OECD countries with a primary focus

funds have an advantage over direct public investments of capital into start-up ventures. State based seed funds in the U.S. and various country wide direct investment funds in Europe can generate over-valuations of investments, and may result in a lack of value-added in the post investment stage. Moreover, the additional terms and conditions required for use of public funds can in fact drive away potential angel investors.[30]

- **Educate angels and entrepreneurs:** Programs that educate angel investors, sometimes known as "Angel Academies," can help potential angels understand the complexities of angel investing and entrepreneurs appreciate the requirements necessary to become investor ready, leading to an increase in both available capital and quality deal flow. These academies can be based on university-private sector partnerships that draw in the appropriate individuals and garner the resources necessary to develop and implement a research-based education program.[31] Small amounts of competitively awarded funding from state and federal sources can have a disproportionately positive impact in generating these partnerships.

d. **Strengthen the Skilled Workforce:** Expanding the skilled workforce through education, training, and retention, through the development of new curricula and delivery methods, and through attracting skilled immigration is needed to promote and encourage innovation. Scientists and engineers are required to develop new ideas and design new processes. Skilled workers and qualified managers are needed to transform those ideas into marketable products and services.

   i. **Support community colleges.** Through their flexibility and proximity to employers and the opportunities they offer, community colleges can and should play an important role in developing industry-relevant skills and training for dislocated

---

on angel investment, see OECD, *Financing High Growth Firms, The Role of Angel Investors*, Paris: OECD, 2011, page 96.
[30] Freear and Jeff E. Sohl "Angles on Angels and Venture Capital: Financing Entrepreneurial Ventures" in *Financing Economic Development in the 21st Century*, 2nd Edition, Z. Kotval and S. White, eds., M.E. Sharpe, Inc: NY (forthcoming).
[31] Amparo San José, Juan Roure, and Rudy Aernoudt, "Business Angel Academies: Unleashing the Potential for Business Angel Investment," *Venture Capital*, Volume 7, Issue 2, 2005.

workers.[32] Initiatives to encourage community colleges train workers in new skills and provide technical credentials should be encouraged and reinforced.[33]

ii. **Encourage worker training**: U.S. workers should be encouraged to engage in life-long learning.[34] Historically, the United States has devoted relatively few resources to worker training compared to other OECD countries, and this contribution is declining.[35] The Federal government should consider providing tax incentives to encourage worker participation in training and skill enhancement programs.[36] Incumbent U.S. engineers with bachelors' degrees should also be encouraged through scholarships to pursue graduate degrees.

iii. **Increase funding and opportunities for dislocated workers**. The Trade Adjustment Assistance (TAA) should be made generic, creating training and therefore opportunities for

---

[32] Michael Greenstone and Adam Looney, "Building America's Job Skills with Effective Workforce Programs: A Training Strategy to Raise Wages and Increase Work Opportunities," Washington, DC: Brookings Institution, September 2011.

[33] Skills for America's Future is an industry led initiative that seeks to "dramatically improve industry partnerships with community colleges and build a nation-wide network to maximize workforce development strategies, job training programs, and job placements." As a part of this effort, the Manufacturing Institute, the affiliated non-profit of the National Association of Manufacturers (NAM), has announced an effort "to help provide 500,000 community college students with industry-recognized credentials that will help them get secure jobs in the manufacturing sector." White House Press Release, June 8, 2011, "President Obama and Skills for America's Future Partners Announce Initiatives Critical to Improving Manufacturing Workforce."

[34] For a review of the need for lifelong learning in the globally competitive economy of the 21st Century, see National Academy of Engineering, *Lifelong Learning Imperative in Engineering, Summary of a Workshop*, D. Dutta, Rapporteur, Washington, DC: The National Academies Press, 2010. See also National Research Council, *Building a Workforce for the Information Economy*, Washington, DC: The National Academies Press, 2001, page 254.

[35] Federal government spending on training and employment as a percent of GDP has fallen steadily since 1979, when it was approximately 0.0047 percent of GDP to 0.0006 percent of GDP in FY 2010, despite increased pressures on the U.S. labor market from international competition. Over the last decade the United States spent less on active labor-market programs, including training, career counseling and job search assistance, as a percent of GDP than almost all OECD countries. See Howard F. Rosen, "Designing a National Strategy for Responding to Economic Dislocation." Testimony before the Subcommittee on Investigation and Oversight House Science and Technology Committee, June 24, 2008. For a review of German policies to foster lifelong learning, see Wilfried Kruse, "Lifelong Learning in Germany –Financing and Innovation: Skill Development, Education Networks, Support Structures," Berlin: Federal Ministry of Education and Research (BMBF), 2003. Access at http://www.bmbf.de/pub/lifelong_learning_oecd_2003.pdf. For an empirical study of the positive impact of German worker training programs on productivity and employment, see Michael Lechner, Ruth Miquel, and Conny Wunsch, "Long-Run Effects of Public Sector Sponsored Training in West Germany," *Journal of the European Economic Association*, Volume 9, Issue 4, 2011.

[36] Brian Bosworth, "Lifelong Learning, New Strategies for the Education of Working Adults." Center for American Progress, December 2007. Access at http://www.americanprogress.org/issues/2007/12/pdf/nes_lifelong_learning.pdf.

workers displaced by shifts in technology and changes brought about by globalization.[37]

iv. **Create incentives to induce retirees and potential retirees to remain active in contributing to the American economy.** Demographic shifts toward an aging population should be mined as an advantage rather than accepted as a weight on society.[38] Citizens of retirement age represent a major investment of skills and knowledge which need to be tapped to train the less-skilled current and potential participants in the workforce. In a web-based global economy, content as well as transmission matters. Much of the needed content can come from this cohort. Entrepreneurship is not confined to the young.[39] Indeed, there is growing evidence that entrepreneurial activity among those over fifty outstrips such efforts by those under twenty-five.[40]

v. **Encourage Immigration of Scientific and Entrepreneurial Talent:**

- **Visas for foreign graduates with advanced U.S. Degrees:** The Congress should immediately establish a special immigration category to allow successful foreign students who have earned advanced science and technology degrees to remain in this country and work in

---

[37] See the discussion in Chapter 2, Section on "U.S. Support for Manufacturing." See Harold F. Rosen, "Strengthening Trade Adjustment Assistance," Policy Brief 08-02, Peterson Institute for International Economics, 2008. Access at http://www.iie.com/publications/pb/pb08-2.pdf. For a review of issues relating to training programs and global competitiveness focusing on the TAA program, see the transcript of the Hearing before the House Ways and Means Committee, "Promoting U.S. Worker Competitiveness in a Globalized Economy." June 14, 2007. Serial No. 110-47, Washington, DC: USGPO, 2008. Access at http://www.gpo.gov/fdsys/pkg/CHRG-110hhrg43113/pdf/CHRG-110hhrg43113.pdf.

[38] The number of retirees today as compared with 1965-70 has nearly doubled, from 13 million to 24 million. This places a great burden on Social Security and Medicare funding, without adequately addressing the contribution to an innovative society and the GDP that this cohort can make. See John Shoven, *Demography and the Economy*, Chicago: University of Chicago Press, 2011.

[39] Stanford University's John Shoven argues that prevailing notions about old age no longer reflect reality. Emerging research is throwing light on a new stage of life between the prime working years and full retirement. See John B. Shoven, ed., Demography *and the Economy*. Chicago: University of Chicago Press, 2011.

[40] According to recent research by the Kauffman Foundation, older Americans are working increasingly beyond their middle years. For 11 of the 15 years between 1996 and 2010, Americans between the ages of 55 to 64 had the highest rate of entrepreneurial activity of any age group. Twice as many founders of U.S. technology companies were over the age of 50 as were under the age of 25. See Robert W. Fairlie, "Kauffman Index of Entrepreneurial Activity, 1996-2010." Kansas City, MO: Ewing Marion Kauffman Foundation, March, 2011.

academia or elsewhere in their areas of expertise.[41] Training of foreign students in this country to the masters or doctoral level involves significant public expenditure; asking such bright and ambitious individuals to leave the country can only hurt U.S. innovation, particularly in light of the significant contributions some make to the creation of new companies.[42]

- **International Agreements on Skilled Worker Mobility.**[43] International agreements ought to be entered into to facilitate the free movement of highly skilled individuals who can contribute to the U.S. economy.
- **Skilled Entrepreneur Visas:** New rules promulgated by the U.S. Citizenship and Immigration Service to attract well-trained entrepreneurs with the funds and experience to found companies are an important positive step with significant potential to add employment.[44] This administrative initiative should be reinforced through legislation that would provide visas, leading to permanent-resident status and ultimately citizenship, for foreign entrepreneurs who have secured financing to start businesses in the United States.[45]

---

[41] See the related Action C-4 called for in National Academy of Sciences et al. report, *Rising Above the Gathering Storm*, op. cit., page 173. The report recommends that "the federal government should continue to improve visa processing for international students and scholars to provide less complex procedures, and continue to make improvements on such issues as visa categories and duration, travel for scientific meetings, the technology alert list, reciprocity agreements, and changes in status."

[42] See discussion in Chapter 2, section on "Hunting for Global Talent."

[43] For a review of the characteristics, trends, and impacts of the global migration of skilled workers, see OECD, International Mobility of the Highly Skilled, Paris: OECD, 2002. Other advanced nations compete for skilled workers. For a Canadian perspective, see Industry Canada, "International Mobility of Highly Skilled Workers: A Synthesis of Key Findings and Policy Implications." Ottawa, April 2008.

[44] Recently, Secretary of Homeland Security Janet Napolitano and U.S. Citizenship and Immigration Services Director Alejandro Mayorkas announced a series of policy and operational initiatives to stimulate investment and firm and job creation by attracting foreign entrepreneurial talent, particularly in the high technology sectors. *Wall Street Journal*, "U.S. to Assist Immigrant Job Creators," August 2, 2011. The impact of well-trained, highly motivated immigrants is not always fully appreciated. As Federal Reserve Chairman Bernanke recently observed, "Contrary to the notion that highly trained and talented immigrants displace native-born workers in the labor market, scientists and other highly trained professionals who come to the United States tend to enhance the productivity and employment opportunities of those already here." See Ben S. Bernanke, op. cit., page 30.

[45] Legislation for "Start-Up Visa", endorsed by the Kauffman Foundation and supported by the Small Business Administration is pending in the U.S. Congress. The House Committee on the Judiciary, Subcommittee on Immigration Policy and Enforcement held a hearing on the "Investor Visa Program" on September 14, 2011.

3. ***Provide a Competitive Tax Framework:***[46] *The United States should assure that the tax framework supports new company creation and investment. In order to be competitive with those of its major trading partners, the U.S. should take measures to address policies that actually disadvantage U.S.-based industry.*

   a. **Benchmark Tax and Regulatory Policy:** Governments at the federal and state levels should engage in regular benchmarking of U.S. tax policies and regulatory costs compared with those of other nations and determine how these differences influence corporate decisions on where to build new industrial capacity and research centers.
   b. **Examine the Tax Code:** The United States should assess the impact of U.S. based innovation and production of tax policies, and consider appropriate adjustments.[47] Where the tax code and regulatory costs are found to be serious impediments to U.S. investment and innovation, the government should seek to narrow or close these competitive gaps, not by abandoning well-grounded regulations, but by fully considering their competitive impact and undertaking measures to reduce the impact and/or provide compensating incentives.[48] Alternatives to current policies should be examined and, if deemed beneficial to the nation, pursued. These alternatives for consideration could include a reduction in the corporate tax rate (now one of the highest nominal rates in the OECD), the limitation of residence based taxation (which may in its present form provide incentives for new companies to incorporate outside the United States), and an increased reliance on consumption taxes (which do affect the location investments.)[49]
   c. **Pursue Prudent Deficit Reduction:** Efforts should be made to ensure that changes in taxation as well as reductions in spending to shrink the federal deficit are allocated with a full understanding of the potential consequences for future growth. Strong and steady public investments are necessary to sustain traditional U.S.

---

[46] See related Finding 3 in Chapter 3.
[47] See the related Action D-3, "Provide Incentives for U.S. Based Innovation" in National Academy of Sciences et al., *Rising Above the Gathering Storm*, op. cit. page 197. The report calls for the examination of alternatives to current economic policies to spur innovation. "These alternatives could include changes in overall corporate tax rates and special tax provisions, providing incentives for the purchase of high-technology research and manufacturing equipment, treatment of capital gains, and incentives for long-term investments in innovation."
[48] Peter R. Merrill, "Corporate Tax Policy for the 21st Century," *National Tax Journal*, December 2010, 63 (4, Part 1), 623-634.
[49] See the discussion in Chapter 2, sections on "Improving Framework Conditions," and "Institutional Support for Applied Research."

advantages in innovation and to accelerate economic growth that will ultimately help address the nation's fiscal challenges.

    d. Make Research and Experimentation Tax Credit Permanent: In 1981, the United States became the first nation to use its tax code to spur innovation. Specifically, the research and experimentation (R&E) tax credit is designed to stimulate company R&D over time by reducing after-tax costs.[50] To capture the benefits that the R&E credit is seen to provide the U.S., many nations have since followed suit with their own, often more competitive tax credits for research and experimentation.[51] According to the OECD (2009) the U.S. tax incentives for R&E now ranks 24th lowest out of 38 countries analyzed.[52] Moreover, the U.S. R&E tax credit has been subject to some 14 renewals, making the fiscal environment for innovation related investments in the United States necessarily uncertain.[53] To draw full benefit from this instrument, The U.S. should make the R&E tax credit permanent to provide greater certainty for long term investments and simplify its administration to make the application process easier (and less expensive).[54] Also, as recommended by National Academies reports, consideration should be given to expanding the credit significantly.[55]

---

[50] Francisco Moris, "The U.S. Research and Experimentation Tax Credit in the 1990s," NSF InfoBrief, NSF 05-316, July 2005.

[51] See Gregory Tassey, "Tax incentives for innovation: time to restructure the R&E tax credit," *The Journal of Technology Transfer*, Volume 32, Number 6 (2007), 605-615. Tassey notes that "in the 25 years since the R&E tax credit was enacted, a steadily increasing number of countries have implemented or expanded competing tax incentives, which in many cases are better structured and larger in size."

[52] Organisation for Economic Co-operation and Development, 2009. OECD *Science, Technology and Industry: Scorecard 2009*. OECD, Paris, France.

[53] Eric Spiegel, Make Permanent the Research and Experimentation Tax Credit, *The Atlantic*, July 12, 2011.

[54] Currently, businesses must choose between using a complex formula for calculating their R&E credit that provides a 20 percent credit rate for investments over a certain base and a much simpler one that provides a 12 percent credit in excess of a base amount. Some analysts argue that the complexity involved in applying for the higher rate effectively lowers the level of credit available. See Robert D. Atkinson, "Expanding the R&E tax credit to drive innovation, competitiveness and prosperity." *Journal of Technology Transfer (*2007) 32:617–628. Recent proposals by the National Economic Council have called for "simplifying its use and expanding its incentive payments by 20%." National Economic Council, Council of Economic Advisers, and Office of Science and Technology Policy, "*A Strategy for American Innovation, Securing our Economic Growth and Prosperity*," Washington, DC: The White House, February 2011.

[55] See the discussion in Chapter 3, Finding 3b-iv. With regard to expanding the R&E tax credit, see Action D-2 in National Academy of Sciences, et al., *Rising Above the Gathering Storm*, op. cit., which calls for an increase in the credit from 20 to 40% of the qualifying increase in research expenditures by companies so that the U.S. R&E tax credit is competitive with that of other countries.

RECOMMENDATIONS   181

4. **Build a 21st Century Innovation Infrastructure:** *The U.S. should increase dramatically investment in state-of-the-art broadband networks and other infrastructure required to maintain American leadership in a 21st century global knowledge economy.*[56]

    a. **Build World Class Infrastructure:** America's physical infrastructure, including road and rail networks, and electric power and natural gas grids—essential for innovation and economic growth—are in many instances aging and in extreme cases falling apart.[57] This comes at a time when other nations are investing significant resources to build modern highways, ports and airports.[58] Using new technological innovations in materials and sensors, *inter alia*, the U.S. should make significant investments to restore and upgrade the nation's infrastructure.[59] To this end, the feasibility of a permanent, national infrastructure bank that could leverage private capital for projects of regional and national significance should be considered.[60]

    b. **Expand Broadband Penetration:** Internet access is an important tool for the development and dissemination of knowledge and is closely linked with economic growth.[61] The U.S. lags other advanced nations in broadband penetration, ranking 15th out of 30 countries in broadband penetration rates.[62] While recognizing the

---

[56] See Related Finding 3 in Chapter 3.

[57] The collapse in 2007 of the Route 35W bridge in Minneapolis aimed a spotlight on the nation's poor infrastructure. See MPR News, "Minneapolis Bridge Collapse," at http://minnesota.publicradio.org/collections/special/2007/bridge_collapse/. In a 2005 survey, the American Society of Civil Engineers issued an average grade of "D" to U.S. infrastructure. See The American Society of Civil Engineers. "Report Card for America's Infrastructure," 2005. Access at http://www.asce.org/reportcard/2005/page.cfm?id=203.

[58] See discussion in Chapter 2, section on "Improving Framework Conditions." Continuing its substantial investments over the past two decades, China plans to invest $1.03 trillion on urban infrastructure during its 12th Five-Year Plan from 2011 to 2015. *China Daily*, "China to invest 7t yuan for urban infrastructure in 2011-15," May 13, 2010.

[59] National Academy of Engineering, Grand Challenges for Engineering, Access at http://www.engineeringchallenges.org/cms/challenges.aspx.

[60] Felix Rohatyn, The Case for an Infrastructure Bank, *Wall Street Journal*, September 15, 2010. In the U.S. Senate, legislation, known as the "BUILD Act, was introduced on May 15, 2011 to fund an infrastructure bank. See also Charles Phillips, Laura Tyson, and Robert Wolf, "The U.S. Needs an Infrastructure Bank," *Wall Street Journal*, January 15, 2010.

[61] A 2006 study by the Economic Development Administration of the Department of Commerce found that "between 1998 and 2002, communities in which mass-market broadband was available by December 1999 experienced more rapid growth in employment, the number of businesses overall, and businesses in IT-intensive sectors, relative to comparable communities without broadband at that time." See EDA, Measuring Broadband's Economic Impact, National Technical Assistance, Training, Research, and Evaluation Project #99-07-13829, February, 2006.

[62] OECD Broadband Portal, Data updated as of June 23, 2011. Access at http://www.oecd.org/document/54/0,3343,en_2649_34225_38690102_1_1_1_1,00.html.

political challenges involved,[63] the United States cannot afford not to make the significant investments needed to upgrade and expand the nation's information infrastructure.[64]

In light of this, a recent National Research Council report found that "further data is required to understand the scope and nature of broadband use by businesses, and more study is required to understand why a significant percentage of households are not linked to the computer and Internet culture that is central to the new, more productive U.S. economy."[65]

c. **Secure *Cross-Border Data Flows*** Many countries want to control, restrict or limit the flow of data and information across international borders.[66] Some countries are trying to affect the international provision of ICT services by favoring domestic interests over international firms or by requiring that international firms provide computing or information services through domestic facilities. Other countries are concerned about access of their citizens to uncensored information or have concerns about privacy or national security.[67] The United States should seek to achieve agreed international norms and rules governing cross-border data flows and the related information technologies.

The following rules and commitments, as recently outlined by U.S. Internet and ICT companies, should form the basis of any international agreements: (1) prohibit restrictions on legitimate cross-border information flows; (2) prohibit local mandates for ICT infrastructure or investments; (3) promote international best practices and standards; (4) improve transparency and predictability concerning regulations of the Internet and the digital

---

[63] In the United States, three industry sectors with different transmission methods --telephony, cable, and satellite--are the primary providers of broadband connections to homes and business. These firms are regulated under dissimilar legal standards and each has opposed changes in rules that could differentially impact their sector. See Adam D. Thierer, "Solving the Broadband Paradox," *Issues in Science and Technology* Spring 2002.

[64] For a review of the FCC's 2010 National Broadband Plan, see Jeffrey Rosen, "Universal Service Fund Reform: Expanding Broadband Internet Access in the United States." *Issues In Technology Innovation*, Number 8, April 2011.

[65] National Research Council, *Enhancing Productivity Growth in the Information Age: Measuring and Sustaining the New Economy,* D. Jorgenson and C. Wessner, eds., Washington, DC: The National Academies Press, 2007.

[66] National Foreign Trade Council, "Promoting Cross-Border Data Flows: Priorities for the Business Community," November 3, 2011.

[67] *Report of the Trilateral Committee on Transborder Data Flows: North American Leaders Summit*, January 2010 at http://web.ita.doc.gov/ITI/itiHome.nsf/0657865ce57c168185256cdb007a1f3a/c444e0c6174952b585 2575d1007eaec2/$FILE/Report%20of%20the%20Trilateral%20Committee.pdf.

economy; (5) ensure that legitimate policy measures (such as those concerning privacy or national security) are not used to restrict legitimate data flows; and (6) ensure that trade agreements increase market access for ICT products and services and cover future technologies and services.[68]

d. **Encourage Energy Conservation**. The United States should pursue substantial improvements energy conservation and efficiency through the accelerated deployment of existing and emerging energy-supply and end-use technologies.[69] U.S. buildings, which alone use more energy than any other entire economy of the world except China, should be an important focus of conservation efforts.[70] U.S. buildings are generally grossly inefficient; it has been widely documented that energy use in new and existing buildings can be cut by 50% or more cost-effectively.[71] Lowering the cost base for location of production in the United States can be fostered by improving conservation, and the techniques learned are themselves marketable globally as innovative services.[72]

  i. **Smart Grid:** It is estimated that making building more intelligent through use of smart grid and active building energy management systems can cut energy use by 20-40%. This would allow buildings to participate in utility pricing schemes that provide large financial incentives to cut peak power use and to enable utilities to reduce energy use demand in response to utility power shortages.[73]

  ii. **Innovative Financing:** Recent studies suggest that potential gains in energy efficiency are significant, opening resources to rapidly expand investments and jobs while strengthening security and effectively addressing important sources of climate change.[74] By some estimates, U.S. businesses and

---

[68] National Foreign Trade Council, *Id.*
[69] See the related Findings in the report of the National Academy of Sciences, et al., *America's Energy Future, Technology and Transformation*, Washington, DC: The National Academies Press, 2009.
[70] U.S. Green Building Council, "Buildings and Climate Change," Accessed on November 3, 2011 at http://www.documents.dgs.ca.gov/dgs/pio/facts/LA%20workshop/climate.pdf.
[71] Greg Kats, *Greening Our Built World, Costs, Benefits, and Strategies*, Washington, DC: Island Press, 2010.
[72] See discussion in Chapter 2, section on "Improving Framework Conditions."
[73] Ibid.
[74] See National Academy of Sciences, *Real Prospects for Energy Efficiency in the United States*, Washington, DC: The National Academies Press, 2010. The NAS panel found that "taking advantage of technologies that save money as well as energy to produce the same mix of goods and services could reduce U.S. energy use to 30 percent below the 2030 forecast level, and even significantly below 2008 energy use. The result would be lower costs and a more competitive

consumers could save as much as two trillion dollars within a decade through more efficient use of energy.[75] Despite these large prospective gains, financing energy efficiency remains a challenge, especially as ARRA funding surge drops off steeply in early 2012. To overcome this investment gap and to capture the very significant savings from improved energy efficiency, the U.S. should substantially raise the level of financing for energy efficiency from its current level of about $20 billion a year. Such investment has the potential to save many billions annually.[76]

5. ***The United States needs to adopt specific policy measures to capture greater economic value from its public investments in research.*[77]** *The America COMPETES Act calls for greater investment in innovation through research and development and Congress should appropriate the funding authorized in the legislation.*[78] *A similarly comprehensive effort needs to be made to exploit the results of the nation's investments in science, technology, and education into more innovative products, business, industries, and well-paying jobs.*

   a. **Strengthen University Links to the Market:** Universities should be encouraged to provide stronger incentives to promote applied research, ease cooperation with industry, and encourage commercialization *in the United States* of research results. Many U.S. universities have already recognized the need for programs that can help to transition good research ideas toward the commercial marketplace, a trend that can be encouraged.[79]

---

economy that uses less fossil fuel, has lower emissions of greenhouse gases, and puts less pressure on environmental quality." Also, see the discussion in Chapter 1, section on "Responding to the Innovation Challenge," and Chapter 2, section on "Strengthening Manufacturing."

[75] See Greg Kats, Aaron Menkin, Jeremy Dommu and Matthew DeBold, "Energy Efficiency Financing - Models And Strategies," Capital E For The Energy Foundation, March 2012.

[76] Ibid.

[77] See related Finding 5 in Chapter 3.

[78] America COMPETES Act was signed by President Bush and became law on 9 August 2007. On January 4, 2011, President Obama signed the America COMPETES Reauthorization Act of 2010. However, as an 'authorization bill', "COMPETES will have teeth only to the extent that the funding levels it lays out are appropriated in practice over the next three years." Eugene Reich, "US Congress passes strategic science bill," *Nature*, December 22, 2010. As this report went into review, the Department of Commerce, in consultation with the National Economic Council released the COMPETES Report, which highlights bipartisan priorities to sustain and promote the pillars of American innovation and economic competitiveness. See Department of Commerce, "The Competitive and Innovative Capacity of the United States," January 2012. Access at http://www.commerce.gov/americacompetes.

[79] See Robert E. Litan and Lesa Mitchell, "A Faster Path from Lab to Market," in *Harvard Business Review*, January/February 2010. See also Krisztina "Z" Holly, "The Full Potential of University Research, A Model for Cultivating New Technologies and Innovation Ecosystems," *Science*

i. **Provide Matching Seed Funds:** Some universities have experimented with seed funds that provide small amounts of capital, e.g., $5,000-10,000, typically to generate a proof of concept based on University research and begin to explore commercial applicability. Where successful, such funds can help to initiate the process of commercialization to the point where private investment can take hold. The federal (and/or state) government should consider providing a small pool of matching funding to help initiate new University seed programs and to defray the costs of starting-up incubation, where those efforts are modeled on successful prior examples. Universities can and should consider emulating those examples that have succeeded. Centers considered to be a success, such as the Deshphande Center at MIT, are managed by professionals, independent of the university.[80]

ii. **Develop University Incubators:** University incubators, which seek to support entrepreneurs as they move their ideas from the laboratory to the marketplace, are a growing phenomenon.[81] In many cases, services provided by university business incubators, including laboratories, equipment, and student employees help nurture new technology based firms.[82] Some universities have experimented with incubators that provide shared office and infrastructure (e.g., wet labs), and in some case shared services (e.g., legal, bookkeeping, etc.). Such incubators can help significantly to defray the cost of a nascent start-up that seeks

---

Progress, June 2010. The author outlines a policy proposal for how the federal government can catalyze economic growth and societal benefit with ideas spawned at major research universities.

[80] For a review of how proof of concept centers can facilitate the transfer of university innovations into commercial applications, see Christine A. Gulbranson and David B. Audretsch, "Proof of concept centers: accelerating the commercialization of university innovation," *Journal of Technology Transfer* (2008) 33:249–258. See also the recommendations of the Hawaii Innovation Council to establish a seed fund to encourage the commercialization of university research. Access this report at http://www.hawaii.edu/offices/op/innovation/council-final-recommendations.pdf.

[81] See for example, Georgia Tech's Venture Lab, which seeks to "move university technologies out of the lab and into the marketplace" and "grow university-based start-up companies in Georgia to create a vibrant industrial base with high-value jobs." For a review of the effectiveness of university incubators, see Frank T. Rothaermela and Marie Thursby, "University–incubator firm knowledge flows: assessing their impact on incubator firm performance." *Research Policy,* Volume 34, Issue 3, April 2005, Pages 305-320.

[82] Mian A. Sarfraz, (2011) "University's involvement in technology business incubation: what theory and practice tell us?" *International Journal of Entrepreneurship and Innovation Management*, Vol. 13, No. 2, pages 113-121.

to commercialize university research. Not all incubators prove successful. Additional research on best practices is required.[83]

iii. **Expand SBIR to Support Commercialization of University Research:** Universities should be encouraged to draw on the potential of competitive innovation award programs like SBIR and STTR.[84] Recent research by the National Academies shows that the SBIR program is "sound in concept and effective in operation." Its awards, particularly through the STTR program, act as a valuable link between universities and the market.[85] In addition, some federal agencies provide commercialization training to SBIR award winners.[86] Systematic efforts to heighten student and faculty awareness would enhance returns on the program while contributing to a culture of innovation.

iv. **Develop Additional Centers of Excellence:** Technological Centers of Excellence can be an effective means to develop specific commercial applications of university research through dedicated government-university-industry cooperation. Positive examples of Centers of Excellence include the Semiconductor industry' four Focus Centers; the National Cancer Institute's designated Cancer Centers; the four Nano-electronics research centers; and the Army funded Federal Display Center at Arizona State University. These centers need to be assured of sustained and sufficient funding,

---

[83] For a recent survey of incubator managers regarding best practices, see David A. Lewis, Elsie Harper-Anderson, and Lawrence A. Molnar, "Incubating Success; Incubation Practices that lead to Successful Ventures," Washington, DC: Economic Development Administration, 2011. Access at http://www.edaincubatortool.org/pdf/Master%20Report_FINALDownloadPDF.pdf.

[84] In testimony before the House Science and Technology Committee on April 23, 2009, Robert M. Berdahl, President of the Association of American Universities noted that "Indeed, SBIR and STTR programs are now widely viewed by many faculty and research administrators as an important tool that can help them transform the research generated in our university laboratories into new industrial products, good, and services. As a result, more and more of our faculty are directly engaged in research funded through these two programs."

[85] Over a third of the respondents in an NRC Survey of four thousand SBIR firms reported some form of university involvement in their SBIR project. Among these, more than 80 percent of respondent companies receiving grants from the National Institutes of Health had at least one founder from academia. The NRC survey also revealed that about one-third of the founders of SBIR firms responding to the survey were most recently employed as academics before founding the company. In addition, about a third of projects surveyed had university faculty as contractors on the project and about a quarter of projects surveyed used universities themselves as subcontractors. See National Research Council, *An Assessment of the SBIR Program*, 2008, op. cit.

[86] These SBIR-related initiatives include, for example, the NIH's Commercialization Assistance Program, and the U.S. Army's Commercialization Pilot Program. These programs have not been subjected to a careful evaluation to date. Also noteworthy is NSF's I-Corps program, which provides a business boot camp and a mentoring service for would-be entrepreneurs. This appears to be a promising initiative but new and small-scale and has not yet been evaluated.

particularly the resources to maintain up-to-date equipment and facilities.[87]

v. **Use of Innovation Prizes:** According to a recent report of the National Academies, inducement prize contests are "thought to have in many circumstances the virtue of focusing multiple group and individual efforts and resources on a scientifically or socially worthwhile goal without specifying how the goal is to be accomplished and by paying a fixed purse only to the contestant with the best or first solution. Inducement prize contests with low administrative barriers to entry can attract a diverse range of talent and stimulate interest in the enterprise well beyond the participant pool."[88] Accordingly, the report recommended that "an ambitious program of innovation inducement prize contests will be a sound investment in strengthening the infrastructure for U.S. innovation. Experimental in its early stages, the program should be carried out in close association with the academic community, scientific and technical societies, industry organizations, venture capitalists, and others."[89]

vi. **Encourage private foundations to take equity positions in start-ups by amending SEC rules:** Investments in innovative startups that balance the focus and discipline of venture investment with the philanthropic missions of private foundations can be an effective way of linking university research to commercial applications for the market. Recent initiatives, such as those by the Gates Foundation, could provide a template for a new approach to financing innovative startups.[90]

---

[87] In this regard, a committee of the National Academies has previously recommended that the Microelectronics Advanced Research Corporation's Focus Centers, in which government and industry jointly support university researchers, be fully funded and, ideally, expanded. See Recommendation b on page 89 of National Research Council, *Securing the Future, Regional and National Programs to Support the Semiconductor Industry*, op. cit.
[88] See National Research Council, *Innovation Inducement Prices at the National Science Foundation*, Washington, DC: The National Academies Press, 2007.
[89] Ibid. Page 2.
[90] See Luke Timmerman, "Gates Foundation Makes First Equity Investment in Biotech Startup, Liquidia Technologies," xconomy.com, March 8, 2011. The Gates Foundation made its first direct equity investment of $10 million in Liquidia Technologies, a for-profit company in March 2011. The goal was to encourage the development and commercialization of technologies that can have a positive impact on global health, a core mission of the Gates Foundation.

b. **Strengthen National Laboratories' Links to the Market**

   i. **Expand use of Research Parks**: The National Laboratories should continue to develop and grow research parks where feasible. The Sandia Science and Technology Park Initiative demonstrates that appropriately structured public-private partnerships can reinforce the mission of the National Laboratories as well as contribute to regional economic development.[91]

   ii. **Expand SBIR to National Laboratories**: The National Laboratories should be encouraged to participate in the SBIR program.[92] Although an important source of technical reviewers for SBIR proposals, the National Laboratories are currently not strongly involved in the SBIR program. As a result, the potentially significant role that they could play as partners with SBIR-award-recipient firms is not being fully realized.[93]

c. **Develop Public-Private Partnerships:** To better capture the economic value of U.S. investments in research, U.S. policy needs to focus on expanding successful U.S. partnership programs including innovation awards, prizes, and research consortia and, where appropriate, consider adopting and adapting successful models from other countries.

   i. **New Initiatives in Early Stage Finance:** New initiatives that are modeled on U.S. programs (such as SBIR) with a proven record of turning new ideas into marketable products might be

---

[91] For a review of the mission and accomplishments of Sandia Park, see Richard Stulen, "U.S. and Global Best Practices, Sandia S&T Park," in National Research, *Understanding Research, Science, and Technology Parks: Global Best Practices*, op. cit. For an early review of Sandia Park, see National Research Council, *A Review of the Sandia Science and Technology Park Initiative*, C. Wessner, ed., Washington, DC: The National Academies Press, 1999. The Park housed 33 companies as of 2011 and (according to an economic impact analysis carried out by SSTP in 2009) directly and indirectly helped create 7725 jobs, providing a significant boost to the region's dynamism. See http://www.sstp.org/about-sstp/economic-impact.

[92] See National Research Council, *An Assessment of the SBIR Program at the Department of Energy*, C. Wessner, ed., Washington, DC: The National Academies Press, 2008. Recommendation C-3 notes that "while the National Laboratories are an important source of technical reviewers for proposals (which is an important component of the administration of the SBIR program at DoE), the Laboratories themselves are not otherwise strongly involved in the SBIR program." Fuller participation includes generation of topics and commercialization and/or adoption of SBIR technologies.

[93] National Research Council, *An Assessment of the SBIR Program at the Department of Energy*, C. Wessner, ed., Washington, DC: The National Academies Press (2008) pages 28-29, 38.

established with a sectoral focus and, perhaps, larger award amounts.

ii. **Support for Industry Consortia:** The federal government should consider assisting new public-private R&D collaborations for sectors, where feasible, such as flexible electronics, , additive manufacturing, photovoltaics, and medical devices modeled after the principles underlying the successful Sematech consortium of the 1980s, and/or well-funded models from abroad, e.g., imec.[94] These consortia should seek to establish industry standards, support pre-competitive R&D cooperation, and provide mechanisms to pool public and private resources in order to accelerate wide-scale commercialization of transformative technologies in multiple industrial sectors.[95] In this regard, recently created cooperative initiatives on manufacturing, nanotechnologies, and the Sunshot initiative for photovoltaics should be sustained and in the case of manufacturing, these efforts should be supported with significantly enhanced resources commensurate with the potential of the sector.

d. **Expand Support for Manufacturing:** Manufacturing is a key source of employment, growth, and a major contributor of R&D. However, individual companies often "cannot justify the investment required to fully develop many important new technologies or to create the full infrastructure to support advanced manufacturing."[96] This comes at a time when many nations are

---

[94] One such example is imec, established by the Flemish government in 1982. imec conducts advanced research in nano-electronics in partnership with leading global companies in information technology, health care, and energy. For a discussion of its operations, see Anton de Proft, "Introduction to imec" in National Research Council, *Innovative Flanders*, C. Wessner, ed., Washington, DC: The National Academies Press, 2008. For a review of Sematech, see Kenneth Flamm and Qifei Wang, "Sematech Revisited: Assessing Consortium Impacts on Semiconductor Industry R&D," in National Research Council, *Securing the Future, Regional and National Programs to Support the Semiconductor Industry*, op. cit.

[95] A National Academies Symposium, "Flexible Electronics for Security, Manufacturing, and Growth in the United States," held on September 25, 2010, discussed possible models for a consortium for flexible electronics. For a review of the potential of a consortium to advance solid state lighting, see , National Research Council, *Partnerships for Sold State Lighting, Report of a Workshop*, C. Wessner, ed., Washington, DC: The National Academies Press, 2002.

[96] The President's Council of Advisors on Science and Technology, "Report to the President on Ensuring American Leadership in Advanced Manufacturing," Washington, DC: The White House, June 2011. The recommendations of this report build on an earlier report submitted to President George W. Bush. See President's Council of Advisors on Science and Technology, "Sustaining the Nation's Innovation Ecosystems, Information Technology Manufacturing and Competitiveness," Washington, DC: The White House, February 2004. See also Stephen Ezell and Robert D. Atkinson, "The Case for a National Manufacturing Strategy," Washington, DC: ITIF, April 26, 2011. Finally, see Gregory Tassey, "Rationales and mechanisms for revitalizing US manufacturing R&D

devoting substantial policy attention to attracting, nurturing, and growing national manufacturing capability, including industry supply chains and required infrastructure.[97]

i. **Provide Incentives for Manufacturing**: The U.S. needs to support the competitiveness of its advanced manufacturing sector by making federal incentive programs permanent and broadening the time horizons of tools such as manufacturing tax credits and loan guarantees so that companies can confidently invest for the long term.[98] Regulatory and tax incentives should be considered and expanded where appropriate to drive downstream demand.[99]

ii. **Expand Manufacturing Support Programs**: Consideration should also be given to the creation or expansion of programs that directly support manufacturing.[100] These programs, in

---

strategies." *The Journal of Technology Transfer*, Volume 35, Number 3, 283-333 (2010). Tassey argues that an advanced economy such as the United States needs a strong manufacturing sector and calls for Increasing "the average R&D intensity of the domestic manufacturing sector to 6 percent"; adjusting "the composition of national R&D to emphasis more long-term, breakthrough research and increasing the amount sufficient to fund a diversified portfolio of emerging technologies commensurate with the size of the U.S. economy"; and improving the efficiency of R&D performance and subsequent technology diffusion."

[97] See the discussion in Chapter 2, section on "Strengthening Manufacturing."

[98] The Department of Energy's loan guarantee programs (through October 2011) committed $35.9 billion in guarantees to 38 projects including renewable and nuclear power generation, renewable energy manufacturing, energy efficiency manufacturing, energy storage, and electric vehicle production.98 This program has now been terminated.

[99] Bronwyn Hall and Beethika Khan note that "the regulatory environment and governmental institutions more generally can have a powerful effect on technology adoption, often via the ability of a government to "sponsor" a technology with network effects." See Bronwyn H. Hall and Beethika Khan, "Adoption of New Technology," NBER Working Paper 9730, 2003. For an illustrative review of the impact of regulatory provisions for energy efficiency on new federal, state, and local policies, programs, and practices across the U.S., see Robert K. Dixon, Elizabeth McGowan, Ganna Onysko, and Richard M. Scheerb, "US energy conservation and efficiency policies: Challenges and opportunities," *Energy Policy* Volume 38, Issue 11, November 2010, Pages 6398-6408.

[100] The Hollings Manufacturing Extension Partnership (MEP) was created in 1988 to offer technical, business, and financial support primarily to small and medium-sized manufacturers in all fifty states. The National Research Council currently has an evaluation of the MEP underway. The assessment will document the achievements and challenges of the program. The Advanced Manufacturing Partnership was developed on the recommendation of the President's Council of Advisors on Science and Technology, which called for a partnership between government, industry, and academia to identify the most pressing challenges and transformative opportunities to improve the technologies, processes and products across multiple manufacturing industries. See President's Council of Advisors on Science and Technology, "Report to the President on Ensuring American Leadership In Advanced Manufacturing." Washington, DC: The White House, June 2011. For a comparative perspective on innovation and manufacturing policy, see Philip Shapira, Building capabilities for innovation in SMEs: a cross-country comparison of technology extension policies and programmes," *International Journal of Innovation and Regional Development*, Volume 3, Number 3-4/2010.

conjunction with programs at the Department of Defense, offer a substantial framework for encouraging and sustaining U.S.-based manufacturing and developing promising new technologies, such as flexible electronics and additive manufacturing.[101]

e. **Sustain Federal Programs to Jump-start new Industries:** Promising federal programs aimed at accelerating commercialization of new technologies, including such areas as photovoltaic cells, advanced batteries, and biomedicine, should be continued with sustainable long-term funding.[102]

f. **Create New Institutions for Applied Research:** The recently announced National Network for Manufacturing Innovation (NNMI)—a private-public partnership program aimed at commercializing and manufacturing U.S. developed technologies—should be fully funded.[103] NNMI calls for precompetitive consortia to conduct applied research on new technologies and design methodologies. Modeled on Germany's Fraunhofer-Gesellschaft, NNMI can help improve the

---

[101] "The Department of Defense Manufacturing Technology (ManTech) Program is a joint program of the armed services and the Defense Logistics Agency. The purpose of the ManTech program is to develop manufacturing technologies for the affordable, low-risk development and production of weapons systems." See National Research Council, Defense Manufacturing in 2010 and Beyond, Washington, DC: The National Academies Press, 1999. See also the discussion in Chapter 2, sections on "Institutional Support for Applied Research," and "Strengthening Manufacturing."

[102] These include the NSF's Engineering Research Centers and Industry-University Cooperative Research program. See also Chapter 5 of this report, which provides detailed case studies of the role of federal programs in the development and commercialization of semiconductor, photovoltaic and advanced battery technologies. For example, the $457 million Sunshot Initiative of the Department of Energy "is a collaborative national initiative to make solar energy cost competitive with other forms of energy by the end of the decade. Reducing the installed cost of solar energy systems by about 75% will drive widespread, large-scale adoption of this renewable energy technology and restore U.S. leadership in the global clean energy race." Department of Energy website: http://www1.eere.energy.gov/solar/sunshot/.

[103] "The proposed Network will be composed of up to fifteen Institutes for Manufacturing Innovation (IMIs or Institutes) around the country, each serving as a hub of manufacturing excellence that will help to make United States (U.S.) manufacturing facilities and enterprises more competitive and encourage investment in the U.S. This program was proposed in the President's fiscal year (FY) 2013 budget and was announced by the President on March 9, 2012. The NNMI program will be managed collaboratively by the Department of Defense, Department of Energy, Department of Commerce's NIST, the National Science Foundation, and other agencies. Industry, state, academic and other organizations will co-invest in the Institutes along with the NNMI program." Federal Register Notice, May 4, 2012. The President's FY 2013 budget requests $1 billion for the NNMI program. See also the discussion in Chapter 2, section on "Institutional Support for Applied Research."

competitiveness of U.S. manufacturers and capture more value from U.S. funded research.[104]

g. **Open Foreign Markets To Business Services.** Services are a large and important component of the U.S. economy and the United States has much to offer the world in high value services. To capitalize on this comparative advantage, the U.S. should be pushing aggressively for services trade liberalization, making common cause with the European Union and other advanced economies to encourage the large, fast-growing developing economies to liberalize their service sectors through multilateral negotiations in the General Agreement on Trade in Services and the Government Procurement Agreement.[105] The infrastructure building boom, particularly in Asia, provides an enormous opportunity for U.S. service firms if the proper policies are in place. Increased trade in services might help rebalance U.S. trade, and both advanced and emerging economies would benefit from the productivity-enhancing gains brought by increased trade in services.[106]

h. **Expand Support for U.S. Manufactured Exports:** The government and private sector should work together to identify and seize new market opportunities and reduce barriers to U.S. exports for products of industries that use advanced manufacturing including electronics, aerospace, and biotechnology.[107] In this regard, resources for export financing should be re-examined to be sure that export financing is fully competitive with foreign export credits.[108] In addition the Department of Commerce should substantially expand the U.S. and Foreign Commercial Service in

---

[104] See Chapter 4 of this report for a description of the Fraunhofer Gesellschaft. See also the presentation by Roland Schindler, Executive Director of Fraunhofer CSE, at the National Academies Symposium on Meeting Global Challenges: U.S.-German Innovation Policy, Washington, DC, November 1, 2010. Germany's Fraunhofer system has established seven research institutes based at U.S. universities, including Michigan State University, Boston University, Massachusetts Institute of Technology, the University of Maryland, the University of Michigan, Johns Hopkins University, and the University of Delaware. These institutes provide research and development services to help translate the fruits of research at U.S. academic institutions into products for the marketplace.
[105] J. Bradford Jensen. "Global Trade in Services: Fear, Facts, and Offshoring" Peterson Institute for International Economics, 2011.
[106] Ibid.
[107] See the discussion in Chapter 2, section on "Strengthening Manufacturing."
[108] Export-Import Bank of the United States, Report to the US Congress on Export Credit Competition and the Export-Import Bank of the United States, June 2010. The U.S. Chamber of Commerce is partnering with the Export-Import Bank of the United States on its Global Access for Small Business initiative to help more than 5,000 small companies export goods and services produced by U.S. workers. For a concise review of the role and performance of the Export-Import Bank in promoting U.S. exports, see Stephen Ezell, "Understanding the Importance of Export Credit Financing to U.S. Competitiveness." Washington, DC: ITIF, June 2011.

order to capitalize on the rapid growth of markets in China, India, and other emerging economies.[109]

i. **Foster Cluster Development:** Recent pilot programs by U.S. federal agencies to align current economic development programs with specific regional innovation cluster initiatives by state and local organizations should be assessed and, where appropriate, provided with greater funding and expanded geographically.[110] Efforts should be made to attract, on a competitive basis, more states and regions, while providing best practice guidance and incentives, preferably with reliance on matching funds.

   i. **Assess Foreign Clusters:** The scope and scale of efforts by foreign governments to develop clusters in new technology areas are impressive. The U.S. needs to assess and draw policy lessons from these efforts that are helping to shape the global competitive landscape. A similar effort should be undertaken for science and technology parks.

   ii. **Support the Development of Science and Research Parks:** In a similar vein, the U.S. should provide competitively awarded federal support to state and regional efforts to develop and sustain modern research parks. These parks can provide valuable means of supporting the missions of national laboratories such as those of the Department of Energy and NASA, national research institutions such as the National Cancer Institute, and university facilities.[111]

j. **Leverage Government Procurement to Establish Early Markets:** As it has done previously in industries such as semiconductors, computers, advanced aircraft, and nuclear power,

---

[109] In remarks at the U.S. Chamber of Commerce, U.S. Commerce Secretary John Bryson has called for restructuring the Foreign Commercial Service to intensify focus on markets where U.S. exports have the best potential for continued growth, including China, Brazil, India, Saudi Arabia and Turkey. See Department of Commerce Press Release, "Commerce Secretary John Bryson Lays Out Vision for Department of Commerce." December 15, 2011.

[110] For a review of recent federal and state efforts, see National Research Council, *Growing Innovation Clusters for American Prosperity*, C. Wessner, rapporteur, Washington, DC: The National Academies Press, 2011.

[111] For a review of the mission and accomplishments of the NASA Ames Research Center, see Simon (Pete) Worden, "NASA Research Park," in National Research, *Understanding Research, Science, and Technology Parks: Global Best Practices*, op. cit. For an early review of the role of the NASA Ames Park, see National Research Council, *A Review of the New Initiatives at the NASA Ames Research Center*, C. Wessner, ed., Washington, DC: The National Academies Press, 2001. For a description of the National Cancer Institute's plan to bring together much of its technology research and development in a park-like setting in Frederick, Maryland, see John Niederhuber, "The National Cancer Institute and NCI-Frederick," in, *Understanding Research, Science, and Technology Parks. Op. cit.*

federal agencies can play a key role in helping build domestic markets for important emerging industries such as electric-drive vehicles, solar power, and solid-state lighting through incentive programs and government procurement.[112] Overall, procurement programs can enable domestic producers to move up the learning curve, push down the cost curve, and enable them to compete successfully in the U.S. and global markets.[113]

i. **Leverage Defense Procurement:** Defense procurement is an important driver of innovation, providing initial markets for products for the military as well as civilian sectors.[114] Traditional defense procurement, however, operates within established and complex sets of regulatory and managerial practices. To shift from a culture of compliance to a culture of results based on performance, cost, and schedule, the U.S. should establish "incentives and rewards for innovation in products and processes that result in continuous performance improvements, at lower and lower costs."[115]

- **Develop a skilled acquisitions workforce:** The government should build an expanded workforce of experienced, smart buyers for the military, including experienced people from industry.[116]
- **Incentivize better performance:** The U.S. should improve its law and regulations concerning procurement practices and export and import controls in order to

---

[112] For a historical perspective of the impact of federal procurement on innovation in the U.S., see Vernon W. Ruttan, 2006, op. cit. For a review of the academic literature, see Jakob Edler and Luke Gerghiou, (2007). "Public procurement and innovation –Resurrecting the demand side." *Research Policy*. 36, 9, 949-963.

[113] Note that federal procurement is generally open to foreign suppliers that are signatories of the World Trade Organization's Government Procurement Agreement.

[114] David Mowery identifies three channels through which public investments in defense-related R&D and procurement affect the innovative performance of sectors or the overall economy. First, defense-related R&D investments can support the creation of new knowledge with defense-related and civilian technology applications. Second, in some cases, defense-related R&D investment can lead to "spinoffs," with civilian and applications. And third, by serving as a "lead purchaser," for early versions of new technologies, defense procurement can enable supplier firms to reduce the costs of their products and improve their reliability and functionality. See David C. Mowery, "National security and national innovation systems," *Journal of Technology Transfer* (2009) 34:455–473. For a contemporary review of the role that Defense procurement can play in advancing new energy technologies, see Ryan Fitzpatrick, Josh Freed, and Mieke Eoyang, "Fighting for Innovation: How DoD Can Advance Clean Energy Technology... And Why It Has To." Washington, DC: The Third Way, June 2011.

[115] Jacques Gansler, "Solving the Nation's Security Affordability Problem," in *Issues in Science and Technology*, Volume XXVII, Number 4, Summer 2011.

[116] Ibid.

reward companies that achieve higher performance at lower costs.

  ii. **Encourage Procurement from Small Business**: Procurement from small firms, such as through the SBIR program, can diversify the supplier base and accelerate innovation through support for early stage funding while addressing the myriad mission needs of government agencies.[117]

6. *Recognize that Trade and Innovation are Closely Linked:*[118]

    a. **Provide a Rules-based Playing Field:** It is the responsibility of the U.S. government to take an activist approach to enforce agreements and provide a rules-based playing field for its industries engaged in competition with foreign industries. Measures that distort foreign trade and investment should be rooted out or offset, especially when these measures risk having a serious adverse effect on U.S. firms continued ability to innovate[119]. This will require support from U.S. industry, but ultimately an independent and well-informed judgment on the part of the U.S. government of policy responses that are in the national interest.

    b. **Develop an Enforceable International Code of Conduct:** Existing international rules have proved to be ineffective in governing the activities of government enterprises engaged in commercial competition. An enforceable international code of conduct is required.

        i. Home governments need to be accountable for their support of their government enterprises as well as the conduct of these enterprises where trade and investment patterns are distorted.

---

[117] For a review of the opportunities as well as challenges small innovative businesses face with respect to federal procurement, see National Research Council, *An Assessment of the SBIR Program.* op. cit., pages 46-49. In recent years, the European Union has sought to expand the use of public procurement to foster innovation, particularly among small and medium sized enterprises. See the Speech of EU Commissioner Geoghegan Quinn to a meeting of IMCO on the role of public procurement policies in supporting EU innovation strategies, 1 February 2011. See also M. Rolfstam W., Phillips, and E. Bakker (2011) "Public procurement and the diffusion of innovations: exploring the role of institutions and institutional coordination." *International Journal of Public Sector Management*, 24 (5).

[118] See related Finding 5 in Chapter 3.

[119] It is far from clear that protectionist measures actually promote innovation in the countries that adopt them.

ii. Going forward, every U.S. government international trade or investment agreement should include a comprehensive code of conduct that includes detailed rules governing government owned, government invested and government supported enterprises (e.g. ensuring enforcement of intellectual property and competition laws and policies), transparency, and dispute settlement.

7. *Capitalize on the globalization of research and innovation:* The United States should capitalize on the globalization of research and innovation to cooperate with other nations to advance innovations that address shared global challenges in energy, environment, health, and security.[120]

   a. **Strengthen International Cooperation:** The United States needs to better capitalize on the new knowledge that is being generated around the world. As one analyst has recently observed, "Gathered from afar and reintegrated locally, knowledge developed elsewhere can be tapped to stoke U.S. innovation."[121] However, international collaboration is supported by only about 6 percent of federal R&D spending, and the United States has no strategy to find and use knowledge from around the world in this regard.[122] The United States should strengthen and expand opportunities for research collaboration by American scientists and entrepreneurs with their counterparts in growing economies such as China, India, Brazil, as well as those in more established countries such as Japan and South Korea and historical partners such as Germany, France and Italy.[123] New initiatives with countries such as Poland, the Czech Republic, Slovakia, Hungary, and Romania should also be undertaken in order to pool resources and talent devoted to solving common challenges in areas such as health, energy, security, and the environment.[124] The need for cooperative efforts is great, and

---

[120] See related Finding 7 in Chapter 3.
[121] Caroline S. Wagner, "The Shifting Landscape of Science," *Issues in Science and Technology*, Volume XXVII, No. 1, Fall 2011.
[122] Ibid.
[123] For a review of opportunities and challenges for further cooperation with India, see National Research Council, *India's Changing Innovation System*, C. Wessner and S. Shivakumar, eds., Washington, DC: The National Academies Press, 2007. For a first hand review of current trends in China's innovation strategies and challenges, see National Research Council, *Building the 21st Century, U.S. China Cooperation in Science, Technology, and Innovation, op. cit.*
[124] To review opportunities for cooperation on innovation with Poland, the National Academies convened two major symposia on "Rebuilding the Transatlantic Bridge: U.S.-Polish Cooperation on Science, Technology, and Innovation" in 2009 and 2010. The meetings reviewed potential for

the potential benefits are substantial. At the same time, considerable care needs to be devoted to the terms and structure of such cooperation, notably to ensure that contributions are comparable and that the fruits of such cooperation are shared in an equitable and sustainable manner. [125]

b. **Expand Exchanges of Scholars and Students:** Expanding exchanges of scholars and students can benefit "both sending and receiving countries, providing access to leading research and training not available in the home country and creating transnational bridges to cutting-edge research."[126] In this regard, self-organized collaborative networks that are steered more by individual scientists linking together across borders for enhanced knowledge creation are found to more often lead to highly cited research articles.[127] This is because researchers are motivated to "compete with each other for collaborations with the most highly visible and productive scientists in their fields, in their own country or abroad. Facilitating this global collaboration could have a considerable impact on knowledge creation and has been promoted, for example, by the EU Framework requirements."[128]

Conclusion

Innovation—from invention through to commercialization—has a vital role to play in maintaining America's position of in the world economy and in addressing the major challenges facing the world today in areas such as energy, climate, health and economic development. There is no single measure, nor even a small number of policy measures that can assure success in preserving

---

cooperative activity in a variety of areas including developing clean coal energy technologies, environmental remediation, and cancer research.

[125] "The costs, complexity, and risk associated with the development of new technologies provide great opportunities for international cooperation in both the public and private sectors....These powerful drivers of cooperation are at the same time a source of greater system friction." See HWWA, IfW, and NRC, *Conflict and Cooperation in National Competition for High-technology Industry,"* Washington, DC: The National Academies Press, 1996, page 46. These challenges include identifying cooperative projects of equal interest to all parties, distributing the costs and benefits in an equitable manner, bridging social and cultural differences and divergent expectations, and ensuring long-term commitment to projects. See page 47. Soundly constructed and effectively managed International cooperation can bring value to all parties; managing expectations and assuring equitable arrangements is often a challenge for national bureaucracies. As the U.S. economic and technological leadership faces increasing competition, we have both greater opportunities for cooperation but greater care is required to ensure that it is mutually beneficial.
[126] National Research Council, *Policy Implications of International Graduate Students and Postdoctoral Scholars in the United States*, Washington, DC: The National Academies Press, 2005.
[127] Caroline S. Wagner. "Network structure, self-organization and the growth of international collaboration in Science." *Research Policy* Volume 34, Issue 10, December 2005, Pages 1608-1618.
[128] National Research Council, *Policy Implications of International Graduate Students and Postdoctoral Scholars in the United States*, Washington, DC: The National Academies Press, 2005.

and enhancing the magnificent record that this country has in innovation. Scientists in industry and government are now accustomed to developing what they call "roadmaps" to identify the challenges that need to be addressed and to bringing forward the next generations of innovative products. Looking abroad to assess what other countries are doing in facing common challenges, this report is designed to contribute to a better understanding of the strengths and weaknesses of the U.S. innovation system. Our purpose is to suggest a path forward. Most importantly, it is essential to understand that this series of cooperative interactions with other countries and efforts to benchmark American policies and measures should be continued in order to help U.S. policymakers improve U.S. performance.

PART II

GLOBAL INNOVATION POLICIES

Chapter 5

The New Global Competitive Environment

America's innovation system has long been the envy of the world. Now the rest of the world is racing to catch up. Virtually every important trading partner has declared innovation to be central to increasing productivity, economic growth, and living standards. They are implementing ambitious, far-sighted, and well-financed strategies to achieve that end. This chapter will describe how different nations studied by the STEP Board are addressing their innovation challenge.

Indeed, just as the global movement toward free markets in the 1990s became known as the Washington Consensus, the first decade of the 21st century has seen the emergence of what could be described as the Innovation Consensus. Governments everywhere have been sharply boosting investments in research and development, pushing universities and national laboratories to commercialize technology, building incubators and prototyping facilities for start-ups, amassing early-stage investment funds, and reforming tax codes and patent laws to encourage high-tech entrepreneurialism. What's more, these efforts are backed by intense policy focus at the highest level of governments in Asia, Europe, and Latin America.

Underlying this trend is an emerging understanding of what makes a nation globally competitive. Carl J. Dahlman of Georgetown University notes that economists traditionally have viewed competitiveness as a function of factors such as capital, the costs of labor and other inputs, and the general business climate. In a more dynamic world in which information technology and communications enable knowledge to be created and disseminated at ever-greater speeds, competitiveness increasingly is based on the ability to keep pace with rapid technological and organizational advances.[1]

---

[1] See presentation by Carl J. Dahlman of Georgetown University in National Research Council, *Innovation Strategies for the 21st Century: Report of a Symposium*, Charles W. Wessner, editor, Washington, DC: The National Academies Press, 2007.

The innovation agendas and precise policies differ from country to country, based on national needs and aspirations. In some cases, governments are implementing policies modeled after those of the United States. In others, they are borrowing from successful models pioneered in Europe and East Asia that leaders regard as more attuned to the competitive realities of the 21st century global economy. In that regard, other nations' experiences offer valuable lessons for policymakers in the U.S. federal government, regions, and states.

To better understand global trends in innovation policy, the National Academies' Board on Science, Technology, and Economic Policy (STEP) conducted an extensive dialogue over the past several years to compare and contrast policies of many nations. This section presents a number of case studies from those symposia and our research. While it is of course difficult to generalize, a number of common policy themes recurred through this extensive dialogue. They include:

- The paramount importance of investment in education to provide the skills base upon which an innovation-led economy is based.
- The value of increasing public and private investment in research and development, with at least 3 percent of GDP generally viewed as a desired target.
- The importance of establishing a far-thinking national innovation strategy that lays out broad science and technology priorities and a policy framework that addresses the entire ecosystem, including skilled talent, commercialization of research, entrepreneurship, and access to capital. Such national strategies require attention of top political leadership, coordination of government agencies, sustained funding, and collaboration with stakeholders at the regional and local level.
- An increasingly prominent role for public-private partnership in which industry, academia, and government pool resources to accelerate the translation of new technologies into the marketplace.
- A recognition that while universities' primary roles are education and research, they also can serve as powerful engines of economic growth if granted greater freedom to collaborate with industry and to commercialize inventions.
- Focus on programs to encourage firms to transform basic and applied research into new products and manufacturing processes.
- Greater policy emphasis on the institutional framework needed to sustain new business creation, such as intellectual property-right protection, competitive tax codes, and an efficient and transparent regulatory bureaucracy.

This chapter will describe how different nations studied by the STEP Board are addressing these and other issues. The chapter describes the innovation policy approaches of nations at three tiers of development.

In the first tier are the emerging economic powers. We looked at China and India in some depth. Both nations have charted ambitious innovation agendas for improving living standards and moving well beyond labor-intensive manufacturing and low-skill services to high-tech and knowledge-intensive industries. They are leveraging their large domestic markets and low-cost workforces to attract foreign investment in next-tier industries and are developing globally competitive corporations. They also are making strategic choices about technologies that address domestic needs and in which they are best positioned to compete globally in the future.

In the second tier are the more mature newly industrialized economies. We focus on Singapore and Taiwan, which have extraordinarily well-educated populations and have attained world standards in industries such as high-tech electronics, biotechnology research, and chemicals. They are striving to develop innovation ecosystems that will allow them to rank among the world's richest nations and compete head-to-head with the West and Japan in next-generation industries.

The third tier represents mature industrialized nations. We devote special attention to Germany because of that nation's ability to remain globally competitive in advanced manufacturing exports despite wages and other costs that are higher than in the United States. Our case studies also include Japan, Finland, Canada, and the Flanders region of Belgium. Each of these nations has revamped their national innovation strategies in order to increase R&D spending, collaboration between industry and academia, and new technology start-ups.

In most cases, it is too early to offer a full assessment of whether the strategies and policy tools selected by other nations will achieve their stated targets. What's more, not all of these policy options are appropriate for America. Yet they offer many valuable lessons for U.S. policymakers and present a picture of the changing global context as America prepares for 21st century competition.

EMERGING POWERS

China's Rapid Rise

After achieving decades of astonishing growth led by export manufacturing and heavy capital investment, China's leadership stresses that the nation's future as a global power rests on its ability to build an innovation-led economy.[2] China has pursued that goal with substantial investment and

---

[2] Government pronouncements on the importance of innovation began earlier. For example, then-President Jiang Zemin declared in the keynote address to the National Innovation Technology Conference on Aug. 23, 1999, that "the core of each country's competitive strength in intellectual

impressive focus. National spending on R&D has risen by an average of 19 percent a year since 1998,[3] and in under a decade has grown from less than one percent of GDP to 1.7 percent.[4] China's share of global R&D spending soared from 6 percent in 1999 to 12 percent in 2010.[5]

By virtually every conventional benchmark—successful patent applications, scientific publications, post-graduate degrees awarded, and global market share in high-tech goods--China's progress in science and technology has been solid. China has emerged as a major exporter of everything from solar cells to high-end telecommunications equipment and has accelerated the construction of high-speed trains. As *R&D Magazine* noted, China's financial commitments and record of generating intellectual property is such that it no longer can be regarded as an "emerging nation" in science and technology.[6] In 2010 alone, for example, China's international patent filings surged by 56.2 percent, to 12,337, compared to average growth worldwide of 4.8 percent.[7] The most visible manifestations of China's innovation push are its sprawling science parks. China's 54 major research parks average 10,000 acres, compared to around 350 acres in the U.S.[8]

China's achievements are a testament to the nation's ability over the past three decades to overhaul a dilapidated science and technology establishment, maintain policy focus at all levels of government, and mobilize immense public resources to invest in higher education, infrastructure, and R&D. That commitment continues to grow. China's long-term plans call for boosting gross R&D spending to 2.5 percent of GDP by 2020 and for science and technology to account for 60 percent of the economy.[9] The government has set an ambitious target of having 2 million patents of inventions, utility models, and designs by 2015.[10]

---

innovation, technological innovation, and high-tech industrialization." Current President Hu Jintao has stressed the importance of innovation in numerous speeches.

[3] UNESCO, *Institute for Statistics Database*, Table 25, Gross Expenditure on Research and Development in constant dollars. Growth rate from 1998 to 2008.

[4] Ministry of Science and Technology of the People's Republic of China, *China S&T Statistics Data Book 2010*, Figure 1-1.

[5] Battelle and R&D Magazine, *2012 Global R&D Funding Forecast*, December 2011

[6] Martin Grueber and Tim Studt, "Global Perspective: Emerging Nations Gain R&D Ground," *R&D Magazine*, Dec. 22, 2009.

[7] Xinhua News Service, "China 2010 International Patent Filings up 56.2%," *China Daily*, Feb. 2, 2011.

[8] Data from Research Triangle Foundation.

[9] State Council of China, *National Medium- and Long-Term Program for Science and Technology Development, 2006-2020*
(http://webcache.googleusercontent.com/search?q=cache:y800l0iQlS8J:www.cstec.org/uploads/files/National%2520Outline%2520for%2520Medium%2520and%2520Long%2520Term%2520S%26T%2520Development.doc+china+National+Medium-+and+Long-Term+Program+for+Science+and+Technology&cd=18&hl=en&ct=clnk&gl=us&client=firefox-a).

[10] State Intellectual Property Office, "National Patent Development Strategy (2011-2020)," (http://graphics8.nytimes.com/packages/pdf/business/SIPONatPatentDevStrategy.pdf).

China's heavy focus on absorbing foreign technology, rather than inventing it, also explains its industrial rise. The U.S. devotes 17.4 percent of its R&D spending to basic research, another 22.3 percent to applied research, and 60.3 percent to R&D development.[11] China invests 82.7 percent of national R&D spending to development of products and manufacturing process, while devoting just 4.7 percent to basic research and 12.6 percent to applied research.[12] [See Figure 5.1]

When it comes to creating truly innovative products, however, China still is regarded as an underachiever.[13] One hurdle is weak R&D spending by Chinese companies, especially state-owned enterprises.[14] Even though business enterprises in China accounted for 73 percent of R&D spending in 2009,[15] a World Bank study of nearly 300,000 Chinese enterprises big and small found that the vast majority did not conduct continuous R&D and described Chinese industry as "manufacturing without innovation."[16]

China's weak protection of intellectual property rights is a serious restraint on innovation, preventing companies from enjoying the full profits of their inventions and making foreign investors wary of conducting sensitive R&D in China.[17] Other often-cited weaknesses are shortages of the right kind of

---

[11] National Center for Science and Engineering Statistics, National Patterns of R&D Resources: 2008 Data Update, Detailed Statistical Tables, NSF 10-314 (March 2010), Tables 1-4.
[12] Ministry of Science and Technology of the People's Republic of China, *China S&T Statistics Data Book 2010*, Figure 1-3 at http://www.sts.org.cn/sjkl/kjtjdt/data2010/cstsm2010.htm.
[13] As a recent National Academy report concluded "China's S&T investment strategy is ambitious and well-financed but highly dependent on foreign inputs and investments. Many of its stated S&T and modernization goals will be unachievable without continued access to and exploitation of the global marketplace for several more decades. China plays a critical role in low- and select high-tech industry production and logistics chains, but it cannot (yet) replicate these processes domestically." National Academy of Sciences, Natural Research Council, *S&T Strategies of Six Countries: Implications for the United States*, Washington, DC: The National Academies Press, 2010, p.23.
[14] Gruber and Studt, ibid.
[15] *China S&T Statistics Data Book 2010*, ibid., Figure 1-2.
[16] Chunlin Zhang, Douglas Zhihua Zeng, William Peter Mako, and James Seward, *Promoting Enterprise-Led Innovation in China*, Washington, DC: The International Bank for Reconstruction and Development/The World Bank, 2009 (http://siteresources.worldbank.org/CHINAEXTN/Resources/318949-1242182077395/peic_full_report.pdf).
[17] For examples of U.S. industry complaints, see John Neuffer, "China: Intellectual Property Infringement, Indigenous Innovation Policies, and Frameworks for Measuring the Effects on the U.S. Economy," written testimony to the United States International Trade Commission Investigation No. 332-514 Hearing on behalf of the Information Technology Industry Council, June 15, 2010. (http://www.itic.org/clientuploads/ITI%20Testimony%20to%20USITC%20Hearing%20on%20China%20%28June%2015,%202010%29.pdf). See also Semiconductor Industry Association, *Maintaining America's Competitive Edge: Government Policies Affecting Semiconductor Industry R&D and Manufacturing Activity*, March 2009, p.31. "Most [semiconductor] companies surveyed indicated that they would not locate their most advanced and critical R&D activities in China, despite encouragement and even pressure by the government to do so, and regardless of the

Box 5.1
Constraints on Innovation in China

China's massive investments in technological infrastructure, science education, and research programs are key elements in laying the foundation for an innovation economy. But these investments in themselves do not mean that China will become a leading innovator in the near term. As China's Vice Minister of Science and Technology, Ma Songde commented in 2006, "most Chinese high-tech products are copies from other countries and that original inventions are rare on the mainland."[18]

In this regard, a recent report by the National Academies noted that "Although the growth in S&T funding is remarkable, there are still institutional issues that must be resolved. In particular, there is a general lack of openness and transparency in funding decisions, which negatively affects the ability of China to recruit first-rate scientists. Additionally, most R&D spending is geared toward development activities, rather than basic research. As a result, the quality and quantity of cutting-edge basic research is still small compared to that of the United States." [19]

The current World Bank report on China observes that notwithstanding China's growing supply of skills and advanced industrial base, most R&D is conducted by the government and state-owned enterprises in a manner that is divorced from the needs of the economy. China has seen a sharp increase in patents and published papers, but few have commercial relevance.[20] The report indicates that "China has relatively few high-impact scientific activities in any field," and that the "quantity [of patents] has not been matched by the quality of the patents."[21]

The centerpiece of China's innovation effort, the so-called 'indigenous innovation" initiative, emphasizes the exertion of commercial leverage against foreign firms to induce the transfer of technology that will be "absorbed, assimilated, and re-innovated" with Chinese intellectual property—arguably not

---

availability quality and size of incentives, due to concerns about the inadequacy of intellectual property protection in that country."

[18] Seminar remarks summarized in Open Source Center Report (July 24, 2006).
[19] National Research Council, *S&T Strategies of Six Countries, Implications for the United States*, Washington, DC: The National Academies Press, 2010, page 30. The report further notes that "although China's university system graduates hundreds of thousands of scientists and engineers each year, a critical shortage exists of highly qualified faculty, many of whom are attracted instead to opportunities in the private sector."
[20] World Bank, *China 2030*, Washington, DC: The World Bank, 2012.
[21] World Bank, Supporting Report 2: China Grows Through Technological Convergence and Innovation. Washington, DC: World Bank, 2012, pages 177-178.

> a program focused on fostering original discoveries.[22] Despite these limitations, developing major new innovations is not the only source of national strength. Programs that focus on acquiring new and established technologies can help develop the technological competitiveness of the Chinese economy and provide the opportunity for commercial success, first within China and next in export markets, thus laying the foundation for steadily higher levels of commercial application of advanced technologies.
>
> To address these challenges to its innovation system, the World Bank recommends that China concentrate on raising the technical and cognitive skills of its university graduates, building a few world-class research universities with links to industry, increasing the availability of patient risk capital for start-ups, and fostering clusters that bring together dynamic companies and universities and allow them to interact without restriction.[23]

human resources, weak linkages between government-funded research institutions and the private sector,[24] a science and technology establishment that prizes the quantity of journal publications and patents over quality and added-value, and over-dependence on government bureaucracy in investing R&D funds. A study by the Chinese Ministry of Science and Technology and the Organization for Economic Co-Operation faulted "deficiencies in the current policy instruments and governance promoting innovation." As a result, the study concluded, the government's heavy investments in R&D have "yet to translate into a proportionate increase in innovation performance."[25] As Deng Wenkui, director-general of the State Council Research Office put it: "Although China is a science and technology country with great skill, it is not a powerhouse." He added that "without reform and innovation, China cannot develop."[26]

---

[22] State Council, "Guidelines for the Medium and Long Term National Science and Technology Program (2006-2020) June 2006.
[23] World Bank, *China 2030*, op. cit.
[24] See Denis Fred Simon and Cong Cao, *China's Emerging Technological Edge: Addressing the Role of High-End Talent,* Cambridge: Cambridge University Press, 2009.
[25] *OECD Reviews of Innovation Policy,* op. cit. This lack of performance is reflected in the innovation component of the World Bank's Knowledge Economy Index (KEI), which ranks China 63rd in the world despite its large absolute spending on R&D. The innovation component of the World Bank's index is based on total royalty payments and receipts, patent applications granted by the U.S. PTO and scientific and technical journal articles. World Bank, *Knowledge Assessment Methodology* at http://go.worldbank.org/JGAO5XE940.
[26] Remarks by Deng Wenkui of the State Council Research Office at the Sept. 19, 2011 National Academies symposium "U.S.-China Policy for Science, Technology, and Innovation" in Washington, DC.

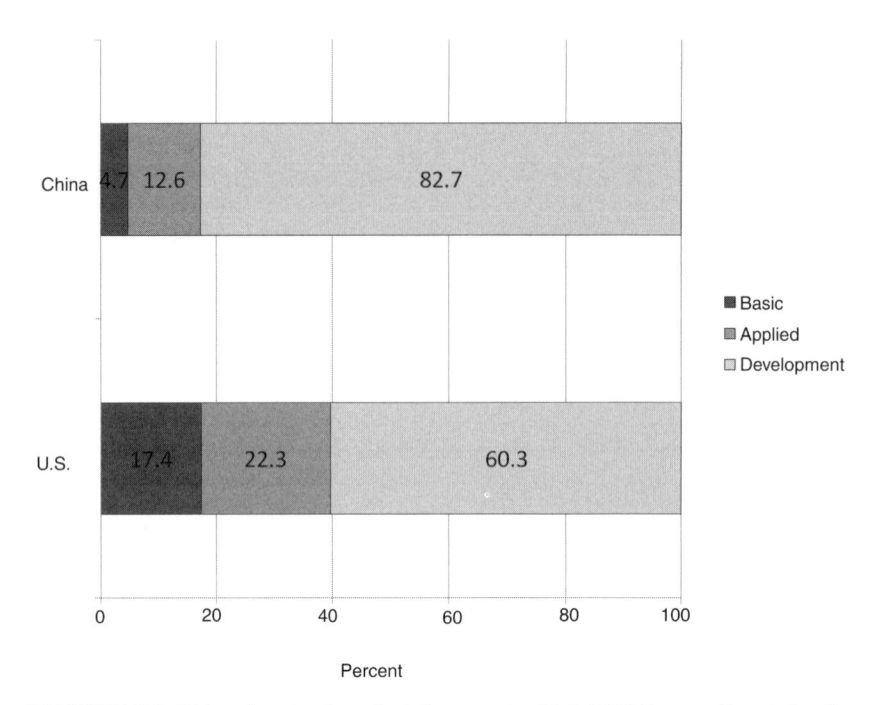

```
           Basic  Applied  Development
China      4.7    12.6     82.7
U.S.       17.4   22.3     60.3
```

Percent

FIGURE 5.1 China devotes less that 5 percent of total R&D spending to basic research.
SOURCE: China: Ministry of Science and Technology of the People's Republic of China, *China S&T Statistics Data Book 2010*, Figure 1-3; for U.S.: National Center for Science and Engineering Statistics, *National Patterns of R&D Resources: 2008 Data Update, Detailed Statistical Tables,* NSF 10-314 (March 2010), Tables 1-4.
NOTES: China data for 2009; U.S. data for 2008.

Determined to correct these shortcomings, the Chinese government over the past five years has launched an ambitious agenda to "transform China's economic development pattern so that it is driven by innovation," in the words of Ministry of Science and Technology official Yang Xianyu.[27] President Hu Jintao has declared that innovation "is the core of our national development strategy and a crucial link in enhancing the overall national

---

[27] From presentation by Yang Xianyu of the Ministry of Science and Technology in National Research Council, *Building the 21st Century: U.S. - China Cooperation in Science, Technology, and Innovation*, Charles. W. Wessner, editor, Washington, DC: The National Academies Press, 2011.

strength."[28] Such pronouncements have been backed with a flurry of initiatives at the central, provincial, and local levels to upgrade the nation's innovation ecosystem. Among other things, the government is greatly increasing spending on R&D, boosting incentives for corporate R&D, urging universities and government research institutes to form stronger links with industry, building immense science parks, investing aggressively in broadband infrastructure, and vowing to improve intellectual property-right protection.

The strategy is embodied in *The National Medium and Long-Term Program for Science and Technology Development, 2006-2020*, a document drafted over two years and that received input from some 2,000 experts.[29] The overarching goal is to make China an "overall well-off society" driven by innovation. Among the key targets for 2020 are to become one of the world's top five generators of invention patents and published scientific papers, and to reduce China's dependence on foreign technology to 30 percent.[30] The document also lists 16 "megaprojects" that will receive heavy government financial backing.

The aspect of the game plan that has generated the most attention overseas is the government's emphasis on "indigenous innovation." The goal is to ease China's dependence on imported technology and to nurture companies that can compete at home and abroad with their own intellectual technology. As outlined in the 15-year science and technology plan and numerous published rules and guidelines over the past five years, the strategy includes compelling foreign companies to transfer core technology as a price for being able to sell into China's immense domestic market.[31]

In addition to generating tension with trade partners, China's innovation strategy seems fraught with internal contradictions. Although the stated goal is to achieve an innovation-driven economy led by market forces and enterprises, the technology drive is built around large state-led projects.

---

[28] Hu Jintao report to the 17th National Congress of the Communist Party of China, Oct. 14, 2007. See *Xinhua*, "Innovation tops Hu Jintao's Economic Agenda," Oct. 15, 2007 (http://news.xinhuanet.com/english/2007-10/15/content_6883390.htm).
[29] Cong Cao, Richard P. Suttmeier, and Denis Fred Simon, "China's 15-Year Science and Technology Plan," *Physics Today*, December 2006 (http://www.levininstitute.org/pdf/Physics%20Today-2006.pdf).
[30] *National Medium- and Long-Term Program for Science and Technology Development*, op. cit.
[31] For an extensive discussion of the controversies surrounding China's indigenous innovation policies, see James McGregor, "China's Drive for 'Indigenous Innovation: A Web of Industrial Policies, "U.S. Chamber of Commerce, Global Intellectual Property Center, APCO Worldwide (http://www.uschamber.com/sites/default/files/reports/100728chinareport_0.pdf). Also see U.S. International Trade Commission, *China: Intellectual Property Infringement, Indigenous Innovation Policies, and Frameworks for Measuring the Effects on the U.S. Economy,* Investigation No. 332-514, USITC Publication 4199 (amended), November 2010 (http://www.usitc.gov/publications/332/pub4199.pdf) and Alan Wm. Wolff, "China's Indigenous Innovation Policy," testimony before the U.S. China Economic and Security Review Commission, Washington, DC, May 4, 2011.

Although the strategy acknowledges that China needs multinational investment and greater international collaboration, it is intends to extract technology from foreign companies to create domestic champions that will eventually compete directly against them. As an extensive study of China's technology modernization drive by CENTRA Technologies concludes: "Caught between a tradition of state planning and the need for markets—and between an interest in foreign technology assimilation of the lure of domestically developed technology—China's innovation system faces an ambiguous future."[32]

Nevertheless, there is little question China has the raw potential—and certainly the determination—to emerge as a 21st century innovation power. China has passed Japan as the world's second-largest spender on R&D.[33] Tertiary enrollment in China rose from 2 percent in 1980 and 22 percent in 2007. As of 2008, China had 27 million post-secondary students, compared to 18 million in the U.S.[34] Forty percent of those students are in engineering, math, and science.[35] China's research workforce that has tripled to some 1.6 million since 1997,[36] and a pool of science and engineering Ph. D's that swelled more than fourfold over that time to 20,000. China has extraordinarily high savings and investment rates of around 40 percent of GDP, double the rate of most other nations. China also has the world's second largest manufacturing base [See Figure 5.2], a surplus labor pool of more than 150 million people, superb trade logistics, the world's fast-growing market for advanced technology products, and the ability to absorb global knowledge through direct foreign investment and an extensive network of overseas Chinese.[37]

---

[32] Micah Springut, Stephen Schlaikjer, and David Chen, "China's Program for Science and Technology Modernization: Implications for American Competitiveness," CENTRA Technology Inc., prepared for The U.S.-China Economic and Security Review Commission, 2011 (http://www.uscc.gov/researchpapers/2011/USCC_REPORT_China's_Program_forScience_and_Te chnology_Modernization.pdf).
[33] OECD, *Main Science and Technology Indicators: Volume 2011/1*, 2011, p. 18. Data comparison based on current U.S. dollars.
[34] UNESCO.
[35] See Carl Dahlman, *World Under Pressure*, op. cit.
[36] UNESCO Science and Technology database.
[37] See presentation by Carl Dahlman of Georgetown University in National Research Council, *Innovation Policies for the 21st Century,* Charles W. Wessner, editor, Washington, DC: The National Academies Press. Also see Carl Dahlman, in *Building the 21st Century: U.S. - China Cooperation in Science, Technology, and Innovation*, op. cit.

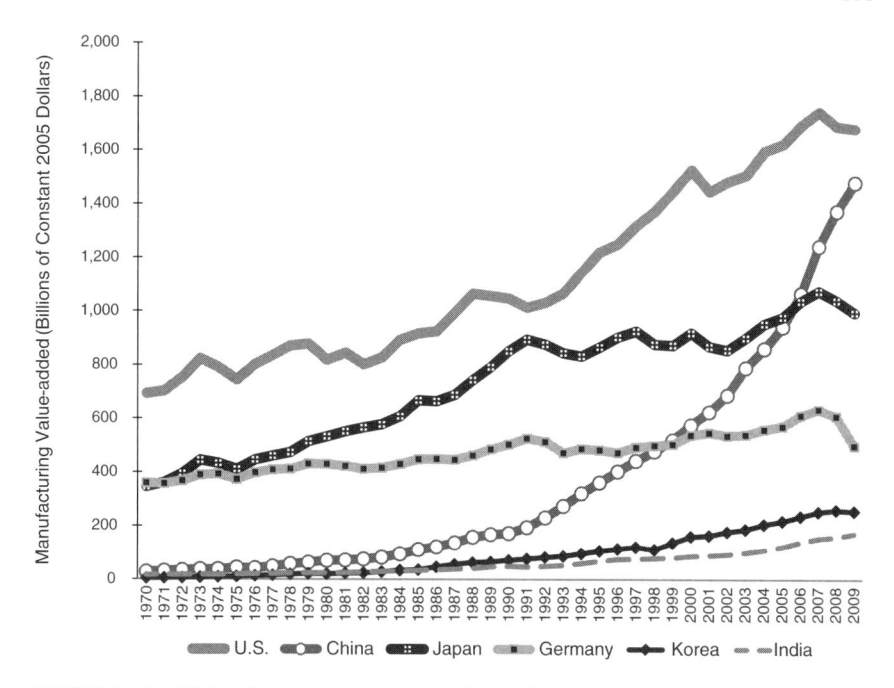

FIGURE 5.2 China is second only to the United States in manufacturing value-added.
SOURCE: United Nations Statistics Division, National Accounts Main Aggregates Database at http://unstats.un.org/unsd/snaama/selbasicFast.asp.

China's Evolving Innovation System

China re-entered the global economy in the late 1970s with a scientific establishment, higher education system, and industrial base that had been crippled by nearly three decades of chaotic rule under Mao Zedong. After its victory in 1949, the Communist Party implemented Soviet-style central planning. Private industrialists fled to Hong Kong and Taiwan, and state took control of the factories left behind. Millions perished in famine as the result of the Great Leap Forward, Mao's disastrous grass-roots industrialization drive. Scientists and academics were purged in an anti-rightist campaign and again during the Cultural Revolution from 1966 to 1976, when educated Chinese were banished to manual work in the countryside and universities were shut to virtually all but workers, farmers, and soldiers. That 10-year period cost China a generation of top scientists and engineers whose absence is still felt.

> **Box 5.2**
> **China's Demographic Challenge**
>
> Driven by the nation's one child policy, China's total fertility rate has fallen over the past 30 years from 2.6, well above the rate needed to hold a population steady, to 1.56, well below that rate.[38] If children of one-child families want only one child themselves, as is typical, China will face a long period of low fertility.
>
> Moreover, China faces a rapid aging of its workforce, leading to a contraction of from 72% to 61% between 2010 and 2050. As the demographic bulge ages, the numbers of those in their early 20s, who are usually the best educated and most productive members of society, will have halved.[39]
>
> As the Economist observes, "The shift spells the end of China as the world's factory. The apparently endless stream of cheap labour is starting to run dry. Despite pools of underemployed country-dwellers, China already faces shortages of manual workers. As the workforce starts to shrink after 2013, these problems will worsen."[40]

China's innovation system, which prior to the revolution featured 210 Western-style universities and 70 research institutes, was remodeled along Soviet lines. The Chinese Academy of Sciences assumed control of basic research. Applied research was the responsibility of thousands of research institutes controlled by central ministries and provincial governments, while state enterprises developed products. Universities focused on human resource development.[41] Although China registered some major achievements, such as development of an atomic bomb and satellites, there was little connection between research and industry.

China's current innovation system began with reforms launched by Deng Xiaoping in 1978. Universities once again admitted students based on

---

[38] See Yong Cai, China's Demographic Reality and Future, *Asian Population Studies*, Vol. 8, No. 1, March 2012. See also Ho Chi-ping, "Demography could threaten China's lead in manufacturing," *China Daily*, April 25, 2012.
[39] Ada C. Mui, "Productive ageing in China: a human capital perspective." *China Journal of Social Work*, Volume 3, Issue 2-3, 2010. See also *The Economist*, "Demography: China's Achilles Heel," April 21, 2012.
[40] *The Economist*, April 21, 2012, op cit. For an analysis of the implications of shifting demographic trends around the world, see Sarah Harper, "Addressing the Implications of Global Aging," *Journal of Population Research*, Vol. 23, No. 2, 2006.
[41] From presentation by Lan Xue of Tsinghua University School of Public Policy and Management at June 28, 2011, Joint Seminar on Comparative Innovation Studies at the Chinese Academy of Engineering in Beijing. This symposium was co-sponsored by the National Academy of Sciences, the Chinese Institute for Strategy Studies in Engineering and S&T.

examination scores, and thousands of China's brightest scholars were allowed to study in the U.S. and Europe. Deng also enshrined science and technology as one of the Four Modernizations, the pillars of the Party's strategy to become a great economic power.[42] The government introduced a wave of programs in the 1980s to advance science and technology and to open the doors to what became a flood of foreign direct investment. The government also shifted much of the implementation of its policies from central ministries to local and provincial authorities.[43]

The first wave of reforms in the 1980s included restructuring and gradual funding cuts of state-run research institutes. Instead, more research funds instead were allocated to specific projects through a competitive process. The State-High Tech Development Plan, better known as the 863 Program, was launched to ease China's dependence on foreign technologies in key areas from satellites to computer processing.[44] A program to build 153 world-class national laboratories in universities and research institutes began in 1984.[45] The National Natural Science Foundation, modeled after the National Science Foundation, was established in 1986 to award peer-reviewed research grants to scientists. The Torch Program was initiated in 1988 to promote industrialization of high technology by developing work forces, organizing science and technology R&D programs to serve national goals, offering preferential access to bank credit for new product development programs, and building 53 high-technology industrial zones.[46] The Spark program targeted rural development. Organizational changes also encouraged different research organizations to establish horizontal linkages and encourage scientists and engineers to become entrepreneurs.

The leadership launched a series of reforms to decentralize, depoliticize, and diversify the higher education system in 1985. Provincial and

---

[42] The Four Modernizations were goals originally promoted by Zhou Enlai in the 1960s and adopted at the Third Plenum of the 11th Central Committee in December 1978.
[43] From presentation by Thomas R. Howell in *Innovation Policies for the 21st Century,* op. cit.
[44] The initial fields covered in the 863 program were biotechnology, space, information technology, automation, energy, and new materials. Other fields, such as telecommunications and marine technology, were added in subsequent five-year plans. An explanation of the program is found on the Ministry of Science and Technology Web site at
http://www.most.gov.cn/eng/programmes1/200610/t20061009_36225.htm.
[45] For a concise explanation of Chinese innovation policies over the past decade, see Can Huang, Celeste Amorim, Mark Spinoglio, Borges Gouveia and Augusto Medina, "Organization, Programme and Structure: An Analysis of the Chinese Innovation Policy Framework," R&D Management 34, 4, 2004
(http://xcsc.xoc.uam.mx/apymes/webftp/documentos/biblioteca/analysis%20of%20the%20Chinese%20innovation%20policy.pdf). Also see Evan Feigenbaum, *Chinese techno-Warriors: National Security and Strategic Competition from the Nuclear Age to the Information Age,* Palo Alto: Stanford University Press, 2003).
[46] An explanation of the Torch program is found on the Web site of the People's Republic of China New York Consulate at http://www.nyconsulate.prchina.org/eng/kjsw/zgkj/t31698.htm.

local governments assumed operating control, and universities were given more management autonomy. Universities also were encouraged to become more commercially viable, compete for faculty and research funding, and cooperate with industry and government.[47] They also were encouraged to form enterprises, incubate new companies, and create science parks.

A second major wave of reforms in the 1990s focused on developing China's national innovation system. Enrollment at universities increased dramatically, and R&D programs were strengthened. Hundreds of universities were merged and restructured, and the number administered by central government ministries dropped from 367 to 120. The National Basic Research Program, better known as the 973 Program, was launched to support 175 chief scientists focusing on "strategic needs," such as agriculture, energy, information, and health.[48] The roles of government research organizations were clarified. After the central government sharply cut its funding, the Chinese Academy of Sciences launched the Knowledge Innovation Program to remake itself as the nation's premier source of basic research and cutting-edge technology in everything from defense and agriculture to health and energy. The CAS hired hundreds of overseas Chinese scientists and consolidated its 120 institutes into 80.[49] As explained further below, thousands of other research institutes controlled by ministries and local governments also were forced to compete for research funds and encouraged to become part of enterprises or go into business themselves.

The most recent innovation push began in 2003 under President Hu Jintao and Premier Wen Jiabao, who elevated innovation to the top of the nation's economic agenda. Coordinated by the Ministry of Science and Technology—which leads development of science policy and overseas many national funding programs to implement projects--and the Chinese Academy of Sciences, the government launched a two-year project to draft a new national strategy for science and technology.[50] The innovation push is part of an overarching strategy to gradually overhaul China's economic model, which Premier Wen described as "irrational" due to its reliance on "the overproduction of low-quality goods, low rates of return, and increasingly severe constraints

---

[47] See Lan Xue, "Universities in China's National Innovation System," prepared for the UNESCO Forum on Higher Education, Research, and Knowledge, 2006 (http://portal.unesco.org/education/en/files/51614/11634233445XueLan-EN.pdf/XueLan-EN.pdf).
[48] The National Basic Research Program, also known as the 973 Program, was approved by the central government in June 1997 and administered by the Ministry of Science and Technology. For an explanation in English of the program, see http://www.973.gov.cn/English/Index.aspx.
[49] For a good analysis of changes in the Chinese Academy of Sciences and reforms of research institutes, see Richard P. Suttmeier, Cong Cao, and Denis Fred Simon, "China's Innovation Challenge and the Remaking of the Chinese Academy of Sciences," *Innovations*, Summer 2006 (http://www.policyinnovations.org/ideas/policy_library/data/ChinasInnovationChallenge/_res/id=sa_File1/INNOV0103_p78-97_suttmeier.pdf).
[50] Xue, "China's Innovation Policy in Context of National Innovation System Reform," op. cit.

resulting from energy and other resource scarcity and severe environmental degradation."[51] The leadership believes that China is overly dependent on export manufacturing of goods that export cheap labor but entail little Chinese value-added. As Lan Xue, dean of Tsinghua University's School of Public Policy and Management explained, the leadership recognized "the need for China to break away from its traditional position in the international division of labor and move up the value chain."[52]

The result was the *Medium to Long-Term Plan for the Development of Science and Technology*. In addition to setting broad goals such as increasing R&D spending to 2.5 percent of GDP by 2020, the lengthy document contained lists of targets for catching up with advanced nations by 2020 in "frontier sciences" such as the study of life processes, earth systems, and the brain; "major scientific programs" that include protein studies, quantum regulation, and nano-scale materials; applied technologies aimed at specific industries such new-energy based vehicles, high-performance computing, sensor networks, high-definition flat-panel displays, high-speed transit, and renewable energies. The 15-year plan addresses framework conditions for a national innovation system, such as the need to put enterprises at the center of innovation, policy support for venture capital, improving protection of intellectual property rights, and investments in infrastructure, human resource development, and promoting public understanding of an innovative culture.[53]

The plan also designated 16 "megaprojects" that would establish China as a global leader in key industries and be backed with significant direct central government funding, bank loans, and policy tools such as tax breaks for companies. The megaprojects include extra large-scale semiconductor manufacturing, next-generation wireless broadband, advanced nuclear reactors, control of AIDS and hepatitis, and large aircraft manufacturing. Beijing has announced more than $100 billion in investments in megaproject schemes since 2008.[54] The megaproject plan had generated active debate over whether central government control over funding for such industrial projects—as opposed to competitive grants allocated through peer review—would lead to financial waste.[55]

A newer government industrial policy initiative calls for nurturing seven "strategic emerging industries"—new-generation information technology, energy efficiency and environmental protection, biology, high-end equipment manufacturing, new energy, new materials, and new energy automotive

---

[51] Wen Jiabao, "Speech at the National Science and Technology Conference," Jan., 9, 2006.
[52] Xue presentation in June 28 Beijing symposium.
[53] *The Medium to Long-Term Plan for the Development of Science and Technology*, op. cit.
[54] A list of government spending announcements for megaprojects is found in Springut, Schlaikjer, and Chen, op. cit.
[55] For an account of internal debates over drafting of the 15-year plan is in McGregor, op. cit.

industries.[56] The goal is for these seven sectors to account for 8 percent of GDP by 2015 and 15 percent by 2020, compared to 4 percent now.[57] To attain these goals, HSBC Global Research calculates that these sectors would have to grow at a compounded annual rate of 35 percent for the next five years and 29 percent over the coming decade and reach between $1.55 trillion and $2.33 trillion in revenue in 2020.[58] The initiative is said to entail an overall investment of $1.5 trillion, with the government planning to account for 5 percent to 15 percent of the funds.[59]

Chinese government bodies offer some of the world's most generous incentives in targeted industries. They include 10-year tax holidays for production plants, exemption from sales tax income earned through technology transferred via foreign investment, low cost or free land, direct equity stakes by government investors, and procurement regulations that favor domestic production. To spur investment in innovation in "high priority" sectors, China offers 1.5 *renmenbi* in tax credits for every *renmenbi* spent on R&D.[60]

TABLE 5.1 Eight Major Innovation Policy Initiatives Resulting from Adoption of the *Outline of the Medium- and Long-Term Plan for National Science and Technology Development*

- Increase investment in R&D
- Tax incentives for investment in STI
- Government procurement policy to promote innovation
- Assimilation of imported advanced technology
- Increase capacity to generate and protect IPR
- Build national infrastructure and platforms for STI
- Cultivate and utilize foreign talents for STI
- Support indigenous innovation

SOURCE: UNESCO, *UNESCO Science Report 2010*, pp. 381-386.

---

[56] The State Council announced Emerging Strategic Industries initiative was released following the Communist Party's 2010 plenary. A Chinese version of the decree, Guo-Fa 2010 No. 32 can be accessed at http://www.gov.cn/zwgk/2010-10/18/content_1724848.htm.
[57] People's Daily Online, "Strategic Emerging Industries Likely to Contribute 8% of China's GDP by 2015," October 19, 2010 (http://english.peopledaily.com.cn/90001/90778/90862/7170816.html).
[58] Steven Sun and Garry Evans, "Emerging Strategic Industries: Aggressive Growth Plans," HSBC Global Research, Oct 19, 2010).
(http://www.research.hsbc.com/midas/Res/RDV?p=pdf&key=lg0uISbcyh&n=280786.PDF
[59] Estimate in Springut, Schlaikjer, and Chen, op. cit.
[60] Yang presentation, op. cit.

The cost of capital is another advantage for Chinese manufacturers. Stephen O'Rourke of Deutsche Bank Securities estimates the Chinese solar cell and module makers pay 3.5 percent interest on average to borrow from government banks. Combined with other incentives, he said, China has an "almost insurmountable" cost advantage over the U.S. as a place to build and operate a factory.[61]

Some government aid to industry has led to friction with trade partners. In December 2010, for example, the U.S. filed a WTO complaint accusing China of providing unfair subsidies to domestic producers of wind-turbines and solar equipment, allegations that China denies.[62] An investigation by the European Commission in February 2011 concluded that Huawei and ZTE received massive subsidies in the form of credit lines from state-owned banks. Huawei and ZTE denied those allegations.[63]

Surging Chinese exports of solar panels also have triggered trade disputes. Seven U.S. manufacturers of solar panels filed an anti-dumping petition with the Department of Commerce in October 2011 alleging that billions of dollars in government subsidies enabled China's largest photovoltaic panel manufacturers to dramatically increase capacity, enabling them to push down prices and dominate the U.S. market. The U.S. manufacturers also accused their Chinese competitors of selling at below-fair value. Chinese manufacturers deny the charges.[64] The China Development Bank reportedly gave $30 billion in low-cost loans in 2010 alone to China's top five manufacturers.[65]

Behind the Indigenous Innovation Push

A steady theme running through the *Medium to Long-Term Plan* is its emphasis on spurring "indigenous innovation." The Chinese term *zizhu chuangxin* roughly translates into "self directed," but has been understood and described in different ways. Many Western commentators have interpreted "indigenous innovation" to mean "self-sufficiency."

Indeed, the *Medium to Long-Term Plan* declares that China must "master core technologies in some critical areas, own proprietary intellectual property rights, and build a number of internationally competitive enterprises." The plan also states that core technologies "in areas critical to the national

---

[61] From presentation by Steve O'Rourke of Deutsche Bank Securities in National Research Council, *The Future of Photovoltaic Manufacturing in the United States: Summary of Two Symposia*, Charles W. Wessner, ed., Washington, DC: The National Academies Press, 2011.
[62] Martin Crutsinger, "U.S. Challenges Chinese Wind-Power Subsidies," *Associated Press* article published in *Seattle Times*, Dec. 22, 2010.
[63] Matthew Dalton, "EU Finds China Gives Aid to Huawei, ZTE," *Wall Street Journal*, Feb. 3, 2011.
[64] Keith Bradsher, "7 U.S. Solar Panel Makers File Case Accusing China of Violating Trade Rules," *New York Times*, Oct. 20, 2011.
[65] Stephen Lacy, "How China Dominates Solar Power," *The Guardian*, September. 12, 2011.

economy and security" should not be purchased from abroad if domestic alternatives are available.[66]

The 15-year plan and other Chinese statements on rules and regulations have heightened fears by foreign companies that the strategy is to reverse-engineer and forcibly extract technology from multinationals as a price for the privilege of selling their products in China. Other policies state that government agencies and government-funded projects—which account for the bulk of important purchases in China due to the government's pervasive role in the economy—should favor products invented in China by Chinese-owned companies over those of foreign companies. The central government and provincial governments issued catalogues to procurement officials specifying which products meet "indigenous innovation" criteria. Few foreign products were on the lists. The indigenous innovation goals also are embedded in Chinese technology standards, anti-monopoly law, patent rules, and tax regulations, according to the U.S. International Trade Commission. "The indigenous innovation 'web of policies' is expected to make it difficult for foreign companies to compete on a level playing field in China," the ITC reported.[67] An American Chamber of Commerce report said "the plan is considered by many international technology companies to be a blueprint for technology theft on a scale the world has never seen before."[68]

Chinese officials and economists have sought to assure foreign companies that China's intent is not to steal foreign technology and shut foreign products out of its market. Rather, the intent is to improve China's ability to create innovative products, add more value to what it produces, and relieve an unhealthy over-reliance on imported knowhow for a country at its stage of development. Mr. Deng of the State Council Research Office noted that in the global supply manufacturing chain, China produces mainly low- and medium-level goods. The core technology and crucial equipment is not made in China. "We need to develop core processes and breakthrough technologies," he said.[69] China's enormous trade surplus with the United States is exaggerated, contended Dr. Xue of Tsinghua University, because conventional trade statistics don't take into account the imported materials that go into exported products and

---

[66] *The Medium to Long-Term Plan for the Development of Science and Technology*, op. cit.
[67] U.S. International Trade Commission, *China: Intellectual Property Infringement, Indigenous Innovation Policies, and Frameworks for Measuring the Effects on the U.S. Economy*, Investigation No. 332-514, USITC Publication 4199 (amended), November 2010 (http://www.usitc.gov/publications/332/pub4199.pdf). Also see For an extensive examination of the implications of Chinese government "indigenous innovation" policies for foreign companies and trade, see Alan Wm. Wolff, "China's Indigenous Innovation Policy," testimony before the U.S. China Economic and Security Review Commission, Washington, DC, May 4, 2011.
[68] McGregor, "China's Drive for 'Indigenous Innovation: A Web of Industrial Policies, U.S. Chamber of Commerce, Global Intellectual Property Center, APCO Worldwide (http://www.uschamber.com/sites/default/files/reports/100728chinareport_0.pdf).
[69] Deng Wenkui presentation, op. cit.

the low value-added of its exports. Dr. Xue estimates that 90 percent of China's trade surplus is in the "processing trade," in which goods are assembled in China from imported parts and materials, and is generated by multinationals and foreign joint ventures.[70]

Dr. Xue said a classic example is the Apple iPhone, which is assembled in China by the Taiwanese contract manufacturer Foxconn. A study by the Asian Development Bank noted that the iPhone, although invented and designed in the U.S., contributed $1.9 billion to the U.S. trade deficit with China in 2009. That is because the $2 billion worth of iPhones shipped from Foxconn's Shenzhen factory contained a little more than $100 million in U.S. parts. Chinese manufacturing accounted for only $73.5 million of value of those $2 billion worth of phones, however. The rest came from imported materials. America's bigger trade deficits from the iPhone, therefore, were with Japan, which supplied $670 million in components, Germany ($326 million), and South Korea ($108 million). [See Figure 5.3] The difference between the $500 selling price of the iPhone and the $179 production cost went to Apple and retailers.[71]

Adding to the sense of urgency over innovation is recognition that rising wages, shipping rates, and other costs are fast eroding China's once-formidable cost advantage as an export-manufacturing base for the world. In 2000, wages and benefits of average Chinese factory workers in the Yangtze River Delta, the nation's leading export region, were one-20th those of comparable workers in Southern U.S. states. By 2015, Chinese wages will be one-quarter of those in the U.S., according to projections by The Boston Consulting Group. Once higher U.S. worker productivity, the actual labor content of a product, logistics costs and other factors are fully accounted for, China's cost advantage will be negligible, BCG predicts.[72] To remain competitive in the years ahead, therefore, China will increasingly have to compete in higher value-added products rather than just on the basis of low labor costs.

---

[70] Xue presentation, op. cit. While it is true that, measured in terms of domestic value-added, China's trade surplus with certain countries such as the United States is overstated, the domestic value-added of Chinese exports has been increasing over time. See Robert Koopman and Zhi Wang, "How Much of China's Exports is Really Made in China? Estimating Domestic Content in Exports When Processing Trade is Pervasive," presented at World Bank Trade Workshop, June 10, 2011.
[71] Yuqing Xing and Neal Detert, "How the iPhone Widens the United States Trade Deficit with the People's Republic of China," ADBI Working Paper 257, Asian Development Bank Institute, December 2010
(http://www.adbi.org/files/2010.12.14.wp257.iphone.widens.us.trade.deficit.prc.pdf).
[72] Harold L. Sirkin, Michael Zinser, and Douglas Hohner, *Made in America, Again: Why Manufacturing Will Return to the U.S.*, Boston Consulting Group, August 2011. This report can be accessed at http://www.bcg.com/documents/file84471.pdf.

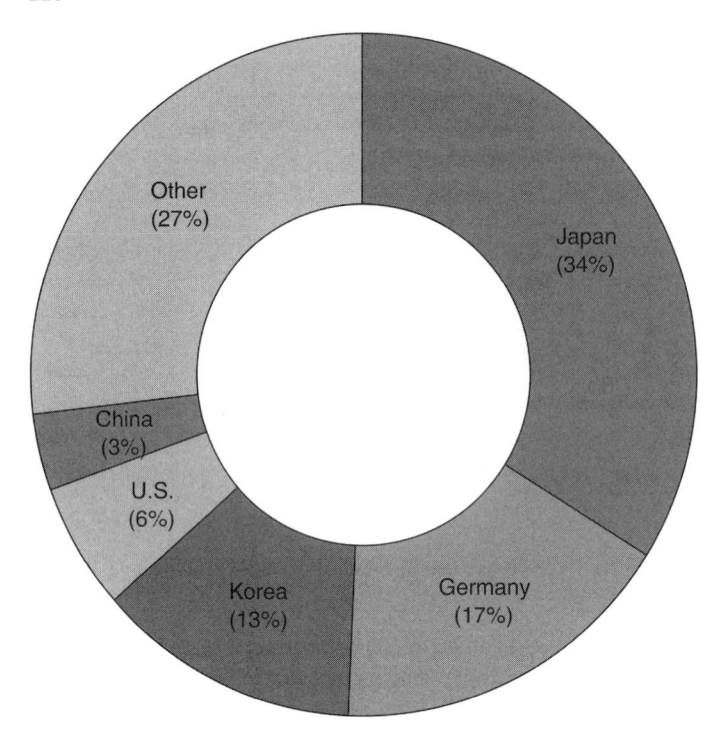

FIGURE 5.3 While trade data indicate that the United States imported $2 billion of iPhones from China in 2009, only an estimated three percent of the value-added was from China.
SOURCE: Yuqing Xing and Neal Detert, "How the iPhone Widens the United States Trade Deficit with the People's Republic of China," *ADBI Working Paper Series*, No. 257, Revised May 2011.

Strategic Priorities

China's innovation push is regarded as integral to achieving a number of top national strategic objectives, such as national security, boosting productivity, addressing what many to believe to be a budding health-care crisis, and meeting future energy needs.

Renewable energy is an especially high priority. China's energy consumption has nearly doubled in five years and is expected to double again in another five years. Currently, the nation relies almost entirely on fossil fuels, especially coal, to generate electricity. "Against this background, renewable energy is our inevitable choice," explained Ren Weimin of the National

Development and Reform Commission.[73] Beijing's target is for a blend of wind, hydro, solar, nuclear, thermal, and other non-fossil fuels to account for 15 percent of consumption by 2020, 20 percent by 2030, and one-third by 2050.[74] That compares to 8.3 percent now. Government also is helping build domestic markets for domestic solar and wind power, energy-efficient solid-state lighting, and electrified vehicles industries through government purchases and generous incentives for consumers. Mr. Ren said China is developing a "comprehensive policy and institutional framework" for renewable energy.

Information and communications technologies (ICT) also are strategically important, not only as promising Chinese growth industries in themselves but also as a means for modernizing the economy. China is becoming a global power in ICT manufacturing and an increasingly important market. In 2011, for example, it produced 140 million PCs and 40 billion ICT-related chips. China has 921 million cell phone and 485 million Internet users.[75] China now is investing heavily to deploy high-speed broadband infrastructure, for example. China views broadband as a catalyst for new growth industries such as software, logistical services, information technology outsourcing, and a wide range of digital devices. Government targets call for 30 percent annual growth for software and information services industry and 28 percent annual growth in software exports.[76] China's domestic electronic commerce industry is estimated to be worth $400 billion industry a year[77] and is growing at around 25 percent a year.

In terms of hardware and software, China is likely to concentrate its R&D efforts on embedded systems, advanced engineering software, large-scale digital control equipment and systems for production lines, integrated IT systems, encryption, virtual reality technologies, and new materials, according

---

[73] From presentation by Ren Weimin of the National Development and Reform Commission in *Building the 21st Century: U.S. - China Cooperation in Science, Technology, and Innovation*, op. cit.
[74] State Council of China, "National Medium- and Long-Term Program for Science and Technology Development, 2006-2020"
(http://webcache.googleusercontent.com/search?q=cache:y800l0iQlS8J:www.cstec.org/uploads/files/National%2520Outline%2520for%2520Medium%2520and%2520Long%2520Term%2520S%26T%2520Development.doc+china+National+Medium-+and+Long-Term+Program+for+Science+and+Technology&cd=18&hl=en&ct=clnk&gl=us&client=firefox-a).
[75] Data from presentation by Xu Jianping of the National Development and Reform at the Sept. 19, 2011, National Academies symposium "U.S.-China Policy for Science, Technology, and Innovation" in Washington, DC.
[76] China's Eleventh Five-Year Plan (2006-2010) also calls for producing around 15 major software enterprises with sales exceeding RMB 10 billion. For a good analysis of China's information technology and communication strategy by Indian software-industry association Indian software-industry association NASSCOM, see "Tracing China's IT Software and Services Industry Evolution," whitepaper prepared by NASSCOM Research, August 2007 (http://www.business-standard.com/general/pdf/082107_01.pdf).
[77] Sirkin, Zinser, Hohner, op. cit.

to a National Research Council assessment. All of these are "areas of weakness and obstacles to autonomy in the IT communications," the report said.[78] Improvement in such areas can "improve and deepen" economic development across industries and the country, explained Xu Jianping of the National Development and Reform Commission's High-Tech Department. Therefore, the government "has made new-generation IT development a core priority," he said.[79]

Actors in China's Innovation System

The Shifting Role of Research Institutions

The main conduits for disseminating technology to China's corporate sector are the some 4,000 research institutes controlled by central ministries, local governments, and the Chinese Academy of Sciences. Compared to applied-research institutes of nations and regions such as Germany, Taiwan, and Finland, the majority of those in China are regarded as having relatively weak linkages with private industry. Reforms since the 1990s, however, have turned several institutes into effective organizations for developing industrial technologies and transferring them to a wide range of enterprises.

Institutes were given several options to cope with funding cuts. They could become the technology-development arms of state enterprises, become contract research organizations for government and industry, or go into business themselves. "When it happened, we were very puzzled, upset, and lost," explained Tian Zhiling, of the China Iron and Steel Research Institute Group (CISRI). "We were being abandoned by the government." Seventy percent of CISRI's employees had master's and Ph. D. degrees and the institute received tax reductions for five years to enter business. But it had little experience with marketing, mass production, finance, and entrepreneurship.[80]

In 1999, 242 central level research institutes under 10 industry bureaus were transferred into enterprises. Local governments transferred another 5,014. Now, these institutes have $17.5 billion in annual revenue and have quadrupled their profits since 2005 to $2.2 billion.[81] The Research Institute of Petroleum Processing, for example, now develops refining and alternative-energy technology for SINOPEC. Zoomlion, formerly a research institute of the Ministry of Construction, now is China's leading manufacturer of construction equipment, with 2010 sales of $7.8 billion. The 242 former state-owned institutes also earn nearly $3 billion in annual income transferring technology to

---

[78] National Research Council, *S&T Strategies of Six Countries*, op. cit., pg. 26.
[79] Xu Jianping presentation, op. cit.
[80] Presentation by Tian Zhiling of the China Iron and Steel Research Institute Group in June 28, 2011, Chinese Academy of Engineering symposium.
[81] Data from presentation by Tian Zhiling of the China Iron and Steel Research Institute Group in June 28, 2011, Chinese Academy of Engineering symposium.

Chinese companies and have earned more than 13,000 patents between 2006 and 2010. CISRI is regarded as a success story. It now leads in the development of new metallurgical technologies for China's steel industry, the world's largest, and its "third-generation steel" makes it a success story. It has developed high-nickel stainless steel, ultra high-strength sheets use in automobiles, and "third-generation steel."[82]

The several thousand research institutes still controlled by government agencies employ around 277,000 R&D staff and focus on applied research and development relating to government missions.[83] The Chinese Academy of Science has numerous institutes that have created some 400 enterprises[84]. The relative role of government institutes in the national innovation system has declined, however. The share of national R&D spending by research institutes has dropped from more than 60 percent a decade ago to around 18 percent of national R&D expenditure, compared to 26.4 percent by universities. Their staffs also have declined. Many state institutes still tend to focus on patents and publishing papers, however, rather than on disseminating technology to industry. Improving these linkages is a strong government priority. Since 2009, institutes have joined more than 40 strategic alliances with industry in areas such as clean coal and solid-state lighting.[85]

Expanding the Mission of Universities

China's higher education system has expanded tremendously in recent decades in size and scope. Between 1980 and 2008, the percentage of Chinese aged 18 to 22 with a college education rose from 2 percent to 23 percent.[86] The number of Ph. Ds. in China, meanwhile, surged from 151,000 in 1999 to 267,000 in 2008, although the rate of growth has slowed to around 5 percent a year compared to annual increases of 20 percent or more a decade ago.[87]

The first mission of universities is "to serve as an engine or driver of a country's core competitiveness," according to Lou Jing of the Ministry of Education's Department of Science and Technology. The government also wants to "markedly raise competitiveness and the quality of higher education,"

---

[82] Ibid.
[83] From Dahlman presentation, *Building the 21st Century: U.S. - China Cooperation in Science, Technology, and Innovation*, op. cit.
[84] The Chinese Academy of Sciences has 12 branch offices, 117 institutes organized as legal entities, over 100 national key laboratories and a staff of over 50,000 people. http://english.cas.cn/ACAS/BI/100908/+20090825_33882.shtml.
[85] From remarks by Mu Rongping of the Chinese Academy of Sciences at the June 28, 2011, symposium at the Chinese Academy of Engineering in Beijing.
[86] Report on the Development of National Education (www.cernet.edu.cn).
[87] National Education Development Statistics cited by Su Jun and Joseph Zhou, "Chinese University in the National Innovation System," presented at June 28, 2011 joint symposium at Chinese Academy of Engineering in Beijing.

she said, and to tighten collaboration among universities, government, and industry.[88]

Chinese universities also have assumed a greater role in government and industrial R&D and creating new businesses. Research funding for Chinese universities has been rising around 20 percent a year, with nearly 40 percent of that now coming from industry. Universities are in charge of some 80 percent of National Science Foundation research programs and 40 percent of national high-technology research-and-development programs. Universities are home to 60 percent of China's "national pilot laboratories," nearly two-thirds of its 140 "national key laboratories," and 26 national engineering laboratories. They also operate 76 science parks. Universities produce more than one-third of Chinese patents for inventions and 60 percent of published science and engineering papers.[89]

Universities also operate 76 science parks in China. The Tsinghua University Science Park, or TusPark, ranks among the largest university science parks in the world. Launched in 1994, TusPark has a 20-building campus in Beijing with 400 companies and 30,000 employees. Google, Sun, Procter & Gamble, and Microsoft are among the multinationals with large R&D centers. There also are 200 innovative local companies—more than half of them established by returnees from overseas.[90] Unlike most Chinese science parks, TusPark also has an active incubator and entrepreneurial-training program for start-ups.

Compared to universities in the U.S., however, most of those in China tend to be ivory towers that put top priority on publishing papers, many of them of questionable quality, in scientific and technical journals. Some scientists blame a research funding system that puts too little emphasis on independent peer review.[91] In an editorial in *Science* magazine, Yigong Shi and Yi Rao, the respective deans of the life sciences programs at Tsinghua University and Peking University, said that major grants often are awarded through personal connections with powerful bureaucrats. Shi and Rao contended that China's research culture "wastes resources, corrupts the spirit, and stymies innovation."[92]

Although 1,354 Chinese institutes of higher learning report R&D programs, less than 50 elite schools dominate important research. Nine

---

[88] Ibid.
[89] Statistics from presentation by Lou Jing of the Ministry of Education's Department of Science and Technology in *Building the 21st Century: U.S. - China Cooperation in Science, Technology, and Innovation*, op. cit.
[90] From presentation by Wu Hequan of Tsinghua University Science Park in June 28 joint symposium.
[91] For a critique of China's scientific research system, see Yigong Shi and Yi Rao, "China's Research Culture," *Science*, Vol. 329, no. 5996, Sept. 3, 2010.
[92] Yigong Shi and Yi Rao, "China's Research Culture, *Science*, Vol. 329, p. 1128, Sept. 3, 2010.

universities, including Peking University, Tsinghua, Zhejiang, and Jiaotong, account for one-quarter of China's scientific papers and citations.[93] The percentage of Chinese researchers at universities has dropped steadily since 1999, to around 15 percent and, although government research grants to universities have grown dramatically, their share of total R&D spending in China has dropped since 1986 to around 8.5 percent, compared 12.8 percent in the United States.[94] Even though more than half of Chinese university research is regarded as applied, there still is a debate at many universities over whether they should focus only on basic research, according to Joseph Zhou of Tsinghua University. "The university role in applied research is a big question mark," Dr. Zhou said, because R&D in China is overwhelming applied.[95]

Chinese universities also have a long way to go to reach world standards. Only eight rate among the world's top 400 schools, according to QS World University Rankings, compared to 86 U.S. institutions. The highest is Peking University at No. 47. Shanghai's Fudan is next at No. 105.[96]

China has launched a number of campaigns to improve this status. Project 211, introduced in 1993, seeks to make 100 universities among the best in the world. The $4.5 billion 985 program, begun in 1998, seeks to raise 39 existing universities to world standards. Central and local governments also are supplying funds for universities to recruit star faculty and establish endowed chairs. A distinguished young scholar program provides cash awards to promising young scientists. The Ministry of Personnel administers a program to identify 100 promising scientists on the frontier of international research, 1,000 leaders of advanced research projects, and 10,000 leaders for academic disciplines.[97]

When it comes to starting companies, one unorthodox aspect of Chinese universities is their propensity to retain ownership or management control. While Chinese universities have spun off 3,665 enterprises, they run or own another 3,569 enterprises.[98] Some of the more significant university-run enterprises include Tsinghua Tongfang, an information technology and environmental technology company owned by Tsinghua University that is listed on the Shanghai Stock Exchange, embedded system company Beida Jada Bird (owned by Beijing University), and information technology company Neusoft (Northeastern University). The majority of firms run and owned by universities are not engaged in science and technology. Dr. Zhou said the large scale, number, and management challenges at university-run enterprises remain

---

[93] Springut, Schlaikjer, and Chen, op. cit
[94] Su and Zhou, op. cit.
[95] From remarks by Joseph Zhou of Tsinghua University at June 28, 2011, joint symposium.
[96] QS World University Rankings 2010/2011 (http://www.topuniversities.com/university-rankings/world-university-rankings).
[97] These programs are described in Springut, Schlaikjer, and Chen, op. cit.
[98] China Statistical Yearbook on Education data cited by Su and Zhou.

significant issues in China.[99] Because only a small portion of university businesses are successful—and can pose serious financial liabilities for universities--the government has been encouraging universities to yield management control at enterprises to professionals so they can be run as modern businesses.[100]

Chinese Corporations as Innovators

According to Chinese statistics, enterprises are the chief drivers of innovation in China. Large and small enterprises account for around 70 percent of R&D investment. They spent nearly $50 billion on R&D in 2009, seven times more than in 2000, and employed nearly 1.5 million R&D personnel, three times the 2000 level.[101]

These investments have enabled China to rapidly become a major global force in a range of advanced industries. Despite all of that activity, however, corporate China can boast few breakthrough products or technologies with the notable exception of internet based e-commerce and social network sites, such as dynamic e-commerce and social network sites such as Tencent, Alibaba, and Baidu. Although China is a leader in some areas of cancer research and genomics,[102] Chinese pharmaceutical companies have marketed few medicines globally except for traditional remedies. China is a leading producer of lithium-ion batteries, but they use decades-old chemistries. China is developing its own narrow-body jet to compete with Boeing and Airbus, but the core systems come from foreign aerospace firms and the body is based on a 1980s design by McDonnell Douglas. China is one of the leading exporters of solar cells and modules, but they use mature polycrystalline silicon technologies.[103] Asked to cite examples of important innovations by Chinese companies in any industry, multinationals executives in China could not come up with any. Said one: "I don't think there is a single success. They have the technology they believe they can scale globally, but if they try to compete on a level playing field they will have problems."[104] Chinese officials agree that corporate innovation remains a significant challenge. China needs to "make enterprises the engines of innovation, as in the United States," stated Li

---

[99] Zhou presentation.
[100] Xue, "Universities in China's National Innovation System," op. cit.
[101] Data cited in Springut, Schlaikjer, and Chen, op. cit.
[102] For a collection of articles that highlight recent cancer research in China, see Cell Research's special issue on cancer research in China. See Cell Research published online on 16 April 2007.
[103] See Dexter Roberts and Pete Engardio, "China's Economy: Behind All the Hype," *BusinessWeek*, Oct. 23, 2009.
[104] U.S. company interview in Beijing (June 2011). (NB: Names and affiliations of this and other interviewees have been withheld pending permission.)

Guoqing, director-general of the State Council Central Finance and Economics Office.[105]

This does not mean Chinese companies are not making rapid progress in innovation. One example is data communications equipment. Huawei Technologies is the world's third-largest makers of network equipment[106] and ranked as one the world's largest network equipment makers, ranking No. 1 in mobile broadband systems, DSL, and global optical networks and No. 3 in routers by various market research firms.[107] Huawei says it spends 10 percent of revenue on R&D, employs 51,000 research staff, and filed for more than 8,000 foreign patents.[108] Although Huawei does not boast breakthrough products, it has a reputation for innovative applications and solutions in wireless communications.[109]

Huawei's top Chinese rival, ZTE, is not far behind. The $10 billion company also invests 10 percent of sales in R&D. It has 30,000 R&D staff and 18 R&D centers, including several in the U.S. Annual revenue have risen from around $3 billion to $10 billion since 2006, with 60 percent of those revenue from overseas.[110] It also contracts out research to more than 20 Chinese universities. Among its innovations is what ZTE calls the world's smallest base station for Long Term Evolution (LTE), a 4G mobile communication standard, which costs half the price of its previous base stations and lowers power consumption by 30 percent. Major research areas include cloud computing and wireless technology beyond 4G. ZTE also has emerged as the world's No. 4 maker of wireless handsets, most of them sold under the private labels of carriers like Vodaphone, T-Mobile, and Verizon. Of the 120 million units it expects to ship in 2012, 18 percent are expected to be smart phones.[111]

Breakthrough innovation remains a challenge for ZTE's handset business. As one ZTE researcher put it, "We see the amazing innovations by Apple." Also, most of the core components ZTE's handsets are imported, such as memory chips, displays, and batteries are from South Korea and Japan. Another challenge is that R&D costs are rising. In China, engineers now earn about $40,000 a year, compared to around $120,000 in Dallas, and job-hopping to other companies has become more intense. As a result, it wants to market its

---

[105] From presentation by Li Guoqing of the State Council Central Finance and Economics Office at the Sept. 19, 2011 National Academies symposium "U.S.-China Policy for Science, Technology, and Innovation" in Washington, DC.
[106] See *The Economist*, "Up, Up and Huawei: China has Made Huge Strikes in Network Equipment," Sept. 24, 2009.
[107] Rankings cited on Huawei corporate Web site.
[108] Huawei data from Web site.
[109] See Huawei press release, "Huawei Receives Innovation Awards for Contribution to CDMA Development," June 17, 2011, Huawei Web site (http://www.huawei.com/en/about-huawei/newsroom/press-release/hw-093167-cdma-award-guangzhou.htm).
[110] ZTE data.
[111] Data supplied by ZTE.

own branded handsets. Market pressures are a much bigger pressure to innovate than government directives. As the ZTE researcher noted, "I don't think about national policies. We look at the market for next year. I just encourage my designers to do fashionable designs."[112]

The multinational research centers cover a vast range of innovation themes. General Electric's China Technology Center (CTC) in Shanghai supports over 20 research labs addressing topics such as digital manufacturing, advanced materials, power electronics, and coal polygeneration.[113] Caterpillar's Wuxi research center, established in 2009, supports the company's Asia-Pacific research needs in areas which include electronics, hydraulics, fuel systems and engine testing.[114] Corning's research center in Shanghai, formed in cooperation with the Chinese Academy of Sciences, is performing research on ceramics, non-metal new materials and lithium cells.[115] In 2011, Intel established a research center in Chengdu with a target staffing of 200 people to develop technology for application in tablet computers and games.[116] In 2010, Toyota announced it would invest $689 million to establish a wholly-owned 200-person research center in China to study energy-efficient and new energy vehicles.[117] Boeing opened an R&D center in Beijing in 2010 to study airplane cabin environment and designs, advanced materials, and computer science.[118]

One hindrance to corporate innovation, in the eyes of some analysts, is the growing domination of state-owned and –supported companies at the expense of smaller, privately held enterprises. China's estimated 8 million small- and medium-sized enterprises account for 60 percent of the nation's industrial output and employ three-quarters of the labor force. They also generate 30 percent more output than state-owned enterprises with the same amount of capital, labor, and materials, according to Renmin University economist Dawei Cheng. Yet they receive little money from China's state-controlled banking system, which primarily lend to government-connected companies. The typical small Chinese enterprise receives only around 10 percent of its working capital from banks, compared to around 40 percent in South Korea and Thailand.[119]

---

[112] Interview with ZTE in Shanghai.(June 2011).
[113] http://ge.geglobalresearch.com/locations/shanghai-china/shanghai-china-featured-labs/.
[114] "Caterpillar Expands China Research Center," Business Daily Update (China) (January 10 2012).
[115] "Corning Sets Up Research Center on the Mainland," Chinadaily.com (June 29, 2011).
[116] "Tenacent, Intel to jointly set up Research Center," SinoCast (April 13, 2011).
[117] Toyota already operated an R&D center in Tianjin. "Toyota rolls out wholly owned Research Center," Chinadaily.com (November 22, 2010).
[118] "Boeing, Tsinghua Open Research Center" Chinadaily.com (October 21, 2010).
[119] Dawei Cheng, "China SMEs: Today's Problem and Future's Cooperation," PowerPoint presentation, School of Economics, Renmin University of China. Presentation can be accessed at http://www.slideshare.net/MIISChina/china-smes-339690.

> **Box 5.3**
> **Innovation with Chinese Characteristics**
>
> Westerners tend to equate innovation with creative ideas and game-changing goods and services. Innovation as generally practiced in China is more modest. The Chinese government actually uses several definitions of innovation. The Chinese government distinguishes "original innovation" (*yuangshi chuangxin*), "integrated innovation" (*jicheng chuangxin*) in which existing technologies are fused together in new ways, and "re-innovation" (*yinjin xiaohua xishou zaichuangxin*), in which imported technologies are assimilated and improved upon.[120] China has put a heavy emphasis on assimilating foreign technology in order to develop indigenous products. However, increasingly China has simply stressed the need to develop (and to favor in procurement) Chinese-owned IP, incorporated in products made by Chinese-owned companies. President Hu Jintao has committed to treat foreign invested enterprises in China as being Chinese for purposes of future procurement. This has not yet translated into complete national treatment at every level of the Chinese government.

China's state enterprises also enjoy many tax advantages, pay lower rates for loans, and do not have to dispense profits to shareholders. As a result, they are under little pressure to generate profits and can amass cash. The average tax burden of 992 state-owned enterprises was just 10 percent, compared to as much as 24 percent for private enterprise, according to the Unirule Institute of Economics, a non-government Chinese think tank. State-owned companies also pay real interest rates of just 0.016 percent for their capital and pay little or nothing for land. A Unirule study found that reported profits of 132 companies under management of the central government's State-Owned Assets Supervision and Administration Commission more than tripled from 2001 to 2008. Yet when low taxes, finance costs, and other special advantages are accounted for, the average real return on equity of state-owned enterprises over that period was negative 6.2 percent.[121]

---

[120] Denis Fred Simon, Cong Cao, and Richard P. Suttmeier, "The Evolution of Business China's New Science and Technology Strategy: Implications for Foreign Firms," *China Currents*, Vol. 6, No. 2, Spring 2007 (http://www.chinacenter.net/China_Currents/spring_2007/cc_simon.htm).
[121] See Jiang Hong, "State-owned Enterprises Research Project Press Release Conference & Academic Seminar Successfully Held in Beijing," Unirule Institute of Economics. 2011. Access at http://english.unirule.org.cn/Html/Events/20110308200838427.html.

Multinational Research Centers

Foreign companies have been key catalysts of China's rise in high-through industries through joint ventures, training programs, and technology-transfer agreements with Chinese partners negotiated in return for access to the domestic market. Foreign companies also have used China as a growing product-development base for their own products, establishing at least 750 R&D centers in Beijing, Shanghai, Guangzhou, Chengdu, and other cities as of 2005.[122] The vast majority of multinational R&D activity in China has been devoted to adapting products and technologies for the domestic market or for products manufactured in China for export.[123]

Such operations continue to grow. Since opening in 2000, General Electric's research center in Shanghai's Pudong district has grown to 1,500 researchers, two-thirds of whom have masters and Ph. D. degrees. The center files around 100 patents a year. Another 700 researchers are in centers in Beijing and Wuxi. The Shanghai center originally was intended to serve as an extension of GE Global Research in Niskayuna, N. Y., to tap lower-cost Chinese talent to help with next-generation products.[124] Although the center has 200 engineers working on long-term research, most of the center's work serves GE's $6 billion annual businesses in China in areas such as aircraft engines, medical equipment, water management systems, rolling stock, oil and gas technology, and home appliances—as well as GE's 26 manufacturing plants in China. GE also is setting up a network of "innovation centers." One in Xian, for example, focuses on light-emitting diodes, coal gasification, and aviation. Another in Chengdu is devoted to rural health care and oil and gas, while one in Shenyang works on manufacturing technology and energy.[125]

Innovations originally for the China market, however, increasingly make their way into products sold around the world. GE Healthcare is one success story. The unit's China operations develop lower-cost and simpler-to-use CT scanners and portable ultrasound equipment for China. Two-thirds of the equipment now is sold in other emerging markets and even in the U.S.[126]

Some Chinese research operations are starting to serve the global needs of U.S. companies. At the IBM Research facility, opened in 1995, has grown to 600 researchers. Virtually all work on global projects. "Originally our (Chinese) researchers were very timid and lacked the confidence and courage to do things," explained a GE representative. "That is completely different today. The experienced ones are really shining, doing extremely well in patents and

---

[122] Data from Lan Xue, "China's Innovation Policy in the Context of national Innovation System Reform," Tsinghua University, Aug. 27, 2007 (http://www.oecd.org/dataoecd/60/62/39310514.pdf).
[123] Company interviews in Beijing and Shanghai.
[124] US Company interview in Shanghai.(June 2011).
[125] US Company Interview in Shanghai.(June 2011).
[126] US Company interview in Shanghai.(June 2011).

contributing to global projects."[127] In all, IBM co-develops products with 10,000 Chinese partners in 350 cities. It also has 100 joint laboratories and technology centers with Chinese universities and offers curricula that have helped trained 860,000 Chinese students and 6,500 teachers.[128]

Microsoft's research center in Beijing also has become integral to development of next-generation products launched around the world. Established in 1998 as basic research laboratory with a couple hundred scientists in fields such as face recognition and motion tracking, the center now is "involved in almost every product Microsoft develops".[129] The center recently opened a new $400 million campus in Beijing's Haidan high-tech district that serves as Microsoft's research hub the Asia Pacific. The some 3,000 staff, including contractors, work in areas such as cloud computing, search tools, hardware development, the mobile Internet, and "natural user interfaces" that enable users to interact with computers using speech, gestures, and expressions.[130] About 95 percent of the work is deployed globally. Since it was established, the lab has published more than 3,000 papers in top international journals and conferences and contributed 260 innovations used in products such as Windows 7, Office 2010, Xbox, and Windows Mobile.[131] The Beijing center has played an especially important role in development of Kinect, the technology that allows users of Xbox 360 game players to control video and music with the wave of a hand or by making sounds.[132] The advantage of being in Beijing is the proximity to major universities such as Tsinghua and Peking University. In China, he said, Microsoft can recruit from among 300,000 computer science graduates a year, about 20 percent of whom are on par with the best in the U.S.

One challenge is that multinationals no longer are the preferred employers of new Chinese graduates, foreign executives said. Several multinationals also said they are losing considerable numbers of seasoned talent to Chinese state-owned enterprises or private Chinese companies willing to double and even triple their salaries, offer senior positions, and provide housing.

Coping with Indigenous Innovation Rules

American companies interviewed in China cited mounting pressures to transfer core technology and discrimination against foreign companies for contracts as their most serious concerns. The government, which has not signed

---

[127] US Company interview in Beijing (June 2011).
[128] From presentation by Mark E. Dean of IBM Research in *Building the 21st Century: U.S. - China Cooperation in Science, Technology, and Innovation*, op. cit.
[129] US Company Interview in Beijing (June 2011).
[130] Descriptions of major research programs at Microsoft Research Asia can be found on the center's Web site at http://research.microsoft.com/en-us/labs/asia/msrabrochure_english.pdf.
[131] Microsoft Research Asia Web site.
[132] US Company Interview. in Beijing (June 2011).

World Trade Organization protocols on government procurement, essentially compels foreign makers of a wide range of advanced products to manufacture in China and transfer technology to domestic companies.[133]

Companies said that such concerns have intensified in recent years. Although China is a major exporter of solar modules to Europe and the U.S., it requires at least 80 percent of equipment for its own solar power plants to be domestically produced.[134] Due to government procurement policies and rapid expansion by Chinese producers, the foreign share of China's annual new purchase of wind power equipment has fallen from nearly 80 percent to around 20 percent between 2004 and 2008.[135] Government bodies essentially require makers of lithium-ion batteries for cars to manufacture in China in order to sell into the growing domestic automobile market.[136] Leveraging its huge market for aircraft, China is using technology transferred by U.S. and European aircraft, engine, and avionics suppliers to achieve its ambitious plans to build a globally competitive commercial aerospace industry.[137] The government also aims to increase the self-sufficiency ratio of integrated circuits used in communications and digital household products to 30 percent and to 70 percent in products relating to national security and defense.[138] The Chinese policies spurred an outcry from American and European companies.[139] Beijing also has reportedly told General Motors that its sales of the Chevrolet Volt plug-in hybrids will not

---

[133] For example, see presentations by James M. Forcier of A123 Systems in *Building the U.S. Battery Industry for Electric-Drive Vehicles*.

[134] Keith Bradsher, "China Builds High Wall to Guard Energy Industry." *International Herald Tribune*, July 13, 2009.

[135] An extensive treatment of China's policies to promote its renewable energy equipment sector can be found in Thomas Howell, William A. Noellert, Gregory Hume, Alan Wm. Wolff,, *China's Promotion of the Renewable Electric Power Equipment Industry: Hydro, Wind, Solar, Biomass*, Dewey & LeBoeuf LLP prepared for National Foreign Trade Council, March 2010.

[136] See presentation by Jason M. Forcier of A123 Systems in forthcoming volume National Research Council, *Building the U.S. Battery Industry for Electric-Drive Vehicles: Progress, Challenges, and Opportunities*, Charles W. Wessner, ed., Washington, DC: The National Academies Press.

[137] See Roger Cliff, Chad J. R. Ohlandt, and David Yang, *Ready for Takeoff: China's Advancing Aerospace Industry*, RAND National Security Research Division for U.S.-China Economic and Security Review Commission, 2011.

[138] Ministry of Information Industry, "Outline of the 11th Five-Year Plan and Medium-and-Long-Term Plan for 2020 for Science and Technology Development in the Information Industry," Xin Bu Ke [2006] No. 309, posted on ministry website Aug. 29, 2006.

[139] A report by the European Union Chamber of Commerce in China said that "industrial-policy interventions and restrictions on foreign investment have been on the rise" and that "European companies are increasingly concerned by the tendency for local companies to be favored over foreign-invested ones." See European Union Chamber of Commerce in China, *European Business in China Position Paper 2009/2010*, executive summary (http://www.euccc.com.cn/images/documents/pp_2009-2010/executive_summary_en.pdf). Also see AmCham-China, "American Business in China: 2010 White Paper," May 22, 2010 (http://web.resource.amchamchina.org/news/WP2010LR.pdf).

qualify for subsidies of up to $19,300 per car available to other hybrids in China unless it transfers core technologies to domestic manufacturers.[140]

In response to high-level complaints by foreign governments, Chinese leaders in 2011 sought to allay major concerns. On a visit to Washington in January 2011, President Hu signed a joint statement with President Barack Obama in which he pledged that "China will not link its innovation policies to the provision of government procurement preferences." The statement also said China will seek to join the WTO Government Procurement Agreement by the end of 2011.[141] At a meeting with U.S. and Chinese businessmen, President Hu said of companies setting up operations in China: "In terms of innovation productions, accreditation, government procurement, (and) IPR protection, the Chinese government will give them equal treatment."[142] On June 29, 2011, China's Ministry of Finance said it would not require companies to transfer patents and other intellectual property to China as a condition for selling equipment and technology to the government. The ministry also said it would rescind other regulations linking government procurement contracts to "indigenous innovation" rules.[143]

Chinese officials have sought to assure multinationals in private meetings as well. An executive of one U.S. corporation with extensive operations in China said an official from the Ministry of Foreign Trade and Cooperation told him that the indigenous innovation policies don't apply to his company because the government regards it as a Chinese company. The executive said his company felt no more discrimination selling products in China than in other nations, such as India, and that it has a fair opportunity to provide input on formation of standards. "Indigenous innovation has been bashed down and killed for now," the executive said. "This is something we've taken off our list as something we have to focus on."[144] Another U.S. executive said that a high-level official of the Ministry of Science and Technology met with multinational representatives in June 2011 and explained that "indigenous innovation" is really about improving China's ability to generate new ideas rather than displacing foreigners, and that China's innovation system is open to multinationals. The MOST official also for the first time discussed ways in which foreign companies could participate in national government-funded research projects, an opportunity many multinationals have long sought.[145]

---

[140] Keith Bradsher, "Hybrid in a Trade Squeeze, *New York Times*, Sept. 5, 2011.
[141] The White House, "U.S.-China Joint Statement," Paragraph 27, Office of the Press Secretary," Jan. 19, 2011 (http://www.whitehouse.gov/the-press-office/2011/01/19/us-china-joint-statement).
[142] The White House, "Remarks by President and Obama and President Hu in a Roundtable with American and Chinese Business Leaders," Office of the Press Secretary, Jan. 19, 2011 (http://www.whitehouse.gov/the-press-office/2011/01/19/remarks-president-obama-and-president-hu-roundtable-american-and-chinese).
[143] *Reuters*, "China Eases Government Procurement Rules After U.S. Pressure," June 29, 2011.
[144] U.S. company interview in China.
[145] U.S. company interview in China

Other American business people based in China, however, said they remain under pressure to transfer core technology to Chinese companies, either to joint ventures or through licenses. One executive that does not want to license its core designs to Chinese companies for fear that they will become future competitors said government officials said it should transfer the knowhow because technology is a "human asset" and should be shared. The company is afraid that if it agrees to license one design, it will become a "slippery slope" in which more technology transfers would be expected. Although there have been "positive comments from individuals" at MOST, the executive said, "the general philosophy there hasn't changed."[146]

U.S. analysts and executives generally regard China's shifting rhetoric on indigenous innovation as a tactical retreat, rather than a fundamental shift in government thinking,[147] and attribute the mixed government messages to the different agendas of different agencies. MOST is regarded as the most dogmatic about enforcing indigenous innovation rules because it spearheads the drive to advance domestic industries. The Ministry of Information Technology has an interest in protecting Chinese IT and telecom companies. The trade ministry, MOFTEC, is more indifferent because its mission is to keep foreign markets open to Chinese products. State-owned industrial companies, meanwhile, tend to be strong advocates of indigenous innovation policies in order to protect their domestic franchises. Private Chinese companies mainly care about being able to buy the best products. The type of foreign business also makes a difference, these executives said. Companies selling expensive high-tech hardware and core components in high-priority Chinese industries are under the most pressure to transfer technology, they said. Companies that offer critical services as well as hardware are under the less pressure as long as most of their products are made in China.[148]

Opportunities for Collaboration

Despite these disputes and the indigenous innovation policy, there are substantial opportunities for scientific and technological collaboration between China and the U.S. Mr. Yang of the Ministry of Science and Technology said China remains committed to international collaboration as a vehicle to "absorb innovation" that can be adapted to "Chinese conditions."[149]

At a government-to-government level, the U.S. and China have signed some 50 cooperative agreements over the past decade. In energy research and

---

[146] U.S. company interview in China
[147] For example, see Adam Segal, "China's Innovation Wall: Beijing's Push for Homegrown Technology," *Foreign Affairs*, Sept. 28, 2010 (http://www.foreignaffairs.com/articles/66753/adam-segal/chinas-innovation-wall) and US-China Business Council, *Issues Brief: China's Domestic Innovation and Government Procurement Policies*, March 2011.
[148] Company interviews in China.
[149] Yang presentation, op. cit.

life sciences, "the United States and China are, in every sense, building a global partnership," noted Deputy Assistant Secretary of State Anna Borg.[150]

Cooperation through universities is also growing. The University of Maryland, for example, has an extensive relationship with China.[151] As the university's former president C. Dan Mote has pointed out, its Institute for Global Chinese Affairs has trained 3,000 Chinese executives since 1995, while 160 Chinese executives have received one-year degrees from Maryland's Executive Master's in Public Administration program. The University of Maryland also has a special "international incubator" that has helped launch 11 Chinese companies in industries such as solar energy and software. In 2002, the Chinese government and Maryland set up a joint research park near campus that now houses facilities of companies from Beijing, Shanghai, and Guangzhou. As Caroline Wagner has pointed out, the growth of such networks creates unprecedented opportunities for cooperation in science to address shared challenges in areas such as energy and health.[152]

As the world's number one and number two economies, the U.S. and China are the two biggest consumers of energy and together emit 40 percent of the world's greenhouse gasses. It is in both nations' interests to accelerate development of clean energy. The National Renewable Energy Laboratory (NREL), based in Boulder, Colo., has a range of collaborations with Chinese companies, research institutes, and government agencies, from long-range planning of wind-power to commercializing specific bio-fuels.[153] Two joint research centers, one focusing on wind power and the other on solar, also have been established. A wide-ranging Sino-U.S. partnership in bio-fuels involves several Department of Energy and Department of Agriculture labs, Chinese research institutes, and mainland companies such as Sinopec, PetroChina, CNOOC, and COFCO.

The Sino-U.S. partnership in medicine is even more deep-rooted. Just as America experienced as its population aged, cancer and other chronic diseases are overtaking infectious diseases in China as the top killers and present a "major health care crisis," according to Anna Barker of the National Cancer Institute.[154] China has 1.6 million cancer deaths a year and reported

---

[150] From presentation by Deputy Assistant Secretary of State Anna Borg in *Building the 21st Century: U.S. - China Cooperation in Science, Technology, and Innovation.*
[151] C. Dan Mote, "Universities as Drivers of Growth in the United States," in National Research Council, *Building the 21st Century : U.S. China Cooperation on Science, Technology, and Innovations,* Washington, DC: The National Academies Press, 2011.
[152] Caroline S. Wagner, *The New Invisible College: Science for Development*, Washington, DC: Brookings, 2008.
[153] From presentation Robin L. Newmark of the National Renewable Energy Laboratory in *Building the 21st Century: U.S. - China Cooperation in Science, Technology, and Innovation.*
[154] From presentation by Anna Barker of the National Cancer Institute in *Building the 21st Century: U.S. - China Cooperation in Science, Technology, and Innovation.*

2.2 million new cases in 2009. The crisis "will get much, much worse in the next 10 to 15 years," she said.

The U.S. needs China's help, too, in order to accelerate the discovery of new treatments and contain skyrocketing drug-discovery costs. New cancer cases in the U.S. are forecast to rise by at least 30 percent by 2020.[155] Annual U.S. spending on cancer treatment is expected to rise from $213 billion to $1 trillion a year. China has immensely valuable data on cancer cases and the largest talent pool of microbiologists, many of them U.S.-trained. China also is a leader in genomics research; its researchers were among the first to identify the SARS genome. The National Cancer Institute is working with Chinese institutes on an ambitious project to sequence genomes of all cancers. It also is partnering with the Beijing Genomics Institute, the world's largest next-generation sequencing center, in brain-tumor research.

China's depth in nanotechnology research, which Dr. Barker said will "touch everything we do in medicine in the next 10 years," is another area of "very strong collaboration." Five thousand scientists at 50 Chinese universities, 20 Chinese Academy of Science Institutes, and 300 nano-technology enterprises focus on the field.[156]

While China needs international cooperation, however, Mr. Deng of the State Council Research Office stressed that it still must develop its internal capabilities. "On the one hand, we have to increase our collaboration and exchange with other countries," he said. "But on the other hand, we have to solve problems with our own efforts. There are a lot of problems that can be solved only with international cooperation." He added that because China is such a large country, it has many "urgent problems" such as water management, energy, and environmental challenges that "we have to solve with self-reliance."[157]

Assessing Chinese Innovation

China's destiny as a science and technology superpower appears to be assured. The nation's steady policy focus and heavy investments in R&D, human capital, infrastructure, and industrial capacity—combined with the world's biggest growth market—all put China in a powerful position to be a leading if not dominant force across a spectrum of emerging advanced industries. China also can play an invaluable role as a research partner in conquering the world's biggest 21st century challenges.

The nation has all of the potential to become a leading force in innovation as well. China's emergence as a source of global patents, for example, demonstrates that it has tremendous inventive capacity key high-tech

---

[155] Ibid.
[156] Data from Science (2005) 309: 65-66.
[157] Deng Wen Kui presentation, op. cit.

sectors such as digital computers and telephone and data transmission systems.[158]

Whether China is on track to achieving its desire to become a giant engine of innovation is less clear. The study by CENTRIC offered a negative prognosis:

> "... (T)he Chinese model of science in its present form is unlikely to deliver the types of creative research on which future high-technology leadership will depend..... China has yet to show that it can meaningfully use the tools of the state to drive the commercialization of discoveries in research labs in a competitive manner. And the nation's drift in a techno-nationalist direction could compromise China's enabling international scientific links.[159]

Mu Rongping of the Chinese Academy of Sciences maintains that the "distance between China and developed nations is still very, very large." Dr. Mu observes that there are "two Chinas"—one that is progressing rapidly in terms of the inputs needed to innovate and another that lags in terms of execution. He notes that China has leapt from 26th place to 17th between 2000 and 2006 in an index of 38 countries measuring national innovation capacity using a model that gives heavy weight to R&D spending and economic growth. In an index of national innovation effectiveness, however, China ranks No. 37, behind Mexico and Romania.[160] And while China has significantly increased its output of scientific publications, the average citation rate for Chinese papers in the Essential Science Indicators database over the period 1998-2008 was still well below the world leaders in science and technology.[161] [See Figure 5.4]

Chinese industries have indeed proved remarkably capable of "catching up" in maturing technologies and driving down prices. With few exceptions, however, they have yet to prove capable of competing at the leading edge. While there has been an explosion of patents, doubts have arisen over the quality of those patents.[162] Although Chinese inventors filed 203,481 patent applications in 2008, according to the World Intellectual Property Organization, more than

---

[158] Evey Y. Zhou and Bob Stembridge, "Patented in China: The Present and Future State of Innovation in China," Thompson Reuters, 2010.
[159] Springut, Schlaikjer, and Chen, op. cit.
[160] See Mu Rongping, Song Hefa, and Chen Fang, "Innovative Development and Innovation Capacity-Building in China," *International Journal of Technology Management,* Vol. 51, No. 2/3/4, 2010.
[161] UNESCO, *UNESCO Science Report 2010*, p. 391.
[162] For example, see Jody Lu, "Who is Making Junk Patents?", *China Daily*, March 6, 2011 (http://ipr.chinadaily.com.cn/2011-03/06/content_12126586.htm). The local government practice of paying patent fees for the first several years also is believed to inflate patent applications. See Zhou and Stembridge, op. cit.

95 percent of those were filed domestically with the State Intellectual Property Office, note Anil K. Gupta and Haiyan Wang, authors of the book *Getting China and India Right*.[163] Chinese inventors accounted for only 473 so-called "triadic" patents filed in the U.S., the European Union, and Japan, the world's prime patent issuers. That compares to 14,525 triadic filings from Europe, 14,399 from the U.S., and 13,446 from Japan. In fact, China accounts for only 1 percent of patent filings and grants by any of the leading patent offices outside of China, even though it accounts for 11 percent of the world's R&D spending. Gupta and Wang also conclude that the vast majority is for "tiny changes on existing designs." Therefore, they label China an innovation "paper tiger" that emphasizes quantity over quality, resulting in "a pandemic of not just incrementalism but also academic dishonesty."[164]

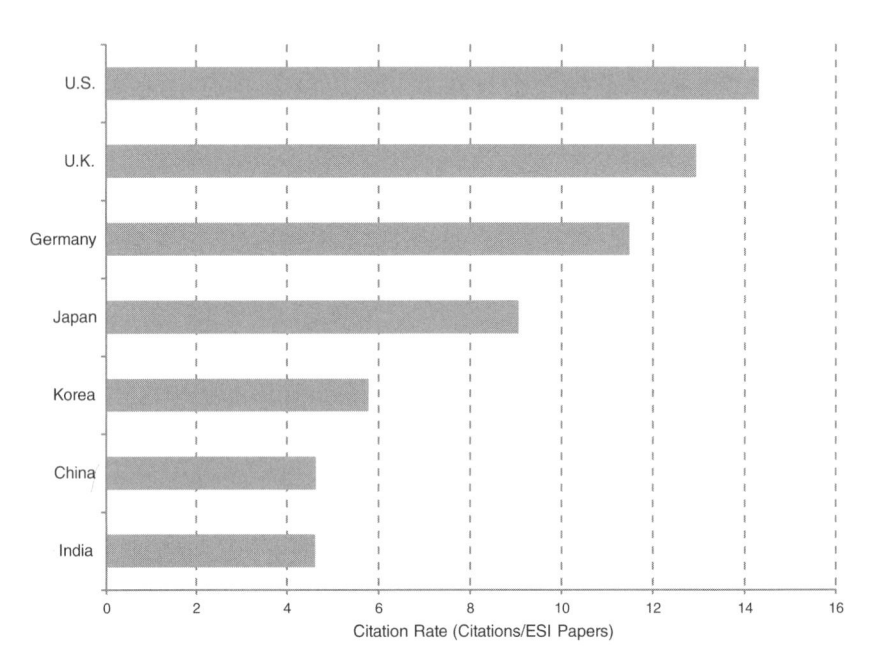

FIGURE 5.4 Citation rate for scientific papers 1998 to 2008.
SOURCE: UNESCO, *UNESCO Science Report 2010*, Paris: UNESCO Publishing, 2010, p.391.

---

[163] Anil K. Gupta and Haiyan Wang, *Getting China Right*, San Francisco, Calif.: Jossey-Bass, 2009.
[164] Anil K. Gupta and Haiyan Wang, "Chinese Innovation is a Paper Tiger," *Wall Street Journal*, July 28, 2011.

A major question is whether a business culture that has focused on scale and market share is ready to shift to a model driven by adding value and creating breakthrough products. Another question is whether state-led policies and programs that try to put national boundaries around intellectual property and curtail foreign competition can succeed in an era when most of the world is moving toward models of open innovation and global cooperation. To the contrary, some analysts warn, such an approach could ultimately make Chinese industry less competitive.[165]

Some Chinese officials agree that fulfilling the high aspirations for innovation will require reform of government institutions and corporate culture. "At present, we believe the innovation of structure is more important than innovation of technology," said Mr. Li of the State Council Central Finance and Economics Office. "Without organizational innovation, there cannot be technological innovation," he said. "We have to learn from the United States."[166]

China's leadership, however, has proved pragmatic and willing to change course if it finds certain policies are retarding economic development. At a time when the U.S. is struggling to maintain funding for current programs, China is providing the financial resources and policy support needed to build a 21st century innovation system. The question now is whether it can devise the right policy framework for China live up to its potential.

While China's leadership has proven to be pragmatic, committed, and willing to spend, China continues to face major challenges in its quest to become an innovator. Even so, as documented in Carl Dahlman's recent work, the sheer scale of China's policies, R&D expenditures, and markets are having an important impact in the U.S. and the rest of the world.[167]

India's Changing Innovation System

The dual faces of its economy define India's great innovation challenges. On the one hand, India is a global leader in information technology and business-process outsourcing services, which account for nearly $60 billion in annual exports and employ more than 2.5 million.[168] On the other, more than one-third of India's 1 billion people live below the poverty line, and three-

---

[165] In her address at a conference on Chinese and U.S. innovation policies hosted by the National Academies, U.S. Assistant Secretary of State Anna Borg warned that Chinese investment barriers or domestic intellectual-property requirements "will ultimately be self-defeating." See Anna Borg presentation in *Building the 21st Century: U.S. - China Cooperation in Science, Technology, and Innovation*, op. cit.
[166] Li Guoqiang presentation, op. cit.
[167] Carl J. Dahlman, "The Innovation Challenge: Drivers of Growth in China and India," National Research Council, Innovation Policies for the 21st Century (Washington, DC: The National Academies Press, 2007) pp 45-60.
[168] NASSCOM, "Indian IT-BPO Industry," 2011. Data can be accessed at http://www.nasscom.in/indian-itbpo-industry.

quarters of those poor live in rural areas.[169] Only 16 percent of India's population has completed high school and 61 percent of the adult population is literate, compared to 97 percent in China. The World Bank estimates that only 4 percent of India's workforce is formally employed in the modern private sector.[170]

For India's world-class technology companies, the goal is to develop more proprietary intellectual property and gain global market share. Private Indian companies such as information-technology giants Infosys and Tata Consulting Services, pharmaceutical producers Piramal Life Sciences and Ranbaxy, and automotive companies Tata Motors and Mahindra & Mahindra and among the many Indian corporations devoting greater resources to R&D, releasing innovative goods and services at home, and striking out into global markets with branded products.

For the Indian government, however, the most urgent priorities in science and technology policy have been basic economic development. Although India's economic growth rate has accelerated sharply since 2003, the benefits of India's dynamic technology sectors have been slow to make a difference in the lives of hundreds of millions of people living in poverty. India is not just focused on improving its capacity to create new products, therefore. The Indian Government also now is paying more attention to what it calls "inclusive innovation," which is defined as "using innovation as a tool to eliminate disparity and meet the needs of the many."[171]

To satisfy the demands of both industry and society, India must dramatically improve its national innovation system.[172] India has enormous potential. It has an immense and growing pool of young English-speaking technology talent, a much younger population than China's, and a large diaspora of overseas Indian technology entrepreneurs and researchers who are rebuilding ties in their homeland. India's economy is projected to grow by more than 7 percent a year for decades. India also has a highly innovative private sector and a number of elite higher-education institutes. India is an important high-tech R&D base for multinationals.

---

[169] From presentation by T. S. R. Subramanian in National Research Council, *India's Changing Innovation system: Achievements, Challenges, and Opportunities for Cooperation*, Charles W. Wessner and Sujai J. Shivakumar, editors, Washington, DC: The National Academies Press, 2007.
[170] World Bank data cited in World Bank, *Unleashing India's Innovation: Toward Sustainable and Inclusive Growth*, Mark A. Dutz, editor, The International Bank for Reconstruction and Development, 2007 (http://siteresources.worldbank.org/SOUTHASIAEXT/Resources/223546-1181699473021/3876782-1191373775504/indiainnovationfull.pdf).
[171] See National Innovation Council, *Towards a More Inclusive and Innovative India*, September 2010 (http://www.innovationcouncil.gov.in/downloads/NInC_english.pdf).
[172] "India suffers from inefficiency in transforming its S&T investments into scientific knowledge (publications) as well as into commercially relevant knowledge (patents)." National Academy of Sciences, *S&T Strategies of Six Countries*, op. cit., p. 43.

While India is becoming a top global innovator, an extensive World Bank study concluded that the country is "underperforming relative to its innovation potential—with direct implications for long-term industrial competitiveness and economic growth."[173] The challenges are numerous. India invests only around 1 percent of GDP in science and technology.[174] [See Figure 5.5] Government controls around 70 percent of national R&D spending,[175] and the biggest recipients have been areas relating to national security, such as atomic energy, aerospace, and ocean exploration. Venture capital is scarce. The talent pool is constrained by the facts that only around 12 percent of college-age Indians are enrolled in higher education, and only 16 percent of Indian manufacturers offer worker training, compared to 42 percent in South Korea and 92 percent in China. India produces only 6,000 Ph. D.s a year in science and 1,000 in engineering.[176] What's more, the legacy of India's obsession with self-sufficiency since independence in 1947 leaves it with some of the highest barriers to product imports, foreign direct investment, and inflows of intellectual property[177] among major trading nations—constraining its access to global innovation.[178]

Linkages between government research institutions and industry are weak. There is little collaboration between India's 400 national laboratories and 400 national R&D institutes and private companies.[179] A European Commission analysis noted that 70 percent of technologies developed by government-funded laboratories remain on the shelf. "A major weakness of the system was the lack of an innovation ecosystem where risk capital and intermediary mechanisms existed to foster and promote technology transfer and the commercialization of public R&D," the report said.[180] India's 358 universities and famed Indian Institutes of Technology, meanwhile, traditionally have played little role in commercializing technology.[181] The World Bank observed that even though recent government policies aimed at generating, commercializing, and absorbing

---

[173] World Bank, *Unleashing India's Innovation: Toward Sustainable and Inclusive Growth*, Mark A. Dutz, editor, The International Bank for Reconstruction and Development, 2007 (http://siteresources.worldbank.org/SOUTHASIAEXT/Resources/223546-1181699473021/3876782-1191373775504/indiainnovationfull.pdf).
[174] GERD = gross domestic expenditures on research and development.
[175] UNESCO data.
[176] Data cited in World Bank, *Unleashing India's Innovation*, op. cit.
[177] The World Bank uses foreign licensing and royalty payments as indicators of intellectual property imports.
[178] Vinod K. Goel, Carl Dahlman, and Mark A. Dutz, "Diffusing and Absorbing Knowledge," World Bank, op. cit.
[179] National Research Council, *S&T Strategies of Six Countries*, op. cit., pg. 38.
[180] European Commission Enterprise Directorate-General, *INNO-Policy Trend Chart Innovation Policy Progress Report: India 2009*. This report can be downloaded at http://www.proinno-europe.eu/trendchart/annual-country-reports.
[181] Martin Gruber, and Tim Studt. "2011 Global R&D Funding Forecast: The Globalization of R&D," R&D Magazine, Dec. 15, 2010.

R&D had achieved some important successes, "their effectiveness has not matched the needs of the Indian economy or been commensurate with the resources invested in them." One reason is that private corporate participation has been minimal. Instead, initiatives are owned and managed by government bureaucracies that "suffer from complex, overlapping structures for policy making and decision making."[182]

India's New Innovation Push

India now is undertaking a number of initiatives to transform its innovation system.[183] As the Planning Commission's steering committee on

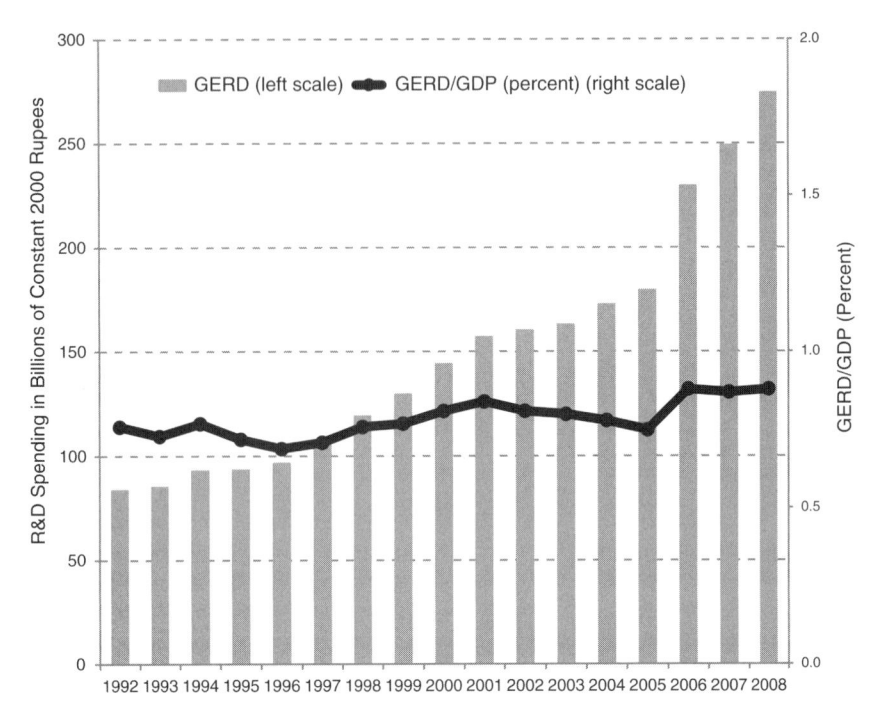

FIGURE 5.5 India invests less than one percent of GDP on R&D spending. SOURCE: UNESCO, *UNESCO Science Report 2010*, Paris: UNESCO Publishing, 2010, p.371.
NOTES: GERD is gross expenditure on research and development. Years refer to fiscal years.

---

[182] Carl Dahlman, Mark A. Dutz, and Vinod K. Goel, "Creating and Commercializing Knowledge," World Bank, *Unleashing India's Innovation*, op. cit.
[183] For a summary of recent initiatives, see 'India Rising' Science (24 February 2012, Vol. 335),

science and technology explained in its report for the current Five-Year Plan, a "strong and vibrant innovation ecosystem" requires an education system that nurtures creativity, an R&D culture and value system that supports both basic research and applied technology, an industry culture that is keen to equity and foreign companies that can be involved.[184]

After doubling national investment in R&D spending between 2002 and 2008 in current Indian Rupees, the government aims to boost research funding by another 220 percent under the current Five-Year Plan for 2007 to 2012. The goal is to boost national R&D investment to 2 percent of GDP by 2020. The government also is both expanding and reforming the nation's higher-education system to strengthen basic research and commercialization. The government's overarching science and technology strategy, as defined in "Technology Vision 2020," puts a heavy emphasis on sectors like agriculture, food processing, health care, electric power, and infrastructure.[185]

The Five-Year plan calls for a number of new universities and greater collaboration between academia, research institutes, and industry.

In terms of research infrastructure, the plan provides for 10 "flagship" programs in areas such as water supply, sanitation, health, and telephony and a national network of globally competitive "centers of excellences" in a range of technologies.[186] To help modernize India's manufacturing sector, the National Council for Skill Development was established to upgrade 5,000 industrial training institutes.

Prime Minister Manmohan Singh, who has pledged that India will embark on a Decade of Innovation, has launched an ambitious effort to formulate a new national innovation strategy. In 2010, Prime Minister Singh established a National Innovation Council charged with formulating a roadmap for the decade ahead that is described as "the first step in creating a cross-cutting system which will provide mutually reinforcing policies, recommendations, and methodologies to implement and boost innovation performance in the country."[187]

Among the Council's early proposals are to set up setting up innovation councils both for states and for different sectors. The council also calls for programs to promote regional innovation clusters, innovation centers at

---

[184] Government of India Planning Commission, "Report of the Steering Committee on Science and Technology for Eleventh Five Year Plan (2007-2012)," December 2006.
[185] "Technology Vision 2020" reports for a number of sectors were prepared by the Technology Information, Forecasting and Assessment Council under the Department of Science & Technology to study and support future technology needs of national importance. Access reports at http://www.tifac.org.in/.
[186] Ibid.
[187] From Introduction on National Innovation Council Web site. See http://www.innovationcouncil.gov.in/index.php?option=com_content&view=article&id=26&Itemid=5.

universities, awards and competitions, outreach programs, and international collaboration.[188]

Focusing on Inclusive Innovation

One of the National Innovation Council's central goals is to foster inclusive innovation[189] that provides "access, affordability and quality, and fosters innovations at the grassroots."[190] The concept builds on the Indian knack for *Jugaad*, or the development of makeshift solutions under conditions of scarcity.[191] The aim, however, is to go beyond relying on informal, makeshift solutions to everyday needs and build a more formal system of low-cost innovation that address the needs of the majority of Indians living at or near poverty.[192] As a council publication explains:

> *India needs more "frugal innovation" that produces more "frugal cost"' products and services without compromising safety, efficiency, and utility of the products. These innovations should also have "frugal' impact on the environment to be sustainable in the long term.*[193]

The National Innovation Council has recently announced a $1 billion Inclusive Innovation Fund to create a "funding platform for solutions aimed at the Bottom of the Pyramid."[194] The government would provide the initial capital for a "fund of funds" that would invest in other intermediate funds and institutes, which in turn would provide seed capital to grassroots innovation projects and that will raise money from companies, banks, insurance companies, and investors. The expectation is that the government contribution will be supplemented by $9 billion in private capital.

---

[188] National Innovation Council, "India Decade of Innovations 2010-2020 Roadmap," March 2011 (http://www.innovationcouncil.gov.in/ideas/ppt1.php#).
[189] For an earlier analysis of inclusive innovation in India and methods to promote it, see Anuja Utz and Carl Dahlman, "Promoting Inclusive Innovation," World Bank, *Unleashing India's* Potential, op. cit.
[190] See National Innovation Council, *Towards a More Inclusive and Innovative India*, September 2010 (http://www.innovationcouncil.gov.in/downloads/NInC_english.pdf).
[191] For a review of the accomplishments and limitations of jugaad, see Navi Radjou, Jaideep Prabhu and Simone Ahuja, , *Jugaad Innovation: Think Frugal, Be Flexible, Generate Breakthrough Growth*, San Francisco: Jossey-Bass, 2012. Some Indian business leaders have criticized jugaad; Anand Mahindra notes that it is "all too often used to excuse cut-price, second-rate answers to his nation's pressing business and social problems." See *Financial Times*, "More with less." May 19, 2012.
[192] See Rishikesha T. Krishnan, *From Jugaad to Systematic Innovation: The Challenge for India*, Indian Institute of Management, Bangalore, self-published, 2010.
[193] National Innovation Council, *Towards a More Inclusive and Innovative India*, op. cit.
[194] *Indian Express*, "$1 bn India Innovation Fund by July, January 17, 2012.

Developing Strategic Sectors

India also has several large initiatives to boost its global standing in strategic science and technologies areas. The government has more than tripled the budget for the Council of Scientific and Industrial Research, which oversees India's national laboratories, in recent years. It also has announced plans to establish 50 centers of excellence in science and technology over six years. Centers will include biotechnology, bio-informatics, nano-materials, and high performance computing, and engineering and industrial design. They will offer doctorate programs and be based at existing institutions.[195]

India has big ambitions in nanotechnology. Under the 10 billion rupee ($220 million) National Science and Technology Nano Mission, created in 2006, three new R&D institutes are being created. Some 50 to 60 science and technology institutes also are to be involved in building nanotech clusters across the country.[196]

In renewable energy, the government announced it aims to quadruple power generation from a range of non-carbon sources to 72.4 gigawatts by 2022, with solar power accounting for 20 gigawatts.[197] India also wants to build on its strength on space research, where it is a world leader in satellite communications, study of the environment, and remote sensing. India has sent 55 satellites into orbit since 1975. In 2009, the National Remote Sensing Center of the Department of Space launched a Web based, three-dimensional satellite imagery tool called *Bhuvban* in August 2009 to offer images of Indian locations superior to that provided by other Virtual Globe software like Google Earth and Wiki Mapia.[198] India also has set a target of a manned space flight by 2016.

Upgrading Higher Education

Improving the quality and quantity of higher education is one of the government's most urgent priorities. The nation's elite science and technology schools are the nine Indian Institutes of Technology, and several strong institutes of information technology, medicine, and science. India also has 10 first-rate graduate business schools, and several Indian Institutes of Management. Seats in these schools are extremely scarce, however. While some of India's 358 universities and more than 20,000 colleges are huge by Western standards, overall quality is poor.[199] India's National Knowledge Commission estimates the

---

[195] Akshaya Mukul, "Govt Plans 50 Centres of Excellence for Science & Tech," *The Times of India*, Jan. 17, 2011.
[196] *Indo-Asian News Service*, "National Mission to Make India Global Nano Hub," Nov. 5, 2007.
[197] Sreejiraj Eluvangal, "Renewable Energy Goal Quadrupled," *DNA Money*, Dec. 30, 2010.
[198] Details on Bhuvan can be accessed on the NRSC Web site (http://bhuvan.nrsc.gov.in).
[199] In his presentation in *India's Changing Innovation System,* former top government official T. S. R. Subramanian said of the nation's more than 500 engineering schools and 600 management institutes, only the IITs and Indian Institutes of Management are world class. The rest greatly need improvement.

nation needs 20 to 30 new "appropriately scaled" universities over the medium term and 1,500 new universities over the long term.[200]

Indian higher education also suffers from a shortage of qualified senior professors, in large part due to poor salaries. Retired IIT-Delhi director P. V. Indiresan, who founded the school's Centre for Applied Research in Electronics and was twice awarded India's highest prize for inventors, said in 2006 that even IIT professors earn roughly as much as an intern at a top Indian company. Partly as a result, India already suffers from acute skill shortages. A study of 25 industrial sectors by the Federation of Indian Chambers of Commerce and Industry in 2007 found there is a 25 percent shortage of skilled personnel in engineering.[201]

Universities also play a small role in the innovation system compared to those in other countries. They account for just 5 percent of India's R&D and interact little with the private sector. The IITs, for example, are renowned for the extremely high caliber of their graduates, who include many of the nation's most famous industrialists, scientists, executives, and business academics. However, the institutes have had few research ties to business, generated few startups, and produce few patents. The constraints on the IITs have included heavy bureaucratic control by the Ministry of Human Resource Development, which some commentators say makes it difficult to respond flexibly to industry needs, expand, and improve their financial base. IITs depend on the government budgets. Only recently have they been allowed to accept donations directly from alumni abroad.[202]

The government is mapping strategies to address all of these shortcomings. It seeks to raise the gross enrollment ratio in higher education, or the number of qualified students who attend, from 11 percent in 2007 to 21 percent in 2017. That would require 8.9 percent annual growth in college and university enrollment.

To accomplish this, the government increased the education budget increased fivefold in the 11th Five year Plan for 2007 to 2012 over the previous five-year plan. The government has established a National Skill Development Mission that hopes to use public-private partnerships to open 1,600 new information technology institutes and polytechnics, 10,000 vocational schools,

---

[200] For an explanation of the National Knowledge Commission's recommendations, see "FAQs on NKC Recommendations on Higher Education" of the commission's Web site at http://www.knowledgecommission.gov.in/downloads/documents/faq_he.pdf.
[201] Federation of the Indian Chamber of Commerce and Industry, "Survey of Emerging Skill Shortages in Indian Industry," 2007 (http://www.ficci-hen.com/Skill_Shortage_Survey_Final_1_.pdf).
[202] From presentation by P. V. Indiresan, retired professor of Indian Institute of Technology-Delhi in *India's Changing Innovation System*, op. cit.

and 50,000 skill-development centers across the country. The goal is to train 10 million new skilled workers a year.[203]

In terms of elite institutions, the government plans to increase the number of Indian Institutes of Technology from nine to sixteen, add five Indian Institutes of Science Education and Research, six Institutes of Management, and 20 Indian Institutes of Informational technology.

Getting universities to play a far bigger role in India's innovation ecosystem and upgrade their standards are other top goals. The government is starting to overhaul the entire system of science and engineering education, explained former Council of Scientific Industrial Research Director General Ramesh Mashelkar.[204] A committee studying reforms of IITs is expected to call for measures to grant them greater management and financial autonomy from the government and to encourage more collaboration with industry.[205] The government also proposes to establish 14 new "innovation universities" that will rank among the best in the world in research.[206]

Yet another initiative involves building interconnections among colleges and universities and to expand their geographic reach. The Indian National Knowledge Network is a government project to build an ultra high-speed broadband network of 10 gigabits and up to connect schools and government agencies across the country. The first phase is operation with a 2.5-gigabit network connecting 96 institutions and 15 virtual classrooms. The plan calls for investing $1.35 trillion over 10 years building more than 1,500 nodes.[207]

Reforming National Laboratories

In a 2005 survey of top executives of Indian manufacturers, 71 percent said that the lack of collaboration between industry and research institutes was the main hurdle to innovation in India.[208] India's national laboratories now are starting to pay more attention to commercialization and linking their research to the greater needs of industry and society. The Council of Scientific Industrial Research, which controls 38 national laboratories and many research institutes, began reforms a decade ago to improve their performance and economic

---

[203] Ministry of Labor and Employment press release, March 5, 2008 (http://pib.nic.in/release/rel_print_page1.asp?relid=36021).
[204] See presentation by Ramesh Mashelkar of the Council of Scientific Industrial Research in *India's Changing Innovation System*.
[205] *Hindustan Times*, "More Autonomy, New Programmes for IITs," Jan. 16, 2011.
[206] For an explanation of innovation universities, see National Innovation Council, "Concept Note on Innovation Universities Aiming at World Class Standards," at http://www.education.nic.in/uhe/Universitiesconceptnote.pdf.
[207] Background on the National Knowledge Network can be found on the Department of Information Technology Web site at http://www.mit.gov.in/content/national-knowledge-network.
[208] Confederation of Indian Industry and Boston Consulting Group, "Manufacturing Innovation: A Senior Executive Survey," 2005.

relevance. Instead of focusing on many small projects and acting like independent entities, CSIR labs now take on larger, networked projects and collaborate more with each other, according to Dr. Mashelkar. Whereas costs had once been no consideration, now time and costs are "sacrosanct," he said. Perfunctory monitoring has given way to stringent monitoring. Rather than being inward-looking, the labs now look outside to harness synergies.[209]

In 1996, CSIR became India's first research institution to manage its own intellectual property. Each laboratory now has marketing teams, and senior staff can serve on boards of private firms. CSIR also introduced financial incentives to motivate scientists, and labs have been allowed to put earnings into reserve funds for carrying out additional research. Patents earned CSIR labs rose from low single digits to more than 200 between 1995 and 2005. Published science papers by CSIR researchers have risen sharply, as have U.S. patents award to the council.

India's Innovative Companies

Although the share of national R&D conducted by businesses in India rose from 19 percent in 2002 to 30 percent in 2008 [See Figure 5.6], industry plays a smaller role in innovation than in many other nations.[210] In China, for example, industry performs around two-thirds of R&D.[211]

Nevertheless, top Indian companies have demonstrated an impressive capacity and desire to innovate in the two decades since they have been freed of the restraints of the country's once-onerous industrial licensing system.[212] Enterprise R&D leapt by seven-fold between 1991 and 2004.[213] A survey of Indian companies in 2006 found that 40 percent had developed a major new product and 62 percent had upgraded an existing product lines, much higher than in China and at about the same level as in the Republic of Korea.[214] In a survey of 83 top manufacturing executives, 82 percent said they believed that generating organic growth through innovation is essential for success.[215]

India's elite corporations are remarkably well-integrated into global innovation networks. The country's information-technology services industry,

---

[209] Mashelkar, op. cit.
[210] Ibid., data from Indian Central Statistics Organization.
[211] Ibid., data from Indian Central Statistics Organization.
[212] Under India's Soviet-inspired planned economy from 1947 through the introduction of reforms in 1991, Indian companies were regulated by an system of licenses and permits derisively known as the License Raj that controlled what and how much companies could manufacture, prices, sources of capital, closing of factories, and firing workers.
[213] Data cited in Dutz and Dahlman, op. cit.
[214] Ibid.
[215] Confederation of Indian Industry and Boston Consulting Group, op. cit.

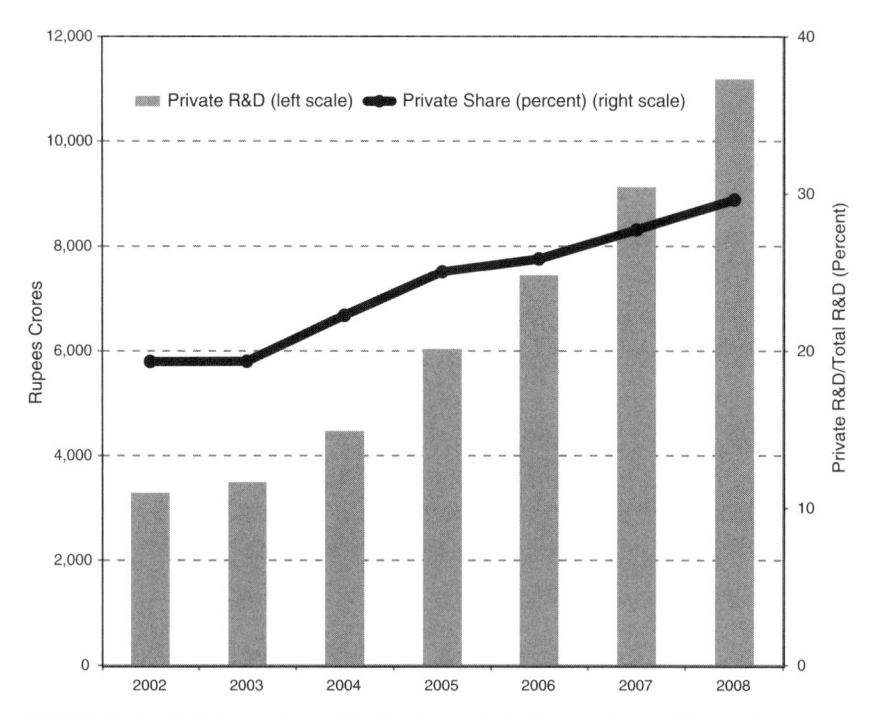

FIGURE 5.6 R&D conducted by business in India: total spending and as a share of national R&D.
SOURCE: Government of India, Ministry of Science and Technology, *Research and Development Statistics 2007-2008* (May 2009), Table 1.
NOTE: Data refer to fiscal years.

for example, has played an integral role in transforming global services industries. Once primarily providers of low-cost outsourced software and call-center services, Indian corporations such as Tata Consulting Services, Infosys, Wipro, and Genpact now help clients ranging from the world's biggest insurance companies and banks to airlines and legal firm develop innovative business processes that boost efficiency, cut cost, and improve customer service.[216] India's biggest IT services companies directly compete with giants such as IBM, Accenture, and Hewlett Packard, who also have major operations in India. NASSCOM, India's IT services industry association, estimates that India accounts for 34 percent of the worldwide business process outsourcing (BPO)

---

[216] Pete Engardio, "The Future of Outsourcing: How it's Transforming Whole Industries and Changing the Way We Work," *BusinessWeek*, Jan. 30, 2006.

market.[217] In 2011, annual revenues of India's IT and business-process outsourcing industry are expected to reach $88.1 billion, with exports accounting for around $59.4 billion of that.[218]

India's pharmaceutical industry, meanwhile, has become an important ally to Western companies that are under mounting financial pressure to get new drugs to market as patents expire on their most valuable products. India's contract drug research industry is estimated to generate $1 billion in revenue a year.[219] By working around the clock with Indian researchers, partners, drug makers hope to slash research time and costs, a crucial consideration given the high risk of failure in explained Eli Lilly executive Robert Armstrong.[220]

Glenmark Pharmaceuticals exemplifies India's prowess in drug research. The company has licensed drug candidates to Eli Lilly and other Western pharmaceutical companies and has new biological entities in clinical testing that are potential treatments for asthma, diabetes, and rheumatoid arthritis. Drug-research firm Piramal Health Care has drug-discovery partnerships with Lilly and Merck, while Ranbaxy has a major collaboration with GSK.[221]

Piramal illustrates the way in which some Indian companies are harnessing the nation's high pool of scientists and engineers and forging strategic alliances in a bid to become global players in innovation. A leading producer of generic drugs, Piramal has expanded manufacturing in the United Kingdom, Canada, China, and the U.S., where it has three plants employing 1,000 workers. But it also has a large and growing early-stage drug-development arm that partners with multinationals. Founder Swati Piramal estimates that her company can develop a new drug for the global market for $50 million, compared to the average of $1 billion spent in the U.S. for every drug brought to market. In India, she noted, Nicholas Piramal can buy "a lot of scientific horsepower" for the money.[222]

India's automotive industry also is leveraging global partners to develop innovative products. To obtain the cutting-edge components needed for Tata Motors' innovative $2,500 small passenger car, the Nano, its affiliate Tata Auto Component Systems (TACO) formed 16 global partnerships, including

---

[217] NASSCOM data can be accessed at http://www.nasscom.in/bpo-0.
[218] NASSCOM data can be accessed at Data can be accessed at http://www.nasscom.in/indian-itbpo-industry.
[219] Data: Zinnov.
[220] From presentation from Robert Armstrong of Eli Lilly & Co. in *India's Changing Innovation System,* op. cit.
[221] Pete Engardio and Arlene Weintraub, "Outsourcing the Drug Industry," *BusinessWeek,* Sept. 4, 2008. Also see Vivek Wadwha, et al, "The Globalization of Innovation: Pharmaceuticals: Can India and China Cure the Global Pharmaceutical Market?" Duke University Pratt School and Engineering and Harvard Labor and Work Life Program, available at SSRN: http://ssrn.com/abstract=1143472.
[222] From presentation by Nicholas Piramal TITLE Swati Piramal in *India's Changing Innovation System,* op. cit.

alliances with Johnson Controls and Visteon. Engineers based in different nations collaborated around the clock. TACO also established four advanced engineering centers, including one in the U.S., and 16 different manufacturing plants for interior plastics, seating systems, exteriors and composites, and other components and modules. Like many Indian companies, TACO regards design as a core strength. TACO executive M. P. Chugh notes that Chinese manufacturers are better at "shoot and ship"—that is, manufacturing a product from a drawings and specifications—while Indian auto manufacturers are better able to design, test, and validate auto parts, as well as manufacture them. The business model, Mr. Chugh explained, is to "not only use the engineering talent in India, but leverage engineering talent in India for a global business market."[223]

The Nano car illustrates another distinct feature of Indian-style innovation: The talent for developing business models that can deliver quality goods and services at extremely low prices. This model also is a crucial element in the government's strategy of meeting the needs of its impoverished population, according to Kapil Sibal, formerly India's Minister of Science and Technology and now Minister of Human Resource Development. To help deliver health care to remote villages, for example, hospitals in Delhi are setting up "medical kiosks" in clusters of villages that enable doctors in Delhi to diagnose patients using satellite technologies. The ministry pays the investment in medical hardware, while hospitals make doctors available. The innovation comes in combining high-tech and very simple technologies to improve the lives of the 500 million people living on less than $2 a day. "The object of technological development is ultimately economic growth and raising the living standard of all, not just a few," Mr. Sibal said.[224]

Public-Private Innovation Partnerships

The government is increasing its incentives for research and development by the private sector. It is reportedly planning to set up an electronics development fund (EDF) to promote R&D in electronics.[225] Minister of Finance Pranab Mukherjee proposed in March 2012 that India's weighted deduction of 200% for R&D expenditures—one of the highest in the world—be extended from 2012 for another five years[226]. The government also trying to bring Indian companies into public-private partnerships aimed at developing new products and tackling national technology needs. The New Millennium Indian Technology Leadership Initiative, funded by the government, involves 60

---

[223] From presentation by M. P. Chugh of Tata Auto Component Systems in *India's Changing Innovation System*, op. cit.
[224] From presentation by Kapil Sibal, then of the Ministry of Science and Technology, ibid.
[225] "Electronics Development Fund to Promote innovation Soon—Official," Indo-Asian News Source (February 21, 2011).
[226] "Union Budget 2012: Full Text of Pranab Mukherjee's Speech," IBN Live (March 16, 2012).

largely networked projects in areas such as agriculture, biotechnology, bioinformatics, pharmaceuticals, materials, information technology, and energy. The initiative involves as least 85 industry partners and 280 R&D programs with 1,750 researchers and has generated cumulative investment of more than $100 million. The program provides small grants to high-risk, low-investment technology projects of research institutions in which India has potential to be a global leader. Projects run by companies can get soft loans at 3 percent interest if Indians or non-resident Indians control them. Projects majority-owned by foreigners get loans at 5 percent interest if they manufacture in India.[227]

New Millennium projects so far have secured 100 international patents and published 150 articles in journals. Products include a system for viewing 3-D images of complex bio-processes, a low-cost embedded computing platform that can replace conventional personal computers for day-to-day office work, an herbal oral psoriasis treatment that is in clinical testing, and an Internet Protocol service that allows users to get television, Internet, and telephone service over telephone lines.[228] The budget for New Millennium projects recently was expanded to $157 million over five years. The program also now includes projects in which industry shares half of costs, that are co-financed with venture capital funds, or that establish innovation centers. Loans can be converted in equity, and foreign companies have greater ability to participate.[229]

Multinationals R&D Centers

The some 300 R&D centers operated by multinationals in India are another powerful force connecting India to global innovation flows. In most emerging markets, multinationals set up research and product-development operations mainly to serve the needs of the local market. In India, however, foreign companies have tended to hire top engineering and design talent to help develop products sold around the world. According to one survey, the biggest reason multinationals invest in China is to access new consumer markets and to tap low-cost labor. In India, foreign companies cited new outsourcing opportunities and access to highly skilled labor as the biggest reason they invest there.[230]

---

[227] Council of Scientific and Industrial Research, "New Millennium Indian Technology Initiative," (http://www.csir.res.in/external/heads/collaborations/Nmitili/NMITLI%20Information%20in%20brief.pdf).

[228] Examples are featured in the brochure Council of Scientific and Industrial Research, "New Millennium Indian Technology Leadership Institute: A Public Private Partnership R&D Programme for Technology Development," which can be accessed at http://www.csir.res.in/external/heads/collaborations/Nmitili/NMITLI%20Brochure%20and%20selected%20achievements.pdf.

[229] Department of Science & Technology press release, "New Millennium Indian Technology Leadership Initiative Scheme," Feb. 27, 2009. This release can be accessed at http://www.dst.gov.in/whats_new/press-release09/new-millennium-scheme.htm.

[230] UNCTAD, *World Investment Report 2005*, United Nations.

General Electric is one multinational that has made Indian talent integral to its global innovation activities. GE's $80 million John F. Welch Technology Center employs 2,500 scientists and engineers. More than 60 percent have advanced degrees and 20 percent with global experience. The 50-acre campus includes state-of-the-art labs for mechanical engineering, electronics, chemical, metallurgy, polymer sciences, new materials, and computer simulation working for GE divisions in everything from health care and energy to aviation and consumer appliances. In its first five years, the center earned 44 patents. They include breakthroughs in computer-tomography, magnetic resonance products, high-performance plastics for automobiles, and next-generation sensors.[231]

Google has set up R&D centers in Bangalore and Hyderabad and regards those operations as on par with those at its headquarters in Mountain View, California, according to Google executive Ram Shriram. Google Finance, which was launched globally, was developed by two researchers in Bangalore.[232] IBM, which employs more than 100,000 in India[233] and has a 100-researcher team IBM Research Laboratory in Delhi, is investing $6 billion to expand its operations.

One topic of growing debate in India, however, is whether the heavy multinational R&D presence is a benefit or a hindrance to the development of a strong national innovation system[234]. Growing competition for top technical talent in India has given rise to concerns that foreign companies are hoarding too much of the nation's most valuable brainpower even though much of the multinational R&D work is oriented toward products sold globally. Some studies suggest, however, that the spillovers will have a positive long-term impact as seasoned engineers leave foreign companies and join domestic ones.[235]

Seeking Global Partnerships

India's national research organizations also are becoming more important global partners. They have joined international mega-science initiatives such as the Large Hadron Collider at the European Organization for

---

[231] From presentation by Kenneth Herd of General Electric, ibid.
[232] From presentation by Ram Sriram of Google, ibid.
[233] Data from Mini Joseph Tejaswi and Sujit John, "IBM is India's second largest pvt sector employer," *Times of India*, Aug. 18, 2010.
[234] For a discussion of the spillover effects of multinational companies R&D centers in India, see R.A. Mashelkar, Technology in Society, April, 008, Vol,30/3-4, Pp 299-308 (Annexure 3); Technonationalism to Technoglobalism by R.A. Mashelkar, Journal of India & Global Affairs, 2009, 90-97. (Annexure 4).
[235] For a discussion of how multinational R&D centers may impact India's domestic innovation ecosystem, see N. Mrinalini and Sandhya Wakdikar, "Foreign R&D Centres in India: Is There any Positive Impact?", *Current Science*, Vol. 94, No. 4, Feb. 25, 2008 (http://www.ias.ac.in/currsci/feb252008/452.pdf).

Nuclear Research, for example, and the International Thermonuclear Experimental Reactor. India has also entered collaborations in agricultural research with the U.S., Brazil, Japan, and South Korea.

India has become a closer partner with the United States in recent years. A 2005 bilateral agreement called for greater cooperation in civilian uses of nuclear, space, and dual-use technology.[236] The two nations also concluded a 10-year framework agreement for defense. The U.S. and India established a new joint science and technology endowment fund to facilitate research collaborations for industrial applications. A $100 million U.S.-India Knowledge Initiative focuses on raising agricultural productivity and increasing agro-industrial business. The U.S. and India also have launched a bilateral dialogue seeking cooperation in oil, gas, nuclear, clean-coal, and renewable energy sources and began discussing cooperation in civilian use of space.

The Challenges Ahead

The government's growing commitments to boost investment research, upgrade higher education, reform its research institutions, and invest in programs and infrastructure to spread the benefits of innovation to the greater population all portend well for India's future as new science and technology power. What remains to be seen is whether the government mobilize and coordinate central and state agencies, universities, and the private sector to execute its ambitious agenda.[237]

An appraisal by the European Commission expressed some skepticism. "(T)he problem is that these innovation policies are rather fragmented among ministries and elite bodies such as the Planning Commission and Prime Minister's Office" and that they "lack coordination and networking." As a result, there is considerable duplication. The report also questioned whether the many discrete programs in areas like telecommunications, information, and pharmaceuticals fit into an overarching framework. "India has not yet articulated a formal national innovation policy as such," the report said.[238]

Another critical issue is political sustainability. If India's booming economy and thriving technology sector do not deliver tangible results for the greater population, political support for expensive science-and-technology programs and universities that seem to benefit the well-off could diminish. Greater participation by India's private sector, both in the form of higher R&D spending and willingness to join public-private partnerships and national programs, also is essential.

---

[236] Presentation by Indian Ambassador to the U.S. Ronen Sen in *India's Changing Innovation System*.
[237] Fir a insightful review of challenges, see R.A. Mashelkar 'Reinventing India', Pune: Sahyadri Publications, 2012.
[238] European Commission Enterprise Directorate-General, op. cit.

As evidenced by recent policies and the growing focus on "inclusive innovation," the government of Prime Minister Singh is well cognizant of these challenges and determined to address them. If such efforts succeed, India appeared destined to be a 21st century innovation powerhouse.

NEWLY INDUSTRIALIZED ECONOMIES

Taiwan

Taiwan's rise from poverty in the 1950s to one of the world's premier high-tech powers has made it a role model of how to use science and technology policy for rapid economic development. Since the 1970s, the government has executed a systematic strategy to absorb advanced technologies from the West and Japan, develop globally competitive products and manufacturing processes, and then transfer the know-how to private companies to create world-class industries. These efforts quickly transformed Taiwan's economy. In 1981, food and textile industries accounted for 40 percent of Taiwan's manufacturing sector, with electronics accounting for less than 15 percent. By 2004, electronics was 35 percent of the island's manufacturing economy, with food and textiles accounting for less than 10 percent. Meanwhile, per-capita income in Taiwan rose from less than $500 in the early 1950s to $18,558 in 2010.[239]

Taiwan's standings in the areas of technology, advanced manufacturing, and knowledge-based industries have risen just as dramatically. Taiwan is the world's leading producer of mask ROMs and optical discs and the world's largest integrated circuit foundry producer and largest packager of integrated circuits.[240] Taiwan is the second-largest producer of large high-definition LCD panels, IC design services and crystalline silicon solar cells.[241] Taiwanese industry is making impressive progress in next-generation industries such as solid-state lighting, thin-film electronics, photovoltaic cells, and biomedical devices using nano-scale materials. The portion of GDP devoted to research and development has risen more than fivefold since the late 1980s, and reached 2.9 percent of GDP in 2009. [See Figure 5.7] Taiwanese companies,

---

[239] International Monetary Fund, World Economic Outlook Database, September 2011.
[240] Republic of China, Council for Economic Planning and Development, *Taiwan Statistical Data Book 2011*, July 2011, Table 4-b2.
[241] *Id.* Taiwanese companies are the world's largest producers of notebook PCs, motherboards, personal navigation devices and LCD monitors, but significant production of products occurs offshore, principally in China. *Id.*, Table 4-b1. See also Xing Yuqing, "China's High-Tech Exports: Myth and Reality," *EAI Background Brief No. 506*, February 25, 2010. "Taiwanese-owned IT companies played a very important role in nurturing the high-tech industries in mainland China. By 2007, they had relocated almost 100% of their production capacities in laptop PC, digital camera, motherboard and LCD monitor for PC into mainland China."

once low spenders on R&D, contributed more than 69.7 percent of total spending on research in Taiwan.[242]

The island is beginning to excel in innovation as well. Taiwan is among the world leaders in U.S. utility and design patents.[243] Indeed, Taiwan generates more patents per 1 million citizens than any other region or nation.[244] Taiwan also has been winning international innovation awards. National research institutes had three winning entries in *R&D Magazine*'s 2010 R&D top 100 Awards, for example. One was for FlexUPD, billed as the first technology to enable the commercialization of paper-thin, low-cost, flexible flat-display panels for electronic products. Taiwan also won awards for a display technology that allows both 2D and 3D information to be viewed simultaneously with the naked eye and for the first non-toxic, fire-resistant composite technology.[245]

What's more, Taiwan's science and technology investments have enabled the economy to meet one of its most crucial strategic challenges: remaining a globally relevant sector in the wake of a rising China. Its giant neighbor has lower costs, vastly more engineers and scientists, and aggressive policies targeting all of the same industries as Taiwan. Despite a massive shift of factory work to the mainland, the value of Taiwanese exports continues to rise. Taiwan had record exports in 2010 of $275 billion, with 42 percent going to China, up from 24 percent in 2000.[246]

Taiwan is reaping the benefits of heavy investments in education and decades of comprehensive science and technology policies aimed at building globally competitive industries. The island of 23 million also has expertly leveraged its strategic geographic location off the coast of China. Estimates of Taiwanese investment in mainland China, including those made through third parties, range from $150 billion to $300 billion.[247] Taiwanese companies control and manage much of the electronics export sector.[248] Taiwan has positioned itself as a global engineering and innovation hub bridging East and West.

---

[242] *Id.*, Table 6-4.
[243] National Applied Research Laboratories, *Yearbook of Science and Technology Taiwan ROC 2010*, May 2011, Table 1-2-4. Taiwan was the 5th in 2009 with respect to both total patents and utility patents.
[244] According to U.S. Patent and Trademark Office data, Taiwan is No. 5 in U.S. utility patents and the third-biggest recipient of U.S. design patents.
[245] *R & D Magazine* "R&D 2010 Winners," July 7, 2010 (http://www.rdmag.com/Awards/RD-100-Awards/2010/07/R-D-100-2010-Winners-Overview/).
[246] Executive Yuan, R.O.C. (Taiwan), Council for Economic Planning and Development, *Taiwan Statistical Data Book 2011*, July 2011, Table 11-9a.
[247] U.S. Department of State, "Background Note: Taiwan," Bureau of East Asian and Pacific Affairs, July 7, 2011.
[248] For instance, Taiwan-owned Foxconn Technology Group, the world's biggest electronics manufacturer, reportedly employed more than 1 million workers in China as of 2010 and plans to increase that workforce to 1.3 million. Frederik Balfour, "IPad Assembler Foxconn Says it Has More Than 1 Million Employees in China," *Bloomberg*, Dec. 10, 2010.

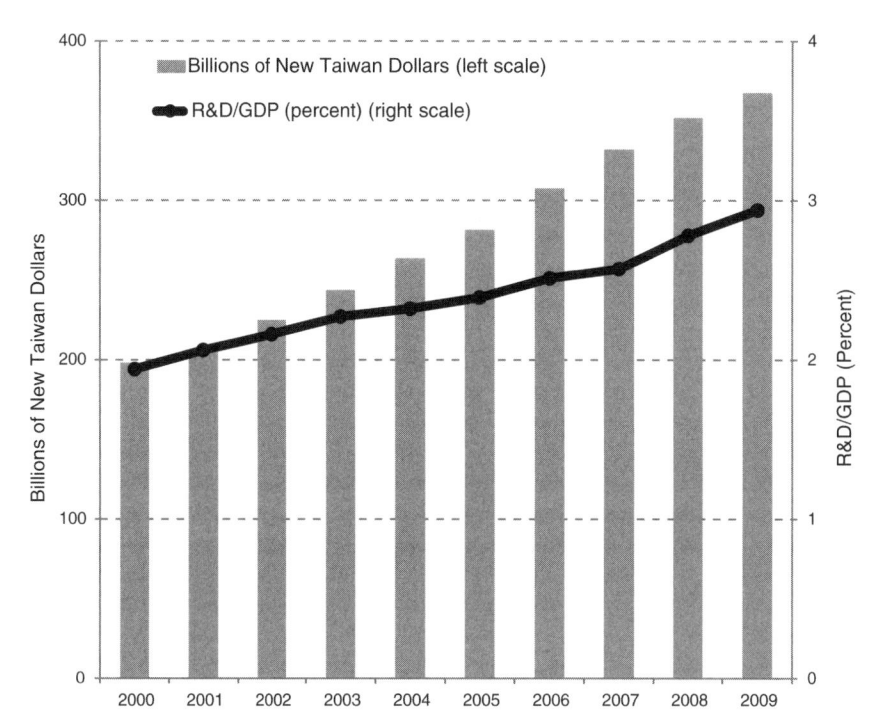

FIGURE 5.7 Taiwan's R&D expenditures increased to 2.94 percent of GDP in 2009.
SOURCE: Executive Yuan, R.O.C. (Taiwan), Council for Economic Planning and Development, *Taiwan Statistical Data Book 2011*, July 2011, Table 6-1.

Fifty-one multinationals have Taiwanese research centers, including Hewlett Packard, Dell, Sony, DuPont, IBM, Fujitsu, Intel, and Dow.[249]

Government planners believe Taiwan needs new economic engines, however, to continue to prosper in a global knowledge economy and amid growing competition from large emerging markets. "Innovation is unquestionably the key to Taiwan's sustained economic growth," states the National Science and Technology Development Plan for 2009 through 2012. To achieve this, "it will be necessary to rethink the country's focus on scientific research, lengthen R&D chains, and strengthen the conversion of R&D results

---

[249]Ministry of Economic Affairs, "Multinational Innovative R&D Center in Taiwan," updated Oct. 5, 2011. Access at http://investtaiwan.nat.gov.tw/matter/show_eng.jsp?ID=433.

into innovative technologies and industrial capabilities."[250] Among other measures, the plan calls for shifting the R&D focus more toward "pioneering" research, strengthening currently weak ties between universities and private industry, building better links between basic research and downstream applications, and reforming Taiwan's education system to encourage more critical thinking and interdisciplinary studies. This will likely mean an attempt to increase spending on basic R&D, which was 10.4 percent of total R&D spending in 2009.[251] [See Figure 5.8]

Chu Hsih-sen of Taiwan's Industrial Technology Research Institute (ITRI) described the goals this way: Taiwan must move from a focus on optimizing existing technologies to exploration, to move from working within single disciplines to integrating multiple disciplines, and to move from developing components to entire systems and comprehensive services. Taiwan also is stressing greater collaboration among its research organizations and industrial and academic partners around the world.[252]

The Taiwan Method

The express purpose of Taiwanese government science and technology policies has always been to establish and sustain domestic industries. The island started in electronics manufacturing with duty-free export zones in the 1960s, when Taiwanese wages were extremely low. In the 1970s, it began investing heavily in industrial technology institutes to stimulate more sophisticated indigenous industries. Ninety-two percent of R&D was devoted to manufacturing as of 2006, compared to 65 percent in the United States and 83 percent in South Korea. Of that, 69 percent was devoted to high-tech manufacturing.[253]

The key elements of the Taiwan method have been to carefully identify industries where the island can make its mark. Rather than attempt to invent new technologies from scratch, Dr. Chu explained, Taiwan's strategy has been to focus on technologies that multinationals already possess and that Taiwanese companies want to apply. Then the government develops the necessary skills base, builds or upgrades common laboratory facilities, and systematically acquires the needed technologies through a combination of licensing, in-house R&D, and partnerships with foreign companies and universities.

---

[250] National Science Council Executive Yuan, "National Science and Technology Development Plan (2009-12), passed July 2, 2009 (http://web1.nsc.gov.tw/public/Attachment/91214167571.PDF)
[251] Id., p. 66. "Promotion of forward-looking, outstanding, interdisciplinary basic research in science and environmental science, biology, and engineering, etc."
[252] From presentation by Chu Hsin-Sen of Industrial Research and Technology Institute in National Research Council, *Innovation Policies for the 21st Century: Report of a Symposium*, Charles W. Wessner, editor, Washington, DC: The National Academies Press, 2007.
[253] From presentation by Chen Choa-Yih, director general of Industrial Development Bureau of Taiwan's Ministry of Economic Affairs, Jan. 5, 2006, symposium 21st Century Innovation Systems for the United States and Taiwan: Lessons from a Decade of Change."

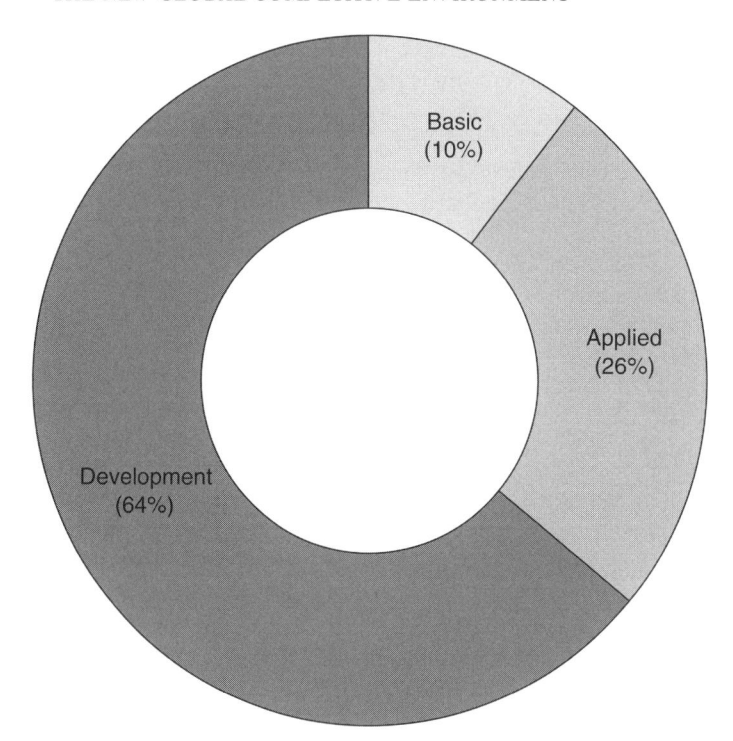

FIGURE 5.8 Taiwan R&D expenditure by type in 2009.
SOURCE: Executive Yuan, R.O.C. (Taiwan), Council for Economic Planning and Development, *Taiwan Statistical Data Book 2011*, Table 6-6.

Working closely with domestic companies, well-staffed industrial research institutes then turn those technologies into prototypes and production processes that are disseminated widely through industry.[254] "Taiwan's miracle is based on government-promoted industries and private domestic firms," observes Massachusetts Institute of Technology political economist Alice Amsden.[255]

To help manufacturing industries take root, government agencies also offer generous assistance, including research grants, early-stage capital, incubators, tax breaks, low-cost access to laboratories and production facilities at world-class science parks, and efforts to build local supply bases of key materials and components. Among other industries, this method has succeeded

---

[254] Chu, op. cit.
[255] From Alice H. Amsden, "Taiwan's Innovation System: A Review of Presentations and Related Articles and Books," submitted for NAS Jan. 4-6 symposium "21st Century Innovation Systems for the U.S. and Taiwan: Lessons From a Decade of Change," Taipei.

with notebook computers, liquid-crystal displays, semiconductor fabrication and design, and bicycles made of carbon composites, an industry Taiwan dominates. The Taiwan government is applying this strategy to a range of new industries, including logistical services.

The National Science Council of the Executive Yuan is the top agency promoting science and technology, receiving 35 percent of the government's $2.9 billion 2008 R&D budget. It funds university research and overseas a network of 11 national laboratories established since 2003. Each lab specializes in developing core technologies with "high societal impact or industrial competitiveness," such as nano-devices, high-performance computing, earthquake simulation, chip implementation, and animal research. Accomplishments include development of biomedical sensor chips, medical visualization products, and what is advertised as the world's first 16 nanometer, single-cell static random-access-memory device. The chip is said to be capable of holding 15 billion transistors that can process 10 times more data than current 45 nm technology and radically reducing the size of circuit boards.[256]

The National Science Council operates a precision instrument development center and a synchrotron radiation center similar to the Max Planck Center in Europe. The National Science Council also operates Taiwan's highly successful and widely imitated Hsinchu Science Park, established in 1980, and several others in southern Taiwan. The park serves as a source of technology development and training for industries like semiconductors, displays, and renewable-energy technologies. Academia Sinica, which conducts research in physical sciences, mathematics, and life sciences, receives around 9 percent of the Executive Yuan's R&D budget.

ITRI's Complex Mission

The backbone of Taiwan's strategy has been its industrial research institutes. ITRI is by far the biggest. Established in 1973, ITRI has grown to a network of 13 research centers that focus on information and communications, advanced manufacturing, biomedical, nanotechnology and new materials, and energy and environmental technologies. More than 60 percent of ITRI's 6,000 employees hold master's or doctorate degrees. ITRI consults with more than 30,000 domestic companies each year. It has helped create 165 start-ups and spinoffs, and generated more than 10,000 patents.[257]

More than 20,000 ITRI alumni work in Taiwan's private sector, around 5,000 of them holding senior executive positions Hsinchu Science Park.[258] According to ITRI official Barry Lo, the institute deliberately seeks an annual

---

[256] National Applied Research Laboratories 2009 annual report.
[257] ITRI, "What is ITRI?" accessed at http://www.itri.org.tw/eng/about/article.asp?RootNodeId=010&NodeId=0101.
[258] ITRI data.

attrition rate of around 15 percent, or about 900 researchers a year, so that they circulate through industry. "If people want to work in a laboratory for life, you don't have the energy to help industry," Lo explained.[259] In addition, ITRI operates a training college that has 3,000 to 5,000 students attended programs lasting one month to one year and an Open Lab that houses some 60 outside companies working on collaborative R&D projects.

ITRI's stated mission is three-fold: to "create economic value through innovative technology and R&D," to "spearhead development of high-value industry in Taiwan," and to enhance the global competitiveness of Taiwanese industry. That gives ITRI a much more complex role, than comparable U.S. agencies, notes Dr. Amsden. "ITRI is not only charged with raising the technological level of Taiwan, but also with increasing the level of its productive capabilities. ITRI has many more tentacles because it has many more jobs to do that are related to industrial diversification and firm formation."[260]

The institute's first big success was launching Taiwan's semiconductor industry. ITRI acquired RCA's technology for 7-micron chips in 1976. Three years later, an experimental lab run by ITRI's Electronics Research and Service Organization (ERSO) was spun off as UMC. Eight years later, ITRI spawned what would become Taiwan Semiconductor Manufacturing Corp., today the world's dominant chip "foundry," which fabricates devices on silicon on a contract basis. Today, TSMC and UMC control some 70 percent of the global chip foundry industry. ERSO also spun off Taiwan Mask Corp., a provider of masking services. ITRI and ERSO also helped launch many of Taiwan's integrated-circuit design companies firms that sprang up around the foundries.

ITRI also was pivotal to the development of Taiwan's personal computer industry. From 1979 through 1991, ERSO began sending teams of engineers to Wang Computer for ten-month training courses in hardware and software design. These engineers helped diffuse the knowhow widely, and helped Acer develop Taiwan's first 16-bit, IBM-compatible computer. Private companies also used ERSO labs to test machines before exporting them, as well as to develop Ethernet, workstations, monitors, and file-management software. The institute transferred some of the first technology that led to the eventual development of Taiwan's liquid-crystal display industry, where companies such as Chi Mei and Au Optoelectronics now are among the world leaders.[261]

The Hsinchu Science Park was another important catalyst for Taiwan because it gave new companies access to first-rate facilities at a low cost. To get into Hsinchu, companies had to meet tough criteria. They had to have the ability to design products for manufacturing according to a business plan, devote a

---

[259] Interview with Barry Lo of ITRI.
[260] Amsden, op. cit.
[261] Ibid.

certain share of resources to high-level R&D, and employ a significant marketing staff within three years. This process enabled the government to "cherry pick" Hsinchu tenants, according to Dr. Amsden. Once admitted to Hsinchu, companies received a full set of subsidies, such as exemption from taxes and import duties, grants, low-cost credit, below-market factory rent, access to government research facilities, good housing, and even bilingual education for expatriates' children.[262]

The government still invests alongside promising companies at the R&D stage. Typically, the government pays for 25 percent of research, explained Mr. Chu. The private company invests half, and the rest comes from government or bank loan.[263] When a company is profitable, it then repays the government's investment.

Dr. Amsden said that government subsidies to companies enjoyed a high rate of success largely because they were tied to "concrete, measurable, and monitorable performance standards." Committees of experts from industry, government, and academia selected winners of government grants. Intellectual property was shared equally with the Ministry of Economic Affairs. Companies had first right of refusal if the ministry wanted to divest. If a company failed to produce a developed product after three years, it not only lost its intellectual property but also had to repay government investments in installments.

Emerging Industries

Now much of ITRI's budget goes toward programs that aim to establish Taiwan in a range of emerging industries. The Taiwanese government is investing $1 billion into clean energy over three years. ITRI priorities include thin-film photovoltaic cells, lighting devices using light-emitting diodes (LEDs), hydrogen fuel cells, offshore wind-power generation, and energy-efficient vehicles. ITRI also is developing flexible electronics products, a service platform for smart living technologies, and cloud computing.

ITRI achieved a major advance by acquiring key technology from Eastman Kodak, the inventor of organic LED (OLED) technology, which the U.S. company was unable to turn into commercially viable products. Among other things, ITRI engineers have used this technology to produce its innovative FlexUPD display. These paper-thin displays, which are "light, malleable, and unbreakable," can be used for rollable mobile phone screens, E-books, e-maps and medical sensors that can be worn or wrapped around the body, according to ITRI. The institute also has developed paper-thin speakers.[264] ITRI is disseminating the technology to domestic opto-electronics manufacturers.

---

[262] Ibid.
[263] Chu presentation, op. cit.
[264] ITRI Web site.

ITRI is leading a similar effort in LED lighting, where it has organized an alliance of 20 Taiwanese manufacturers. The companies are developing LED products and materials for street lighting. The goal is to establish a "vertically functioning" LED industrial chain within Taiwan.[265] The institute also has an open laboratory to make 8-inch wafers for microelectromechanical systems (MEMS). ITRI engineers help companies design, test, package, and manufacture MEMS components such as biomedical devices. Other next-generation ITRI R&D programs include technology for wireless multimedia systems on a chip, wireless sensor networks, and wireless broadband.

An example of ITRI's cultural shift toward more creative, knowledge-based industries is the Creativity Lab, which is developing technology concepts for new consumer lifestyles. Launched in 2005 and based in Hsinchu, the program is a collaboration with the MIT Media lab and has hired staff with psychology degrees and from the arts and media.[266]

ITRI also is playing a role in the government's $1.8 billion program to improve the island's information and communications infrastructure. The initiative seeks to raise industry competitiveness, improve government efficiency, improve quality of life, and increase the number of broadband users to 6 million.[267]

Taiwan's Innovation Challenges

Many of these new programs reflect initiatives adopted by Taiwanese economic planners to shift toward more knowledge-based industries[268] and to address perceived shortcomings in the island's innovation ecosystem. The National Science Council challenges were enunciated in the 2009-2012 plan.

One flaw in Taiwan's innovation system cited by the Council is weak technology-transfer and commercialization efforts by universities. Small-business incubators and entrepreneurial training are relatively new in Taiwanese universities. As a result, universities launch few start-ups. The Council also faults the Taiwanese teaching system. Because students must focus on either liberal arts or science and technology at an early age, many do not get broad interdisciplinary education. Engineering courses, meanwhile, are criticized for not training students to think creatively. As a result, the plan states, "the education system does not provide students with the knowledge, skills, and

---

[265] ITRI Web site.
[266] Bruce Einhorn, "A Creativity Lab for Taiwan," *BusinessWeek*, May 16, 2005.
[267] For details on the e-Taiwan program, see http://www.etaiwan.nat.gov.tw/content/application/etaiwan/egenerala/guest-cnt-browse.php?cnt_id=779).
[268] Taiwan has identified six emerging industries to help develop an optimal industrial structure. These are green energy, biotechnology, tourism, medical care, cultural creativity and quality agriculture. National Applied Research Laboratories, *Yearbook of Science and Technology Taiwan ROC 2010*, May 2011, "Report I: Six Emerging Industries – Recreating Prosperity."

attitudes that they will need to confront and deal with the problems of a fast-changing society."[269]

The plan calls for making universities more business-friendly and more open to outside collaboration. Among the recommended measures are establishment of more incubators, better incentives for academics to commercialize research, expanded entrepreneurial training, programs to "broaden students' knowledge of practical innovation design skills," and curricula that promote interdisciplinary knowledge.[270] To secure sufficient manpower, Taiwan should recruit more talent from abroad, especially from mainland China, the plan says. The document also calls for universities to establish stronger links with science parks and national laboratories. The Ministry of Education plans to invest $1.7 billion over five years in first-rate universities. The goal is that at least one will be rank among the top 100 in the world and among the top 10 in the Asia-Pacific.

Another perceived flaw in Taiwan's innovation system is that researchers at public institutes are treated as civil servants and therefore may not work for private companies. Since such a large share of Taiwanese R&D is conducted at Academia Sinica, national laboratories, and industrial research institutes, the Council regards such rules are major obstacles to commercialization and recommends that they be eased.

The fundamental approach to R&D by government and industry almost must change, the Council said. "It will not be enough to merely improve technologies and raise efficiency, as has been done over the past few decades," the plan states. Instead, R&D should focus on "pioneering technology," and policy should be "shaped by demand pull and vision of Taiwan's future."

Opportunities to Collaborate

International collaboration is likely to become a more important aspect of Taiwanese innovation strategy. ITRI already has extensive overseas ties. In addition to the relationship with the Media Lab, ITRI works with MIT's artificial intelligence lab. ITRI has joint research programs with the University of California at Berkeley in nanotechnology and clean energy, five labs at Carnegie Mellon University, and a strong relationship with Stanford Research Institute. Among its many other collaborations are projects with Japan's RIKEN, the University of Tokyo, the Netherlands' Organization for Applied Scientific Research, Russia's Ioffe Physical-Technical Institute, and Australia's Commonwealth Scientific and Industrial Research Organization. ITRI's long list of multinational partners includes Corning, Broadcom, Sun Microsystems, Hewlett Packard, Bayer, BASF, ARM, GSK, and Nokia.

---

[269] National Science and Technology Development Plan 2009-2012, op. cit.
[270] Ibid.

The National Science Council calls for expanding Taiwanese collaborations with international research institutes and industry consortia. It also recommends attracting more multinationals to use the island as a global innovation base.

Singapore's Focus

Science and technology policy has been central to Singapore's emergence as one of the world's wealthiest nations. Since separating from Malaysia in 1965, per-capita income has soared from a mere $512 to $42,653 in 2009.[271] Like Taiwan, Singapore's takeoff was fueled first by labor-intensive manufacturing in the 1960s. Singapore then thrived as an Asian hub for trade, services, manufacturing, and corporate product development. Now the island of 5.1 million aspires to become one of the world's premier innovation zones for 21st century knowledge industries. As the government's science and technology plan for 2006-2010 stated, "The critical success factor for Singapore will be its ability to become an international talent node—nurturing its own talent as well as drawing creative and talented people from all corners of the world to live and work in Singapore."[272]

Singapore is making impressive progress. The nation's heavy investments in higher education and R&D infrastructure and ability to execute visionary and comprehensive innovation policies has enabled the country to reinvent itself as a magnet for multinational research labs and top-notch international talent in fields such as genomics, infectious diseases, advanced materials, and information technology. R&D manpower more than doubled between 1998 and 2009 to 41,388, research organizations increased from 604 to 854, and total R&D spending more than doubled to S$6.04 billion, despite contracting by 15 percent from 2008 levels because of a sharp decline in private sector R&D as a result of the global recession.[273] [See Figure 5.9]

Singapore ranks No. 2 worldwide in global competitiveness, according to the World Economic Forum.[274] Singapore's innovation system is built upon a strong foundation in education. The share of university graduates in the population leapt from 4.5 percent in 1990 to 23 percent in 2010,[275] and the

---

[271] International Monetary Fund data.
[272] Ministry of Trade and Industry, *Sustaining Innovation-Driven Growth, Science, and Technology*, Government of Singapore, February 2006.
(http://app.mti.gov.sg/data/pages/885/doc/S&T%20Plan%202010%20Report%20(Final%20as%20of%202010%20Mar%2006).pdf).
[273] Department of Statistics, Ministry of Trade & Industry, Republic of Singapore, *Yearbook of Statistics Singapore 2010*, July 2010 and Agency for Science, Technology and Research Singapore, *National Survey of R&D in Singapore 2009*, December 2010.
[274] World Economic Forum, *The Global Competitiveness Report 2011-2012*, op. cit.
[275] Singapore Department of Statistics, Census of Population 2010.

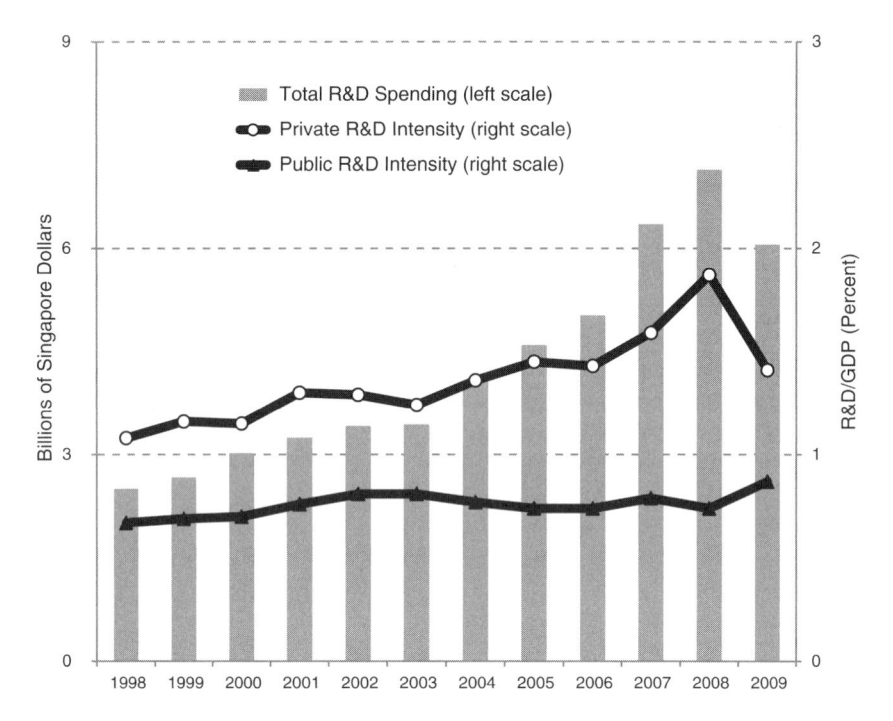

FIGURE 5.9 Singapore R&D expenditures declined in 2009 on a sharp drop in business R&D intensity.
SOURCE: Agency for Science, Technology and Research Singapore, *National Survey of R&D Singapore 2009*, December 2010.

portion of the resident workers with degrees jumped from 14.6 percent to 27.8 percent between 1999 and June 2010.[276] More than 153,000 students were studying at the nation's universities and polytechnics as of 2009.[277] Singapore grade-schoolers perennially rank at or near the top in math and science scores.[278] The government's strong commitment to science and technology encourages students to pursue those fields, and the highly skilled workforce in turn enables Singapore to frequently transform itself, explained Yena Lim of the Singapore Agency for Science, Technology, and Research.[279]

---

[276] Ministry of Manpower, *Report on Labour Force in Singapore 2010*.
[277] *Yearbook of Statistics Singapore 2010*.
[278] Source: *Trends in International Mathematics and Science Study* (TIMMS).
[279] From presentation by Yena Lim of Singapore Agency for Science, Technology, and Research in National Research Council, *Understanding Research, Science, and Technology Parks*, op. cit.

In terms of international patents, start-ups, and the dynamism of domestic companies, Singapore is still far from an innovation powerhouse.[280] The government has charted an ambitious agency to push its innovation system to a higher level. In 2004, the Ministerial Committee on Research and Development was formed to review the nation's R&D strategies and direction and compare them with those of other nations. The panel concluded that Singapore needed to "refocus its research and innovation agenda to keep up with international developments." Singapore's position as an open innovation zone also is connected to its national defense strategy as a small country surrounded by large Southeast Asians with which it sometimes has been at odds. Singapore's strategy "is based on establishing itself as a valuable partner in the information age and on making an attack on its territory prohibitively expensive for potential enemies," according to a National Academies assessment.[281] To attract multinational manufacturing and R&D centers in targeted industries, Singapore offers incentives such as 10-year tax holidays, fast one-stop shop regulatory approvals, and subsidized vocational training.

To transform Singapore into an in "innovation-driven economy," the panel recommended the government boost R&D resources, select areas of national importance on which to focus, increase private R&D, and strengthen linkages between universities and business. It also recommended that "investigator-led research" be balanced with "mission-led research."[282]

A New Strategy

The government unveiled a new strategy in 2006. The Ministry of Trade and Industry announced $10 billion in R&D spending over five years, triple the level of the previous five-year plan. It set a target of raising national R&D spending to 3.5 percent of GDP by 2015. The plan gave special focus to life sciences, environmental and water technologies, and interactive and digital media, sectors where jobs in a "whole spectrum of research capacities" are expected to double to 80,000 and value-added to triple to $27 billion by 2015. The plan also declared Singapore must become "a global talent hub, attracting talent here by providing a vibrant environment and an open society that offer opportunities for communities of creative and talented people."[283]

---

[280] Fewer than 500 U.S. patents originated in Singapore in 2009, according to U.S. Patent and Trademark Office, not a major improvement over the previous five years, and compared to nearly 7,800 from Taiwan. The WEF's *Global Competitiveness Report* recommended that Singapore take measure to improve the "sophistication" of domestic companies.
[281] National Research Council, *S&T Strategies of Six Countries: Implications for the United States*, Committee on Global Science and Technology Strategies and Their Effect on U.S. National Security, Washington, DC: The National Academies Press, 2010.
[282] *Sustaining Innovation-Driven Growth, Science, and Technology*, op. cit.
[283] Ibid.

To lead the national drive, Singapore set up a high-level Research, Innovation and Enterprise Council. Prime Minister Lee Hsien Loong chairs the council, which also includes several cabinet ministers and international science and technology experts such as Stanford University President John Hennessy, Harvard Business School professor Clayton M. Christensen, Novartis International corporate research head Paul Herrling, and DuPont Chief Innovation Officer Thomas M. Connelly Jr. The council oversees the $3.9 billion five-year budget of the National Research Foundation, which funds and coordinates research within the national framework.

The Search for Talent

The Agency for Science, Technology, and Research (A*STAR) leads many of the programs aimed at making Singapore a global R&D base. A*STAR spearheads efforts to develop clusters in high value-added manufacturing, such as microelectronics, new materials, chemicals, and information and communications equipment, and the rapidly growing biomedical sector.[284] The agency also manages Singapore's ambitious new multibillion-dollar science parks, Biopolis and Fusionopolis, which combine a high concentration of public and corporate research organizations in a contemporary urban setting.

A*STAR also leads Singapore's aggressive efforts to recruit top international scientists and to develop homegrown talent. Its policy is described as "pro-foreign and pro-local." It runs a Graduate Academy that aims to train 1,000 Singaporean science and engineering Ph. Ds. The list of star foreign scientists recruited to Singapore's well-funded research labs is impressive. To cite a few examples: Former National Cancer Institute clinical research director Edison Liu now is executive director of the Genome Institute of Singapore. Sir David Lane, discoverer of the p53 gene, is executive director of the Institute of Molecular and Cell Biology. Nobel laureate Sydney Bremmer of the Salk Institute chairs Singapore's Biomedical Research Council and leads the Genetic Medicine Laboratory. Leading cancer geneticists Neal Copeland and Nancy Jenkins left the National Cancer Institute to lead a research team using the mouse genome to study human diseases. MIT professor Jackie Ying is executive director of Singapore's Institute of Bioengineering and Nanotechnology, while University of Texas at Austin professor Dim-Lee Kwong is executive director of the Institute of Microelectronics.

Singapore also is upgrading higher education to meet the demands of a 21st century knowledge economy. While schools such as the National University of Singapore and Nanyang Technological University are strong in science and technology, the government wants them to become world-class research institutions and to become fonts of entrepreneurialism. The government also

---

[284] S. Chaturvedi, "Evolving a National System of Biotechnology Innovation, Some Evidence from Singapore," Science Technology & Society, 2005.

wants its universities to become much more globally connected and to train more students for the kind of multidisciplinary, creative industries the government wants to develop.

New Institutions

Underscoring its commitment to educating a creative class, the government established the new Singapore University of Technology and Design. Developed in collaboration with MIT and China's Zhejiang University, the university will have a multi-disciplinary curriculum and research programs. It is expected to open in 2012. Prime Minister Lee said the university will "teach students to be creative" and will "stimulate students to go beyond the book knowledge, to apply it to solving problems."[285] The university will house an International Design Center modeled after a smaller facility at MIT and intends to "become the world's premier hub for technologically intensive designs." MIT will help design programs to encourage innovation and entrepreneurship.

A Focus on Entrepreneurship

The government also is moving to address another perceived weakness in its innovation system: A shortage of entrepreneurialism and breakthrough innovation by Singapore companies, which some analysts blame on a cultural aversion to risk and the heavy government role in the corporate sector.[286] In 2008, Singapore launched the National Framework for Innovation and Enterprise. For those who believe government should play an active role as an investor to spur innovation, Singapore sets a high benchmark. The initiative calls for spending $275 million over five years in the following areas to promote entrepreneurialism. Initiatives include—

- **Establishment of a high-level Enterprise Board** and innovation fund at each university. The fund supplements the universities' own resources to finance entrepreneurship education, technology incubators, entrepreneur-in-residence programs, and commercialization of university technologies.
- **Grants to companies accepted into incubators** covering 85 percent of the cost of developing a proof of concept, up to S$250,000. The National Research Foundation gets an equity stake in exchange that co-investors may buy out in the next round of financing.

---

[285] Singapore Ministry of Education press release, Jan. 25, 2010.
[286] For example, see Richard W. Carney and Loh Yi Zheng, "Institutional (Dis)incentives to Innovate: An Explanation for Singapore's Innovation Gap," *Journal of East Asia Studies* 9(2):291-319
(http://www.thefreelibrary.com/Institutional+(dis)incentives+to+innovate%3A+an+explanation+for...-a0202704740). Also see Patrick Lambe, "The Engineer's Dilemma: Innovation in Singapore," *Straits Knowledge*, 2002 (http://www.greenchameleon.com/thoughtpieces/engineer.pdf).

- **Seed money for early-stage venture capital funds.** The NRF will match capital raised by venture capitalists in these funds, which will be managed by professional investors and must invest only in Singapore-based high-tech startups.
- **An incubator devoted to disruptive innovation.** The NSF will fully match funds for start-ups, which will be assessed based on their "potential to disrupt a current industry and create new ones," according to the disruptive innovation methodology of Harvard Business School professor Clay Christensen.[287]
- **Grants for polytechnics to perform translational research** on R&D conducted by universities and public research institutes. The aim is to get polytechnics and universities to become "strategic partners to bring research breakthroughs to the marketplace."
- **"Innovation Vouchers" for small and midsized enterprises** to produce R&D and other services from higher-educational institutes and national laboratories.
- **A national center for innovation studies** that will propose policies and initiatives to encourage innovation in the public and private sectors.[288]

Gauging Singapore's Success

It is too early to fully assess the success of Singapore's innovation initiatives. A number of companies have been spun off of Singapore research laboratories in areas such as nanotechnology, medical devices, water-purification, and ultra-low power electronics.[289] But Singapore still is regarded as falling short in generating startups that grow to globally recognized companies. Richard W. Carney and Loh Yi Zheng blame institutional "disincentives," such as heavy government financial and management control over major Singapore corporations and small capital markets that make it hard for entrepreneurs to raise private risk capital and publicly float successful companies. Carney and Zheng also cite a business mind-set that is risk-averse and that focuses on "incremental innovation" rather than "radical innovation."[290] A sharp drop in private research spending that was blamed on global economic

---

[287] Based on concepts described in Clayton M. Christensen, *The Innovator's Dilemma: The Revolutionary Book that Will Change the Way You do Business*, Cambridge: Harvard Business School Press, 1997.
[288] National Research Foundation, "National Framework for Innovation and Enterprise," Prime Minister's Office, Republic of Singapore, 2008 (http://www.nrf.gov.sg/nrf/otherProgrammes.aspx?id=1206).
[289] S&T Strategies in Six Nations, op cit.
[290] Carney and Zheng, op. cit.

conditions pushed Singapore's R&D spending-to-GDP ratio down to 2.3 percent as of 2009, well short of the 3 percent goal for 2010.[291]

Singapore's small scale is another perceived handicap in generating domestic innovation, especially in science-based industries such as biotechnology. Singapore's big investments in life sciences and incentives for pharmaceutical multinationals has resulted in a large biomedical manufacturing base, whose output tripled to S$21.7 billion from 2000 through 2009.[292] They also have created high-paying jobs and spurred development of suppliers of materials and R&D services. But progress in luring high-value multinational R&D and stimulating collaboration between foreign companies and domestic ventures has been disappointing, contends Joseph Wong in his book *Betting on Biotech*.[293] The reason, Wong contends, is that Singapore still lacks the critical mass of researchers, biotech commercialization expertise, and companies with which to collaborate needed to get multinationals to transfer R&D operations to the country. Some critics also have said Singapore's science strategy depends too much on recruiting aging foreign star scientists rather than grooming domestic talent and younger, foreign-trained Singaporeans to lead research programs.[294]

As a research base, however, Singapore clearly has progressed. The nation is becoming a global leader in research in infectious diseases and environmental technologies such as water and waste treatment. Singapore also is poised for "industrial domination" in Asia in digital gaming and virtual reality technologies, as well as in networked command and control of traffic and unmanned aerial vehicles, according to the NAS assessment.[295] Multinational corporations continue to expand and open new Singapore R&D centers. By constantly improving infrastructure, higher education, and investment incentives, Singapore hopes that this growing research activity will eventually translate into homegrown innovation.

INDUSTRIALIZED NATION CASE STUDIES

Germany

Germany is proving that even a high-wage nation can compete globally in manufacturing. Exports of everything from kitchen equipment and industrial

---

[291] Ministry of Trade and Industry, "Economic Survey of Singapore 2010,"
[292] Economic Development Board data cited in Wong Siew Ying, "Biomedical Manufacturing Output Grew to S$21b in 2009," *channelnewsasia.com*, March 17, 2010.
[293] Joseph Wong, *Betting on Biotech: Innovation and the Limits of Asia's Developmental State*, Ithaca, N.Y.: Cornell University Press, 2011.
[294] Alice S. Huang and Chris Y. H. Tan, "Achieving Scientific Eminence Within Asia," *Science*, Vol. 329, p. 1471-2, Sept. 17, 2010.
[295] *S&T Policies in Six Nations, op. cit.*

machinery to high-speed trains and wind turbines by small and large firms alike[296] surged by 18.5 percent in 2010 to €951.9 billion ($1.3 trillion),[297] leading the country out of recession. German net exports of goods contributed 1.4 percentage points to its 3.6 GDP growth in 2010, or 40 percent of the total increase.[298] German exports to China soared by 44 percent, which could become Germany's biggest export destination overall by 2015.[299] Unemployment in Germany fell to an 18-year low in January 2011.[300]

Innovation and a system for efficiently converting new technologies into marketable products and large-scale production are keys to this success.

Germany's innovation system is characterized by heavy corporate and government investment in research, innovative small- and medium-sized enterprises, extensive workforce training, and strong institutions such as Fraunhofer-Gesellschaft that collaborate with Germany industry. The government also works to assure that the nation is a "lead market" for important, emerging technologies through methods such as consumer incentives, government procurement, and standards.[301]

Such policies have enabled Germany to become the world's leading exporter of research-intensive products, according to the German Institute for Industrial Research (DIW Berlin).[302] More than 12 percent of Germany's exports are research-intensive, double the level of the U.S.[303] Programs

---

[296] Anthony Faiola, "Germany Seizes on Big Business in China," *Washington Post*, Sept. 18, 2010.
[297] Statistisches Bundesamt Deutschland, "German Exports in 2010: +18.5% on 2009," Press Release No.052/2011-02-09. These data refer to German exports to both other EU members as well as countries outside the EU. Exports to countries outside the EU increased by 26 percent.
[298] World Trade Organization, "Trade Growth to Ease in 2011 But Despite 2010 Record Surge, Crisis Hangover Persists," WTO Press/628, April 7, 2011.
[299] Estimate by UniCredit Markets and Investment economist Andreas Rees cited in Jeff Black, "Germany's Future Rising in East as Exports to China Eclipse U.S.," *Bloomberg*, April 6.
[300] German Federal Labor Agency data. Germany's unemployment rate was at 7.3 percent as of March 2011, compared to an average of 9.7 percent for the previous two decades.
[301] A "lead market" is a regional market that can establish the early commercial success of an innovation and large-scale production, increasing the chances of global diffusion. A discussion of Germany's strategy of establishing a lead market in photovoltaic cells and other technologies can be found in Klaus Jacob, et al, *Lead Markets for Environmental Innovations,* ZEW Economic Studies, Volume 27, Heidelberg: Physica-Verlag, 2005.
[302] DIW Berlin's definition of research-intensive industries includes automobiles and parts, chemicals, pharmaceuticals, machinery, and engines. In 2009, $670 billion in research-intensive products, compared to $561 by the United States and $388 by Japan. DIW Berlin data cited in presentation by Rainer Jäkel of the Federal Ministry of Economics and Technology in May 24-25, 2011, NAS symposium "Meeting Global Challenges in Berlin. Also see presentation by Stefan Kuhlmann, Fraunhofer ISI in National Research Council, *Innovation Policies for the 21st Century*, Charles W. Wessner, editor, Washington, DC: The National Academies Press, 2007, and DIW Berlin, "Germany is Well Positioned for International Trade with Research-Intensive Goods," *DIW Berlin Weekly Report*, No. 11/2010, Volume 6, March 26, 2010.
[303] Heike Belitz, Marius Clemens, Martin Gornig, Florian Mölders, Alexander Schiersch, and Dieter Schumacher, "After the Crisis: German R&D-Intensive Industries in a Good Position," *DIW*

promoting wide dissemination of environmental technologies have enabled German companies to capture 16 percent of world trade in that sector, which employs 1.5 million in Germany. Germany is a world leader in optics, a €2 billion industry that also has received significant public support. German machine tool makers are the world market leaders with a share of 19 percent. The nation has some 500 biotechnology companies, and the nanotechnology sector boasts 740 companies and 50,000 industrial jobs.[304] Germany also ranks No. 4 in the world in patents granted.[305]

Over the past decade, the German government has implemented an ambitious agenda designed to maintain the strength of Germany's global competitiveness. Chancellor Angela Merkel's government has increased investments in R&D, which rose by one-third to €12 billion ($17.1 billion) from 2005 through 2008.[306] Germany spent €80 billion in economic stimulus during the financial crisis, followed by a further €11 billion in stimulus that went to education and science and technology. Coming at a time when other nations were cutting back such spending in the face of recession, the major commitment to innovation represented "a paradigm shift of some importance" for Germany, explained Rainer Jäkel, director general for technology and innovation policy at the Federal Ministry of Economics and Technology (BMWi).[307]

The government also has been implementing a wide range of policies and programs to improve its innovation system. They include initiatives to—

- upgrade basic science,
- boost private R&D spending,
- strengthen collaboration between universities and business,
- improve the environment for high-tech start-ups, and
- nurture regional innovation clusters.

The government also has unveiled what it describes as Germany's first comprehensive national innovation framework, High-Tech Strategy 2020, which seeks to consolidate public programs around well-defined missions.[308]

Some of these efforts appear to be bearing fruit. Corporate investment

---

Economic Bulletin 2, 2011. This paper can be accessed at http://www.diw.de/documents/publikationen/73/diw_01.c.377100.de/diw_econ_bull_2011-02-1.pdf.
[304] Federal Ministry of Education and Research, *Research and Innovation for Germany: Results and Outlook*, (http://www.research-in-germany.de/dachportal/en/downloads/download-files/34648/research-and-innovation-for-germany-results-and-outlook-2009-116-pages-.pdf).
[305] World Intellectual Property Organization data as of 2008.
[306] Federal Ministry of Education and Research, data.
[307] Jäkel presentation, op. cit.
[308] Federal Ministry of Education and Research, *Ideas. Innovation. Prosperity. High-Tech Strategy 2020 for Germany*, Innovation Policy Framework Division, 2010.

in R&D surged from €2.3 billion in 2005 to an estimated €9.4 billion in 2008.[309] Total R&D spending in Germany reached 2.82 percent of GDP in 2009, the highest level since reunification with Eastern Germany. [See Figure 5.10] The number of people employed in the German R&D sector rose by 15 percent, to 162,000, over that period, with further growth expected.[310] A 2009 study by the German Association of Chambers of Industry and Commerce found that around 30 percent of all companies attribute their innovations to improved federal policy.[311] The World Economic Forum ranks Germany No. 6 among 142 nations in global competitiveness, including No. 7 in innovation and No. 4 in business sophistication.[312] In rankings by the Innovation Union, Germany places a close fourth behind Sweden, Denmark, and Finland, among the European Union's 27 members, and No. 1 in terms of "innovators."[313]

German innovation still faces a number of serious challenges, however. They include a scarcity of venture capital and bank loans for innovative companies, declining momentum in sectors such as electronics and aircraft, and weak performance in eastern Germany and Berlin, which consume a large share of federal research spending but produce relatively little innovation.[314] Germany ranks below most other industrialized nations in researchers as a percentage of total employment [See Figure 5.11], measures of international collaboration in research, and venture capital as a percentage of GDP.[315] There also are fears of a looming skills shortage due declining university enrollment as the population ages and disinterest in science and technology fields grows among German youth.[316] The Expert Commission on Research and Innovation, known by its German acronym EFI, reports an "urgent need to expand education, research and innovation" and warns that Germany's global competitiveness is under threat. The EFI also contends that Germany's tax system must become more innovation-friendly.[317]

---

[309] German Federal Ministry of Education and Research data.
[310] Ibid.
[311] German Association of Chambers of Industry and Commerce data cited in BMBF, "High-Tech Strategy 2020," op. cit.
[312] World Economic Forum, *The Global Competitiveness Report 2011-2012*, Klaus Schwab, editor, 2011.
[313] Pro Inno Europe InnoMetrics, *Innovation Union Scoreboard 2010: The Innovation Union's Performance Scoreboard for Research and Innovation*, Feb. 1, 2011.
[314] Kuhlmann presentation, op. cit.
[315] Jäkel presentation, op. cit.
[316] Expert Commission on Research and Innovation (*Expertenkommission Forschung und Innovation*), "Research, Innovation and Technological Performance in Germany Report 2010," http://www.kompetenznetze.de/service/bestellservice/medien/kn2010_englisch_komplet.pdf.
[317] Expert Commission on Research and Innovation, "Research, Innovation and Technological Performance in Germany Report 2009," http://www.e-fi.de/fileadmin/Gutachten/2009_engl_kurz.pdf.

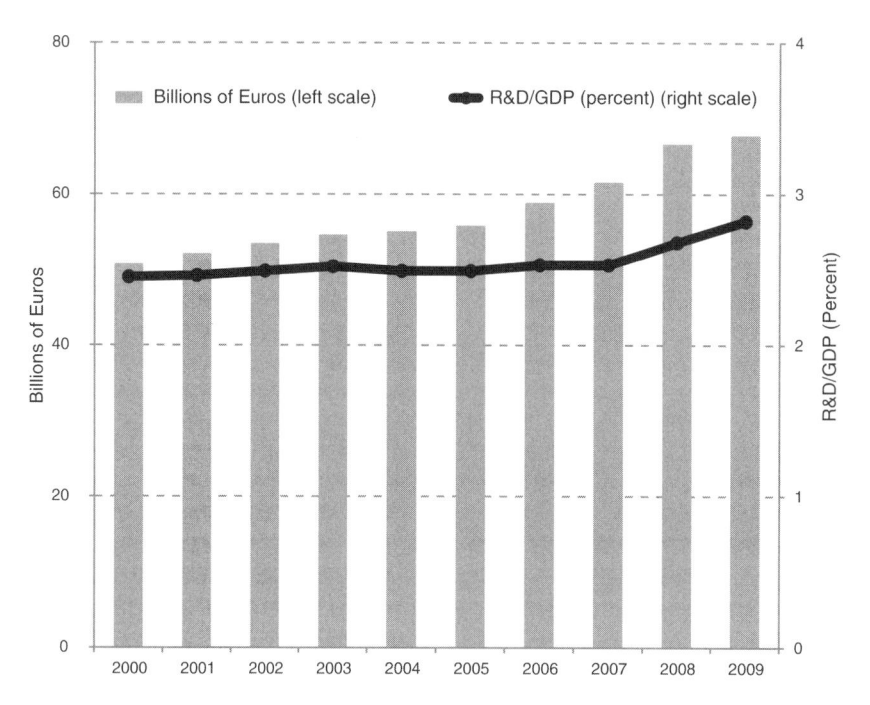

FIGURE 5.10 Germany's R&D expenditures reached 2.82 percent of GDP in 2009.
SOURCE: Eurostat, date of extraction was 10/29/2011.

For these reasons and others, a study by DIW Berlin rated Germany lower than did the WEF and Innovation Union—at only No. 9 among 17 leading industrialized nations in innovation capacity. While giving Germany strong marks in high technology and in research institutions, the study cited financing of innovative projects in particular as a "major barrier to innovation." Aversion to risk is another constraint. The study noted that 42 percent of Germans think one should set up a business if there is a chance of failure, compared to nearly 70 percent of Koreans and Irish and 74 percent of Americans. The study rated Germany No. 11 in innovation policy, primarily for spending just 4.5 percent of GDP on education in 2005 compared to 6 percent in countries like Finland and Sweden.[318] And while R&D-intensive products are important in Germany's

---

[318] German Institute for Economic Research (*Deutsches Institut für Wirtschaftforschung*), "Innovation Indicator for Germany 2009," Deutsche Telekom Foundation and Federation of German Industries. The DIW uses 180 different data items to measure innovation capacity. The summary of the report can be accessed in English at

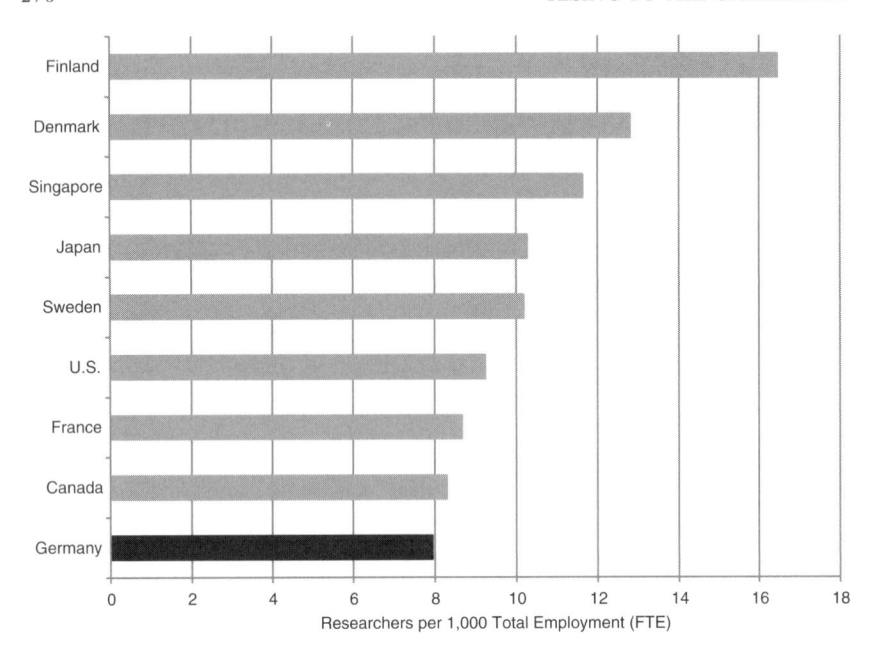

FIGURE 5.11 Germany ranks below most other industrialized nations in researchers per 1,000 total employment.
SOURCE: UNESCO, UNESCO Institute for Statistics, Science and Technology, Table 19.
NOTE: Data are 2009 or most recent year.

foreign trade, "Germany has a marked weakness in foreign trade in the area of cutting-edge technologies."[319] German R&D-intensive products overall make a positive contribution to Germany's foreign trade, but not in the most R&D-intensive areas. [See Table 5.2]

The government's new national innovation strategy aims to address these shortcomings.

---

http://www.innovationsindikator.de/fileadmin/user_upload/Dokumente/summary2009.pdf. The full report in German can be accessed at http://www.innovationsindikator.de/fileadmin/user_upload/Dokumente/innovationsindikator2009.pdf.

[319] The Commission of Experts on Innovation, *Research, Innovation and Technological Performance in Germany Report 2011*, February 2011, p. 130.

TABLE 5.2 Net Contribution of R&D-Intensive Products to Germany's Foreign Trade

	R&D-Intensive Products	High-Technology	Medium High-Technology
1995	70	-24	94
2000	49	-36	85
2005	50	-33	83
2009	47	-19	66

SOURCE: The Commission of Experts on Innovation, *Research, Innovation and Technological Performance in Germany Report 2011*, February 2011, Table C7-6.
NOTE: Net Contribution = Contribution to Exports – Contribution to Imports.

The German System

Germany's innovation system differs from that of the U.S. is several fundamental ways. While the U.S. has an "entrepreneurial economy," explained Engelbert Beyer of the Federal Ministry of Education and Research's Directorate for Innovation Strategies, Germany's model is more oriented toward "solid, high-quality progress."[320] While labor and skilled talent easily move to other jobs in the U.S., mobility is more limited in Germany. In terms of federal science and technology policy, programs are dispersed across many agencies in the U.S. In Germany, the Federal Ministry of Education and Research (*Bundesministerium für Bildung und Forschung*), better known as the BMBF, has a broad portfolio that includes most federal R&D activities and programs to promote commercialization. The Federal Ministry of Economics and Technology, known by its German acronym BMWi, also has a range of technology and innovation programs.

The "innovation rhetoric" differs in Germany and the United States, too, Mr. Beyer said. In the U.S., it is generally believed that government should play a limited role in industry and commerce. In Germany, "it is quite common to refer to government as a problem solver," Mr. Beyer said. Dr. Jäkel of BMWi pointed out that the German government has no qualms about providing "cradle to grave" financial assistance for R&D and commercialization efforts by small- and medium-sized enterprises in the case of "market failure" by private lenders. "The government has the right to intervene," he said. "It is well known that the banks are not so supportive."[321]

The German system also distributes its R&D investments very differently than the United States. While the U.S. innovation system seeks breakthroughs in a broad spectrum of sciences and technologies, most German

---

[320] Comments by Engelbert Beyer of the Federal Ministry of Education and Research in Nov. 1, 2010, "Meeting Global Challenges" symposium in Washington, op. cit.
[321] Jäkel presentation, op. cit.

R&D spending is on industries in which the country already is established, such as automobiles and machinery. The U.S. is great at "disruptive technologies and radical inventions," explained German State Secretary for Education and Research Georg Schütte. "Germany is strong in gradual change." The chief beneficiaries of Germany's high focus on applied research are the *Mittelstand*, the small and medium-sized export manufacturers that are the nation's "hidden champions" and that pursue market leadership in niches, he said. State Secretary Schütte noted that there is wide discussion in Germany whether this model is sustainable in light of growing shortages of skilled workers as the population ages and intensifying competition in manufacturing industries from East Asia. "Some people think we are sitting in the dining car," he said.[322]

The emphasis on manufacturing and exports, however, has served Germany well over the past few years of global turbulence. The government's response to the 2008-2009 global financial crisis and recession highlighted the importance that Germany places on preserving its manufacturing sector, which was hit hard. In the U.S., manufacturers laid off workers, who then sought public unemployment benefits. Germany, by contrast, subsidized manufacturing salaries so that staff could stay on payrolls while working part time. As a result, consumer spending and service industries remained robust through the recession, according to Klaus F. Zimmerman, former president of DIW Berlin. When recovery came, German exporters were able to quickly increase production and gain market share.[323]

There are strong linkages between government-funded research and public-private commercialization activities. Nearly half of federal R&D funds go to national research institutes. They include the Max Planck Society, which conducts basic research, and the Helmholz-Gemeinschaft, which has a network of 17 major research centers on long-term scientific challenges such as health, energy, the environment, and transportation. The Leibniz Association, an umbrella organization of 87 independent state-controlled institutes, performs research on scientific issues of strategic importance, such as life sciences, natural sciences, and social sciences. Fraunhofer- Gesellschaft acts as a "technology bridge" to industry, in the words of Executive Director Roland Schindler.[324]

Funding of research organizations is shared between the federal government and the states. This reflects changes to Germany's constitution in 1946, when it was under Allied control. The constitution was designed to prevent the central government from having sole control over education.

---

[322] From May 24, 2011, remarks by Georg Schütte at "Meeting Global Challenges" symposium in Berlin, op. cit.
[323] From presentation by Klaus F. Zimmerman of the German Institute for Economic Research in Nov. 1, 2010, "Meeting Global Challenges" symposium.
[324] See presentation by Roland Schindler, executive director of Fraunhofer, in November 1, 2010, "Meeting Global Challenges: U.S.-German Innovation Policy" symposium in Washington, op. cit.

Overall, the German research system "is a system that works well, with a clear division of labor, varying degrees of autonomy and state control, and distinct research organizations whose activities partly overlap and partly are different," explained Leibnitz Association President Karl Ulrich Mayer. One challenge, Mr. Mayer noted, is that many German state governments have weak tax bases, making it difficult for them to carry the educational funding load.[325]

One downside of Germany's large science and technology bureaucracy is that it also is considered unwieldy. The two agencies that dominate policy, the BMBF and BMWi, collaborate in many realms but have tended to compete and duplicate each other's work, as Stefan Kuhlmann of Fraunhofer ISI observed. He added that the BMBF's immense array of technology and innovation programs so broad that they are difficult to track.[326] The same is true of BMWi whose extensive portfolio sometimes overlaps with BMBF programs. Coordination among German's big research institutions also has traditionally been spotty. In addition, the Länder—Germany's 16 states--have their own research and technology programs.[327]

Forging a Common Strategy

To address these challenges, German leaders have been seeking to better coordinate the efforts of government, research organizations, and industry to improve innovation and the translation of technology into marketable products. In October 2003, Chancellor Gerhard Schröder summoned representatives of public agencies, policy circles, major companies, small business associations, major research organizations, and other stakeholders to debate challenges to German innovation. The forum led to formation of the Partnership for Innovation initiative, which aimed to define a national innovation framework.

In 2010, the German government unveiled High-Tech Strategy 2020, which it described as the country's "first broad national concept in which the key stakeholders involved in innovation share a joint vision." The primary goals are to build lead markets, bridge industry and science, and improve framework conditions such as access to early-stage finance, intellectual property protection, public procurement, and the ability of universities and research institutions to commercialize research. The plan aims to "stimulate Germany's enormous scientific and economic potential in a targeted way and find solutions to global and national challenges".[328]

The strategy is to coordinate government-funded research and

---

[325] Presentation by Leibnitz Association President Karl Ulrich Mayer in May 24-25 "Meeting Global Challenges" symposium in Berlin.
[326] Kuhlmann presentation, op. cit.
[327] Ibid.
[328] Federal Ministry of Education and Research, *High-Tech Strategy 2020 for Germany*," op. cit. Framework Division, 2010.

innovation activities across all departments around five broad themes: climate and energy, health and nutrition, mobility, security, and communication. The government defines "forward-looking projects" in science, technology, and social development and detailed roadmaps to achieve each of the overarching missions over 10 to 15 years. These "projects" include developing technologies and services to make model regions "carbon-dioxide neutral," development of smart grids and large power-storage systems, technologies that help people live well into old age, deploying electric vehicles on Germany roads by 2020, and developing new work organization models that allow people to remain productive longer.[329]

The national strategy incorporates a number of national innovation programs and initiatives to achieve these goals. They include programs to develop specific technologies, promote regional innovation clusters, upgrade higher education and scientific research, forge "innovation alliances" among corporations and universities, support small enterprises, and launch technology startups. German activities in energy and the environment, information and communications technology, and transportation illustrate how high-minded policy goals are integrated with strategies to develop globally competitive export industries.

Energy Efficiency: By 2020, the federal government proposes to reduce Germany's greenhouse gas emissions by 40 percent below 1990 levels and to double energy productivity. It also has set a target of having renewable energy meet 18 percent of gross energy consumption and provide for 35 percent of electricity by 2020. By 2050, greenhouse gasses are to be 80 percent below 1990 levels, while renewable sources will account for 60 percent of total consumption and 80 percent of electricity generation.[330]

To achieve those goals, there are detailed "fields of action." They include investing aggressively in renewable sources such as offshore wind power and bioenergy; expanding nuclear energy and promoting clean coal-fired plants; upgrading the national power grid; research and incentive programs to promote energy efficiency in industry, buildings, and households; aggressively promoting green transportation; and expanding R&D in innovative new energy technologies. Recognizing that the global market for energy-efficiency and environmental products is projected to reach €2.2 trillion by 2020, the BMBF,[331] BMWi, Fraunhofer, and other organizations also have ambitious programs to commercialize technology through regional innovation clusters, public-private

---

[329] Ibid.
[330] Federal Ministry of Economics and Technology and Federal Ministry for the Environment, Nature Conservation and Nuclear Safety, *Energy Concept for an Environmentally Sound, Reliable and Affordable Energy Supply*, Sept. 28, 2010
(http://www.bmu.de/files/english/pdf/application/pdf/energiekonzept_bundesregierung_en.pdf)
[331] Data cited in Federal Ministry of Economics and Technology, *Research and Innovation in Germany*, op cit.

research alliances,[332] and small-business assistance.

Government measures to establish Germany as a "lead market" for new energy technologies also help German industry attain scale in emerging industries. High feed-in tariffs for all renewable energy were instrumental in making Germany a leader in photovoltaic manufacturing, for example.[333] Even though Germany only receives about as much sunlight as Alaska, its photovoltaic manufacturing industry outperforms that of the U.S. by a factor of six, noted John Lushetsky, director of the Solar Energy Program of the U.S. Department of Energy. As a result of the scale of investment, solar modules in Germany cost around half of what they do in the United States.[334] The BMBF predicts that the renewable-energies sector could employ up to 500,000 people by 2020, and that exports of renewable-energy products will rise from €500 million in 2007 to €9 billion by 2020.[335]

Partnerships for Transportation Technology: Germany is a global leader in automobiles, high-speed train systems, and other transportation equipment. By setting a target of having 1 million electric vehicles by 2020, the government wants to make Germany a lead market in electric mobility and associated information systems. As the center of Europe's automotive industry, Germany in a good position to provide momentum for new technologies, the marketability of innovative vehicles, light-weight construction methods for aircraft, and development of global standards.

To this end, German public-private partnerships are investing in a wide range of transportation-related technologies. There is a national plan to develop lithium-ion technologies for electric-car batteries and for hydrogen fuel cells, for example.[336] Concerted research and product-development programs are underway in new drive systems, fuels, satellite navigation systems, traffic-control networks, mobile electronic services, and logistic concepts, among other fields.[337]

Information and communication technologies: The Federal Government's new ICT five-year plan sets out a broad agenda to "better harness

---

[332] Federal Ministry for the Environment, Nature, Conservation and Nuclear safety, for instance, has announced it will invest €250 million in four "innovation alliances" for climate-protection technologies. Press release, Nov. 11, 2008.

[333] See the Photovoltaic Industry Case Study in Chapter 6 of this volume.

[334] See presentation by John Lushetsky of the U.S. Department of Energy in Nov. 1, 2010, "Meeting Global Challenges" symposium in Washington.

[335] *Research and Innovation in Germany*, op. cit.

[336] See German government "National Hydrogen and Fuel Cell Technology Innovation Programme," May 8, 2006 (http://www.nkj-ptj.de/datapool/page/3/NIP-en.pdf).

[337] A comprehensive explanation of German programs to develop next-generation transportation technologies can be found in *German Federal Government's National Electromobility Development Plan*, August 2009 (www.bmvbs.de/cae/servlet/contentblob/27978/publicationFile/9729/national-electromobility-development-plan.pdf).

the large potential of ICT for growth and employment in Germany."[338] Broad goals include wiring the country with high-performance broadband networks of at least 50 megabits per second that will reach three-quarters of the population by the end of 2014 and the entire country as soon as possible. They also include deploying state-of-the-art, smart IT networks and services across industries and enabling paper-free government by 2012.

The ICT Strategy for 2015 calls for accelerating development of flagship projects, such as a super high-speed Internet service, digital data protection technologies, and intelligent networks for education, energy, mobility, public administration, and tourism. Public-private research projects, "innovation alliances," and research contracts are devoted to virtual reality, cloud computing, intelligent autonomous devices, connected household appliances, smart national identity cards. Various Germany government agencies also support programs to support ICT startups, provide expert help for small- and medium-sized businesses, vocational training, and export promotion for ICT-enabled services. The government envisions that such services can add 30,000 jobs.[339]

Improving Germany's Innovation System

Beyond the broad technology goals and industrial-development targets, Germany is pursuing a range of initiatives aimed at improving the nation's ability to innovate and disseminate new technologies more quickly and efficiently. The intent is to strengthen the linkage between the creation of knowledge and creation of products, explained Bernhard Milow, director of the German Aerospace Center's energy program. "Our view is that if we continue with the innovation process we have today, we cannot achieve our goals because it is too slow."[340] These initiatives include upgrading the quality of university research and commercialization, regional innovation clusters, public-private research alliances, promotion of small and midsized business, and improving the environment for enterprises.

Unleashing Universities and Research Institutes: The federal government and Länder—which control funding for higher education--have taken a number of steps over the past decade to upgrade the quality of university research, break down barriers between academia and industry, and commercialize university R&D. "A high-wage country needs to invest in education, universities, and research to foster innovation," explained former DIW president Zimmerman. "This is what you have to do to maintain

---

[338] Federal Ministry of Education and Research, *ICT Strategy of the German Federal Government: Digital Germany 2015*, November 2010 (http://www.bmwi.de/English/Redaktion/Pdf/ict-strategy-digital-germany-2015,property=pdf,bereich=bmwi,sprache=en,rwb=true.pdf).
[339] Ibid.
[340] From presentation by Bernhard Milow of German Aerospace Center energy program in Nov. 1, 2010, "Meeting Global Challenges" symposium in Washington.

competitiveness."[341]

A turning point was the Knowledge Creates Markets initiative in 2001 to create a "broad-based patenting and exploitation infrastructure."[342] Among other things, universities were given ownership of intellectual property created by academics, along the lines of the U.S. Bayh-Dole Act of 1980. The reform made it easier for universities to create larger portfolios of technologies and compete with top U.S. universities.[343] To help universities commercialize research and negotiate contracts, Patent Marketing Agencies were set up in each state, with the federal and Länder governments splitting the costs. These agencies also pooled resources into a national network called TechnologieAllianz e. V., which provides services to 200 scientific institutions with more than 100,000 scientists.[344]

Major research institutes such as Max Planck, meanwhile, have been given greater leeway to commercialize their research. The Freedom of Science Act gives these institutions more latitude to manage their own financial resources. Also, compensation for researchers no longer is restricted by government civil-service rules.[345] However, these institutes are expected to orient more of their research around missions defined by the national High-Tech Strategy.

Research universities have received substantial funding increases under several major programs. The Initiative for Excellence, for example, is a €1.9 billion, five-year federal program run by the German Research Foundation that allocates funding on a competitive basis to promote cutting-edge research, collaboration among institutions and disciplines, and international research alliances. One goal is to create a set of elite Germany universities.[346] So far, the initiative has granted funds to 44 graduate schools to expand research by young scientists, "clusters of excellence" based at universities, and eight "universities of excellence" that have developed "future concepts" for high-level research.[347] The initiative is expected to create 4,000 new positions. As a result of the program, "universities have moved to the center of the German science system,"

---

[341] Zimmerman presentation, op. cit.
[342] See Federal Ministry of Education and Research and Federal Ministry of Economics and Technology, *Knowledge Creates Markets: Action Scheme of the German Government*, March 2001 (http://www.bmbf.de/pub/wsm_englisch.pdf).
[343] Engelbert Beyer presentation, op. cit.
[344] TechnologieAlliance Web site, http://translate.google.com/translate?hl=en&sl=de&u=http://www.technologieallianz.de/&ei=EJ-sTfHgJ8Tk0QGujdmWDQ&sa=X&oi=translate&ct=result&resnum=6&sqi=2&ved=0CD4Q7gEwB Q&prev=/search%3Fq%3DGermany%2BTechnologieAlliance%2Be.%26hl%3Den%26prmd%3Divns.
[345] Federal Ministry of Education and Research, *Research and Innovation in Germany*, op. cit.
[346] Gretchen Vogel, "A German Ivy League Takes Shape," *Science Magazine*, Oct. 13, 2006.
[347] Excellence Initiative data from German Research Council (Deutsche Forschungsgemeinschaft) Web site at http://www.dfg.de/en/magazine/excellence_initiative/index.html.

according to a BMBF report. The federal government and Länder have approved another €2.7 billion in funding through 2017.[348]

Another goal is to sharply increase enrollment in higher education. Under the Higher Education Pact 2020, the federal government and Länder will provide €26,000 per place until 2015 to help create new positions for up to 275,000 first-semester students in tertiary institutions.[349]

Regional Innovation Clusters: Research programs in Germany have tended to be dispersed across the country, making it difficult to develop regional innovation clusters that commercialize new technologies.[350] Several public-private initiatives have sought to form regional innovation clusters in emerging industries. The Fraunhofer institutes are leading a government effort to help consolidate research activities into 16 innovation clusters. An emerging bioenergy cluster based in North Rhine-Westphalia district, for example, has 17 regional partners from industry and academia. Other innovation clusters that the Fraunhofer institutes are helping to organize include one in optical technologies based in Jena, electronics for sustainable energy based in Nuremberg, turbine-production technologies based in Aachen, and digital production based in Stuttgart.[351]

The BMBF also has a program to support regional innovation clusters. In 2007, the ministry launched the Top Cluster competition in which industry-led strategic partnerships around Germany vied for €200 million in BMBF funds. The first five winners were an aviation cluster forming in the Hamburg region, Solar Valley in Mitteldeutschland (Middle Germany), energy-efficiency innovations in Saxony, and electronics and cell- and molecular-based medicine in the Rhine-Neckar metropolitan region.[352]

Innovation Alliances: New forms of German public-private partnerships are being encouraged to advance new technologies. One initiative is called "innovation alliances." Under the program, corporations must decide at the board level to co-invest with government. The German government is investing €500 million and private industry €2.6 billion in nine such alliances. Government funds are typically leveraged five-fold through private investment. An initiative for a cluster in molecular imaging for medical engineering, for example, includes Bayer Schering Pharma, Goehringer Ingelheim Pharma, and Siemens. The alliance, which has a €900 million research budget, seeks to create

---

[348] Details on the Excellence Initiative can be found on the German Research Council (Deutsche Forschungsgemeinschaft) Web site at http://www.dfg.de/en/magazine/excellence_initiative/index.html.
[349] Data from Research in Germany Web site of the BMBF http://www.research-in-germany.de/research-landscape/r-d-policy-framework/60122/higher-education-pact.html
[350] Schindler presentation, op. cit.
[351] Information on Fraunhofer innovation cluster initiatives is found on the Fraunhofer Web site at http://www.fraunhofer.de/en/institutes-research-establishments/innovation-clusters/.
[352] Federal Ministry of Education and Research Web site.

new diagnostic products and imaging procedures for clinics at the molecular and cell level. Other such alliances focus on automotive electronics, energy-efficient lighting using organic light-emitting diodes, organic photovoltaic cells, and lithium-ion batteries for energy storage.[353]

BioIndustry 2021 is another such initiative. Launched by the BMBF in 2008, the program allocates funds to strategic partnerships between scientific organizations and industry aimed at speeding up the translation of ideas and research findings into marketable products. The federal government, industry, and Länder contribute funds. Five clusters have been selected. One is Hamburg-based Biocatlaysis2021, devoted to manufacturing chemicals, cosmetics, foods, and detergents. It includes 19 small- and medium-sized companies, 22 academic research groups, and seven agencies. [354]

Reinvigorating Small Business: More than one-quarter of innovation expenditure in German manufacturing and nearly half in knowledge-intensive businesses are by *Mittelstand* businesses. Small and medium-sized manufacturers are regarded as the backbone of Germany's advanced industrial sector. Many have been run by the same families for three or four generations, frequently reinventing themselves to keep up with the times, and tend to be "be long-term minded when it comes to research and developed," noted BWMi official Jäkel.[355] *Middelstand* companies also tend to be anchored in their regions and maintain close ties with local universities and research institutes, and therefore "won't easily relocate from one country to another." Small and medium-sized enterprises are especially important in Germany's biotechnology, optical, nanotechnology, and information technology sectors. Still, the vast majority of SMEs spend little on regular R&D, according to Germany's Center for European Economic Research.[356] Mr. Jäkel cited difficulty raising funds for R&D from banks as a major reason.

The government is seeking to increase R&D by smaller companies by connecting them to federal research programs and through expanded financial subsidies. In 2008, several SME-related activities within the BMWi were consolidated into the Central Innovation Programme SME, known by its German acronym ZIM.[357] One of those BMWi programs, Pro Inno, was established in 1999 and had distributed research grants to thousands of firms and more than 240 research organizations. An evaluation by Fraunhofer praised Pro

---

[353] Information on Germany's Innovation Alliances is found on the Research in Germany Web site, http://www.research-in-germany.de/coremedia/generator/research-landscape/rpo/networks-and-clusters/41832/10-3-innovation-alliances.html.
[354] BioCatalysis2021 Web site, http://www.biocatalysis2021.net//?page=Partner.
[355] Jäkel presentation, op. cit.
[356] Centre for European Economic Research, "Monitoring and Evaluation of 'KMU-Innovativ' Within the High-tech-Strategy" (http://www.zew.de/en/forschung/projekte.php3?action=detail&nr=814).
[357] In 2008 and 2009, the programs PRO INNOII, INNO NET, NEMO, and INNO-WATT were restructured and integrated into Central Innovation Programme.

Inno's high transparency, easy access, and relative lack of bureaucracy and found that some three-fourths of participating firms would not have conducted the R&D had it not been for the program.[358]

ZIM has an annual budget of around €300 million, and it received an additional €900 million through Germany's economic stimulus program in 2009 and 2010. It also expanded its services to include larger companies with up to 1,000 employees. ZIM's stated goals are to encourage SMEs to dedicate more efforts to innovation, reduce the risks of technology-based projects, and rapidly commercialize research.[359] ZIM has different programs to support cooperative research projects between enterprises and research organizations, R&D commercialization projects by individual SMEs, and innovative networks involving at least six companies. ZIM offers grants of up to €350,000 covering 35 percent to 50 percent of R&D costs, depending on the company's size and location. The program is not limited to technologies or sectors. Research institutions that cooperate with these firms can receive grants covering their entire costs up to €175,000.[360] As of late 2010, ZIM had allocated €1.4 billion in grants that were matched by €1.5 billion in SME contributions.[361] The program receives around 6,000 applications a year.

The BMBF has its own SME innovation program, KMU-Innovativ. It focuses on biotechnology, information and communications, nanotechnology, optics, energy, and production research. KMU-Innovativ also offers research grants and helps SMEs get better access to federal research funding. One objective of the program is to lower the technological and financial track record requirements that had prevented many smaller enterprises from competing for research funds.[362]

Early-Stage Capital

The BMBF declares that "Germany needs to go back to being a country of start-ups."[363] The paucity of venture capital in Germany is a "major bottleneck" to achieving this goal, explained Dietmar Harhoff, chairman of the Commission of Experts for Research and Innovation, whose annual reports have detailed shortcomings in Germany's entrepreneurship environment. In the U.S., Dr. Harhoff noted, young inventors seeking to start a company often go to angel

---

[358] Khulmann presentation, op. cit.
[359] Federal Ministry of Economics and Technology, *Central Innovation Programme (ZIM)*, January 2011 (http://www.zim-bmwi.de/download/infomaterial/informationsbroschuere-zim-englisch.pdf).
[360] Ibid.
[361] For an analysis of ZIM, see European Commission, *ZIM, the Central Programme for SMEs (Zentrales Innovationsprogramm Mittelstand)*, PRO INNO Europe, INNO-Partnering Forum, Document ID: IPF 11-005, 2010.
[362] Centre for European Economic Research, op. cit.
[363] *High-Tech Strategy 2020*, op. cit.

investors. In Germany, they often go to the government—or abroad.[364]

To address this goal, the government has taken several moves to help make more capital available to start-ups. In 2005, the BMBF formed a €272 million public-private capital partnership called the High-Tech Start-ups Fund. Participants include BMWi, KfW Bank Group, BASF, Deutsche Telekom, Siemens, Robert Bosch, and Daimler. The fund invests up to €500,000 in new, promising companies. The goal is to support young companies for up to two years from the R&D stage through development of proof of concept and even market entry, by which time it is hoped that private financing will be available. In its first five years, the fund has pledged to take holdings in 177 technology companies.[365]

The BWMi has its own program to aid start-ups, called EXIST. The program helps universities build infrastructure to assist technology- and knowledge-based ventures. EXIST also provides stipends of up to €2,500 a year for equipment, materials, coaching, and childcare to scientists and students who wish to develop business ideas. EXIST's Transfer of Research program provides grants for up to 18 months for technology startups.[366] BMWi official Jäkel said this public-private model has proved very successful and that private ventures "have really gone on board." Now BWMi is setting up a second fund that will have contributions from at least 10 German companies.[367]

International Cooperation

Recognizing that Germany cannot attain its broad technology goals on its own, the government's innovation policies devote considerable attention to international cooperation. There are more than 50 bilateral agreements between German and U.S. institutions.[368] In February 2010, Germany and the U.S. signed their first umbrella science-and-technology agreement and signed memorandums of understanding in the fields of energy and cancer research.[369] The German government also recently opened the German Center for Research and Innovation in New York and broke ground on the Max Planck Florida Institute in Florida. German and American scientists collaborate on numerous programs, such as the German Electronic Synchrotron, the Large Hadron Collider at the European Organization for Nuclear Research, and on what is described as the world's most powerful spallation neutron source SNS under

---

[364] Remarks on May 24, 20111, by Dietmar Harhoff of the Commission of Experts for Research and Innovation at "Meeting Global Challenges" symposium in Berlin, op. cit.
[365] Data from the Web page http://www.hightech-strategie.de/en/879.php.
[366] Data from BWMi Web site.
[367] Jäkel presentation, op. cit.
[368] For a comprehensive explanation of bilateral cooperation in science and technology, see Federal Ministry of Education and Research, "Germany and the United States Increase Their Cooperation," March 24, 2011 (http://www.bmbf.de/en/6845.php).
[369] See presentation of John Holdren at Nov. 1, 2010, "Meeting Global Challenges" symposium in Washington.

construction at Oak Ridge National Laboratory. The German government is seeking to expand such partnerships.[370] German Minister of State Werner Hoyer noted that Germany and the U.S. face several immense common technological challenges, such as the need to develop renewable energy, improve energy efficiency, and safeguard their nations from terrorism and other asymmetrical threats. "We are more likely to succeed if we combine our resources and technology," Minster Hoyer said.[371]

Remaining Challenges

Despite the abundance of innovation initiatives and programs, there are concerns that Germany isn't moving fast enough in some areas. DIW Berlin projects that the nation will have a shortfall of 270,000 skilled workers by 2020,[372] for example. A recent study by the Cologne Institute for Economic Research estimated that Germany's skills shortage costs the economy up to €20 billion a year, or one percentage point of GDP.[373] In addition to spending more on education, the authors assert Germany should be more open to immigration. Non-EU residents wishing to work in Germany must have a yearly income of €80,000, for example. The study estimates Germany's GDP could increase by up to €100 billion by 2020 if it relaxed immigration rules for skilled workers.

The Experts Commission on Research and Innovation calls for dispensing with income thresholds and instead linking admission to immigrants' qualifications. The commission also notes that many German scientists leave the country after they graduate, while not many foreign scientists move to Germany. It calls for the government to find ways to retain and recruit top talent. The commission also says the government must do more to encourage more German youth to study mathematics, engineering, and science, where college enrollment is declining.[374]

Germany's tax policies also are cited as a disincentive to investment. Germany cut its corporate tax rate from 38.65 percent to 29.83 percent in 2007, placing it near the median point of European economies. But the Experts Commission notes that Germany is one of few industrial nations that do not offer a tax credit for R&D. The commission blames tax policies for falling R&D investment by small- and medium-sized enterprises and the scarcity of private

---

[370] From presentation by German Ambassador to the United Sates Klaus Scharioth in Nov. 1, 2010, symposium "Meeting Global Challenges: U.S.-German Innovation Policy" in Washington, DC.
[371] Remarks by Minister of State Werner Hoyer at May 24, 2011, "Meeting Global Challenges" symposium in Berlin.
[372] German Institute for Economic Research data cited in Juliane Kinast, Christian Reiermann, and Michael Sauga, "Labor Paradox in Germany: Where have the Skilled Workers Gone?," *Spiegel Online*, June 22, 2007.
[373] Cologne Institute for Economic Research data cited in Bertrand Benoit, "German Gap Costs €20 bn," *Financial Times*, Aug. 20, 2007.
[374] Expert Commission on Research and Innovation, "Research, Innovation and Technological Performance in Germany Report 2010," op. cit.

risk capital. The commission asserts that shortages of angel funding and venture capital could worsen unless Germany adopts an "internationally competitive, growth-promoting tax framework."[375]

Although technology transfer from universities and research institutes has improved in recent decades, there still is room for improvement. The Expert Commission calls for strengthening the autonomy of universities and research institutes, freeing scientists from public-service regulations, and allowing professors to use their time more flexibly, such as by making teaching requirements less rigid.[376] The commission also suggests that universities create performance-related incentives for scientists and transfer of team members and to ease constraints on university and research institution participation in spin-offs companies.

Flanders

Openness and an emphasis on public-private partnership pervade the innovation system of Flanders, the Dutch-speaking region of Belgium. This approach has made Flanders, with its population of only about six million, an influential voice in technology policy in Europe.

Flanders' official strategy is to become "a region where businesses establish their research centers and where high-tech companies can develop."[377] To accomplish this, Flanders has adopted a cohesive strategy that combines strong public funding for science and technology, high-level guidance, and strong bottom-up input from industry. Its assets include a well-educated and multi-lingual workforce, strong higher education system, first-rate transportation and logistical infrastructure, and central location in Europe.[378]

The government is investing heavily in its universities and a range of new organizations to develop human resources and spur commercialization of knowledge. The government also provides early-stage financing for small and midsized enterprises and spreads the innovation message relentlessly through schools and the media.

---

[375] Ibid.
[376] Expert Commission on Research and Innovation, "Research, Innovation and Technological Performance in Germany Report 2009," op. cit.

[377] Greta Vervliet, *Science, Technology, and Innovation,* Ministry of Flanders, Science and Innovation Administration, 2006.
[378] See presentation by Peter Spyns of the Flanders Department of Economy, Science, and Innovation in National Research Council, *Innovative Flanders: Innovation Policies for the 21st Century—Report of a Symposium*, Charles W. Wessner, editor, Washington, DC: The National Academies Press, 2008. This volume summarizes proceedings from a symposium convened by the NAS STEP Board in Leuven in the Flanders region of Belgium in September 2006 titled "Synergies in Regional and National Innovation Policies in the Global Economy."

Box 5.4
The German Fraunhofer Institutes

Fraunhofer-Gesellschaft has been a major factor behind Germany's continued export success in advanced industries despite high labor costs. Established in 1949 as part of the effort to rebuild of Germany's research infrastructure,[379] the non-profit organization is one of the world's largest and most successful applied technology agencies. Fraunhofer's 80 research institutes and centers in Germany and around the world employ some 17,000 people—4,000 of them with Ph. Ds and master's students—and has a $2.3 billion (€1.62 billion) annual budget.[380]

The mission, in the words of Executive Director Roland Schindler, is to act as a "technology bridge" to German industry.[381] Although Fraunhofer researchers publish scientific papers and secure patents—they filed 685 applications in 2009—their primary mission is to disseminate and commercialize technology. Most of the organization's remarkable range of applied-research programs, which span microsystems, life sciences, communications, energy, new materials, and security, focus on clearly identified market opportunities and collaboration with German manufacturers.

Fraunhofer institutes offer a broad portfolio of services to its 5,000 corporate clients. Fraunhofer engineers develop intellectual property on a contract basis, hone product prototypes and industrial processes, and work with manufacturers on the factory floor to help implement new production methods. The institutes also can conduct market research and offer consulting services. Some *Mittelstand* manufacturers—the small and medium-sized enterprises that are the backbone of Germany's high-value export sector—have been Fraunhofer clients for generations. Nearly one-third of clients have 250 or fewer employees.[382]

A major source of Fraunhofer's strength and durability has been its diverse funding base, which enables its institute to perform their own in-house cutting-edge research, remain engaged in strategic national innovation programs, and collaborate with industry. Federal government and state funding, which covers one-third of its budget, has been stable and has steadily increasing, enabling Fraunhofer to plan for the long term, Dr. Schindler explained. Another third of Fraunhofer's revenue comes from manufacturing clients. Fraunhofer's

---

[379] For a history of the organization, see, *60 Years of Fraunhofer-Gesellschaft*, Munich: Fraunhofer-Gesellschaft, 2009. The publication can be accessed at http://www.germaninnovation.org/shared/content/documents/60YearsofFraunhoferGesellschaft.pdf.
[380] Fraunhofer data.
[381] See presentation by Roland Schindler, executive director of Fraunhofer, in *Meeting Global Challenges: U.S.-German Innovation Policy,* op. cit.
[382] Fraunhofer data.

59 Institutes of Applied Research in Germany collaborate closely with manufacturers in 16 different clusters. The federal government generally matches funds raised from industry. Half of the industry contracts are with small and medium-sized enterprises. When contracted to perform research, institutes agree to meet deadlines, milestones, and deliverables. Customers own the intellectual property.

The remaining third comes from publicly funded research projects that it wins on a competitive basis from the German government and the European Union. These keep Fraunhofer at the forefront of developing technologies meeting national and European Union priorities. Fraunhofer institutes own the intellectual property resulting from German government-funded research.

The diversity also enables Fraunhofer to use different approaches to commercialize technology. One way is by helping develop specific technologies for companies. Schott Solar, for instance, contracted with Fraunhofer to develop technology for absorber tubes used in solar receivers that now are being exported out of Schott's factory in Albuquerque, N. M. Fraunhofer earned Industrial research revenue of $654 million in 2009.[383]

Fraunhofer commercializes technology developed through in-house research or through its numerous R&D collaborations in Germany and abroad. Recent Fraunhofer lab inventions for industry include touch-controlled organic light-emitting diode (OLED) lighting, artificial animal tissue for drug testing, lightweight bicycle seat posts, new steel-cutting techniques for car manufacturers, micro-helicopters, and ultra-efficient gem-cutting tools.[384] One of the most lucrative Fraunhofer success labs is an algorithm co-developed with AT&T Bell Labs and other collaborators to reduce the size of audio files used in MP3 players. The institute earned several hundred millions from licensing the digital-compression technology.[385] In 1999, the organization set up Fraunhofer Ventures, a consulting service for start-ups credited with assisting 150 spin-offs.[386]

The technological diversity of Fraunhofer institutes also enables them to pursue promising niches in hybrid industries, such as the integration of multimedia technologies with medical devices. The Fraunhofer Heinrich Hertz Institute, which specializes in information and communications technology, has been developing 3-dimensional imaging, sensor, and communication network technologies with many possible medical applications. Engineers have developed immersive displays that can allow a surgeon to view a 3-D image of a heart without wearing special glasses, for example, and to manipulate the

---

[383] Fraunhofer-Gesellschaft, *Annual Report 2009: With Renewed Energy* (http://www.fraunhofer.de/en/Images/Annual-Report_2009_tcm63-60137.pdf).
[384] Explanations of these are examples are found in Fraunhofer-Gesellschaft 2009 annual report, ibid.
[385] See Mary Bellis, "The History of MP3," *About.com*, (http://inventors.about.com/od/mstartinventions/a/MPThree.htm).
[386] Fraunhofer Ventures Web site, http://www.fraunhoferventure.de/en/

continued

images by moving his or her fingers. The institute also is developing a handheld devise using Terahertz waves to probe for cancer cells inside the body.[387]

Another element of Fraunhofer's success is its ability to strike a balance between coordination of its German institutes with management autonomy. The Fraunhofer Group for Microelectronics, which has a $285 million annual budget, coordinates 12 Fraunhofer institutes working on topics such as automation and smart-system integration. The Information and Communication Technology group has 14 member institutes, including digital media, e-business software, and traffic and mobility. A production group coordinates R&D activities of seven institutes. Other institutes collaborate on the emerging field of flexible electronics, another Fraunhofer strength. The Institute for Photonic Microsystems and Institute for Electron Beam and Plasma Technology, for example, have demonstrated what it is billed as the first successful roll-to-roll manufacturing system that deposits OLED materials on sheets of aluminum, a process with wide potential applications for solar cells, memory systems, sensors, lighting, and other devices.[388]

At the same time, Fraunhofer's headquarters tries to give its institutes the latitude needed set their own direction in developing technologies and responding to opportunities. "Headquarters does not tell institutes what to do. We try to give them as much autonomy as possible," explained Anke Hellwig, liaison office for Fraunhofer USA's seven U.S. centers. "We try to keep a very delicate balance between the institutes having their own culture and with a Fraunhofer culture and branding."[389]

Headquarters appoints institute directors and imposes guidelines. For example, each institute director must also serve as the department chair at a local university in order to maintain strong links with academia. Headquarters allocates funding to each institute based on performance, using a formula that is heavy weighted toward their ability to raise funds from industry and research work. "We support the institutes that are successful in the market and want them to grow further," Ms. Hellwig explained. Revenue from industry must range from between 25 percent to 70 percent. If the institute operates below that number for several years or run steady deficits, headquarters could dissolve an institute or transfer operations to another organization.[390]

Retaining talent also is a challenge because the pay scales of scientists and engineers are dictated by German civil service rules. On the one hand, a

---

[387] From presentation by TK of Fraunhofer Heinrich Hertz Institute in May 25-26 symposium "Meeting Global Challenges: German-U.S. Innovation Policy," organized by the German Institute for Economic Research and the National Academies, Berlin.
[388] Printed Electronics World, "Smoothing the Way for Economic Flexible OLEDs," April 20, 2010.
[389] Interview with Anke Hellwig of Fraunhofer in Berlin.
[390] Ibid.

certain level of staff turnover is good for Fraunhofer as well as German industry. For engineering graduates, the opportunity for landing jobs at top German technology collaborating with Fraunhofer is a major assure of joining the organization. Fraunhofer alumni at German companies also help provide the institutes with relationships for new business. German industry, meanwhile, gets a pipeline of talent in emerging technologies. The abundance of Fraunhofer engineers with experience in solar cell and module manufacturing helped Germany establish itself in that industry. "Part of the secret of the photovoltaic industry is that the workforce was present when this industry developed," Dr. Schindler explained.[391]

On the other hand, Fraunhofer also needs experienced engineers and executives.

Institute directors coming from private industry general take steep pay cuts, but they can earn outside income as consultants. The institutes also offer some flexibility for researchers who want to try their hand at becoming entrepreneurs but are wary of completely abandoning the security of their Fraunhofer positions. Researchers can take leaves of absence to join a company. If they decide to return within two years, they can have their jobs back.[392]

One of Fraunhofer's new strategic thrusts is to extend its brand overseas. Fraunhofer has opened a number of centers in the U.S. specializing in different fields. In Plymouth, Mich., Fraunhofer USA has a state-of-the-art center to develop laser technologies, components, and systems. It also has joined with Michigan State University to open a center for advanced coatings and laser technology applications. The Fraunhofer Center for Sustainable Energy Systems (CSE), based in Cambridge, Mass., is a non-profit applied research and development lab dedicated to commercializing clean energy technologies. Another center in Brookline, Mass., focuses on manufacturing innovation. A Fraunhofer center in Delaware focuses on molecular biotechnology. A center at the University of Maryland develops software.[393] In these centers, Fraunhofer engineers work with nearby manufacturers to develop prototypes. It also evaluates company research projects and provides some funding.[394] Fraunhofer also recently opened a testing center for solar panels in Albuquerque, N. M., that it says will speed entry of North American manufacturers in the world market.[395]

Fraunhofer also is rolling out a program in the U.S. to help energy startups. Called TechBridge, the program aims to bridge the gap between

---

[391] Schindler presentation, op. cit.
[392] Hellwig, op. cit.
[393] Details of Fraunhofer centers in the U.S. are found on the Fraunhofer USA Web site, http://www.fraunhofer.org/.
[394] Schindler, op. cit.
[395] Fraunhofer USA press release April 7, 2011.

> *continued*
>
> laboratory research to wide-scale production without having to sacrifice intellectual property rights.[396] It offers design and modeling expertise, equipment, and access to Fraunhofer facilities in Germany. Fraunhofer's U.S. institutes earned $25 million in revenue in 2010 and expect to bring in $30 million in 2011.[397] "I see a huge market for this kind of research in the United States," Ms. Hellwig said.

Collaboration is a central element in public policy. Some 10 percent of R&D expenditures as of 2005 in Flanders involved industry partnerships with academia, compared to 6.9 percent in the EU and 6.3 percent in the U.S.[398] Flanders has initiated a variety of new public-private partnership programs in recent years to promote collaboration further.[399] They include regional innovation "cooperation networks," centers for collective research that serve traditional industries, and "competency poles" that often are located near universities. Finland also has strategic research centers for microelectronics, biotechnology, energy and environment, and broadband technology.[400] Each center has a mandate to work with the private sector, and is allowed to collaborate with foreign companies as long as they contribute to the Flemish economy. Chambers of commerce and labor unions also are involved in regional efforts.

Re-orienting higher education to add commercialization to its traditional functions of teaching and research has been another key thrust of Flemish policy. The region has seven universities, 22 non-university institutions of higher education, and five university-based institutes of higher education designed specifically to diffuse knowledge. The new emphasis on commercialization represents "a very big sea change" to centuries-old institutions, noted Free University of Brussels professor Bruno de Vuyst.[401] A study by Van Looy and Koenraad Debackere found that technology-transfer activities do not detract from the amount or quality of basic scientific research. In fact, technology transfer tends to support publication of more research papers.

---

[396] Fraunhofer USA Website, http://cse.fraunhofer.org/about/.
[397] Hellwig, op. cit.
[398] European Commission, *Third European Report on Science and Technology Indicators 2003*.
[399] Flanders' numerous innovation partnership programs are described in Vervliet, op. cit.
[400] See the presentation by Peter Spyns of the Department of Economy, Science, and Innovation in *Innovative Flanders*, op. cit.
[401] Presentation by Bruno de Vuyst of Free University of Belgium in *Innovative Flanders*.

Groups that collaborate reinforce their scientific research because industrial partners present academics with real problems. [402]

To incentivize universities, the government provides block grants to institutions that strive to meet performance metrics such as increased numbers of spin-offs, patent applications, and contracts. "Performance-based funding is the key," explained Fientje Moerman, former Minister for Economy, Enterprise, Science, Innovation, and Foreign Trade.[403] To speed up the process, the Flemish Innovation Agency was set up in 1991. The agency acts as a "one-stop shop for innovation," offering direct financing for technology-related R&D and coordinating other innovation efforts of the Flemish government. It also provides services for new business.

In 2003, the government drew up an Innovation Pact between academia and industry. The pact urged all parties to boost R&D investment to meet the EU target of 3 percent of GDP.[404] After a study by the Flemish Science Policy Council two years later found that old barriers between academia and business remained, the government introduced two new innovative mechanisms. In 2004, it established an €11 million Industrial Research Fund to encourage universities to hire post-doctoral staff to conduct further research on findings deemed to have high potential for near-term market application. Each university creates its own portfolio of industry-oriented projects. The government also instructed each university to set up a technology-transfer office. The Flemish government introduced a program to place young academic researchers into industry and to support Ph. D. students wishing to launch their own companies.[405]

The Katholieke University Leuven, or K. U. Leuven, has an especially interesting system to encourage industry collaboration. The university's 50 research divisions, which include faculty from different departments, can reinvest proceeds from industrial involvement into equipment, infrastructure, and stipends of post-doctoral students. The university also has 40 staff offering management, information-technology, and advisory help to start-ups. Leuven Inc., as it is called, has spun off more than 100 companies.[406]

To strengthen the flow of knowledge from universities to business, the Flanders government in 2006 launched a large €232 million program for strategic basic research of benefit to industry, the non-profit sector, and government policy objectives. The biggest investments in this program go to

---

[402] Bart Van Looy, Koenraad Debackere et al, *Research Policy,* 2004. The researchers used data based on ISI-SCIE figures.
[403] See remarks by Fientje Moerman, former Minister for Economy, Enterprise, Science, Innovation, and Foreign Trade, in *Innovative Flanders*.
[404] This target challenges all EU nations to raise their total investment in research and development to 3 percent of GDP by 2010. According to the independent web portal EurActiv, however, this target is increasingly unlikely to be met (www.euractiv.com).
[405] Moerman presentation, op. cit.
[406] From presentation by Koenraad Debackere of K. U. Leuven in *Innovative Flanders.*

four high-level research institutes: Interuniversity Micro-Electronics Centers (IMEC), the Flemish Interuniversity Institute for Biotechnology (VIB), the Flemish Institute for Technological Research (VITO), and the Research Center for Broadband Technology (IBBT).

IMEC is the most globally prominent of these Flemish research organizations. Established in Leuven in 1984, IMEC is one of the largest semiconductor research partnerships in the world and strives to be a "worldwide center of excellence," according to IMEC Chairman Anton de Proft.[407] About half of IMEC's revenue, which reached €275 million in 2009, comes from contracts with international industry. The Flemish government and the European Commission also are big contributors.[408] IMEC has more than 1,750 staff and more than 550 resident and guest researchers from around the world.[409]

The center emphasizes pre-competitive R&D by bringing together researchers from industry and academia related to areas such as chip design, processing, packaging, microsystems, and nanotechnology that may not meet industry needs for three to 10 years.[410] [See Table 5.3] As a result, IMEC enables its partners, who include Texas Instruments, ST Microelectronics, Infineon, Micron, Samsung, Panasonic, Taiwan Semiconductor Manufacturing, Intel, and a number of equipment makers to undertake risky research they may not do on their own. Given the high cost and long time horizons of chip research, partnerships like the one with IMEC are essential to sustaining the semiconductor industry, according to Allen Bowling of Texas Instruments.[411] In August 2010, Intel announced it was investing in a new ExaScience Lab in Leuven with IMEC, the Agency for Innovation by Science and Technology, and five Flemish universities that aim to achieve breakthroughs in power-reduction software to run on future computers delivering 1,000 times the performance of today's machines.[412]

Flanders' biotech facility, the Interuniversity Institute for Biotechnology (VIB), has an equally ambitious mandate to translate research into innovative industry. Until VIB was founded, "we had a lot of activity, but no translation from the universities to the economic growth of Flanders," explained Lieve Ongena, VIB's senior science advisor. "VIB was given a compound mission designed to overcome that problem."[413] The €62 million institute, formed in 1995, invests in basic research, training of researchers, commercialization of discoveries, and explanation of science to the public.

---

[407] From presentation by imec Chairman Anton de Proft in *Innovative Flanders*.
[408] Vervliet, op. cit., p. 57.
[409] Data: IMEC.
[410] Mission Statement.
[411] Presentation by Allen Bowling of Texas Instruments in *Innovative Flanders*.
[412] Press release issued Aug. 6, 2010 (http://www2.imec.be/be_en/press/imec-news/flandersexasciencelab.html).
[413] From presentation by Lieve Ongena of VIB in *Innovative Flanders*.

TABLE 5.3 IMEC—Major Areas of Research

- sub-22nm CMOS
- Heterogeneous integration
- Electronics for healthcare and life sciences
- Wireless communication
- Imaging systems
- Organic electronics
- Energy
- Sensor systems for industrial applications

VIB now has 60 research groups in nine departments, and 50/50 cost- and profit-sharing partnerships with its four universities. It focuses on work of "strategic importance," such as cancer, cardiovascular biology, neurodegenerative disorders, inflammatory diseases, growth and development, proteomics, and bioinformatics. The VIB supports 850 scientists and technicians, of whom 300 are Ph. D candidates. It has helped launch companies that have developed microscopic worms for drug discovery, a drug-targeting tool using camel antibodies and another using a bacterium as a living drug-delivery tool. Start-ups had raised more than €220 million in venture capital as of 2006.

Another organization that partners with industry is the Flemish Institute for Technology Research, Belgium's premier research center for energy, the environment, and materials.[414] The Research Centre for Broadband Technology, established in 2004 as a "virtual" center to help Flanders become a leader in information and communications technology. Its mission is to develop multidisciplinary talent and perform demand-driven research for industry and government, with an emphasis on health care. Business partners include Philips, Siemens, and Alcatel. The center hopes to recoup its investment through licensing and spinoffs, in which IBBT typically retains a 5 percent interest. The institute's goal is "to stimulate economic activity," according to General Manager Wim de Waele.[415]

The government has a raft of programs in addition to these institutes to spur innovation and commercialization. One subsidizes industry-initiated cooperative ventures that aim to commercialize or add value to corporate research. Another supports economic networks that encourage innovation. The Flemish Innovation Cooperative Ventures program supports collective research,

---

[414] Presentation by VITO Managing Director Dir Fransaer in *Innovative Flanders*.
[415] Presentation by IIBT General Manager Wim de Waele in *Innovative Flanders*.

technological services, and projects that foster innovation for particular issues or in sub-regions.

Flanders also has programs aimed at addressing an aversion to entrepreneurial risk on the part of the domestic financial sector and business community, which is regarded as a serious obstacle to innovation.[416] The government created a program in 2001 called Arkimedes, which provides government guarantees and tax credits for investments in certain small-denomination bonds. Money raised in the bond offerings goes into a "pool of pools" that is invested in several R&D funds. As with a venture capital fund, the risk is spread among a number of companies. The program is too young to draw conclusions about its effectiveness.[417]

One question about Flanders' approach is whether its open attitude toward R&D generates enough domestic industrial activity. Although IMEC plays an important role in international semiconductor research, for example, there is debate over whether it is establishing a semiconductor cluster in Belgium, which was the center's original mission in 1984. There have been at least 20 spinoffs from IMEC through 2002, noted Kenneth S. Flamm of the University of Texas at Austin, only a few related to devices, materials, or equipment manufacturing. None were major players in their sectors, Dr. Flamm said.[418]

IMEC Chairman de Proft noted while the economic impact so far is hard to measure directly, it is several times the level of government funding. He also said that the institute's concentration of 300 top researchers and 200 Ph.D. students from around the world are likely to make an impact as they develop networks and rise through their organizations.[419] Another indication of success, he noted, is that other nations have mimicked the public-private model of IMEC and other Flemish research institutions.

Finland

Despite its population of just 5.4 million, Finland has emerged as a global leader in innovation, consistently ranking the near top of the World Economic Forum's annual Global Competitiveness Index.[420] Finland has been

---

[416] From remarks by Rudy Aernoudt, then Secretary-General of the Flemish Department of Economics, Science, and Bruno de Vuyst of the Free University of Brussels in *Innovative Flanders*.
[417] See presentation by Rudy Aernoudt of the Department of Economic, Science, and Innovation in *Innovative Flanders*.
[418] From presentation by Kenneth Flamm of University of Texas at Austin in National Research Council, *21st Century Innovation Systems for Japan and the United States: Report of* a Symposium, Sadao Nagoka, Masuyuki Kondo, Kenneth Flamm, and Charles Wessner, editors, Washington, DC: The National Academies Press, 2009.
[419] Presentation by de Proft, op. cit.
[420] Finland ranked No. 3 in innovation and No. 4 in overall competitiveness in the World Economic Forum Global Competitiveness Index for 2011-12.

rated as Europe's most innovative business environment.[421] This has enabled the nation to restructure an economy that depended on pulp and paper for two-thirds of its exports in the 1960s to one dominated by electronics, most notably telecommunications equipment. Finland's economy also has grown faster than the OECD average both before and after the 2008 recession.[422]

Much of the credit goes to far-sighted government technology policies initiated in the 1980s that focus both on scientific research and on disseminating new technologies to industry. As a result, a close "Triple Helix" relationship has developed among Finnish universities, private industry, and government funding agencies.[423] In 1981, R&D accounted for around 1.2 percent of Finland's GDP. R&D intensity increased significantly in the mid-1990s and by 2009 had risen to 4 percent of GDP, one of the highest levels in the world, before falling slightly to 3.9 percent in 2010.[424] [See Figure 5.12] Private companies accounted for 70 percent of Finnish R&D spending in 2009, or €4.85billion.[425] Between 1992 and 2008, Finland's annual exports of high-tech products leapt more than five-fold, to €11.4 billion.[426] But high-technology exports fell sharply in 2009 and 2010 as electronics and telecommunications products fell dramatically, primarily mobile phone sales. [See Figure 5.13] In addition to electronics and telecom equipment, Finland achieved dramatic export growth in energy technologies and chemicals.[427]

Finland's innovation system is guided by the Science and Technology Council, which issues broad technology investment recommendations every three years that other ministries and agencies use as guidelines for setting funding priorities. The council is chaired by Finland's prime minister and includes five cabinet ministers and representatives from industry, unions, and academia. There is a high degree of coordination between the Academy of Finland, which funds basic research, and Tekes, a Ministry of Trade and Industry agency that funds applied-research collaborations between the public and private sectors.

Finland's high competitiveness rankings are attributable to the close link between national research programs and industry, according to Tekes Deputy Director General Heikki Kotilainen. Even though the government

---

[421] The Lisbon Council & Allianz Dresdner Economic Research, "The Lisbon Review, 2008."
[422] Eurostat data and ETLA calculations.
[423] See presentation by Heiki Kotilainen of Tekes in National Research Council, *Comparative National Innovation Policies: Best Practice for the 21st Century*, Charles W. Wessner, ed., Washington, DC: The National Academies Press, 2005.
[424] Finnish Science and Technology Information Service and Statistics Finland, *Research and Development 2010*, October 27, 2011.
[425] Ibid.
[426] Eurostat and Statistics Finland data.
[427] Data from Tekes. See http://www.tekes.fi/en/community/Statistics_and_comparisons/790/Statistics_and_comparisons/1740.

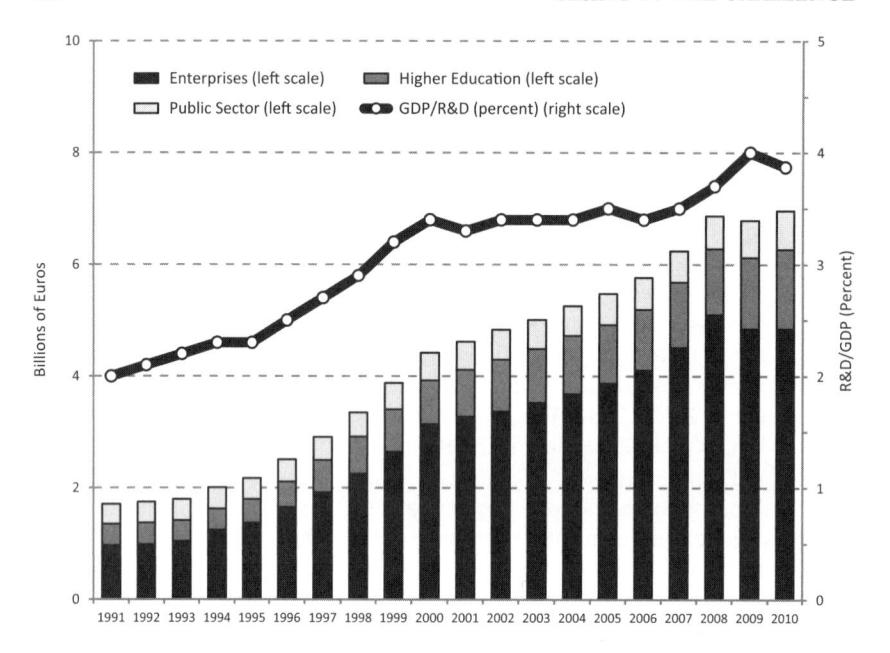

FIGURE 5.12 Finland's R&D intensity reached 4 percent in 2009 before declining slightly to 3.9 percent in 2010.
SOURCE: Statistics Finland, Science and Technology Statistics, Accessed at <http://www.research.fi/en/resources/R_D_expenditure/R_D_expenditure_table>.

accounts for only 10 percent of total R&D spending, it has been essential to stimulating R&D investment by companies. "You cannot jump from pure science to innovation immediately," Dr. Kotilainen said.[428]

Tekes estimates that government investments in research have yielded a return of around 20 times for the Finnish economy. In 2009, Tekes estimated its R&D investments contributed to more than 900 new products and services, the introduction of 328 production processes, 709 patent applications, and 775 academic theses.[429] According to Tekes customer surveys, more than half of small and mid-sized companies and 60 percent of large companies said research projects completed in 2006 led to commercial success.[430] Outcomes of collaborations with industry include Finnish companies that have developed lactose-free milk products, high-end computer monitors, recyclable bio composites that are used in everything from furniture to musical instruments,

---

[428] Kotilainen presentation, op. cit.
[429] Tekes data.
[430] Tekes 2009 customer surveys.

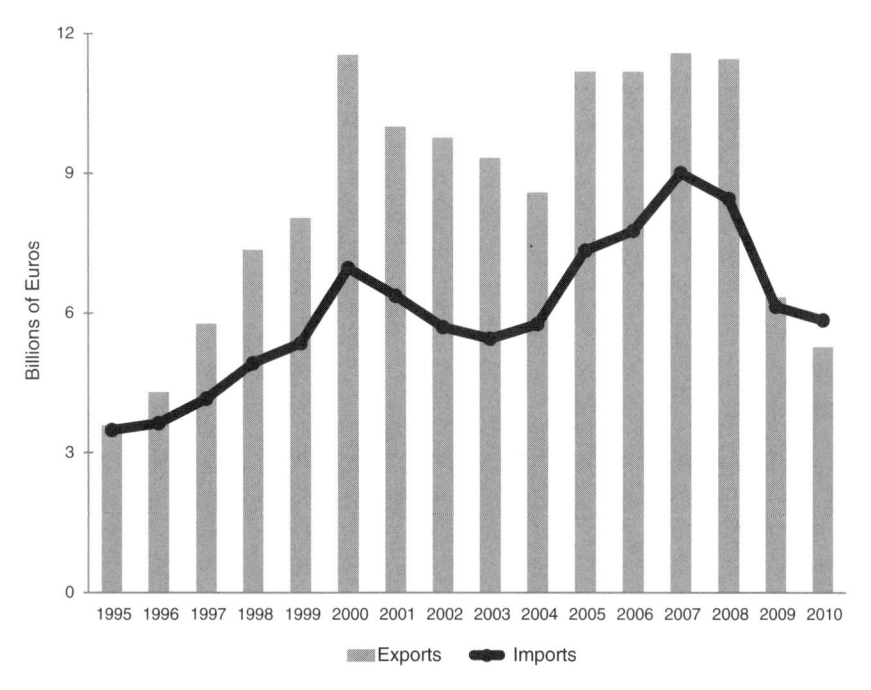

FIGURE 5.13 Finnish exports of high-technology products fell sharply in 2009 and 2010.
SOURCE: National Board of Customs, Finland, (Tullihallitus, Tilastoyksikkö), March 21, 2011.

equipment for recovering oil from offshore spills, and bio-carbon derived from wood and agro biomass that is said to be equal to high-quality coal as a fuel source.[431]

Tekes' approach to R&D funding illustrates the Triple Helix method. Rather than act as a regulator and coordinator of Finnish innovation, the agency views itself as a partner, networker, and investor, Dr. Kotilainen explained. Of the €579 million Tekes invested in 2009, €343 million went directly to enterprises. Small and midsized companies received 61 percent of those funds, and 87 percent of those companies have fewer than 500 enterprises. Funding applications by Finnish companies leapt by 40 percent in 2009.[432] Other funds

---

[431] Examples of successful Tekes investments can be found on
http://www.tekes.fi/en/community/Success%20stories/416/Success%20stories/666.
[432] Data from Tekes Annual Review 2009
(http://www.tekes.fi/en/community/Annual%20review/341/Annual%20review/1289).

go to universities, national research universities, and early-stage financing for start-ups.

Whether the applicant is a university or private company, Tekes favors projects that involve cooperation between the two sectors. Private companies are required to provide matching funds when participating in university research. Companies receive credit if they invite universities to join their own research projects. Sometimes companies and universities pool their R&D personnel. Tekes tends to divide its funds between established R&D projects that can involve multiple companies and universities and unsolicited project proposals. The agency also promotes international collaborations.

Integration of basic and applied research is an important feature of Finland's innovation system. Tekes and the Academy of Finland, for example, fund university and corporate-led programs simultaneously to help insure that basic research leads to technology development. Tekes also consults Finnish companies on their immediate and long-term needs. The goal is to make sure Tekes' limited resources are invested in technology that companies can absorb and that is relevant to the economy. Current focus areas include information and communication technology, renewable energy, new materials, and health and wellbeing.

Several studies have found that Finland's investments in R&D have had a significant impact. According to a 2006 study by the Research Institute of the Finnish Economy (ETLA), public subsidies by Tekes have improved productivity in small and medium-sized companies and in "companies near the frontier in productivity."[433] A study by Finland's National Audit Office found that Tekes funding allowed companies to implement R&D projects more quickly and broadly. It is also found that 57 percent of projects in the study would not have been undertaken without support from Tekes.[434]

Canada

Among industrialized nations, Canada ranks very high in education and in living standards, boasting the second-highest per-capita income among G7 nations. Yet Canada is not among the leaders in most benchmarks of innovation, ranking 12th in the World Economic Forum's latest Global Competitiveness Index.[435] In part, this is due to low R&D spending by business, which has

---

[433] Hannu Piekkola, "Knowledge and Innovation Subsidies as Engines of Growth—The Competitiveness of Finnish Regions," Research Institute of the Finnish Economy (ETLA), Sarja B 216 Series, Helsinki: Taloustieto Oy, 2006.
[434] Findings of the National Audit Office and other studies of Tekes' performance can be found in Markus Koskenlinna, "Additionality and Tekes," *Impact Analysis*, Nov. 25, 2003 (http://www.taftie.org/Files/PDF/MarkusKoskenlinna.pdf).
[435] World Economic Forum, *The Global Competitiveness Report 2011-2012*, op. cit. Canada was 11th in the innovation ranking.

declined in inflation-adjusted terms since 2001.[436] Canada's BERD intensity is among the lowest of industrialized economies. [See Figure 5.14] Some analysts attribute this paradox to Canada's abundant natural resources[437] and close integration with the United States, which keep its industries at the technological forefront even though domestic companies spend relatively little on research and development. "In many, many sectors, there is one economy," explained Peter J. Nicholson, president of the Council of Canadian Academies. "A great deal of technical sophistication in the Canadian economy is embodied in imported capital."[438]

The Canadian government has promoted domestic innovation much more actively in the past decade. One reason is that resolution of serious government fiscal problems in the mid-1990s freed up public resources. Another was realization that innovation would have to propel a greater share of future growth. "If we wanted to have something that was home-grown and that could give us a degree of independence, we had to build our innovation capacity from the ground up," Dr. Nicholson explained.[439] Another source of motivation was alarm over slowing productivity growth, which lagged that of the U.S. After Canadian productivity reached 91.4 percent of the U.S. level in 1984, it fell steadily. By 2006, Canadian productivity was at 73.7 percent of the U.S. level, the lowest level since the 1950s.[440] The International Institute for Management Development ranks Canada 24 among 33 advanced economies in productivity growth.[441] A report by the Council of Canadian Academies concluded that "Canada has a serious productivity growth problem."[442] As a result, "economists are increasingly focusing on a lack of innovation in Canada as a contributor to poor productivity performance," reported the Science and Technology and Innovation Council in its report *State of the Nation 2010*.[443]

---

[436] Council of Canadian Academies, *Innovation and Business Strategy: Why Canada Falls Short*, Report by Expert Panel on Business Innovation, 2009. This report can be accessed at http://www.scienceadvice.ca/uploads/eng/assessments%20and%20publications%20and%20news%20releases/inno/(2009-06-11)%20innovation%20report.pdf.
[437] Freedman shows that BERD intensity in Quebec and Ontario is much higher than in the other, more resource dependent provinces. Ron Freedman, "Re-Thinking Canada's BERD Gap," *The Impact Group*, January 2011.
[438] See remarks by Peter J. Nicholson in *Innovation Policies for the 21st Century*, op. cit. At the time, Dr. Nicholson represented the Office of the Prime Minister.
[439] Ibid.
[440] Andrew Sharpe, "Lessons for Canada from the International Productivity Experience," Centre for the Study of Living Standards, Research Report 2006-02, 2006.
[441] IMD data cited in Science, Technology and Innovation Council, *State of the Nation 2010*, June 2011. This report can be accessed at http://www.stic-csti.ca/eic/site/stic-csti.nsf/eng/00043.html.
[442] Council of Canadian Academies, op. cit.
[443] Science, Technology and Innovation Council, *State of the Nation 2010*, op. cit.

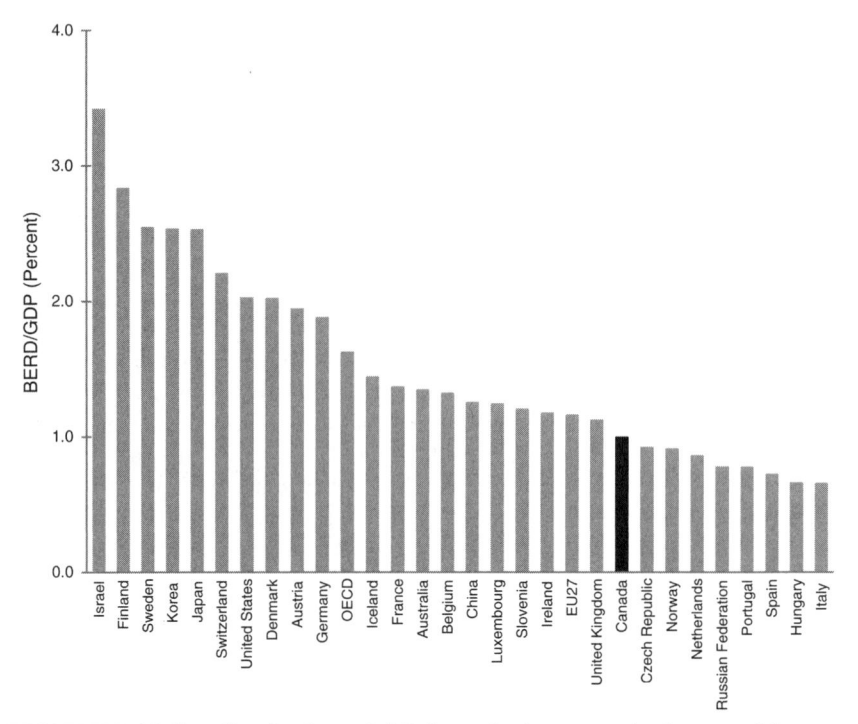

FIGURE 5.14 Canadian business R&D intensity is among the lowest of the industrialized countries.
SOURCE: OECD, Main Science and Technology Indicators Database, June 2011.
NOTE: Data refer to 2009 or most recent year available.

The nation's education system gives Canada a strong base to build upon. Forty-six percent of Canadians aged 25 to 64 are post-secondary graduates, the highest rate among OECD nations.[444] Canada spent 63 percent of GDP on higher education R&D as of 2007, by far the highest level among G7 nations.[445] It also had the most citizens aged 25 to 64 with a tertiary education-- 49 percent.[446] Canada has an extensive network of 24 national research institutes under the National Research Council, which employ 4,500 and hosts 1,200 guest researchers.

---

[444] Human Resources and Skills Development Canada data.
[445] OECD Science and Technology Indicators 2009.
[446] OECD data.

The government's approach to science and technology shifted significantly in the 1990s. Public funding for basic research rose sharply. R&D spending by Canadian universities and research hospitals nearly tripled between 1998 and 2004, to around $2.3 billion.[447] There also was a healthy increase in "intramural" funding. Direct subsidies for industry were curtailed, and a model of sharing risk between the public and private sectors was adopted.[448] The government also established a number of programs to build world-class research institutions, encourage companies to invest more in R&D, and disseminate technology more widely through the economy. Canadian policies were in part influenced by studies of the experiences of such nations as Sweden and the United Kingdom and by the innovation system of the European Union.[449]

Canada introduced several institutions in the late 1990s to lead an innovation drive. The Canada Foundation for Innovation was established and given the mission of transforming research and technology development, fostering strategic research planning at universities, attracting and retaining world-class researchers, and promoting collaborative and cross-disciplinary research. The Department of Finance and the Department of Industry began to formulate an innovation framework for the country in 1998. The government also launched an initiative in the mid-1990s to establish a network of "centers of excellence" to create research partnerships in advanced technologies, engineering and manufacturing, life sciences, environmental technologies, and natural resources.

The national innovation strategy, presented in a 2001 report called *Achieving Excellence*, set ambitious benchmarks. The document called for Canada to rank among the world leaders in share of private-sector sales attributable to innovations, match the U.S. is per-capita venture-capital investment, improve recruitment of foreign talent, and increase graduate student admissions by 5 percent each year. To make the business environment more globally competitive, the strategy called for regulatory reform, lower taxes, and high-speed broadband that is widely accessible to Canadian communities. The document set a target of developing at least 10 internationally recognized technology clusters.[450] Minister of Finance Paul Martin, who later became Prime Minister, announced a goal that Canada would move from No. 15 in the world in

---

[447] Science, Technology, and Innovation Council of Canada, "State of the Nation, 2010." Access at http://www.stic-csti.ca/eic/site/stic-csti.nsf/vwapj/10-059_IC_SotN_Rapport_EN_WEB_INTERACTIVE.pdf/$FILE/10-059_IC_SotN_Rapport_EN_WEB_INTERACTIVE.pdf .
[448] Ibid.
[449] For a good analysis of the evolution of Canadian innovation policy, see Thomas Liljemark, "Innovation Policy in Canada: Strategies and Realities," Swedish Institute for Growth Policy Studies, A2004:24 (http://www.vinnova.se/upload/EPiStorePDF/InnovationPolicyInCanada.pdf) .
[450] Industry Canada, Achieving Excellence: Investing in People, Knowledge and Opportunity—Canada's Innovation Strategy, 2001. (http://dsp-psd.pwgsc.gc.ca/Collection/C2-596-2001E.pdf).

government R&D spending as a percentage of GDP to No. 5 by 2010. That would require research investment to triple.

Canada has made especially strong progress in strengthening its infrastructure for basic research. Canada ranks No. 6 among OCED nations in scientific publications per capita and fifth in quality of publications.[451] Much of the credit goes to measures launched in the mid-1990s. The Foundation for Innovation and the Canada Research Chairs program have had an especially broad impact. In 2007, the government unveiled a new science and technology strategy. It stated that Canada "must be connected to the global supply of ideas, talent, and technologies." Among other things, the plan called for focusing on research relating to the environment, natural resources and energy, health, and information and communication technologies. It included an initiative called Knowledge Advantage to build on Canada's research strengths to generate innovation and another called People Advantage aimed at developing and recruiting knowledge workers.[452]

The Foundation for Innovation, established in 1997, awards grants covering up to 40 percent of the cost of university R&D projects. The competitive application process led to a sharp improvement in the quality of research projects, according to Dr. Nicholson. The foundation also allocates funds to upgrade research facilities, spur international collaborations, and help first-time researchers. The foundation's board includes some government appointees but operates independently.

As of September 2009, the foundation had committed nearly $5.2 billion to 6,300 projects at 130 research institutions across Canada. The program had attracted 8,050 new faculty members to Canadian universities, with nearly 3,200 from other nations. Forty-four percent of the 1,806 new researchers were recruited internationally, and nearly 80 percent of project leaders said the availability of foundation-funded infrastructure was important to their decision to join the institution. More than 21,000 post-doctoral fellows and graduate students used the infrastructure for their research, and the foundation had supported more than 1,600 collaborative research agreements. The foundation also was credited with creating nearly 4,700 jobs and at least 54 new companies. [453] A $61 million round of investments in 245 projects announced in January 2011 offers a flavor of the research that is supported. The projects include design of innovative molecules to treat breast cancer at the University of

---

[451] Organization for Economic Co-operation and Development, *OECD Science, Technology and Industry Outlook 2010: Country Profiles,*
[452] See Industry Canada, *Mobilizing Science and Technology to Canada's Advantage—2007,* 2007 (http://www.ic.gc.ca/eic/site/ic1.nsf/vwapj/SandTstrategy.pdf/$file/SandTstrategy.pdf).
[453] Canada Foundation for Innovation, *2009 Report on Results: An Analysis of Investments in Infrastructure*
(http://www.innovation.ca/docs/accountability/2009/2009%20Report%20on%20Results%20FINAL EN.pdf).

Guelph, a project to improve understanding of the brain and spinal cord at Dalhousie University in Halifax, and monitoring of ecosystems in the Canadian Arctic at the Université du Québec à Rimouski.[454]

Canada Research Chairs complements the foundation by funding development of world-class research capacity at universities and a cadre of researchers. The program has a $300 million annual budget to recruit and retain top-flight academics. Since it began operation a decade ago, the program has established 2,000 chairs at degree-granting institutions. Thirty percent of the 1,845 chairs filled so far are occupied by academics recruited from outside Canada.[455]

Each degree-granting school receives allocations of chairs based on research grants they win in national competitions, with special consideration for small institutions. Universities nominate academics whose work complements their strategic research plans. Academics recognized by their peers as "world leaders" in their fields are paid $200,000 annually for seven years with indefinite renewal. "Exceptional emerging scholars" receive $100,000 for five years with one renewal. Together, the foundation and chairs program have "powerfully boosted Canada's research capacity at the front end," Dr. Nicholson said.[456]

To spur innovation among small businesses, Canada operates the Industrial Research Assistance Program (IRAP). The program, managed by the National Research Council, has a $281 million budget and employs 240 industrial technology advisors in 147 sites across the country. These advisors work with nearly 8,000 companies, dividing their time between consulting small businesses and supporting projects such as feasibility studies, pre-competitive R&D, hiring, and international sourcing. Seventy-five percent of advisors have masters or Ph. D. degrees, 45 percent had run their own R&D facilities, and 35 percent have been entrepreneurs. IRAP advisors improve business proposals and help connect small and midsized enterprises to national and global innovation networks. Help from the agency often gives small companies credibility in the financial community, making it easier to raise capital.[457]

IRAP also provides financial support to help small and midsized Canadian enterprises develop technologies for competitive advantage. The program provides up to $1 million a year to some 1,400 firms, 80 percent of which have fewer than 50 employees, with the average receiving around $100,000. Companies receiving funds typically contribute half of a project's cost. Small contributions can be approved in as little as two weeks.

---

[454] Canada Foundation for Innovation press release, Jan. 21, 2011.
[455] Data: Canada Research Chairs Web site.
[456] Nicholson presentation, op. cit.
[457] National Research Council, "NRC-Industrial Research Assistance Program," Power Point presentation, March 2010, (http://acamp.ca/alberta-micro-nano/images/docs-conventional-energy/Finance/Generic%20IRAP%20PPT_FINAL%20ENG_March-2-2010.pdf).

In most cases, IRAP agrees to work with companies over time, rather than only provide one-time help for specific projects. IRAP charges companies for advisory services, but doesn't break even. The help contributed to the Canadian economy, however, Dr. Nicholson explained. A 2007 evaluation of IRAP suggested that the approach of supporting development of firms is one reason behind its success. Sales of IRAP client firms averaged 28 percent growth in revenues and 30 percent growth in employment over the previous five years. For each 1 percent increase in IRAP contribution and advisory services, firms exhibited an 11 percent increase in sales, a 12 percent jump in productivity, a 13 percent rise in R&D spending, and a 14 percent increase in employment. The analysis found that the program contributed between $2.3 billion and $6.5 billion to the Canadian economy between 2002 and 2007, meaning that benefits equaled four to 12 times IRAP's costs.[458]

Low business investment in R&D has received growing emphasis in recent years. The 2007 federal science and technology plan included a program called Entrepreneurial Advantage to translate knowledge into practical applications. The plan called for improving investment incentives and allowances for capital costs, establishing centers of excellence in commercialization and research, and expanding support for small and midsize companies.[459]

The Networks of Centers of Excellence program, meanwhile, has been broadened. In 2007, the government committed $46 million to fund large collaborative networks that support private-sector innovation headed by business consortia. Seventeen new Centers of Excellence for Commercialization and Research have opened since 2008 to promote stronger partnerships between researchers and industry.[460] They include the Perimeter Institute for Theoretical Physics at the University of Waterloo, the Brain Research Centre at the University of British Columbia, and the National Optics Institute in Quebec City. To date, the Centers of Excellence program is credited with creating more than 100 spin-off companies and training 36,000 personnel. Each year, it generates more than 100 patents and leverages $71 million in added investment.[461]

Canada also has made generous use of tax credits to entice corporations to build R&D centers and advanced manufacturing facilities in Canada. Under the Scientific Research and Experimental Development incentive scheme, companies can get 30 percent rebates from the government on their R&D spending. The credits are awarded regardless of a company's size, industry

---

[458] National Research Council, "Impact Evaluation of the NRC Industrial Research Assistance Program (NRC-IRAP)," Executive Summary, 2008 (http://www.nrc-cnrc.gc.ca/eng/evaluation/evaluation-irap.html).
[459] *Mobilizing Science and Technology to Canada's Advantage*, op. cit.
[460] Networks of Centres of Excellence Web site.
[461] Ibid.

sector, or technology area. Companies can deduct the full cost of R&D machinery and equipment. Large Canadian and foreign corporations can claim 20 percent credits that can be used to offset federal taxes due within the next 20 years. An estimated $3.5 billion in benefits were awarded in 2009.[462]

Despite all of these programs, striking the right balance with public support of private companies has proved challenging in Canada. In 1996, a program called Technology Partnerships Canada, which had focused on the defense industry, began covering 25 to 30 percent of companies' cost of industrial research, prototype development, and testing in other industries. Investments targeted "enabling" technologies such as biotech, materials, and information and communications technology. Companies were to repay the funds when they became profitable. As of 2006, only about 3 percent of funds invested by the Technology Partners had been repaid by companies. The program also was criticized for taking too long to approve projects.[463] Technology Partnerships was discontinued in 2006 and absorbed into the Industrial Technologies Office, which no longer offers such subsidies.[464]

Challenges faced by the Technology Partners program offered some lessons regarding public investments. One is that it is difficult to design repayment terms that properly reflect risk and reward of a specific research project, Dr. Nicholson acknowledged. Technology Partners was criticized for taking too long to approve projects. The program's broad objectives also made it difficult for Technology Partners to maintain a consistent approach. "That tends to invite a lot of objections from people who were disappointed," Dr. Nicholson said. "Someone can always find a precedent and say, 'but you approved that one, so what's wrong with me?'" It also was sometimes hard to demonstrate a direct impact to the Canadian economy from public investments: Recipients of Technology Partners funds included companies like IBM and Pratt & Whitney that operate in a world of global supply chains.[465]

Business spending on R&D also continues to lag in Canada, falling in 2010 for the third year in a row, to $14.8 billion.[466] Although business funding of university research has risen sharply since 2001,[467] corporations still account for only about 50 percent of total R&D spending in Canada, one of the lowest among major economies. A 2009 survey of 6,233 Canadian enterprises in 67 industries found that only 18.8 percent said their strategic focus is to regularly

---

[462] For details of how Canadian R&D tax credits work, see Foreign Affairs and International Trade Canada, "Invest in Canada: We Take Care of Business," September 2010 (http://investincanada.gc.ca/download/142.pdf).
[463] Nicholson presentation, op. cit.
[464] Industrial Technologies Office Web site.
[465] Nicholson, op. cit.
[466] Statistics Canada at http://www40.statcan.ca/l01/cst01/econ151a-eng.htm, accessed November 1, 2011.
[467] Science, Technology and Innovation Council, *State of the Nation 2010*, op. cit.

introduce new or significantly improved goods and services.[468] The Science, Technology, and Innovation Council said in a 2008 report that R&D spending by Canadian firms is "falling behind our major competitors and the gap is growing."[469] Business R&D spending equaled around 1 percent of GDP in 2009, compared to a 1.6 percent average for OECD nations.[470] [See Figure 5.14] Milway, executive director of the Institute for Competitiveness and Prosperity, recently remarked that this performance "is another bit of evidence that our businesses are not competing on the basis of innovation, value-added and sophistication."[471] Total R&D intensity in Canada has thus been trending downward for the past decade, to 1.81 percent of GDP in 2010. [See Figure 5.15]

There also are concerns that Canada is falling short of its goal of building a sufficient base of knowledge workers. A report by the Canadian Council on Learning in August 2010 said Canada lags in early childhood education. While science, math, and reading test scores still are relatively high in secondary school, other nations are advancing faster.[472] Canada ranks 20th among OECD nations in terms of natural science and engineering degrees as share of total degrees and 17th in the number of people in science and technology occupations.

Such challenges have not slowed Canada's commitment to investing in the science and technology foundations of an innovation-led economy. It is early to pass judgment on Canada's efforts to stimulate private investment in R&D, since many of the new programs were implemented just prior to the 2008-2009 recession, which forced companies to cut back. To address challenges in R&D investment and with the skilled workforce, the Canadian government also remains committed to expanding research collaborations with foreign companies and universities, to improving incentives to attract direct foreign investment, and to recruiting top talent.

---

[468] Industry Canada, Foreign Affairs and International Trade Canada, and Statistics Canada, "Survey of Innovation and Business Strategy," 2009. A summary of the survey's findings can be found on the Industry Canada Web site at http://www.ic.gc.ca/eic/site/eas-aes.nsf/eng/h_ra02118.html.
[469] Science, Technology, and Innovation Council, *State of the Nation 2008* (http://www.stic-csti.ca/eic/site/stic-csti.nsf/eng/00019.html).
[470] OCED, *Main Science and Technology Indicators*, 2010.
[471] Rebecca Lindell, "Canadian R&D Spending Continues Downward Spiral: StatsCan," *Postmedia News*, Dec. 8, 2010.
[472] Canada Council on Learning, *Taking Stock of Lifelong Learning in Canada (2005-2010): Progress or Complacency?* Aug. 25, 2010.

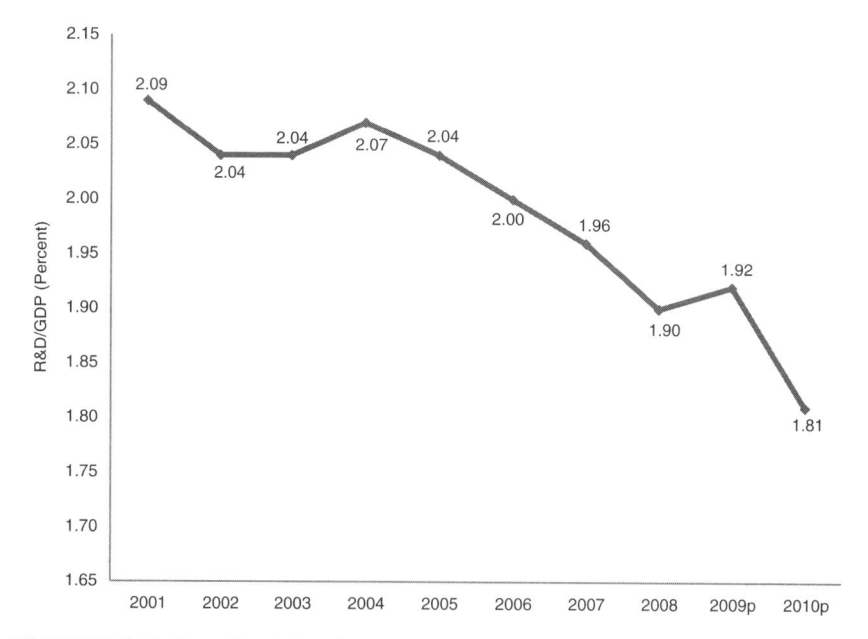

FIGURE 5.15 Canadian R&D intensity has been trending downward in the past decade.
SOURCE: Statistics Canada, CANSIM, tables 358-0001 and 380-0017 and Catalogue nos. 88-001-XIE and 88F0006XIE.
NOTE: Data for 2009 and 2010 are preliminary.

Japan

Japan has taken a number of actions since the mid-90s to improve its innovation system, many of them inspired by the United States.[473] Japan has strengthened protection of intellectual property, overhauled science and technology policy institutions, enacted its own version of the Bayh-Dole Act to make it easier for universities and research laboratories to commercialize technology, and bolstered industry and academic science partnerships.[474] Japan

---

[473] A National Academy report recently concluded, however, that Japan has still not adequately addressed some longstanding weaknesses in its S&T system "which include immobility of personnel, inadequate entrepreneurialism, insufficient opportunity for younger researchers, and abiding problems with industry-university-government collaboration." National Academy of Sciences, *S&T Strategies of Six Countries*, op. cit., p. 43.
[474] See Sadao Nagaoka and Kenneth Flamm, "The Chrysanthemum Meets the Eagle— The Co-evolution of Innovation Policies in Japan and the United States," in National Research Council, *21st Century Innovation Systems for Japan and the United States: Lessons from a Decade of Change*,

also undertook a number of initiatives to increase entrepreneurialism, including a small-business loan program similar to America's Small Business Innovation Research program.

To spur corporate R&D spending, Japan grants generous tax credits. Largely as a result, Japanese spending on research and development surged from 2.77 percent of GDP in 1994 to 3.8 percent in 2008 before declining slightly to 3.62 percent in 2009.[475] [See Figure 5.16] Japanese companies account for three-quarters of that spending, the highest ratio among OECD nations.[476]

Driving this change was the realization that innovation would be central to restoring growth to the Japan's stagnating economy in the wake of the financial crash of 1990. Even though Japanese R&D investment and output of patents remained quite strong on world standards throughout the 1990s, Japanese companies stumbled as they tried to make the transition from products derived from well-developed technologies to the creation of more fundamental breakthroughs.[477] Japan's competitiveness in industries such as semiconductors and consumer electronics waned with the rise of new rivals in South Korea and Taiwan. Japan had largely missed out on the U.S.-led booms in biotechnology and software.[478] Japan's commercial scene, dominated by large conglomerates, was not producing many dynamic start-ups. The rapid pace of change ushered in by the information technology revolution and globalization did not play to the strengths of Japan's large industrial conglomerates.

Japan's policy shift began in earnest with passage of the Basic Law on Science and Technology in 1995.[479] Under that plan, the government spent ¥17 trillion ($206 billion in current U.S. dollars) from 1996 through 2000 on science and technology programs. During the subsequent five-year basic plans, another ¥49 trillion were invested. These funding increases helped Japanese universities and national laboratories upgrade laboratories that had become outdated.[480]

Sadao Nagaoka, Masayuki Kondo, Kenneth Flamm, and Charles Wessner, Eds., Washington, DC: The National Academies Press, 2009.
[475] Japanese Ministry of Internal Affairs and Communications, Statistics Bureau at http://www.stat.go.jp/english/data/kagaku/index.htm. Data refer to fiscal years.
[476] OECD, *OECD Science, Technology and Industry Scorecard 2011*, Figure 2. 5.2.
[477] Lee Branstetter and Yoshiaki Nakamura, "Is Japan's Innovation Capacity in Decline?" National Bureau of Economic Research, Working Paper 9438, January 2003.
[478] Some analysts attribute Japan's decline as a leader in consumer electronics, characterized by innovative products such Sony's Walkman audio devices, to increased importance of embedded software, an industry dominated by U.S. companies, rather than hardware design. See Ashish Arora, Lee G. Branstetter, and Matej Drev, "Going Soft: How the Rise of Software-Based Innovation Led to the Decline of Japan's IT Industry and the Resurgence of Silicon Valley," National Bureau of Economic Research, Working Paper 16156, July 2010.
[479] For an unofficial translation of the Science and Technology Basic Law (Law No. 130 of 1995) see http://www.mext.go.jp/english/kagaku/scienc04.htm.
[480] National Science Foundation, "The S&T Resources of Japan; A Comparison with the United States," Access at http://www.nsf.gov/statistics/nsf97324/intro.htm.

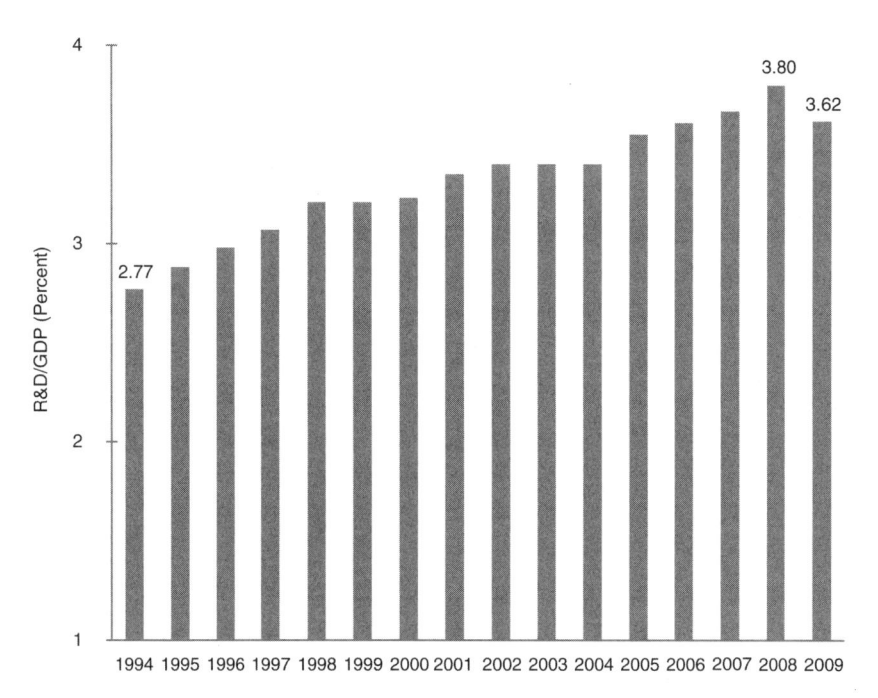

FIGURE 5.16 Japanese R&D intensity peaked at 3.8 percent of GDP in FY2008 before declining slightly in FY2009.
SOURCE: Japan Ministry of Internal Affairs and Communications, Statistics Bureau, Accessed at <http://www.stat.go.jp/english/data/kagaku/index.htm>.
NOTE: Data refer to fiscal years.

Japan also strengthened national coordination of its innovation strategy. The Council for Science and Technology Policy, established in 2001, became part of the Prime Minister's Cabinet. The council drafts comprehensive science and technology policies to respond to national and social needs, advises on how to allocate resources, and evaluates major projects. Funding focused on life sciences, nanotechnologies and new materials, information and communication, and environmental technologies.[481]

The government did not, however, assume greater central control over research. To the contrary, in 2004 it gave national universities and research institutes more autonomy to allocate resources, collaborate with industry, and set

---

[481] For an extensive discussion of changes in Japanese innovation policies, see Akira Goto and Kazuyuki Motohashi, "Technology Policies in Japan: 1990 to the Present," in *21st Century Innovation Systems for Japan and the United States*.

their own research priorities by separating them from the civil-service system. These institutions were transformed into non-profit corporations. Because they account for the bulk of scientific and technological research, the independence given universities and national labs is expected to allow resources to be used more flexibly and efficiently. In another crucial institutional reform, government agencies have begun to allocate much greater shares of R&D funds on the basis of peer-reviewed competition.[482]

The greater focus on innovation has led to dramatic increases in scientific research in strategic areas.[483] In 1992, the government set a goal of tripling investment in life sciences over the next decade. By 2001, the number of biotech companies had risen from a few dozen to 250; the goal was to have 1,000 biotech companies by 2010. In nanotech, Japan was spending almost as much on research as the United States--$940 million—as of 2004. Fuel cells, an important technology not only for portable electronic devices but also for future electrified vehicles, also received heavy emphasis.

Robotics is another top Japanese research priority. The government is especially interested in developing technologies used in core components that can be applied across the industry, such as power sources, control systems, mechanics, software, and structures. Two of Japan's biggest investments in science were the $1 billion Spring-8, one of the world's largest synchrotron radiation facilities, and the Earth Simulator, a $450 million scientific computer billed as the world's fastest when it opened in 2003.

Japan also has resuscitated R&D consortia, a key element of industrial policy until the 1980s. The government cut funds for consortia in areas like semiconductors following trade friction with the U.S., but began to renew such programs after Sematech started to benefit U.S. producers and Japanese chipmakers' fortunes declined.[484]

Strengthening University-Industry Partnerships

Japan has moved to strengthen universities' collaboration with industry. In 1999, Japan enacted a law that gave universities and research institutes the ability to patent investments derived from publicly funded research, similar to the Bayh-Dole Act of 1980. Since then, these institutions have established technology-transfer organizations. The government also helped universities set up Collaborative Research Centers that compete for government grants for joint

---

[482] A concise analysis of Japan's shift in innovation policy is found in National Research Council, *S&T Strategies of Six Countries: Implications for the United States*, Committee on Global Science and Technology Strategies and Their Effect on U.S. National Security, Washington, DC: The National Academies Press, 2010.
[483] See presentation by David K. Kahaner of the Asian Technology Information Program in *Innovation Policies for the 21st Century*, op. cit.
[484] Nagaoka and Flamm, op. cit.

university-industry research, small-business incubators, and a network of 45 Venturing Business Laboratories, which help young researchers commercialize their work. In addition, the government relaxed rules that had barred university faculty from serving on the boards of private companies.

These efforts led to significant results. University-industry research collaborations surged from around 1,500 in 1995 to more than 6,000 in 2003. Companies spun out of universities increased to around 150 a year as of 2003, nearly half of them in life sciences and information and communication technologies.[485]

While it is too early to assess the full impact of Japan's reforms, there have been noticeable improvements. The World Economic Forum ranks Japan 9th overall in its most recent Global Competitiveness Index and 4th in innovation.[486] Patent applications by universities and technology-licensing offices increased from 641 in 2001 to 8,527 in 2005, a comparable level to the United States. University-industry joint research projects jumped from less than 1,500 annually in 1995 to more than 10,000 in 2005. Spinoffs from Japanese universities also rose sharply.[487] And overall, Japanese patent applications have been increasing in recent years. [See Figure 5.17]

Such data suggest that university-industry partnerships have become "important for science-based innovation in Japan," said Masayuki Kondo of Japan's National Institute of Science and Technology. "They narrow the gap between Japanese high science and technology potential and low industrial performance to help strengthen the innovation capability of Japanese industry." However, Mr. Kondo said, Japanese universities bring in only a fraction of the licensing revenues of American universities. Only a handful of Japanese spinoffs so far have gone public.[488]

Stronger protection of intellectual property rights has improved Japan's innovation system since the early 1990s. Initially, the Japanese government responded to pressure from the U.S. to strengthen enforcement of violations. The World Trade Organization's Trade-Related Aspects of Intellectual Property Rights (TRIPs) agreement in 1995 also had a major impact. The government enacted a series of other reforms since then, including the Basic Law on Intellectual Property in 2003 and establishment of the Intellectual Property High Court in 2005, which is modeled after the U.S. Court of Appeals of the Federal Circuit. Criminal sanctions have been raised, and the scope of invention that is patentable has been greatly broadened.[489]

---

[485] Presentation by Masayuki Kondo of Japan's National Institute of Science and Technology Policy in *21st Century Innovation Systems for Japan and the United States*.
[486] World Economic Forum, *The Global Competitiveness Report 2011-2012*, op. cit.
[487] Presentation by Masayuki Kondo, op. cit.
[488] Ibid.
[489] See presentation by Sadao Nagaoka of Hitotsubashi University in *21st Century Innovation Systems for Japan and the United States*.

316 *RISING TO THE CHALLENGE*

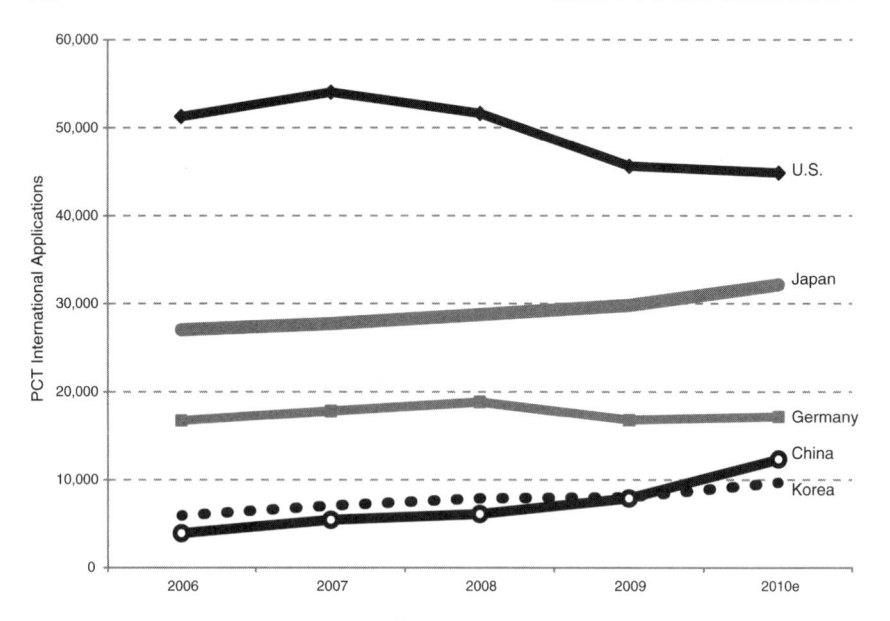

FIGURE 5.17 Japanese patent applications have been increasing in recent years.
SOURCE: WIPO, "International Patent Filings Recover in 2010," February 2, 2011, PR/2011/678.
NOTE: 2010 data are estimated.

IPR protection in Japan is now widely recognized to be very high. According to Business Software Alliance, Japan has the third-best record of enforcement following the U.S. and New Zealand. Patent-infringement claims have increased sharply. The overall impact on Japanese innovation is more difficult to assess because there are concerns that the IPR system's complexity and overburdened judiciary may hinder the ability of companies to commercialize technologies efficiently and raise transaction costs.[490]

Rediscovering Small Companies

Small business played a big role during Japan's post-war economic takeoff. But starting in the 1970s, new company formation began to fall to the point where entrepreneurship was perceived as stagnant, explained Takehiko Yasuda of Japan's Research Institute of Economy, Trade, and Industry. One reason was that Japanese policy tended to protect small enterprises from large

---

[490] Ibid.

firms, rather than see them as sources of innovation and job creation.[491] Policymakers also viewed large corporations as bigger contributors of wage and labor productivity. By the 1990s, however, the government recognized that start-ups were providing major stimulus to the economies of the U.S. and England.[492]

The government began introducing policies to encourage more start-ups in 1999. It enacted the Small and Medium Enterprise Basic Law to promote their growth. Two years later, the government launched the Start-up Doubling Plan, which set a goal of increasing the number of start-ups from 180,000 in 2001 to 360,000 in five years. Japan removed minimum capital requirements for new limited-liability companies, established the National Startup and Venture Forum to educate entrepreneurs, reformed the bankruptcy code, and launched a start-up loan program through the government-owned National Life Finance Corporation. The loans required no collateral, guarantors, or personal guarantees. In 2008, this unit was folded into the Japan Finance Corporation, whose small- and medium-sized business unit provided ¥20 trillion in support in 2009.[493]

Japan also established its own Small Business Innovation Research program, modeled after the one run by the U.S. Department of Commerce. The program aims to enhance the ability of small and midsized enterprises to develop technology and innovative products. As with the U.S. SBIR program, Japanese agencies that make research grants set aside a certain portion of their funds for small and midsized enterprises.

Removing the minimum capital requirement of ¥10 million for joint-stock companies in 2004 had an immediate impact. Between Feb. 1, 2004, and Jan. 21, 2006, there were 24,639 confirmed applications with 20,211 notification completions. Based on the success of this policy, the Japanese government enacted the Corporate Law in 2005 to remove the minimum capital requirement for establishing firms in general, which is consistent with the U.S. joint-stock corporation policy.

Remaining Challenges for Start-Ups

One of Japan's most pressing challenges is to create new companies. A 1997 survey by Japan's Ministry of Public Management, Home Affairs, Post and Telecommunications found that only one in 50 employed people aspired to become entrepreneurs, a very low level on world standards, and that only half of them were actually preparing to become self-employed.[494] The environment has

---

[491] *S&T Policies in Six Nations*, op. cit.
[492] See presentation by Takehiko Yasuda of the Research Institute of Economy, Trade, and Industry in *21st Century Innovation Systems for Japan and the United States*.
[493] Japan Finance Corporation Web site.
[494] Employment Status Survey by the Ministry of Public Management, Home Affairs, Post and Telecommunications, 1997.

not improved dramatically since then. Of 59 nations studied by the Global Entrepreneurship Monitor, Japan ranks second to the bottom, behind only Italy, in entrepreneurial activity.[495]

A lack of capital is a major reason. A survey of start-ups found that 49 percent of Japanese entrepreneurs reported that "procuring funds for entry" is a major problem, well ahead of finding customers and hiring high-quality employees.[496] To remedy this problem, the National Life Finance Corporation set up a new program to lend up to ¥10 million to start-ups without requiring collateral, guarantors, or personal guarantees. Between 2002 and 2006, the number of recipients rose from 2,975 to 7,942.[497]

Some Early Progress

Japan's new innovation system has begun to change the dynamics of the national economy. Patenting and technology transfer from Japan's top public research institutes have increased sharply. That system is still evolving, however, and inefficiencies remain. America's National Institutes of Health, for example, coordinates all government-funded biomedical research. In Japan, similar activity is dispersed among many funding agencies that do not share information on researchers, according to a 2006 analysis by Yosuke Oka, Kenta Nakamura, and Akira Tohei.[498] Nor are there guiding principles of peer review across agencies. "This could explain why a small number of star scientists receive a large share of research funds from multiple funding agencies," the authors noted. Government research funding also tends to flow to a handful of top schools. The top 10 universities garner half of research grants in Japan.[499]

Even though patent filings increased, technology transfer from Japanese research universities was not impressive when measured licensing revenue, according to Dr. Oka, Dr. Nakamura, and Dr. Tohei. Among other things, they attributed the poor performance to rudimentary technology-transfer contract practices and overly restrictive rules on using research funds. University researchers prefer "informal collaborations" to get around red tape. What's more, despite relaxed rules allowing academics to work in the private sector, most university researchers remain at their jobs rather than circulate

---

[495] Donna J. Kelley, Niels Bosma, Jóse Ernesto Amorós, "Global Entrepreneurship Monitor 2010 Global Report," Global Entrepreneurship Research Association, 2011, pg. 23.
[496] Applied Research Inc., "Survey of Environment for Start-ups," November 2006.
[497] Data cited in Yasuda presentation, op. cit. For an explanation of the National Life Finance Corporation program, see Jun-ichi Abe, "Small Business Finance & Support for Startups in Japan (Case of NLFC)," National Life Finance Corporation, December 2004 (http://www.afdc.org.cn/upload/18/downloads/JUN-ICHI%20ABE.pdf).
[498] Yosuke Oka, Kenta Nakamura, and Akira Tohei, "Public-Private Linkage in Biomedical Research in Japan: Lessons of the 1990s," in *21st Century Innovation Systems for Japan and the United States.*
[499] Ibid.

through industry. Sadao Nagaoka and Kenneth Flamm suggest that Japan still may lack the complementary institutions needed to make U.S.-style industry-university partnerships more effective, such as infrastructure for supporting high-tech startups, availability of risk capital, and professional services.[500]

A number of reforms have been proposed in Japan to address many of these shortcomings. While it is too early to measure progress, the changes implemented over the past decade in Japan's innovation ecosystem have provided a much stronger institutional framework for success in the 21st century global knowledge economy.

---

[500] Nagaoka and Flamm, op. cit.

Chapter 6

National Support for Emerging Industries

The appropriate role of public policy in promoting specific industries has been a source of passionate debate in the United States since the founding of the Republic.[1] Many nations in Europe and Asia have not hesitated to use the full force of government to attain commercial competitive advantage in industries they regarded as strategic. In the United States, however, the idea of proactive government help for private industry in the name of economic development has sometimes raised concerns about distorting market forces and the wisdom of letting public servants "pick winners." The debate began with Alexander Hamilton, who was an early advocate of "bounties" to encourage desirable industry, continued through the 19th century, and has resurfaced many times in the post-war era as U.S. industry confronted new competitive challenges. These policy debates have to some extent obscured actual practice, both in the United States and abroad.

In reality, the U.S. federal government has played an integral role in the early development of numerous strategic industries, not only by funding research and development but also through financial support for new companies and government procurement. Telecommunications, aerospace, semiconductors, computers, pharmaceuticals, and nuclear power are among the many industries that were launched and nurtured with federal support.

The intensifying global race to dominate an array of emerging high-tech industries once again has focused attention on the role of public policy. As China, South Korea, Germany, and Taiwan target industries such as renewable energy equipment, solid-state lighting, electric vehicles, and next-generation

---

[1] The link between national security and the need to develop key domestic industries was identified by Adam Smith, a contemporary of Hamilton, who noted that "if any particular manufacture was necessary, indeed, for the defense of the society it might not always be prudent to depend upon our neighbors for the supply." Adam Smith, An Inquiry into the Nature and Causes of the Wealth of Nations, 1776.

displays with comprehensive strategies and generous subsidies, the U.S. has struggled to compete. The financial crisis of 2008 has made it even more difficult for U.S. technology companies to raise the capital needed to turn designs into prototypes and prototypes into products made in large volumes.

In recent years, the Science, Technology, and Economic Policy Board of the National Academies has extensively studied the competitive challenges facing a number of important high-tech industries. The STEP board also has studied the policies adopted other nations and compared them to those of the United States.

This chapter explores the major policy issues in four of these industries—semiconductors, photovoltaic products, advanced batteries, and pharmaceuticals. Each of these industries can be regarded as strategic to the United States. Integrated circuits are the building blocks of all electronics products and have enabled the breathtaking advances in information technology that drive productivity gains across all industries. American leadership in semiconductors also is vital to the technological superiority of the U.S. military. Photovoltaic cells are the enabling technology of solar power, a key source of renewable energy that can serve America's national interests in reducing dependence on petroleum and cutting greenhouse gas emissions. Advanced batteries and their electrical management systems are the core components of hybrid and electric vehicles, much as internal combustion engines have been to conventional gasoline-powered cars and trucks. A strong domestic battery industry, therefore, is regarded as crucial to the future competitiveness of the U.S. auto industry. Lightweight, long-lasting, rechargeable energy-storage systems also are required for advanced weapons systems being developed by the U.S. military and for storing renewable energy for utility power grids. The pharmaceuticals industry is likewise strategic, producing medicines and vaccines that are essential to the well-being of Americans and indeed the world's people. U.S. leadership in this sector has been secured through enormous federal investments, though the industry faces numerous challenges in terms of litigation, regulatory pressure, and counterfeit drugs.

Each of these three industries shares another characteristic. The core technologies are the fruits of decades of research at U.S. universities and national laboratories at considerable American taxpayer expense. Many of the early U.S. companies that pioneered these industries, moreover, were supported over the years through federal research grants, small-business loans, and government and military procurement.

As they reached the point of large-scale commercial production, each of these U.S. industries encountered severe global competitive challenges[2]. Concerted Japanese government policies to facilitate joint R&D, transfer

---

[2] See Glenn Fong, "Breaking New Ground, Breaking the Rules—Strategic Reorientation in U.S. Industrial Policy," International Security 25:2 pp 152ff.

commercial technology to companies, protect domestic producers from imports helped Japanese companies in the 1970s and 1980s seize a commanding global market share in dynamic random-access memory chips, sending the U.S. semiconductor industry into crisis. U.S. companies dominated the nascent photovoltaic industry through the 1980s. Leadership in mass production of cells and modules, however, was assumed by Japan in the 1990s—and then Germany, Taiwan, and China—after each of these nations or regions enacted policies to build domestic markets for solar power or to promote manufacturing. The lithium-ion industry is one of several high-tech sectors that grew from U.S.-invented technology but was never industrialized domestically. Instead, Japanese companies were the first to mass-produce rechargeable lithium-ion batteries for electronic devices and notebook computers because of their large-scale production of consumer electronics. South Korean and Chinese manufacturers followed their lead. Asian producers, therefore, have a huge advantage in the small but extremely promising market for rechargeable batteries for cars and trucks.

The four industries illustrate different aspects of the public policy debate. **The U.S. semiconductor industry** is a case study in how a strategic sector that had lost competitive advantage in production and a once-dominant market share was able to regain global leadership through cooperation on pre-competitive R&D and public policy initiatives with responsive government actions. The public-private research consortium SEMATECH and assertive U.S. trade policies in response to Japanese dumping and protectionism enabled the industry rebound.

The photovoltaic industry is an example of a U.S. high-tech sector that has lost global share but has a solid opportunity to re-emerge as a leader with the right mix of federal and state policy support. In the case of solar power, a deciding factor will be whether the United States will become a big enough market to support a large-scale, globally competitive manufacturing industry. Federal and state incentives will be essential for the next few years, until the cost of solar energy can compete against electricity generated from fossil fuels without subsidies. Another question is whether U.S. companies that focus on products incorporating promising new technologies will be able to survive surging imports of low-cost photovoltaic cells and modules based on mature technologies long enough to attain economies of scale. What's more, because technologies are still evolving rapidly, and there are not yet commonly accepted manufacturing standards, the global race for future leadership remains wide open. Public-private research partnerships will be essential to ensure that the U.S. can be a leader in the race for global market share.

The emerging **U.S. advanced battery industry** represents a bold experiment by the federal government in direct financial support of private companies to establish a domestic manufacturing industry. Prior to 2008, the

U.S. had a number of lithium-ion battery start-ups but virtually no production plants.[3] It now has dozens of battery-related factories that are beginning to ramp-up, thanks in part to $2.4 billion in grants and support under the American Recovery and Reinvestment Act. Like photovoltaic cells, however, prices of lithium-ion auto batteries are too high, making hybrid and electric vehicles expensive for most consumers compared to conventional gasoline-powered vehicles. Larger demand, in turn, is required for the industry to attain the economies of scale that will bring prices down, in turn generating higher demand. In addition, further innovation is required to improve battery performance and reduce cost. Federal policies to support expansion of the market and public-private R&D collaboration will likely be required for the foreseeable future, but the long-term gain to the economy and national security can be significant.

The ascent of the U.S. pharmaceutical industry has been driven by massive federal support for life sciences R&D, primarily by the National Institutes of Health (NIH). During the decade of 2001, U.S. firms developed 57 "new chemical entities" (NCEs) compared with 33 by European firms and nine by Japanese firms, erasing the European lead which existed in prior decades. Despite the spectacular successes of past two decades, the U.S. pharmaceutical industry's future prospects are uncertain. Many of the blockbuster drugs that drove the industry's success have gone off patent or will do so soon, including first-generation biotechnology drugs, and branded producers face growing competitive pressure from generic drug makers. The costs and risks of developing new drugs and bringing them to market are rising, while the productivity of the industry's R&D appears to be declining. In light of key developments, especially in emerging markets, a key challenge is to sustain the productivity and competitiveness of this strategic U.S. industry.

SEMICONDUCTORS

A little more than two decades ago, the U.S. semiconductor industry appeared to be going the way of the U.S. consumer electronics industry. Japanese companies had seized a commanding world market share and technological lead in memory devices and were rapidly adding more production capacity. Struggling U.S. chipmakers were abandoning a large segment of the industry that made memory products, an essential part of computers and other leading semiconductor technologies of the eighties. There was widespread concern that erosion of America's semiconductor industry posed not only economic challenges, but national security risks as well. Even after the U.S. government had begun to mount a strong policy response to bolster U.S.

---

[3] "In 2009, the U.S. made less than 2 percent of the world's lithium-ion batteries." Jon Gertner, "Does America Need Manufacturing?" *The New York Times*, August 24, 2011.

competitiveness, a defense task force warned in 1987 that a dependence on foreign suppliers for state-of-the-art chips for weapons was an "unacceptable situation" because it would undermine the U.S. military strategy of maintaining technological superiority.[4] This national security concern and the willingness of the semiconductor industry to collectively seek policy help from Washington were instrumental in reversing the loss of market share and technology lead that seemed irretrievably lost.

Remarkably, as recounted below, the U.S. semiconductor regained global leadership by the early -1990s and —despite the dramatic rise of new competitors in South Korea, Taiwan, and China—remains today a top semiconductor producer. Even though the U.S. market accounts for only 18 percent of the global sales for integrated circuits, sales by U.S. companies accounted for 48 percent of the world market in 2010.[5] [See Figure 6.1] While only one U.S. company is still a major player in memory chips, the U.S. semiconductor industry dominates the lucrative market for logic devices such as microprocessors and analog mixed signal products.[6]

Moreover, despite rapid growth in outsourcing to Asian foundries (wafer fabrication factories that produce integrated circuits on a contract basis for other firms), the vast majority of production and R&D by U.S. semiconductor companies remains in the United States.[7] Seventy-seven percent of capacity owned by America semiconductor companies is located in U.S. and 74 percent of compensation and benefits is paid to U.S.-based workers.[8] And while the vast majority of chip companies now outsource fabrication of the devices they design to foundries located in Asia, approximately 500 of the world's 1,200 so-called "fabless" design firms—including most of the industry leaders—are headquartered in North America.[9]

---

[4] See U.S. Department of Defense, *Report on Semiconductor Dependency*, Office of the Undersecretary of Defense for Acquisition, prepared by the Defense Science Board Task Force, Washington, DC, February 1987.
[5] Source: Semiconductor Industry Association citing data from based on World Semiconductor Trade Statistics data.
[6] Micron Technologies, headquartered in Boise, Idaho, is the leading U.S. producer of computer memory chips.
[7] For an analysis of semiconductor R&D has remained in the U.S. despite outsourcing of production, see Jeffrey T. Macher, David C. Mowery, and Alberto Di Minin, "Semiconductors," chapter 3 in National Research Council, *Innovation in Global Industries: U.S. Firms Competing in a New World*, Jeffrey T. Macher and David C. Mowery, eds., Washington, DC: The National Academies Press, 2008.
[8] Semiconductor Industry Association (SIA), *Maintaining America's Competitive Edge: Government Policies Affecting Semiconductor Industry R&D and Manufacturing Activity*, March 2009. This report can be accessed at http://www.sia-online.org/galleries/default-file/Competitiveness_White_Paper.pdf.
[9] Global Semiconductor Alliance, *Industry Data* at http://www.gsaglobal.org/resources/industrydata/facts.asp. The largest fabless companies include QUALCOMM, Broadcom, AMD, NVIDIA, and LSI.

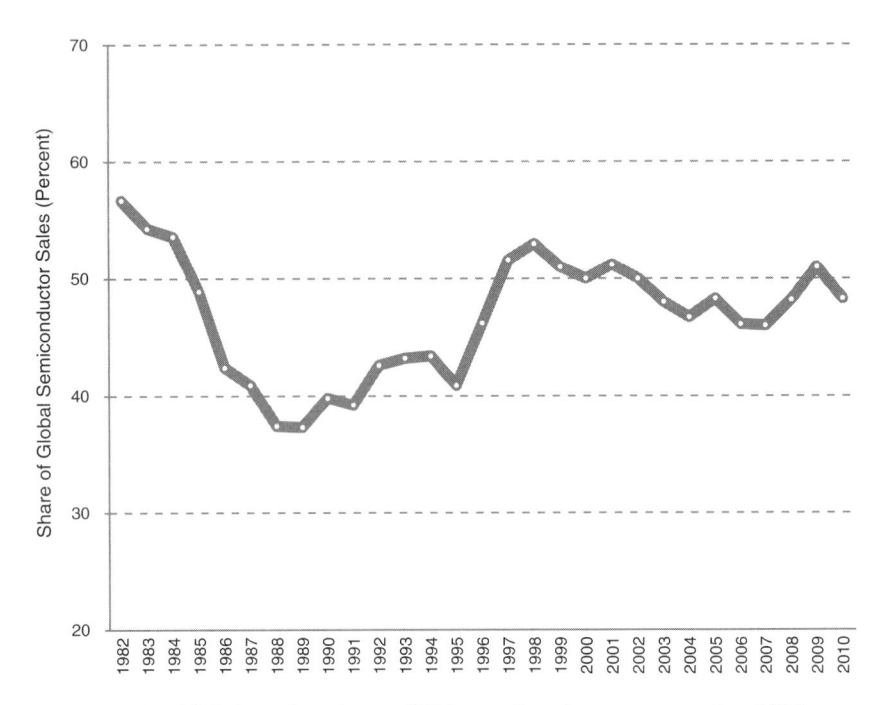

FIGURE 6.1 Global market share of U.S. semiconductor companies, 1982-2010.
SOURCE: Semiconductor Industry Association.
NOTE: Share data based on nationality of company.

This turn of fortunes is primarily due to strategic moves and investments in new technologies by U.S. semiconductor manufacturers. Yet, their success also rests on the important contributions of U.S. policy that was driven by an engaged industry. There were two additional interrelated elements to the U.S. success:[10] The research consortium SEMATECH, a $200 million-a-year research effort co-funded by the federal government and most large American chip companies, accelerated productivity and innovation in semiconductor manufacturing based on a common technology roadmap and

---

[10] The recovery of the U.S. industry has been described as a three-legged stool. It is unlikely that any one factor would have proved sufficient independently. Trade policy, no matter how innovative, could not have met the requirement to improve U.S. product quality. On the other hand, by their long-term nature, even effective industry-government partnerships can be rendered useless in a market unprotected against dumping. Most importantly, neither trade nor technology policy can succeed in the absence of adaptable, adequately capitalized, effectively managed, technologically innovative companies.

enabled a rapid decline in prices.[11] Persistent trade negotiations and enforcement of previous agreements won commitments from Japan to open its market to U.S. semiconductors and curtail dumping in any world market.[12] This was deemed essential to prevent the United States from becoming a high-priced island in a sea of underpriced semiconductors. Had that occurred, it would have severely disadvantaged downstream American electronics equipment producers compared with competitors producing abroad utilizing lower-priced dumped chips.[13]

The decline and resurgence of the U.S. semiconductor industry offers many useful lessons for policymakers and industrialists grappling with how to bolster other American high-technology sectors facing intense international competitive pressure. It shows that erosion of U.S. leadership in manufacturing is not irreversible as long as both industry and government are committed to cooperative action, both on trade policy and in well-designed research programs that will lead to innovation. In a comprehensive analysis of the semiconductor experience, the National Research Council concluded that overcoming competitive challenges requires "continued policy engagement and public investment through renewed attention to basic research and cooperative mechanisms such as public-private partnerships."[14]

---

[11] For analysis of the contributions of SEMATECH, see presentation by Kenneth Flamm of the University of Texas in National Research Council, *Innovative Flanders: Innovation Policies for the 21st Century—Report of a Symposium*, Charles W. Wessner, editor, Washington, DC: The National Academies Press, 2008. For a more extensive treatment, see Kenneth Flamm, "SEMATECH Revisited: Assessing Consortium Impacts on Semiconductor Industry R&D," in National Research Council, *Securing the Future,* OP. CIT. See also, Peter Grindley, David C. Mowery and Brian Silverman. "SEMATECH and Collaborative Research: Lessons in the Design of High Technology Consortia, *Journal of Policy Analysis and Management*, 13(4) 1994, pp. 723-758.

[12] In the U.S.-Japan Semiconductor Trade Agreement, signed on Sept. 2, 1986, Japan agreed to eliminate dumping of semiconductors following a U.S. Department of Commerce finding that Japanese producers sold memory chips in the U.S. at below the cost of production. Japan also agreed to open its market to foreign-made chips and to cease dumping in any market. In 1990, Japan signed a second bilateral trade agreement that provided U.S. producers with a "fast-track" process for addressing dumping allegations and promised to fulfill an earlier pledge that foreign producers achieve a minimum 20 percent share of the Japanese semiconductor market. This figure was chosen because it would give foreign producers access to the customer base of the six giant vertically integrated Japanese companies that controlled the Japanese market. The trade agreement was remarkable in that it did not close the U.S. market, but instead opened the previously closed Japanese markets and stopped dumping in third markets.

[13] For a full description of the how Japan closed its market for all foreign semiconductor producers, see Thomas R. Howell, William A. Noellert, Janet H. McLaughlin, and Alan Wm. Wolff, *The Microelectronics Race,* Boulder, Colo., and London: Westview Press, 1988.

[14] National Research Council, *Securing the Future: Regional and National Programs to Support the Semiconductor Industry*, Charles W. Wessner, editor, Washington, DC: The National Academies Press, 2003.

The Strategic Importance of Semiconductors

The importance of semiconductors to the United States is difficult to overstate. As an industry, the semiconductor sector directly employs over 180,000 Americans and has consistently ranked as either America's No. 1 or No. 2 export industry.[15] Semiconductors represent the core technology of the modern electronics revolution, enabling products from smart phones and computers to advanced weapons systems. More importantly, semiconductors have made possible the rapid advances in information technology that drive productivity gains across other industries. As one National Academies study noted—

> "...often called the 'crude oil of the information age,' semiconductors are the basic building blocks of many electronics industries. Declines in the price/performance ratio of semiconductor components have propelled their adoption in an ever-expanding array of applications and have supported the rapid diffusion of products utilizing them. Semiconductors have accelerated the development and productivity of industries as diverse as telecommunications, automobiles, and military systems. Semiconductor technology has increased the variety of products offered in industries such as consumer electronics, personal communications, and home appliances."[16]

The impact of semiconductor-based information technology has been so pervasive that many economists regard it as the catalyst behind the acceleration in productivity growth in the U.S. economy since the mid-1990s.[17] Meeting critical national needs such as increased energy efficiency, lower-cost and improved health care services, and ubiquitous access to high-speed broadband data communications will depend on further advances in

---

[15] Patrick Wilson, Director of Government Affairs, Semiconductor Industry Association, "Maintaining US Leadership in Semiconductors," AAAS Annual Meeting, February 18, 2011.

[16] This excerpt is taken from Jeffrey T. Macher, David C. Mowery, and David A. Hodges, "Semiconductors," *U.S. Industry in 2000: Studies in Competitive Performance*, David C. Mowery, ed., Washington, DC: National Academy Press, 1999, p. 245.

[17] For an analysis of the role of new information technologies in recent high productivity growth, often described as the New Economy, see Dale W. Jorgenson, "The Emergence of the New Economy" in *Enhancing Productivity Growth in the Information Age*, Dale W. Jorgenson and Charles W. Wessner, eds., Washington, DC: National Academy Press, 2007. Also see National Research Council, *Measuring and Sustaining the New Economy, Report of a Workshop,* D. Jorgenson and C. Wessner, eds., Washington, DC: National Academy Press, 2003, and Council of Economic Advisers, *Economic Report of the President,* H.Doc.107-2, Washington, DC: USGPO, January 2001.

semiconductors.[18] Semiconductors also remain vital to national security, observes the Industrial College of the Armed Forces, because "they are the building blocks of the nation's infrastructure and the space, communications, and weapons systems that allow the projection of American diplomatic, information, military, and economic power."[19]

A New Set of Challenges

Continued American leadership in semiconductors certainly cannot be taken for granted, however. The industry faces a range of technological, financial, and competitive challenges. Among the most prominent—

- **Declining share of capacity**: U.S. semiconductor companies still invest billions of dollars in wafer fabrication facilities in the United States. But investment by manufacturers in Asia is expanding faster. The share of global installed wafer fabrication capacity in the United States declined from 42 percent in 1980 to about 16 percent in 2007.[20] American semiconductor companies are investing a proportionately larger share of their total worldwide fabrication capacity spending outside of the United States. The share of spending in the United States for wafer manufacturing capacity has dropped by 14.6 percentage points between 1997-1999 and 2005-2007, from 78.5 percent to 63.9 percent.[21] The Semiconductor Industry Association (SIA) expects the U.S. share to decline by another 9.3 percentage points by 2013.[22] What's more, only 14 percent of leading-edge capacity (300 mm wafers) is located in the United States. The largest market for state-of-the-art manufacturing equipment is in Asia, principally South Korea, Taiwan and Japan.[23]
- **Business and capital costs:** As the cost of building new leading-edge wafer fabrication plants reach some $4 to $6 billion, factors such as tax rates and government incentives now heavily influence corporate

---

[18] The RAND Corporation, for example, estimates that application of information technology in the health care sector could result in annual efficiency savings of $77 billion. See RAND Corporation, *Health Information Technology: Can HIT Lower Costs and Improve Quality?*, 2005, (http://www.rand.org/pubs/research_briefs/RB9136/index1.html). Also see Jorgenson, "The Emergence of the New Economy," op. cit.
[19] Industrial College of the Armed Forces, *Electronics 2010,* Industry Study Final Report, National Defense University, Spring 2010, (http://www.ndu.edu/icaf/programs/academic/industry/reports/2010/pdf/icaf-is-report-electronics-2010.pdf).
[20] SIA, *Maintaining America's Competitive Edge*, op cit.
[21] Ibid.
[22] Ibid.
[23] SEMI Industry Research and Statistics Group data.

decisions on where to build capacity. Countries such as Malaysia, India, Singapore, China, and Israel and regions such as Taiwan offer tax holidays or significantly reduced rates. Germany offers grants and loans to chip manufacturers. Federal and state tax breaks and other benefits offered in the U.S. are often either insignificant or non-competitive,[24] according to the SIA.

- **Talent:** The American semiconductor industry is becoming increasingly dependent on foreign-born R&D staff at a time when immigration rules have tightened and opportunities abroad are growing. More than 50 percent of students graduating from U.S. universities with master's degrees and 70 percent of doctorates in science and engineering disciplines applicable to semiconductors are foreign nationals.[25] Meanwhile, nations and regions such as India, China, and Taiwan are rapidly increasing their supply of semiconductor engineers. An inability of industry to hire top talent in the U.S. could lead to a greater shift of R&D offshore.

- **Offshore R&D:** Even though U.S. semiconductor companies conduct most of their R&D onshore, that proportion has declined by 8.4 percent points from 1997-1999 to the 2005-2007 period. Most of the work is going to Europe, Israel, and Singapore, and increasingly to Romania. Meanwhile, the outsourcing by American companies of chip fabrication to Asian foundries—plants that fabricate chips on a contract basis—means that semiconductor design can go to any place that has the best supply of engineers.[26]

- **Competing Consortia:** While federally funded U.S. research is under budget pressure, other nations have learned from the accomplishments of SEMATECH and have formed their own public-private partnerships aimed at becoming the first to commercialize next-generation semiconductor technologies. At the same time, the ability to continue improving the performance of integrated circuits along the path predicted by Moore's Law[27] through current transistor technology may be nearing its physical limits.[28] The U.S. faces growing competition to develop technologies to replace silicon-based, CMOS semiconductors,

---

[24] The U.S. currently offers a 9 percent manufacturing tax credit and a temporary R&D tax credit, although states such as New York offer sizeable incentives.
[25] SIA, *Maintaining America's Competitive Edge*, op cit.
[26] Ibid.
[27] Moore's Law is based on the prediction by Intel co-founder Gordon Moore in 1965 that the number of transistors that can be placed inexpensively on an integrated circuit doubles every two years.
[28] One recent development that could alter this view is Intel Corp.'s recent announcement that it had successfully demonstrated the world's first 3-D transistor, called Tri-Gate, used in a 22nm microprocessor. Intel claimed its technology will "advance Moore's Law into new realms."

a challenge that Nanotechnology Research Institute Director Jeffrey Welser says is as dramatic as the replacement of vacuum tubes by semiconductors in the 1940s.[29]

These challenges must be addressed. "At some point," the SIA warns, "without sufficient U.S. government support of basic R&D and supportive tax, immigration, and education policies, it may well prove to be very difficult if not impossible to reverse current trends."[30]

Industry Growth and U.S. Policy

The federal government was at the outset deeply involved in the U.S. semiconductor industry. Indeed, as economist Laura Tyson observed in 1992: "The semiconductor industry has never been free of the visible hand of government intervention."[31]

The U.S. Signal Corps was the prime funder of the R&D that led to development of the transistor and semiconductors for three decades and purchased most of the initial output. The military funded the first pilot production lines of Western Electric, General Electric, Raytheon, and Sylvania and construction of production capacity far in excess of demand. From the late 1950s through the early 1970s, the federal government funded between 40 to 45 percent of U.S. R&D in semiconductors.[32] Military purchases of semiconductors enabled the industry to establish the scale that led to a dramatic drop in prices between 1962 and 1968,[33] making them more practical for commercial use.

Japan's entry into the dynamic random-access memory (DRAM) industry, backed by low-cost capital and a protected home market, resulted in dramatic increases in capacity and dumping of product on third-country markets. Some U.S. companies also lagged the Japanese competition in quality and productivity using the same equipment sets. The result was a reduction of the U.S. global share in this market from around 90 percent to less than 10 percent by 1985, and producers such as Intel, Advanced Micro Devices, and National

---

[29] Testimony by Jeffrey Welser, Nanoelectronics Research Initiative director, before the House Committee on Science, Space, and Technology's Subcommittee on Research and Science Education, April 14, 2011, http://science.house.gov/sites/republicans.science.house.gov/files/documents/hearings/Welser%20Testimony%20FINAL.pdf).
[30] SIA, *Maintaining America's Competitive Edge*, op. cit.
[31] Laura D'Andrea Tyson, *Who's Bashing Whom? Trade Conflict in High Technology Industries*, Washington, DC: Institute for International Economics, 1992.
[32] A concise history of U.S. government involvement in establishment of America's electronics industry is found in Kenneth Flamm, *Mismanaged Trade?: Strategic Policy and the Semiconductor Industry,* Washington, DC, Brookings Institution, 1996. pp. 27-38.
[33] Defense Science Board, "High Performance Microchip Supply," 2005.

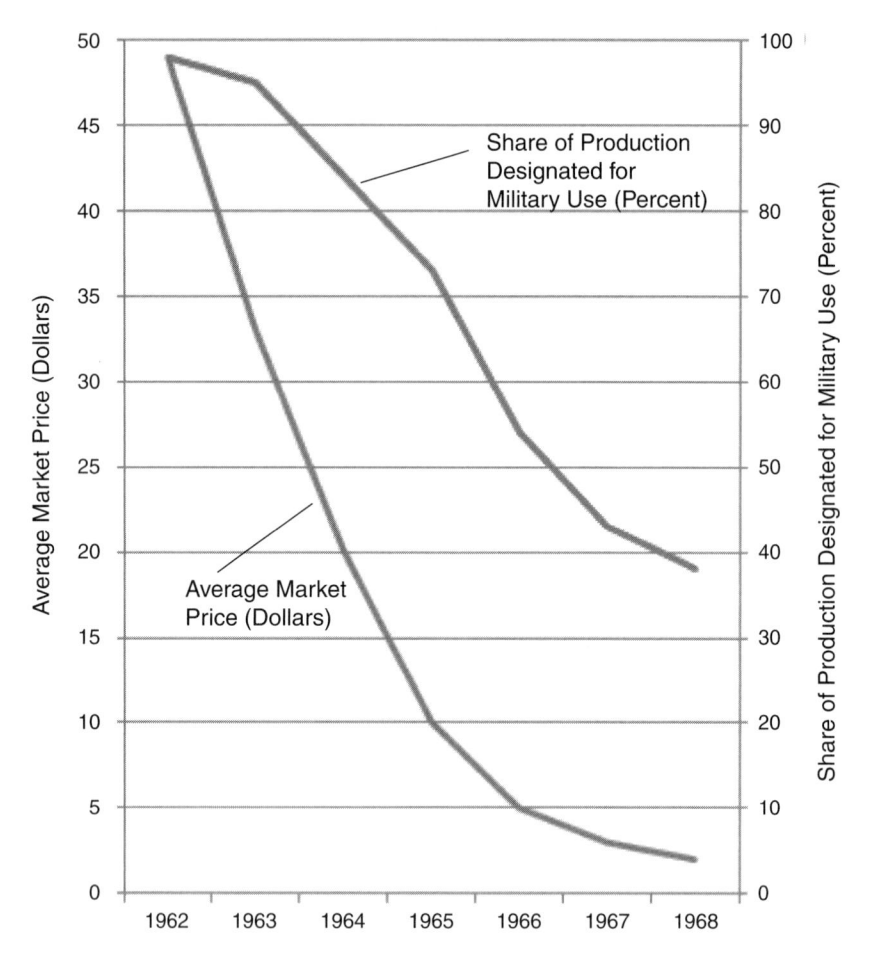

FIGURE 6.2 Government procurement as a catalyst for semiconductor development
SOURCE: Defense Science Board, "High Performance Microchip Supply," 2005.

Semiconductors were driven from the DRAM business.[34] The loss of market leadership in DRAMs was considered a major setback for the U.S. industry,[35]

---

[34] For a discussion of American competitiveness challenges in the 1980s, See Laura Tyson, *Who's Bashing Whom? Trade Conflict in High Technology Industries*, Washington, DC: Institute for International Economics, 1992. Also see Clyde Prestowitz, *Trading Places: How We are Giving Away our Future to Japan and How to Reclaim* It, New York: Basic Books, 1988.

especially because the high-volume memory devices were process technology drivers for the industry. The scale of production of the high-volume commodity DRAM chips justified investment in new process technologies and wafer fabrication facilities that could then also be used for lower-volume integrated circuits.

The impact of these policies and trade practices convinced the industry that it needed government policy support. By the early 1980s, the U.S. industry was in crisis and reached out to the federal government for help. The industry argued that Japan violated rules of the General Agreement on Tariffs and Trade as a consequence of trade and industry policy coordinated by Japan's Ministry of International Trade and Industry (MITI) and supported by NTT.[36] The industry also blamed Japanese government toleration of anticompetitive practices of Japanese companies. Reflecting growing concern for the health of the industry, the Defense Science Report in 1987 cited declining U.S. market share in semiconductors as a national security concern.[37] By that time, the U.S. government had put into place the measures that were to improve the competitive position of U.S. producers to counter Japan government's industrial policies.

The first step was to shore up research and enable U.S. companies to collaborate. In 1982, the semiconductor industry formed and funded the Semiconductor Research Corporation, an independent affiliate of the SIA, to conduct silicon-based research at universities. Two years later, President Ronald Reagan signed the National Cooperative Research Act, which reformed U.S. antitrust law to encourage joint R&D consortia.[38] The Microelectronics and Computer Technology Corp., a privately funded industry consortium, was established in response to Japan's government-funded "Fifth Generation" R&D program that aimed to put Japanese computer makers at the leading edge of technology. This first U.S. semiconductor consortium had a menu of projects that members could choose to fund and participate in, but was viewed as a failure and shut down in 2001.[39]

SEMATECH was the second and more successful consortium. At the

---

[35] See Andy Procassini, *Competitors in Alliance: Industry Associations, Global Rivalries, and Business-Government Relations,* New York: Greenwood Publishing, 1995.

[36] For an account of Japanese trade practices, see Prestowitz, op. cit. MITI and Nippon Telephone and Telegraph had worked with the large vertically integrated Japanese producers to move at least a generation ahead of their Western competitors in the production of DRAMs.

[37] Department of Defense, *Report on Semiconductor Dependency,* op. cit.

[38] For an account of the evolution of U.S. semiconductor research policy, see Kenneth Flamm presentation in National Research Council, *21st Century Innovation Systems for Japan and the United States: Lessons from a Decade of Change,* Sadao Nagaoka, Masayuki Kondo, Kenneth Flamm, and Charles W. Wessner, eds., Washington, DC: The National Academies Press, 2009. Also see Kenneth Flamm and Qifei Wang, "SEMATECH Revisited: Assessing Consortium Impacts on Semiconductor Industry R&D," in National Research Council, *Securing the Future,* op. cit.

[39] Flamm, ibid.

recommendation of industry and the Defense Science Board, Congress in 1987 voted to match industry contributions for precompetitive research in a non-profit consortium. SEMATECH corporate members consisted of all of the largest device makers at the time, including IBM, Intel, Motorola, Texas Instruments, Hewlett Packard, and National Semiconductor. Former Intel chairman Gordon Moore described the organization as unique in that industry made sure that U.S. companies assigned top people to a public-private partnership.[40] The strategy was to have SEMATECH focus on fabrication equipment and processes so that semiconductor companies could focus on design, quality, and innovation. The consortium included major initiatives in critical processing technologies, such as lithography, furnace and implant, plasma etch, and deposition. The SIA also coordinated government, industry, and academia to produce a roadmap guiding research and development and oversaw implementation of research.

SEMATECH is widely perceived as effective in accomplishing its goals and making a contribution to the U.S. semiconductor industry's resurgence. By 1993, the U.S. industry had regained leadership in world market share in semiconductors.

A National Research Council analysis found that the consortium "played an integral role in promoting effective manufacturing technology in the semiconductor industry."[41] SEMATECH also helped the equipment industry develop reliable, standardized chip-manufacturing tools, particularly in lithography. SEMATECH is credited with reducing R&D duplication by its members, thus lowering costs and freeing funds for additional investment.[42]

SEMATECH also helped achieve the original goals of the DOD to preserve access to state-of-the-art, low-cost chips from domestic commercial sources.[43] In a subsequent review, a defense task force labeled the consortium "a resounding success."[44]

---

[40] For a first-hand account of the formation of the SEMATECH consortium, see Gordon Moore, "The SEMATECH Contribution," in National Research Council, *Securing the Future: Regional and National Programs to Support the Semiconductor Industry*, C. Wessner, ed., Washington, DC: The National Academies Press, 2003. Also see Larry D. Browning and Judy C. Shetler, *SEMATECH: Saving the U.S. Semiconductor- tor Industry,* College Station: Texas A&M University Press, 2000. For a view from the Semiconductor Industry Association at that time, see also Procassini, op. cit.
[41] *Securing the Future*, op. cit. In particular, see Gordon Moore presentation in that volume.
[42] Flamm and Wang, op. cit.
[43] See Jacques Gansler, *Defense Conversion: Transforming the Arsenal of Democracy*, Cambridge, MA: MIT Press, 1995. See also the presentation by Paul Kaminski, then Under Secretary of Defense for Technology and Acquisition, in National Research Council, *International Friction and Cooperation in High-Technology Development*, Washington, DC: National Academy Press, 1997. Dr. Kaminski points out that tighter linkage with commercial markets shortens cycle time for weapons-systems development and reduces the cost of inserting technological improvements into DoD weapons systems. By placing greater reliance on commercial sources, the DoD can field technologically superior weapons at a more affordable cost.
[44] Department of Defense, "SEMATECH 1987-1997: A Final Report to the Department of Defense," Defense Science Board Task Force on Semiconductor Dependency," February 21, 1997.

Rapid advances in semiconductors, in turn, enabled dramatic innovation in information technology that resulted in robust industries and higher productivity growth.[45] The *Securing the Future* report observed: "SEMATECH's record of accomplishment was achieved in no small part through the flexibility granted its management and the sustained support provided by DARPA, the public partner, complemented by the close engagement of its members' senior management and leading researchers."[46]

Perhaps the clearest measure of SEMATECH's success is that corporate members in 1994 agreed to continue the consortium without further government financial help, except for a $50 million grant by the DoD. Foreign companies have since joined SEMATECH, which became an international consortium in 1999, and other governments have established similar programs—often on a larger scale with greater political support. (See descriptions of several of these programs below).

International SEMATECH remains active, and has broadened its activities to design, materials, testing, and packaging technologies. Among other activities, it funds development of new 300-mm tools and continues to pursue technology roadmaps. Initiatives include mask-making tools and next-generation lithography using very-short-wavelength violet light from a special laser. Other U.S. industries, such as optoelectronics and nanotechnologies, also have emulated the SEMATECH model.[47]

The Role of U.S. Trade Policy

State-of-the-art manufacturing process technologies and yield improvements were not the only elements that helped restore the U.S. semiconductor industry to health. An assertive U.S. response to Japanese trade practices that began in the mid-1980s also helped stem and then reverse the decline of the American semiconductor industry. In response to Japanese dumping and protection of its own market,[48] the United States and Japan

---

[45] Council of Economic Advisers. *Economic Report of the President*, Washington, DC: Government Printing Office, 2001.
[46] *Securing the Future*, op. cit.
[47] Flamm and Wang, op. cit. In April 2011, the school received a $57.5 million Department of Energy grant to become the base of the U.S. Photovoltaic Manufacturing Consortium, a partnership that includes SEMATECH and the University of Central Florida. See College of Nanoscale Science and Engineering news release, April 5, 2011 (http://www.albany.edu/news/12770.php).
[48] See Prestowitz, op. cit, for an inside account of early U.S.-Japan trade conflicts over semiconductors. Also see Kenneth Flamm, *Mismanaged Trade? Strategic Policy and the Semiconductor Industry*, Washington, DC: Brookings Institution Press, 1996. Prestowitz co-chaired a U.S. Japan High Tech Work Group set of discussions, a largely fruitless exchange of views between the U.S. and Japan during his term of government service, but this allowed time for further industry research into the nature of Japan's market closure and was a useful step in obtaining U.S. government understanding of the problem and action several years later. He was also instrumental

initiated a bilateral working group on high technology in 1983 to address trade conflicts. Two years later, the two nations agreed to completely eliminate tariffs on imported semiconductors. The SIA filed a Section 301 petition alleging that the Japanese government kept out imported chips through non-tariff barriers. In 1986, the U.S. Department of Commerce concluded that Japanese semiconductor firms were selling memory chips in the U.S. market at prices substantially below the cost of production. Together with the injury caused to U.S. industry, this warranted a finding of dumping. The further finding by the U.S. Trade Representative in 1987 that Japan had still not opened it market for foreign products and had breached its antidumping commitment prompted President Ronald Reagan to impose a 100 percent duty on $300 million worth of Japanese goods.[49]

The two nations reached an unprecedented agreement in 1986 under which Japan pledged that imported chips would account for 20 percent of its domestic market.[50] The number was chosen because Japan's integrated producers of semiconductors, who were at the same time large semiconductor consumers, accounted for only 13 percent of Japanese consumption of semiconductors. A 20 percent goal required that Japanese producers and the Japanese government allow access to a customer base beyond the big vertically integrated Japanese producers. Japan also agreed to a "fast-track" approach to resolving dumping allegations. In return, the U.S. dropped anti-dumping duties and its Section 301 case. By late 1992, the Japanese market was open to competitive foreign products, and foreign chips did indeed account for 20.2 percent of Japan's market.[51]

The series of U.S. Japan Semiconductor Agreements "was a pivotal point in the recovery of the U.S. semiconductor industry and its return to global leadership," said Semiconductor Industry Association President George M. Scalise.[52] Antidumping cases provided a means for companies like Intel to stay in the production of erasable programmable read only memories (EPROMS), which allowed it to progress to the production of flash memory. The U.S.-Japan Semiconductor Agreements also enabled Texas Instruments and Micron Technologies to stay in the DRAM business and gave South Korea and Taiwan

---

in getting the Department of Commerce to self-initiate an antidumping case that provided needed leverage to obtain an end to the dumping of chips by Japan.
[49] Proclamation 5631 by President Ronald Reagan, "Increase in the Rate of Duty for Certain Articles from Japan," April 17, 1987. The details of penalties were provided in an April 22, 1987 annex to the Federal Registry.
[50] The original target amount committed to was in a side letter to the agreement.
[51] For a discussion of the Semiconductor Trade Agreement, see National Research Council, Hamburg Institute for Economic Research, and Kiel Institute for World Economics, *Conflict and Cooperation in National Competition for High-Technology Industry*, Washington, DC: National Academy Press, 1996. Andrew A. Procassini, *Competitors in Alliance: Industry Associations, Global Rivalries, and Business-government Relations*, Westport, CT: Quorum Books, 1995.
[52] Interview with Semiconductor Industry Association President George Scalise.

an opportunity to enter the memory market. Creation of a competitive multiple vendor base, in turn, spurred the production of ever more powerful personal and mainframe computers at diminishing cost and fueled the information technology revolution. Also, the agreements allowed Intel and other companies to pursue more attractive opportunities in devices such as microprocessors.[53] The trade pacts with Japan are widely credited with giving the U.S. and foreign industries breathing room to adjust and regain the profitability needed to invest in advanced capacity and new technologies. Notably, by 2010, five of the top 10 semiconductor producers in the world were based in the United States, compared to two from Japan.[54] [See Table 6.1] The agreements also enabled some U.S. manufacturers to make the transition from commodity memory products to new types of highly specialized products. In short, intervention to end Japan's market closure and the restoration of the U.S. industry produced a worldwide burst of innovation that has never slowed.

New U.S. Research Consortia

The United States has a number of other public-private research collaborations addressing technological challenges under the umbrella of the Semiconductor Research Corp. Since its founding, the SRC has managed more than $1.2 billion in research funds, supported 2,000 faculty and 9,000 students at 257 universities, and produced 373 patents.[55] One of the most extensive programs is the Nanotechnology Research Initiative (NRI), which seeks to advance technologies that ultimately can replace complementary metal-oxide semiconductor (CMOS) technology,[56] the digital design style and set of

---

[53] Dale A. Irwin, "The Politics and the Semiconductor Industry," in Anne O. Kreuger, editor, *The Political Economy of American Trade Policy*, Chicago: University of Chicago Press, 1996. Few if any academics understood the dangers posed for the IT revolution due to Japan's dumping and market closure. Nor could the participants in the bilateral U.S.-Japan Semiconductor Agreement negotiations foresee how these agreements would expand to cover all major semiconductor-producing countries and industries, create a tariff-free global trade environment for semiconductors, and encourage full cooperation toward shared environmental and energy-saving goals. For a short description of these new arrangements, see the World Semiconductor Council Web site and the series of conclusions reached by the six-nation Government and Association meeting on Semiconductors, and the tariff agreement on multi-component chips (MCPs) announced by then USTR and now Ohio Senator Rob Portman.
[54] Two of the five U.S. companies (Qualcomm and Broadcom) are "fabless" producers, companies that develop and design integrated circuits but contract the production out to "foundries," or contract fabrication facilities run by other companies.
[55] Welser testimony, op. cit.
[56] Complementary metal-oxide-semiconductor (CMOS) refers to a style of digital circuitry design and process used to implement the circuitry. CMOS is the most common technology used in very-large-scale integrated circuits, such as microprocessors, static random-access memory devices, and microcontrollers. In CMOS devices, power is drawn by switching transistors between on and off stages. The devices have gates, typically of polysilicon or metals. The technology allows a high density of logic functions.

processes used in very large-scale integrated circuits such as microprocessors. Industry experts say that at some point, the extreme miniaturization of transistors—the basic building block within an integrated circuit--results in undesirable quantum effects that inhibit performance of the device.[57] Today's most advanced semiconductors contain billions of transistors.[58]

The Nanotechnology Research Initiative: The NRI, which receives funding through the National Science Foundation and NIST, supports four institutes—each based at universities—that pursue high-risk, pre-commercial research on technologies that are likely to result in commercial products within the next decade. Each institute, which brings together its own partnerships of universities, focuses on different approaches to developing devices cable of replacing CMOS in logic chips by 2020.[59] Corporate members GlobalFoundries, IBM, Intel, Micron Technology, and Texas Instruments, as well as the states where the centers are based, also contribute funds.

The Western Institute of Nanoelectronics (WIN), for example, is led by the University of California at Los Angeles and includes UC Berkeley,

TABLE 6.1 Top Ten Semiconductor Companies in 2010 by Sales

Rank	Company	2010 Revenue (Billions of Dollars)	Country
1	Intel	40.4	U.S.
2	Samsung	27.8	Korea
3	Toshiba	13.0	Japan
4	Texas Instruments	13.0	U.S.
5	Renesas	11.9	Japan
6	Hynix	10.4	Korea
7	STMicroelectronics	10.3	France/Italy
8	Micron Technology	8.9	U.S.
9	Qualcomm	7.2	U.S.
10	Broadcom	6.7	U.S.

SOURCE: iSuppli, "Samsung Closes in on Intel for Semiconductor Market Leadership in 2010," April 19, 2011.
NOTE: Sales based on vendor. Foundries not included.

---

[57] The physical limits of transistor size was described in Paul A. Packan, "Pushing the Limits: Integrated Circuits Run Into Limits Due to Transistors," *Science,* September 24, 1999.
[58] The coming generation of advanced chips will have line widths of 22nm.
[59] The mission statement and research objections of the Nanotechnology Research Initiative are found on the Semiconductor Research Corporation Web site at http://www.src.org/program/nri/about/mission/.

UC Santa Barbara, and UC Irvine. WIN focuses on nano-magnetic circuits, spin wave devices, spin torque logic, and SpinFET. The Institute for Nanoelectronics Discovery (INDEX) based at the University of Albany in New York, partners with schools such as MIT, Purdue, and Harvard. INDEX conducts research on a wide range of topics, such as new nanomaterials and atomic-scale fabrication technologies. Among other things, the INDEX consortium is studying the use of graphene to transmit electrons. Graphene is a strong, flexible atom-thick carbon material that are capable of carrying 1,000 times the density of electric current as copper wires, which researchers believe could lead to a new generation of super-fast, super-efficient electronics.[60] The Midwest Institute for Nanoelectronics Discovery (MIND), based at Notre Dame, concentrates on energy-efficient devices and systems. The Southwest Academy of Nanoelectronics (SWAN), led by the University of Texas at Austin, focuses on the Bilayer Pseudospin Field Effect Transistor, which the SRC describes as a promising graphene-based device in terms of power consumption and speed.[61]

The Focus Center Research Program: The SRC oversees a number of other semiconductor-related research initiatives. The Focus Center Research Program, funded by $40 million from the DoD and industry contributions and run by SIA affiliate Microelectronics Advanced Research Corp. (MARCO), is devoted to pushing CMOS technology to its limits. The Focus Center program, supported by the Defense Advanced Research Projects Agency, involves 41 universities, 33 faculty, and 1,215 doctoral students.[62] The guiding philosophy of MARCO is to have universities control research projects, back them with significant funding, train top students, and encourage "out of the box" approaches to technical problems.[63] The Global Research Collaboration, another initiative of the SRC, funds R&D projects that address everything from sub-32 nm mixed-signal manufacturing processes and computer-aided design to advanced circuit and systems design. The new National Institute for Nanoengineering, based at Sandia National Laboratories, explores nano-enabled solutions to technologies that address various critical national challenges.

Today's Competitive Challenges

The competitive landscape has changed dramatically since the 1980s. The market is increasingly global, as are the locations of supply among the U.S., Japan, South Korea, Taiwan, the EU and China. Important new pools of

---

[60] Holly B. Martin, "Miracle Material: Two-Dimensional Graphene May Lead to Faster Electronics, Stronger Spacecraft and Much More," National Science Foundation Web site, May 19, 2011, accessed at http://www.nsf.gov/discoveries/disc_summ.jsp?cntn_id=119493&WT.mc_id=USNSF_1.
[61] Semiconductor Research Corp. Web site.
[62] Semiconductor Research Corp. data.
[63] See *Securing the Future*, op. cit.

engineering talent are emerging. Decisions on where to build capacity are heavily influenced by government incentives. In addition to commodity memory chips, the new market-share battles are also fought on the basis of design and innovation. The coming technology transition has launched a new global R&D race. Government policy will loom large in determining the winners and losers. The following are some of the new challenges facing policymakers.

Declining U.S. Share of Global Capacity in the United States

The share of global production capacity located in the United States continues to decline. In 1980, 42 percent of worldwide fabrication capacity was located in the United States. That dropped to 30 percent in 1990 and reached 16 percent in 2007.[64] IC Insights, a market research firm for the semiconductor industry, estimated that the share of installed wafer fabrication capacity in the Americas (primarily the United States) was 14.7 percent in 2010.[65] [See Figure 6.3] Japan and Europe also lost share over the same period.

The rapid expansion of Asian semiconductor companies and offshore investment by U.S. companies are behind the shift. South Korea and Taiwan have been the largest gainers, led by Samsung and Hynix for South Korea and Taiwan Semiconductor Manufacturing Corp and UMC for Taiwan.[66] Both Taiwanese companies are foundries. Samsung, one of the largest integrated device manufacturers, also entered the foundry business in 2005. Significantly, the vast majority of new leading-edge 300mm wafer fabrication capacity is being installed in Asia, an estimated 80 percent in 2011 and a forecasted 70 percent in 2012.[67]

The U.S. is drawing some important new investment. In 2009, GlobalFoundries, the former manufacturing operations of AMD and Chartered Semiconductor and 86 percent owned by Abu Dhabi's Advanced Technology Investment Co., began construction of a $4.6 billion 300mm fab in Malta, NY,

---

[64] SIA analysis of data from SEMI Industry Research and Statistics Group and Robert C. Leachman and Chien H. Leachman, "Globalization of Semiconductors," in Martin Kenney and Richard L. Florida (eds.), *Locating Global Advantage: Industry Dynamics in the International Economy*, Palo Alto, Calif.: Stanford University Press, 2004.

[65] IC Insights, "Taiwan to Pass Japan as Largest Source of IC Wafer Fab Capacity," *Research Bulletin*, November 11, 2010.

[66] The history of ITRI's role in establishing Taiwan's semiconductor industry is addressed below. The U.S. investigated brought a countervailing duty case against South Korean DRAM producer Hynix in response to allegations that the South Korean government had subsidized the company's exports by orchestrating a financial bailout. The dispute was dropped without punitive duties being assessed.

[67] Paul Dempsey, "Foundry Overcapacity – Yes, It Could Happen," *Tech Design Forum*, June 20, 2011. The data cited in the article are from Gartner Dataquest. Article at http://www.techdesignforums.com/eda/eda-topics/design-to-silicon/foundry-overcapacity---yes-it-could-happen/.

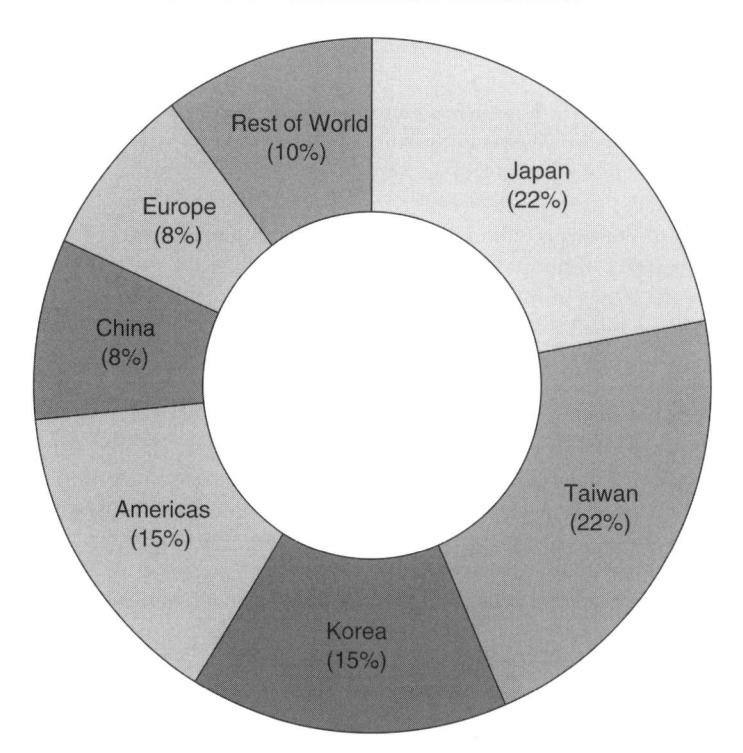

FIGURE 6.3 Estimated integrated circuit wafer fabrication installed capacity by region – July 2010.
SOURCE: IC Insights, "Taiwan to Pass Japan as Largest Source of IC Wafer Fab Capacity," *Research Bulletin*, November 11, 2010.

not far from the College of Nanoscale Science and Engineering at the University of New York at Albany. This facility will be able to produce 60,000 wafers per month with line widths of 28nm and below.[68] The plant will deploy a technology called High K Metal Gate developed with IBM, Samsung, Infineon, and other partners that it claims far exceeds the capabilities of competing foundries. The company is seeking further financial aid from the state of New York to expand the plant.[69] Intel announced in 2009 that it intends to invest $7 billion to upgrade

---

[68] GlobalFoundries Web site. See also Chapter 7 of this report.
[69] Drew Kerr, "GlobalFoundries Seeks New Incentive Money From State to Expand Operations," *PostStar,* March 26, 2010.

existing plants in Oregon, Arizona, and New Mexico to produce next-generation 32nm chips.[70]

Capital spending by U.S. semiconductor companies on new or upgraded wafer plants rose by 10.6 percent from 1997-1999 to 2005-2007.[71] Yet the portion of total investment in the United States slid from 78.5 percent to 63.9 percent over that period.

China is rising fast as a semiconductor consumer and producer, although the vast majority of production in China is still carried out by foreign semiconductor firms.[72] Sales of integrated circuits produced in China reached 144 billion yuan ($21.3 billion) in 2010,[73] which represented about 7.6 percent of total world integrated circuit sales in 2010.[74] Because labor constitutes a small share of semiconductor manufacturing cost, China's low wages are not a significant advantage. Rather, its advantages are access to low-cost capital and government policies aimed at leveraging China's immense domestic market.[75] Chinese consumption of semiconductors has grown at a 25 percent compound annual rate since 2001, four times faster than total worldwide consumption, and has represented 43 percent of global sales growth since 2003.[76] Since 2009, China has become the largest consumer of semiconductors because approximately one-quarter of the world's electronic products are assembled there by foreign-invested enterprises. Most of these products, once assembled, are then exported by foreign-invested factories as finished goods.[77] Thus, approximately two-thirds of chips sold in China go into electronics products that are exported, such as mobile phones, personal computers, color TVs, and digital cameras.[78] Most chips have to be imported because China does not produce many of these sophisticated semiconductor devices. This has led to a large and growing Chinese trade deficit in integrated circuits, which reached $128 billion in 2010. [See Figure 6.4] From 2008 through 2010, China's imports of integrated circuits have exceeded its oil imports. Domestic demand is growing

---

[70] Nicholas Kolakowski, "Intel Investing $7 Billion in Manufacturing Facilities," *eWeek*, Feb. 10, 2009.
[71] SIA, *Maintaining America's Competitive Edge,* op. cit.
[72] PricewaterhouseCoopers estimates that "there is no Chinese company within the top 50 suppliers to the Chinese semiconductor market." PricewaterhouseCoopers, *Global Reach: China's Impact on the Semiconductor Industry 2010 Update*, November 2010, p. 14, (http://www.pwc.com/gx/en/technology/assets/china-semicon-2010.pdf).
[73] "China to Boost IC Sector as 'State Strategy'," *Xinhua*, April 16, 2011. China IC sales data from MIIT.
[74] WSTS IC sales for 2010 were $278.52 billion. Total semiconductor sales (ICs plus discretes) were $298 billion. WSTS, "WSTS Projects Semiconductor Market to Grow by 7.6 Percent to $338.4 Billion in 2012," *Press Release*, June 7, 2011.
[75] SIA, *Maintaining America's Competitive Edge*, op. cit.
[76] PricewaterhouseCoopers, *Global Reach*, Ibid.
[77] Ulrich Schaefer, "Semiconductor Market Forecast 2010-2013," WSTS European Chapter, EECA-ESIA & WSTS, December 1, 2010.
[78] PricewaterhouseCoopers, *Global Reach*, Ibid.

rapidly as well, including advanced devices required for weapons systems and telecommunications.

Most Chinese wafer fabs are several generations behind those of Japan, South Korea, Taiwan, and the United States. Only 27 percent of the new or committed capacity in China is for 300mm wafers, compared to a global average of 45 percent. Most will produce 6-inch or 8-inch wafers.[79] As businesses, moreover, many of China's semiconductor manufacturers have met with mixed success. In the first quarter of 2009, capacity utilization in China sank to 43 percent, the lowest level since 2000 and dramatically below the 92 percent utilization rate of mid-2004.[80] Most Chinese fabs are foundries. Because most use mature technology, they cannot fabricate the most advanced chips and instead make thin-margin, commodity devices. As a result, many Chinese chip manufacturers have not earned the high profits required to invest in next-generation wafer fabs. Some analysts believe that China's strategy has "collapsed."[81]

It is important to note that there is more to China's semiconductor strategy than just investment in Chinese-owned fabs or inducing foreign manufacturers to produce chips in China. The government also has introduced programs to deploy Chinese-owned intellectual property. The Ministry of Information Industry has announced a goal that China become 70 percent self-sufficient in integrated circuits used for information and national security and 30 percent for those used in communications and digital household appliances.[82] One of the government's goals, to have all Chinese supercomputers use Chinese-made central processors, reached a milestone in late 2011 when China's National Supercomputer Center in Jinan unveiled its first supercomputer, the Sunway BlueLight MPP, based entirely on Chinese microprocessors.[83]

The Chinese government still regards developing a globally competitive semiconductor industry as a high strategic priority. As part of its "indigenous innovation drive," the government also is offering generous incentives to convince multinationals to build advanced capacity in China. In 2007, Intel agreed to build a 300 mm wafer fab in the coastal city of Dalian for chip sets. China's glut in capacity also means that it is in a strong position to gain substantial share in chips and other silicon-based devices that do not require

---

[79] Ibid.
[80] Dylan McGrath, "China's Fab Utilization Sinks to 43%, says iSuppli," *EE Times*, April 20, 2009.
[81] See for example, iSuppli analyst Len Jelinek quoted in McGrath, Ibid.
[82] Ministry of Information Industry, "Outline of the 11th Five-Year Plan and Medium-and-Long-Term Plan for 2020 for Science and Technology Development in the Information Industry," Xin Bu Ke [2006] No. 309, posted on ministry website August 29, 2006.
[83] John Markoff, "China Has Homemade Supercomputer Gain," *The New York Times*, Oct. 28, 2011. See also Alan Wm. Wolff, testimony before the U.S. China Economic and Security Review Commission, Washington, DC, May 4, 2011.

344 RISING TO THE CHALLENGE

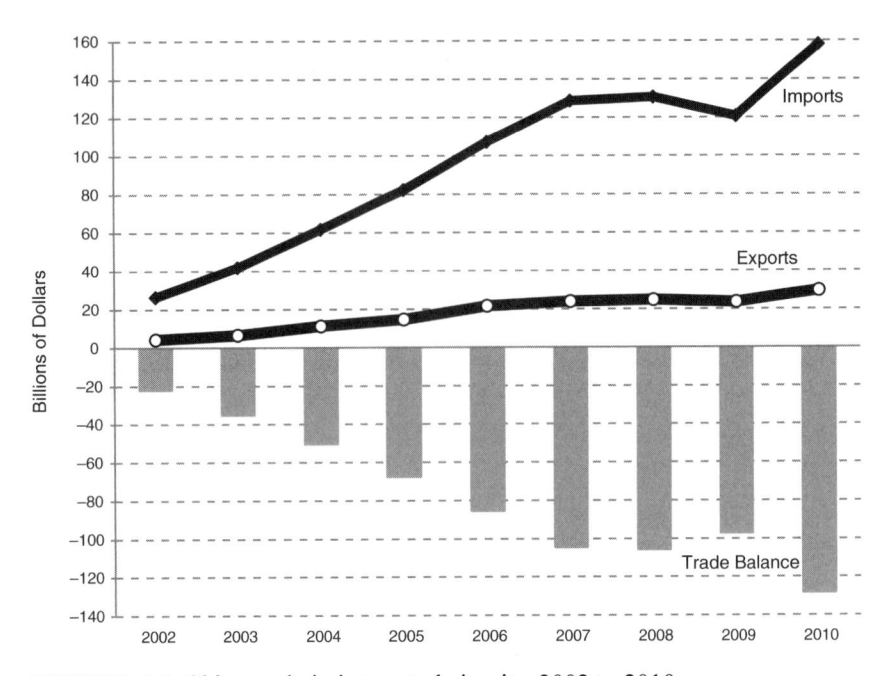

FIGURE 6.4 China trade in integrated circuits, 2002 to 2010.
SOURCE: United Nations, UN Comtrade database. Accessed at <http://comtrade.un.org/db/dqQuickQuery.aspx>.
NOTE: Commodity code HS 8542 used to calculate trade values.

the most advanced technology, such as photovoltaic cells and light-emitting diode chips for solid-state lighting.[84]

Asia will likely remain the largest market for leading-edge semiconductor manufacturing equipment. Because process R&D and wafer fabrication are closely linked, moreover, the continued erosion of U.S. market share in wafer fabrication capacity could eventually give the technological advantage to nations that are investing more aggressively in state-of-the-art capacity. For this reason, U.S. industry leaders say, it is important that tax and regulatory measures be taken to encourage chip companies to build new and next-generation wafer fabs in the United States.[85]

---

[84] China had an estimated 62 manufacturers of light-emitting diode chips as of 2010. *LEDinside*, "Ranking of LED Chip Manufacturers in China—Report on China's LED Epitaxy Industry," 2009.
[85] SIA, *Maintaining America's Competitive Edge*, op. cit.

Competition for Financial Incentives

The soaring cost of fabricating chips has made financial incentives an important determinant of where new capacity is built. Tax breaks, grants, low-cost loans, free land and other incentives typical defray $1 billion of a plant's cost over a 10-year period. The SIA maintains—

> *"As a practical matter, any U.S. semiconductor management answerable to its shareholders must establish a new fab in a location that offers this type of incentive package or risk becoming less competitive vis-à-vis a competitor who receives such incentives. In other words, government incentives play a decisive role in determining the geographic location of advanced wafer fabrication facilities, and thus indirectly determine the location of the process R&D associated with that facility."*[86]

Nations and regions such as India, Israel, Malaysia, China, Taiwan and Singapore offer complete five- to 10-year tax holidays for corporate profit taxes or sharply reduced rates for R&D and for plant construction spending. Germany and other governments offer direct grants, project equipment, and central and state government loans and loan guarantees to semiconductor manufacturers. The German federal government and the state of Saxony, for example, covered the total construction cost of AMD's "Fab 36" to produce 45nm and 65nm 300mm wafers in Dresden in 2004. Government agencies also provided $798 million in cash and allowances, a loan guarantee of 80 percent of losses sustained by lenders, and further funds for expansion.[87] The Israeli government offered more than $1 billion in aid, including a $525 million grant, for Intel's 300 mm plant in Kiryat Gat in 2005, plus $660 million in the form of tax benefits to upgrade another fab. Intel said the grants were pivotal in deciding to build the plant in Israel.[88]

Some U.S. states have offered generous incentives nearly matching those of foreign governments. New York, for instance, awarded incentives worth $660 million over 10 years to persuade IBM to build a new $2.5 billion wafer fab in Fishkill N.Y., in 2001.[89] The state also awarded $1.2 billion in cash and tax incentives to GlobalFoundries, 86 percent owned by Abu Dhabi's Advanced Technology Investment Co., to build a $4.6 billion fab in Malta, N.Y.

---

[86] Ibid.
[87] Ibid.
[88] Ibid.
[89] Jack Lyne, "IBM's Cutting-Edge $2.5 Billion Fab Reaps $500 Million in NY Incentives," *Site Selection*.

The deal amounted to the largest private-public investment in the state's history.[90]

Such U.S. state incentives are awarded case by case, however, and remain highly controversial--especially at a time when budget deficits are forcing states to slash public services. What's more, semiconductor manufacturers still must pay federal corporate taxes. In 2006, Intel CEO Craig R. Barrett testified that it cost $1 billion more to "build, equip, and operate" a $3 billion chip plant in the United States than it does outside the U.S., with 90 percent of that difference due to government policies.[91]

The Dispersion of Design

The United States remains the world leader in semiconductor design. Three-quarters of what American chip companies invest in R&D is spent in the U.S. America's continued dominance of semiconductor design cannot be taken for granted, however. The chip-design industries in Taiwan, India, and China have grown tremendously, either as outsourcing destinations or as development bases for domestic industries. The share of research by U.S. companies performed in the United States declined from 86.2 percent in 1997-99 to 77.8 percent in 2005-2007, according to the SIA. By 2013, the portion invested in the United States is projected to drop by another 9.3 percentage points, with most of that activity going to Europe.[92]

The growing importance of foundries, wafer fabrication plants dedicated to contract manufacturing, has brought about a significant structural shift in the semiconductor industry that has accelerated the global dispersion of design work. By outsourcing manufacturing to large foundries, even small chip companies can gain access to state-of-the-art wafers and production processes without having to raise the billions of dollars required to build their own modern production capacity. Instead, they can focus their resources on design around standardized parameters. What's more, major foundries offer "IP libraries" so that companies with only specialized proprietary designs can develop entire "systems on a chip."[93] As a result, the industry has been undergoing a process that D. A. Hodges and R. C. Leachman describe as "vertical disintegration."

---

[90] Empire State Develop Corporation, "Empire State Development Corporation: A Description of the Corporations Operations and Accomplishments,"
(http://www.siteselection.com/ssinsider/incentive/ti0011.htm).
[91] Craig R. Barrett, testimony before the Subcommittee on Select Revenue Measures of the House Ways and Means Committee, June 22, 2006.
[92] SIA, *Maintaining America's Competitive Edge*, op. cit.
[93] D.A. Hodges and R.C. Leachman, "The New Geography of Innovation in the Semiconductor Industry." For the full presentation, see <http://web.mit.edu/ipc/www/hodges.pdf>.

Even though the dedicated foundry industry is almost entirely based in Asia and is dominated by two Taiwanese companies—TSMC[94] and United Semiconductor Corp. (UMC)—the U.S. design industry has thrived.

Seventeen of the top 25 "fabless" semiconductor companies in the world and nine of the top 10 are based in the United States, led by Qualcomm, AMD, and Broadcom.[95] Because chip designs can be transmitted digitally, design R&D does not need to be close to wafer production plants. Indeed, an SIA survey found that location of fabrication capacity is not a key factor in a company's decision of where to locate design R&D.[96]

By the same token, however, the shift to the foundry model means that design can be based any place with the best available talent. A number of governments are targeting semiconductor design and development for rapid development. India, already a major R&D base for companies such as Intel and Texas Instruments, has a plan to increase the nation's share of the very large integrated-circuit market from 0.5 percent to 5 percent and to boost annual revenue to $1 billion.[97] The India Semiconductor Association predicts that annual revenue of India's semiconductor development industry will grow from $7.5 billion in 2010 to $10.6 billion in 2012. It also advocates a strategy to incubate at least 50 fabless semiconductor companies, each with annual revenue of $200 million or more, by 2020.[98] Eastern Europe, Russia, Brazil, and Israel are growing centers of semiconductor design as well.[99]

As a world technology leader in computers, displays, and smart phones, Taiwan also has become a major factor in semiconductor design. In 2002, the Taiwanese government launched the Si-Soft Project, which stands for "silicon and software." The objective is to push the island's industry beyond contract manufacturing and to become a major player in design of very large-scale

---

[94] Taiwan Semiconductor Manufacturing Co., led by former Texas Instruments executive Morris Chang, was formed in 1987 as a joint venture between the Taiwan government and Philips Electronics NV. It was the first company dedicated entirely to the foundry business. United Microelectronics Corp. was spun off of the Industrial Technology Research Institute in 1980 as Taiwan's first semiconductor manufacturer. UMC evolved into a dedicated foundry and became to first to fabricate 300mm chips on a contract basis.
[95] See David Manners, "Top 25 Fabless Companies," *Electronics Weekly*, January 19, 2010, (http://www.electronicsweekly.com/Articles/2010/01/19/47816/top-25-fabless-companies.htm).
[96] SIA, op. cit.
[97] Department of Information Technology, *Special Manpower Development Programme in the Area of VLSI Design and Related Software* (http://www.mit.gov.in/content/special-manpower-development-programme).
[98] India Semiconductor Association, *Study on Semiconductor Design, Embedded Software and Services Industry*, prepared by Ernst & Young, April 2011, (http://www.ey.com/Publication/vwLUAssets/Study_on_semiconductor_design_embedded_software_and_services_industry/$FILE/Study-on-semiconductor-design-embedded-software-and-services-industry.pdf).
[99] SIA, op. cit.

integrated circuits.[100] Initiatives include establishment of a science park modeled after the Hsinchu Science and Industrial Park dedicated to design of systems on a chip. Sci-Soft also established six university research consortia in fields such as mixed-signal design, digital IP, electronic design automation, and system on a chip.[101]

China also is becoming a major location for chip design. Multinationals such as Intel and Freescale have opened Chinese design centers and a number of fabless design companies have opened in Shanghai and Beijing. China's lack of intellectual property protection, however, has prevented the country from attracting more foreign investment. In an SIA survey of U.S. chip companies, a majority indicated they would not locate their most advanced and critical R&D activities in China, "despite encouragement and even pressure by the government to do so, and regardless of the availability, quality and size of incentives, due to concerns about the inadequacy of intellectual property protection."[102] If China follows through on commitments to protect intellectual property, however, the fact that it has the fastest growing market for semiconductors indicates that it has enormous potential to grow in chip R&D.

Workforce Issues

Perhaps the biggest threat to long-term U.S. leadership in semiconductor R&D is availability of talent. Foreign nationals comprise half of the master's degree candidates and 71 percent of the PhD candidates graduating from U.S. universities in the engineering fields needed to design and manufacture integrated circuits and other semiconductor devices.[103] One indicator of this foreign dependence is to look at where engineering Ph. D. graduates from U.S. universities receive their bachelor's degrees. Only one U.S. school—MIT—ranked among the top 10. The leading university, Tsinghua University in Beijing, had 421 students who went on to earn Ph. D's from U.S. universities in 2006, which was more than the 241 graduates from all California universities combined.[104]

The ability of companies to hire this talent in the United States has been complicated by tightened immigration procedures and a sharp reduction in temporary H-1B work visas. Taken together, these restrictions serve to inhibit U.S. semiconductor firms from growing research programs in the United States

---

[100] For an explanation of the objectives of Si-Soft, see Chun-Yen Chang and Charles V. Trappey, "The National Si-Soft Project," National Chiao-Tung University, (http://web.eecs.utk.edu/~bouldin/MUGSTUFF/NEWSLETTERS/DATA/si-soft-speech.pdf).
[101] See Chang Chun-Yen and Wei Hwang, "Development of National System-on-Chip (NSoC) Program in Taiwan," National Chiao Tung University, November 18, 2004 (http://www.cs.tut.fi/soc/Chang04.pdf).
[102] SIA, *Maintaining America's Competitive Edge*, op. cit.
[103] Ibid.
[104] Ibid.

NATIONAL SUPPORT FOR EMERGING INDUSTRIES 349

that depend on being able to hire the best and the brightest talent," says the SIA.[105]

Other nations, meanwhile, are expanding their pools of semiconductor engineers and expanding efforts to woo émigrés back home. India, which has an available semiconductor engineering workforce of 160,000,[106] has a number of programs to increase the supply further. The VLSI Manpower Initiative[107] of the Department of Information Technology operates programs to expand semiconductor engineering training through the master's and doctorate level at universities and the nation's famed Indian Institutes of Technology and Indian Institutes of Information Technology.[108] The India Semiconductor Association calls for boosting semiconductor manpower 20 percent a year and for India to have 500,000 in five years.[109]

Other Research Consortia

The perceived success of SEMATECH and other U.S. public-private partnerships have encouraged other nations and regions to expand semiconductor research collaborations among government, industry, and academia. For example—

- **Japan**: After curtailing heavy government industrial policies in the 1980s, the Japanese government and industry established a number of new consortia when the industry slumped in the 1990s.[110]

  The Association of Super-Advanced Electronics Technology (ASET) is completely funded by the government and focuses on equipment and chip R&D. ASET has produced more than 100 patents and completed a number of projects with industry, including ones that developed technology for X-ray lithography and plasma physics and diagnostics. It recently has launched the Dream Chip Project, which focuses on 3-D integration technology, and another relating to next-generation information appliances.[111]

  The Semiconductor Leading Edge Technology Corp. (SELETE), by contrast, is a joint venture funded by 10 large Japanese semiconductor

---

[105] Ibid.
[106] India Semiconductor Association, op. cit.
[107] VLSI is an acronym for "very large scale integrated" circuits.
[108] Department of Information Technology, op. cit.
[109] India Semiconductor Association, op. cit.
[110] For an overview of Japanese semiconductor consortia, see Shuzo Fujimura presentation in *21st Century Innovation Systems for Japan and the United States*, op. cit.
[111] Association of Super-Advanced Electronics Technologies Web site, http://www.aset.or.jp/english/e-link/e-link_index.html.

companies with no government contributions. Established in 1996, the joint venture conducts precompetitive R&D for production technologies using 300mm wafer equipment. Currently, SELETE is nearing completion of a research collaboration to develop 45nm to 32nm technologies.[112]

Other Japanese research consortia include the Millennium Research for Advanced Information Technology (MIRAI) program, which concentrates on alternative materials for future large-scale integrated circuits. MIRAI's R&D base is the $250 million Tsukuba Super Clean Room. In 2002, Japan's Ministry of Economy, Trade and Industry (METI) launched a five-year industry-government R&D project to develop extreme ultraviolet lithography for 50-nm device manufacturing in conjunction with 10 Japanese device and lithography equipment purchasers.

- **Flanders:** The Interuniversity Micro-Electronics Centers (imec) in Flanders, the Dutch-speaking region of Belgium, is one of the world's largest semiconductor research partnerships and strives to be a global "center of excellence," according to Chairman Leuven Anton de Proft.[113] The organization, which received around half of its €285 million in revenue in 2010 from company research contracts and most of the rest from the Flemish government and the European Commission, has a staff of 1,900 and more than 500 industrial residents and guest researchers. It also has research partnerships in the Netherlands, Taiwan, and China.[114] has "core partnerships" with Texas Instruments, ST Microelectronics, Infineon, Micron, Samsung, Panasonic, Taiwan Semiconductor, and Intel, and "strategic partnerships" with major equipment suppliers.[115]

imec emphasizes pre-competitive research that is three to 10 years ahead of industry needs, and therefore takes on risky projects that partners cannot afford to do on their own.[116] Researchers from academia and industry work together under the same roof. Subject areas include chip design, processing, packaging, microsystems, and nanotechnology. In July 2005, imec produced its first 300mm silicon

---

[112] Semiconductor Leading Edge Technology Corp. Web site, http://www.selete.co.jp/?lang=EN&act=selete_message.
[113] Presentation by Anton de Proft of imec in National Research Council, *Innovative Flanders*, op. cit.
[114] Interuniversity Micro-Electronics Centers data.
[115] Greta Vervliet, *Science, Technology, and Innovation,* Ministry of Flanders, Science and Innovation Administration, 2006.
[116] imec Mission Statement.

disks with working transistors, using its second clean room, a new, 3,200-square meter facility. A production ASML lithography system installed in 2006 offered capabilities that at the time were beyond those available at the U.S.-based SEMATECH.

imec has been of "great value" to its members, according to Texas Instrument executive Allen Bowling, who noted that moving a new material or device into production requires seven to 12 years of precompetitive research.[117] In 2010, Intel announced it was investing in a new ExaScience Lab in Leuven with, the Agency for Innovation by Science and Technology, and five Flemish universities that aims to achieve breakthroughs in power reduction software that can deliver 100 times the performance of today's computers.[118] Because of its multinational membership, some analysts question whether is actually delivering on its mission to help develop a domestic semiconductor industry in Flanders.[119], however, maintains that it is building a large research base that will eventually lead to the growth of domestic companies and the location of related industry.[120]

- **Taiwan:** Public-private research programs have been instrumental to Taiwan's rise as a semiconductor power since the mid-1970s, when the government-funded Industrial Technology Research Institute acquired 7-micron chip technology from RCA and spun off chip manufacturer UMC. In the 1980s, ITRI helped launch TSMC, now the world's dominant foundry.[121] ITRI continues to operate substantial semiconductor-related R&D partnerships. The institute's Electronics and Optoelectronics Research Laboratories, for example, include programs in fields such as next-generation memories and chips for lighting and 3D imaging.[122]
- **France:** After a previous semiconductor research consortium involving

---

[117] Presentation by Allen Bowling of Texas Instruments in National Research Council, *Innovative Flanders*, op. cit.
[118] Press release issued Aug. 6, 2010 (http://www2.imec.be/be_en/press/imec-news/flandersexasciencelab.html).
[119] For example, see remarks by Kenneth Flamm in *21st Century Innovation Systems for Japan and the United States*, op. cit.
[120] de Proft, op. cit.
[121] For a concise history of the role ITRI played in launching Taiwan's semiconductor industry, see Alice H. Amsden, "Taiwan's Innovation System: A Review of Presentations and Related Articles and Books," submitted for NAS Jan. 4-6 symposium "21st Century Innovation Systems for the U.S. and Taiwan: Lessons From a Decade of Change," Taipei.
[122] ITRI , "3D System and Application Division," Web site of ITRI Electronics and Optoelectronics Research Laboratories, accessible at http://www.itri.org.tw/eng/EOL/research-and-development-category-detail.asp?RootNodeId=020&NavRootNodeId=02042&NodeId=0204222.

ST Microelectronics, Philips, and Freescale folded in 2007,[123] the French government has launched an initiative called Nano 2012. Billed as the nation's largest industrial project, the aim is to make the Grenoble region a world center for developing 32nm and 22nm CMOS technologies.[124] The program involves nearly €4 billion in funding from the national, state, and local governments for R&D and equipment. Among the initiative's partners are the CEA-Leti Institute for Micro- and Nanotechnology Research; IBM's Fishkill, N. Y., semiconductor production complex; ST Microelectronics; the University of New York at Albany; ASML Holdings of the Netherlands; and Oregon-based and ST Mentor Graphics of Wilsonville, Oregon.[125]

The program is based at MINATEC, a campus in Grenoble that has become an important European center for semiconductor innovation.[126] MINATEC also is diversifying into biotechnology and clean-energy technologies to complement its strength in micro-systems. MINATEC brings academic programs from four universities. Its state-of-the-art facilities include a 300mm silicon wafer center that operates around the clock, a 200mm micro-electro-mechanical systems (MEMS) prototyping line for fast development of new products, and one of Europe's best facilities for characterizing new nano-scale materials. The campus is home to 2,400 researchers and 600 technology transfer experts. MINATEC's 200 industrial partners include Mitsubishi, Philips, Bic, and Total, and two-thirds of its annual €300 million annual budget comes from outside contracts. It also receives funding from the French and local governments, the French Atomic Energy Commission, and private investors. Researchers have filed nearly 300 patents and published more than 1,600 scholarly papers.[127]

Lessons

The decline and resurgence of the U.S. semiconductor illustrates that government policy can help retain a high-tech manufacturing industry to keep America at the technological forefront. Government financial and policy support

---

[123] Anne-Francoise Pele, "Freescale Eases Out of Crolles2 Alliance," *EE Times Europe*, June 26, 2007.
[124] MINATEC Web site (http://www.minatec.com/en/actualites/07/07/2009/nano-2012-underway).
[125] Anne-Francoise Pele, "Mentor Joins 2012 R&D Alliance," *EE Times*, March 16, 2010.
[126] See presentation by David Holden of Minatec in National Research Council, *Understanding Research, Science, and Technology Parks: Global Best Practices: Report of a Symposium*, Charles W. Wessner, editor, Washington, DC: The National Academies Press, 2009.
[127] MINATEC data.

for the original SEMATECH research consortium is widely regarded as a successful experiment and has influenced subsequent public-private partnerships in other U.S. industries and in other nations. The determination of the U.S. government to challenge unfair trade practices and to take action within international law at the time helped stem Japanese dumping and provided inroads into the Japanese market, providing U.S. semiconductor companies with an opportunity to make the large investments needed to diversify into other, more lucrative products.

The continued leadership of the U.S. semiconductor industry cannot be taken for granted, however. A new set of competitive challenges has arisen, such as America's declining share of leading-edge manufacturing capacity, possible skilled talent shortages, and China's drive toward "indigenous innovation" that envisions a diminishing foreign share in its huge and growing semiconductor market. Each of these elements requires American policy attention at a time of intensifying global competition. As the industry heads into historical technology transition, government funding of basic research is critical to maintaining U.S. semiconductor leadership. The U.S. partnership between industry, academia, and government is unrivaled in developing and implementing leading-edge semiconductor technology and in training talent. The U.S. should continue to nurture areas in which it leads the world today and compete in areas such as tax and regulatory policy that determine where companies build new production and R&D capacity.

THE PHOTOVOLTAIC INDUSTRY

Photovoltaic cells represent a classic case of technology developed in the United States with heavy federal support where high volume manufacturing developed largely offshore because of more supportive foreign government policies. Bell Laboratories scientists invented the first silicon-based cell capable of converting sunlight directly into electricity in 1954. Solar panels were first deployed in U.S. satellites. The U.S. government funded most of the pioneering research to develop photovoltaic cells as a source for clean energy in response to the oil shocks of the 1970s and backed the first successful start-ups. In the 1980s, the U.S. accounted for more than half of global production.[128] The U.S. maintained its global manufacturing leadership position until 1999. But in the 21st century, the United States has fallen behind other nations in both installing and manufacturing solar-energy systems. [See Figure 6.5]

---

[128] PV News and Navigant Consulting data cited in presentation by Minh Le of the Department of Energy at the May 24-25, 2011, symposium "Meeting Global Challenges in Berlin hosted by the National Academies' Science, Technology, and Economic Policy board and DIW Berlin.

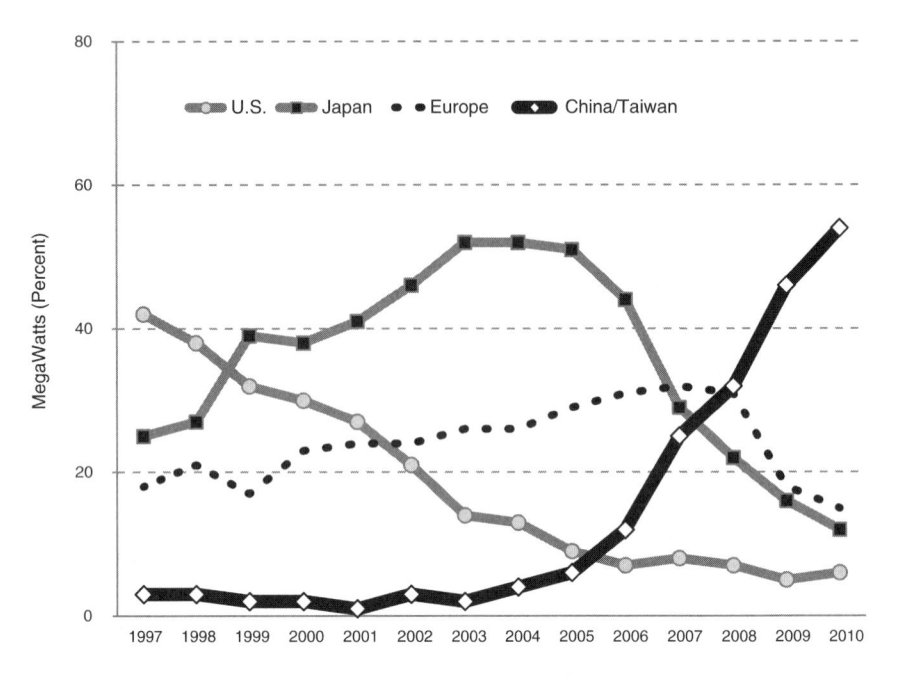

FIGURE 6.5 Global share of PV shipments by region, 1997 to 2010. SOURCE: Paula Mints, "Reality Check: The Changing World of PV Manufacturing," *Electro IQ*, October 3, 2011.

Government policies have been a major reason behind the changes in global manufacturing leadership passing from the United States to Japan to Europe and now to China. Other nations have done more to promote adoption of solar energy and to encourage development of large-scale manufacturing. Even though Germany receives far less sunlight than any U.S. state except for Alaska, for example, it had 17.2 GW of installed solar capacity in 2010 compared to just 2.5 GW in the United States.[129] The U.S. accounted for just 5.3 percent of new global photovoltaic capacity installations in 2010. Because U.S. capacity additions have lagged those in other countries, the U.S. share of worldwide cumulative installed capacity has fallen from 9.5 percent in 2000 to

---

[129] European Photovoltaic Industry Association, *Global Market Outlook for Photovoltaics Until 2015*, April 2011. Data from EPIA for U.S. capacity differ slightly from data released by the Solar Energy Industry Association (SEIA) in the United States. SEIA indicates that the United States had 2.593 GW of installed capacity by 2010, while EPIA indicates an installed capacity of 2.528 GW. EPIA data are used here for international comparison purposes because it is a consistent data source for all countries.

6.4 percent in 2010.[130] [See Figure 6.6] The future of photovoltaic power generation as a significant source of electric power will depend on innovation in device and process technology which reduces the cost of PV relative to other sources of electricity production.[131]

Yet the United States has considerable opportunities to re-emerge as a global leader in solar and other clean energies. Among the world leaders in the two dominant photovoltaic technologies are SunPower in polysilicon and First Solar in thin film. Both of these companies are based in the United States, as are some of the industry's premier producers of raw materials and manufacturing equipment.[132] GE recently announced that it would build the largest solar factory in the United States in Aurora, Colorado. The factory, to start up in 2012, will use thin-film technology to produce solar panels that will be "more efficient, lighter-weight and larger than conventional thin film panels."[133] A U.S. federal solar-power production tax credit of 30 percent, introduced in 2008, has sparked a boom in large-scale commercial systems.[134] Billions of dollars in private investment has flowed into more than 100 U.S.-based solar firms since 2006,[135] while federal grants and loan guarantees have enabled cell and panel manufacturers to build major domestic plants.[136] These policy measures helped

---

[130] Ibid.

[131] See Powell, Buonassisi, et al, Crystalline Silicon Photovoltaics, a Cost Analysis Framework, Energy and Environmental Science (February 2012), 5, 5874, http://pubs.rsc.org/en/journals/journalissues/ee, which is discussed in, Kevin Bullis, Technology Review: http://www.technologyreview.com/energy/39771/, 2012.

[132] While SunPower's corporate headquarters are in San Diego, California, it manufactures solar cells in the Philippines and Malaysia and assembles solar panels in the Philippines and, through third-party contract manufacturers, in China, Mexico and Poland. Sunpower will soon assemble solar panels in California with a contract manufacturer. SunPower Corporation, *Annual Report 2010*, February 2011, p. 10. First Solar, headquartered in Tempe, Arizona, manufactures thin-film solar modules in Ohio, Germany and Malaysia. It plans to add manufacturing facilities in Malaysia, Germany, France, Vietnam and the United States. First Solar, Inc., *Annual Report 2010*, February 2011, p. 6.

[133] "GE Plans to Build Largest US Solar Factory in Colorado, Expand Solar Innovation in New York and Deliver Lighter, Larger, More Efficient Thin Film Panels," *GE Press Release*, October 13, 2011.

[134] Peter Asmus and Clint Wheelock, "Clean Energy: Ten Trends to Watch in 2011 and Beyond," Pike Research, 2011 (http://www.pikeresearch.com/wordpress/wp-content/uploads/2011/05/CE10T-11-Pike-Research.pdf).

[135] See presentation by Robert Margolis of the National Renewable Energy Laboratory in National Research Council, *The Future of Photovoltaic Manufacturing in the United States: Summary of Two Symposia*, Charles W. Wessner, editor, 2011. This volume summarizes presentations in two symposia on the U.S. photovoltaic industry convened in Washington April and July 2009 by the National Academies' Science Technology and Economic Policy board.

[136] The loan guarantee program that was launched in 2009 as part of President Obama's stimulus initiative expired in 2011. In April 2012, the Department of Energy indicated that it would offer a smaller volume of loan guarantees to solar, wind, and geothermal energy products pursuant to a loan guarantee program established under the Energy Policy Act of 2005. "Energy Dept. to Revitalize a Loan Guarantee Program," *New York Times* (April 5, 2012).

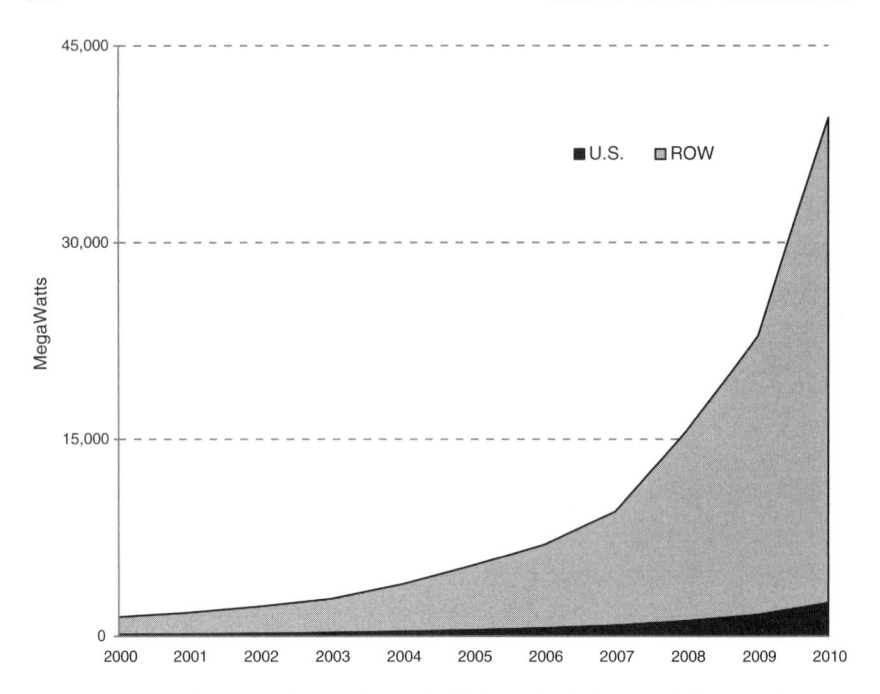

FIGURE 6.6 The U.S. share of worldwide installed photovoltaic capacity, 2000 to 2010.
SOURCE: EPIA, *Global Market Outlook for Photovoltaics Until 2015*, April 2011.

more than double the number of U.S. photovoltaic installations in 2010 compared to 2009.[137]

Perhaps the most important development is that solar-power is steadily nearing "grid parity," the point at which solar-generated electricity costs the same as power generated by fossil fuels offered by utilities without subsidies.[138]

---

[137] Solar Energy Industries Association, *U.S. Solar Market Insight 2nd Quarter 2011: Executive Summary*, 2011.
[138] A number of different definitions for grid parity are used in the industry. The point at which solar-generated power is regarded as cost-competitive with conventional power offered through the grid differs depending on electricity costs in a given region. Parts of Europe, where electricity rates are much higher than in the U.S., grid parity can be reached sooner. The DoE's SunShot Initiative measures grid parity in terms of "the installed system as a whole," including costs associated with permitting. A more common understanding views grid parity in terms of electricity cost of between 8 cents and 12 cents per kilowatt hour in the U.S., about the price of power from a natural gas-fired plant. If defined as capacity cost, grid parity is generally defined as $1 per watt during peak demand.

Although there is considerable debate over how soon grid parity can be achieved,[139] progress is unmistakable. The cost of installing photovoltaic systems connected to the power grid in the U.S. dropped from an average of $11 per watt in 1998 to $6.20 in 2010. Costs dropped by 17 percent in 2010 alone and by a further 11 percent in the first half of 2011.[140] In 1990, solar energy cost between 50 cents and around 65 cents per kilowatt-hour, compared to around 5 to 8 cents utilities charged for conventional power. In 2011, residential rates for solar power were around 21 cents to 27 cents without federal subsidies.[141] In some parts of the U.S., such as northern California, solar power already is regarded as economically viable, although further cost reductions and efficiency improvements are needed before it can compete against conventional electricity production in most of the U.S.[142] The Department of Energy's SunShot Initiative sets a target of reducing the total installed cost of utility-scale solar electricity to a "grid parity" rate of around 6 cents per kilowatt-hour without subsidies by 2020, a development that it predicts "will result in rapid, large-scale adoption of solar electricity across the United States."[143]

While there are many positive trends, a number of challenges still must be overcome to achieve wide-scale adoption of solar energy in the United States and for the U.S. photovoltaic industry to resume global leadership—

- **Inadequate scale**: Because the United States accounts for less than 6 percent of global photovoltaic cell and module production, many U.S.-based manufacturers lack the scale to compete on cost with high-volume producers in Asia and Europe.[144] Scale applies to installation costs as well. It costs only $3.83—about 60 percent less than in the U.S. -- to add one watt of capacity for a residential solar system in

---

[139] The nonprofit Prometheus Institute, for example, predicts that two-thirds of the U.S. will have achieved grid parity by 2015. One criticism of grid-parity data presented by solar-industry advocates is that market prices for electricity in most nations ultimately are distorted by government policy, taxes, and subsidies. For a contrarian view of the progress toward reaching grid parity, see Lux Research, "The Slow Dawn of Grid Parity," 2009.
[140] Galen Barose, Naïm Darghouth, Ryan Wiser, and Joachim Seel, "Tracking the Sun IV: An Historical Summary of the Installed Cost of Photovoltaics in the United States from 1998 to 2010," Lawrence Berkeley National Laboratory, September 2011. This report can be accessed at http://eetd.lbl.gov/ea/emp/reports/lbnl-5047e.pdf.
[141] Michael Woodhouse, et. al, "An Economic Analysis of Photovoltaics Versus Traditional Energy Sources: Where Are We Now and Where Might We Be in the Near Future?," National Renewable Energy Laboratory, presentation at the 37th IEEE Photovoltaic Specialists Conference, Seattle, Wash., June 19-24, 2011.
[142] Ibid. Also see presentation by Eric Daniels of BP Solar in National Research Council, *The Future of Photovoltaic Manufacturing in the United States*, op. cit.
[143] Department of Energy SunShot Initiative Web site, http://www1.eere.energy.gov/solar/sunshot/.
[144] See comments by Ken Zweibel of the George Washington University Solar Institute and First Solar CEO Michael J. Ahearn in *The Future of Photovoltaic Manufacturing*, op. cit.

Germany, for example, mainly due to greater construction efficiencies.[145]

- **Excess global capacity:** Explosive growth in the production of solar cells and modules, especially in China, is pushing down world prices for commodity devices.[146] While that makes solar power systems less expensive, it is even harder for U.S.-based manufacturers deploying next-generation technologies to compete with low-priced imports using mature technologies.
- **Dependence on Subsidies**: The relatively high prices of panels and installation means that solar power is not yet cost-competitive with fossil fuels for power generation without public subsidies such as feed-in tariffs.[147] Those subsidies can change due to policy shifts, making demand hard to predict.
- **Intense International Competition:** Other nations are investing aggressively in R&D and manufacturing capacity to attain global leadership. With more than 100 manufacturers of photovoltaic cells and more than 400 makers of panels,[148] China accounts for more than half of global production capacity.[149] Germany has invested more than €2 billion in public-private photovoltaic R&D and €5 billion in support for manufacturing.[150]
- **Technical challenges**: Some industry experts maintain that another technological leap in materials and process technologies is required before solar power can become cost-competitive with carbon-emitting energy.[151] Due to the high costs and risk of such R&D, public-private

---

[145] Presentation by Minh Lee of the U.S. Department of Energy in May 24-25, 2011, symposium "Meeting Global Challenges: German-U.S. Innovation Policy" in Berlin, jointly organized by the Germany Institute for Economic Research and the National Academies.

[146] For a discussion of the impact of excess Chinese capacity on the world market, see National Foreign Trade Council, "China's Promotion of the Renewable Electric Power Equipment Industry: Hydro, Wind, Solar, Biomass," prepared by Dewey & LeBoeuf LLP, March 2010 (http://www.nftc.org/default/Press%20Release/2010/China%20Renewable%20Energy.pdf).

[147] Lux Research, op. cit. The same is true for Germany, even though utility electricity prices are higher than in the U.S. and the cost of solar-energy systems lower. See Thilo Grau, Molin Huo, and Karsten Neuhoff, *Survey of Photovoltaic Industry and Policy in Germany and China*, Climate Policy Initiative Report, DIW Berlin and Tsinghua University, March 2011.

[148] ENF counted 102 Chinese manufacturers of cells and 473 panel manufacturers in 2010. ENF, Market Survey: Chinese Cell and Panel Manufacturers, 4th Edition, December 2010. Synopsis on ENF Web site.

[149] GTM Research, "U.S. Solar Energy Trade Assessment 2011: Trade Flows and Domestic Content for Solar Energy-Related Goods and Services in the United States," prepared for Solar Energy Industries Association, August 2011. This report can be accessed at http://www.seia.org/galleries/default-file/Solar_Trade_Assessment.pdf.

[150] Grau, Huo, and Neuhoff, op. cit.

[151] See presentations by John E. Kelly of IBM and Steven C. Freilich of E. I. du Pont de Nemours and Co. in *The Future of Photovoltaic Manufacturing*.

consortia of industry, universities, and government agencies may be required.
- **Lack of technological standards**: The market for photovoltaic products remains divided among several competing technologies with different materials and production processes and no industry-wide roadmap similar to the one adopted by the semiconductor industry in the 1970s. That makes it difficult for companies to decide where to make big investments in R&D and capital equipment with a long-term payoff.

Solar Power's Strategic Importance

Solar power is among a portfolio of renewable energy sources upon which many nations are counting to reduce their dependence on petroleum and coal and to reduce greenhouse-gas emission. In the United States, these energy goals are regarded as important for the environment, national security, and economic growth.[152] President Barack Obama has set a target of boosting the portion of energy consumed in the U.S. coming from renewable sources from 7 percent in 2007 to 25 percent by 2020. Other U.S. targets are to conserve 3.6 million barrels of oil within 10 years and to cut U.S. greenhouse-gas emissions by 83 percent by 2050.[153]

Although solar power now accounts for just 2 percent of non-hydroelectric renewable energy in the U.S., capacity is expected to increase more than five-fold by 2035.[154] "For a long-term, sustainable energy source," notes the National Academy of Engineering, "solar power offers an attractive alternative" because energy transmitted from the sun is abundant, environmentally clean, and free.[155] A strong domestic manufacturing industry for photovoltaic cells and modules is vital in order to dramatically lower the costs of installing solar-power systems in the United States and to keep the U.S. at the technological forefront in new materials and high-tech production

---

[152] See National Academy of Sciences, *Electricity from Renewable Sources: Status, Prospects, and Impediments*. Washington, DC: The National Academies Press, 2009. See also National Research Council, *The National Academies Summit on America's Energy Future: Summary of a Meeting*, Washington, DC: The National Academies Press, 2008. Also see National Research Council, *Hidden Costs of Energy: Unpriced Consequences of Energy Production and Use*. Washington, DC: The National Academies Press, 2009.

[153] For explanation of the role of solar energy in meeting U.S. energy targets, see July 29, 2009, presentation by former Under Secretary of Energy Kristina Johnson in *The Future of Photovoltaic Manufacturing*.

[154] U.S. energy Information Administration, *Annual Energy Outlook 2011* (http://www.eia.gov/forecasts/aeo/pdf/0383(2011).pdf).

[155] National Academy of Engineering, *Grand Challenges for Engineering*, Washington, DC: The National Academies Press, 2008, (http://www.engineeringchallenges.org/Object.File/Master/11/574/Grand%20Challenges%20final%20book.pdf).

processes. A large-scale domestic manufacturing and installation industry for solar power and other renewable energies also is a potential source of millions of new jobs.[156]

The Industry's Origins

Although physicists had experimented with materials to achieve the "photovoltaic effect" of converting light to electricity since the mid-19th century,[157] the photovoltaic industry didn't emerge until the U.S. space race with the Soviet Union. Researchers at Bell Laboratories were the first to develop a working photovoltaic using silicon in 1954. The Signal Corps of the U.S. Army recognized the potential of solar-powered energy for satellites. The California Institute of Technology and the Jet Propulsion Laboratory led early development of photovoltaic cells, with the National Science Foundation as the lead funding agency.[158] In 1958, solar cells were first deployed on the Vanguard I, which operated for eight years.

Serious research aimed at developing commercially viable solar power for energy began soon after the 1973 Arab oil embargo, when more than 100 representatives from government, industry, and academia convened at a conference in Cherry Hill, N. J., to develop a 10-year technology roadmap for crystalline silicon photovoltaic technology.[159] Attendees called for $295 million for crystalline silicon technology research. Silicon Valley replaced Los Angeles as the base of the leading U.S. solar-energy cluster, with national laboratories and other public research institutions playing a heavy role in development of the U.S. industry.[160]

---

[156] John Lushetsky of the U.S. Department of Energy noted that solar energy has created around 200,000 jobs in Germany. If solar energy would be adopted on a similar scale in the United States, a much larger market, it therefore would create an estimated 1 million jobs. From remarks at Nov. 1, 2010, "Meeting Global Challenges: U.S.-German Innovation Policy" symposium in Washington organized by the National Academies and DIW Berlin. Also see July 29, 2009, remarks by U.S. Senator Mark Udall (D-Colo.) in *The Future of Photovoltaic Manufacturing*.

[157] French physicist Alexandre Edmond Baquerel is credited with discovering the photovoltaic effect in 1839 when he observed that illumination increases the conduction of electricity from metal electrodes and electrolyte. The first solar cell, made with selenium, was developed in 1877.

[158] For a history of the development of photovoltaic cells, see John Perlin, *From Space to Earth: The Story of Solar Electricity*, Ann Arbor: AATEC Publications, 1999. Also see Steven S. Hegedus and Antonio Luque, "Status, Trends, Challenges and the Bright Future of Solar Electricity from Photovoltaics," in Antonio Luque and Steven Hegedus, editors, *Handbook of Photovoltaic Science and Engineering*, Chichester, England: John Wiley & Sons, 2003. The Department of Energy also offers a concise timeline of the history of solar technology, (ttp://www1.eere.energy.gov/solar/pdfs/solar_timeline.pdf).

[159] For an account of the Cherry Hill conference, see Henry W. Brandhorst, Jr., "Photovoltaics—The Endless Spring," NASA Technical Memorandum 83684. (http://ntrs.nasa.gov/archive/nasa/casi.ntrs.nasa.gov/19840023712_1984023712.pdf).

[160] For a good overview of the role of U.S. public research institutions in the origins of the photovoltaic industry, see "Phech Colatat, Georgeta Vidican, and Richard K. Lester, "Innovation

In 1978, Congress introduced tax credits to spur installation of solar panels and other renewable energy sources as part of the National Energy Act. The following year, President Jimmy Carter proposed a solar strategy to "move our Nation toward true energy security and abundant, readily available energy supplies."[161] Measures included installing 350 solar systems on government facilities and buildings, establishment of a Solar Bank, and $1 billion in federal investment in the form of tax credits, loans, and grants.[162] President Carter set a goal of the sun meeting 20 percent of U.S. energy needs by 2000, and even had solar panels installed on the White House roof. The Reagan Administration dismantled much of the Carter Administration's solar programs on the grounds that the government should limit involvement in programs that should be led by the private sector.[163] A sharp drop in oil prices in the early 1980s also undermined political support for large investments in renewable energy.

U.S. companies still had a commanding world lead in the nascent photovoltaic industry through the mid-1990s. Then Japanese companies developed solar panels that could be installed on residential rooftops, which SunPower CEO Dick Swanson described as the "killer app."[164] The Japanese government also created a large market for photovoltaic panels by introducing financial incentives through the Residential PV System Dissemination Project in 1994. Japan became the global market leader in 1999 as the U.S. share steadily declined.[165] [See Figure 6.5] Sanyo acquired a leading U.S. photovoltaic producer, Solec International, while another leader, Solar Technology International, was sold to Atlantic Richfield and then to Siemens. European nations such as Germany and Spain then took the lead by introducing high feed-in tariffs,[166] driving the second wave of industrial expansion. Companies such as Suntech, Q Cells, and Solarworld became new leaders in a market that had been dominated by Sharp, Sanyo and Kyocera. The rapid market expansion in

---

Systems in the Solar Photovoltaic Industry: The Role of Public Research Institutions," Massachusetts Institute of Technology Industrial Performance Center, Working Paper Series, MIT-IPC-09-007, June 2009 (http://web.mit.edu/ipc/research/energy/pdf/EIP_09-007.pdf).

[161] Carter Administration initiatives included enlarging the budget for NREL, which was established in 1974 as the Solar Energy Research Institute.

[162] President Jimmy Carter, "Solar Energy Message to the Congress," June 20, 1979, http://www.presidency.ucsb.edu/ws/index.php?pid=32503&st=foreign+oil&st1=#axzz1OmYKbnIb.

[163] For a discussion of the Reagan Administration's solar-energy policy, see J. Glen Moore, "Solar Energy and the Reagan Administration," Mini Brief Number MB81265, Science Policy Research Division, Congressional Research Service, the Library of Congress, archived Sept. 23, 1982 (http://digital.library.unt.edu/ark:/67531/metacrs8799/m1/1/high_res_d/MB81265_1982Jul26.pdf).

[164] From presentation by SunPower CEO Dick Swanson in *The Future of Photovoltaic Manufacturing*.

[165] See presentation by Robert Margolis of the National Renewable Energy Laboratory in *The Future of Photovoltaic Manufacturing*.

[166] Under a feed-in tariff system, utilities are required to purchase electricity generated by renewable sources under long-term contracts at premium rates high enough to guarantee a financial return for developers of power systems. The costs are generally passed on to rate-payers.

Europe also triggered a surge of venture capital and private equity investment in U.S. photovoltaic companies over the past decade, although they located most of their initial large-scale manufacturing in Asia and Europe. The volume of venture capital investment in clean energy technology has increased year-over-year in every year since 2005, except 2009 when VC investment fell off sharply. Venture capital investments in U.S. cleantech companies totaled $4.3 billion in 2011, an all-time high, although this figure is comprised substantially of ongoing investments, the funding of start-ups having declined.[167]

Competing Technologies

There currently are two main types of solar power technologies: flat plates and concentrators. The latter technology uses mirrors or lenses to concentrate solar thermal energy onto a small area. Solar thermal plants transfer the heat from concentrated sunlight into a hot working fluid, which powers a generator that produces electricity. Concentrated photovoltaic systems concentrate sunlight onto a small, highly efficient PV semiconductor device. Because mirrors or lenses can only concentrate an image of the sun, their use tends to be limited to cloudless regions with abundant, direct sunlight, such as deserts in the U.S. Southwest.

Flat plates are the far more widely used. The most common photovoltaic cells use polycrystalline materials to absorb and release photons that then are converted into electrical current. Polycrystalline cells, which account for 90 percent of the market, typically are laminated on large glass panels. Because their weight and rigidity, panels with polycrystalline cells tend to be manufactured close to the end market, and installation accounts for around half of the system cost. The other main type of photovoltaic cells use materials such as cadmium telluride or gallium arsenide to absorb light that are deposited in ultra-thin layers on more flexible materials, such as thin sheets of metal or polymers. Thin-film cells on the market yield less power, but are far lighter and easier to install than rigid polycrystalline cells, so their overall cost can be lower.[168] Thin-film is expected to rise from around 50 megawatts of generating capacity installed in 2007 to around 4.5 GW (gigawatts) by 2012.[169] Other competing technologies, such as dye-sensitized and nano-particle photovoltaics,

---

[167] "There is No Cleantech Venture Bust, Sorry Wired," Cleantech (February 14, 2012); "Busting the Myth of the 'Clean-tech' Crash," Notes (February 15, 2012).
[168] For a description and explanation of tradeoffs, benefits, and costs of each type of solar technology, see National Academy of Sciences, *Electricity from Renewable Sources: Status, Prospects, and Impediments*. Washington, DC: The National Academies Press, 2010.
[169] From presentation by Mark Hartney of FlexTech Alliance in *The Future of Photovoltaic Manufacturing*, 2010.

are at an early stage of development. Commercialization will require much more technology development.[170]

U.S. Advantages

The United States still has considerable advantages that could enable it to regain global leadership. The U.S. remains a global leader in photovoltaic research, with at least 11 public-private collaborative R&D consortia involving universities, industry, and government.[171] The U.S. photovoltaic industries includes some 2,000 companies spanning the photovoltaic supply chain, including manufacturers of polysilicon, polymers, wafers, cells, modules, invertors, glass, and production equipment in 17 states. They include First Solar, the world's leading producer of thin-film photovoltaic modules and a top provider of complete solar power systems. First Solar plans to boost production capacity to 2.8 gigawatts by the end of 2012.[172] San Jose-based SunPower,[173] a major producer of polycrystalline cells, also is regarded as the world technology leader.[174]

In fact, although the United States is a major net importer of solar modules, it enjoyed a $1.9 billion trade surplus in solar products in general in 2010, led by shipments of polysilicon—the feedstock for crystalline silicon photovoltaics—and capital equipment.[175] SunPower President Emeritus Richard Swanson estimates that 70 percent of the content in a SunPower solar module is American, even though the device itself is manufactured in the Philippines. Most of the polysilicon, for example, comes from Hemlock Semiconductor Corp. in Saginaw, Mich., the world's largest producer. Most of the equipment used to make wafers is made by U.S. companies such as Applied Materials. As production becomes more automated, Mr. Swanson said more work can shift to the United States if there is a sufficient market.[176] Many U.S. plants have

---

[170] Ibid.
[171] For an extensive description of collaborative photovoltaic research programs in the U.S., see Charlie Coggeshall and Robert M. Margolis, *Consortia Focused on Photovoltaic R&D, Manufacturing, and Testing: A Review of Existing Models and Structures*, National Renewable Energy Laboratory, Technical Report NREL/TP-6A2-47866, March 2010, (http://www.nrel.gov/docs/fy10osti/47866.pdf).
[172] First Solar Web site.
[173] In April 2011, SunPower agreed to sell a 60 percent stake to Total of France for $1.38 billion but says its headquarters will remain in San Jose.
[174] SunPower modules boast the highest efficiency rate in the industry, 22.4 percent, according to the DoE.
[175] GTM Research, "U.S. Solar Energy Trade Assessment 2011: Trade Flows and Domestic Content for Solar Energy-Related Goods and Services in the United States," prepared for Solar Energy Industries Association, August 2011. This report can be accessed at http://www.seia.org/galleries/default-file/Solar_Trade_Assessment.pdf.
[176] Swanson presentation, op. cit.

struggled in the past year, however, due to plunging prices caused by a dramatic expansion of capacity in China.[177]

Although a number of U.S. module plants have closed because they could not compete on costs, others have opened or are expanding.[178] As of 2009, a study by MIT counted 46 solar-cell manufacturing establishments in California alone, and half of those are in the Bay Area. MIT estimated 100 start-ups that had received some funding.[179] Of $2.3 billion in venture capital and private-equity investment in solar companies in 2010, the U.S. accounted for 76 percent.[180]

Another big U.S. advantage is ample sunlight, the basic resource of solar energy. Most territory in the Western and Southern states receives as much sunlight as Spain or more. Parts of the Southwest can receive the equivalent of more than 2,000 kilowatts of energy per square meter each year.[181] The "sunniest" part of Germany receives 60 percent of the energy that reaches the "sunniest" spot in the U.S.[182]

Finally, the U.S. market has growth momentum at a time when new solar installations in Germany, Italy, and Spain have slowed due to reductions and caps in feed-in tariffs.[183] In 2010, photovoltaic demand in Germany reached nearly 8 gigawatts, compared to less than 1 gigawatts in the U.S. Demand for new capacity in the U.S. is projected to leap fivefold by 2013, however, while investment in Germany is set to decline below U.S. levels.[184]

The New U.S. Solar Policy Thrust

Unlike many other nations, the U.S. does not have a feed-in tariff system requiring utilities to purchase solar and other renewable energies at a premium rate. Instead, the U.S. allows companies to accept either a tax credit or cash grant to cover 30 percent of investment in solar power-generation systems.[185] The extension of tax credits to utilities has led to a dramatic increase

---

[177] Swanson presentation, op. cit.
[178] Solar Energy Industry Association and GTM Research, "U.S. Solar Market Insight: 2010 Year in Review," Executive Summary, 2010 (http://www.seia.org/galleries/pdf/SMI-YIR-2010-ES.pdf.
[179] Colatat, Vidican, and Lester, op. cit.
[180] Source: Department of Energy based on *Bloomberg NEF data*.
[181] National Renewable Energy Laboratory data.
[182] Presentation by John Lushetsky of the U.S. Department of Energy at first Germany symposium.
[183] Lux Research, op. cit.
[184] Source: Department of Energy citing 2011 data from Barclays Capital, Citigroup Goldman Markets, Goldman Sachs, Jeffries & Co., and other sources.
[185] Section 1603 of the American Recovery and Reinvestment Act of 2009 (H. R. 1) allows companies to claim either a cash grant or tax credit to cover portions of investments in renewable energy technologies. For solar energy projects, the grant is equal to 30 percent of investment in solar-energy property. The program has been extended through 2011.

in solar power systems for electrical grids. Pike Research predicts utility-scale capacity will surpass 10,000 megawatts by 2016.[186]

The federal government also supports the photovoltaic industry with R&D funding, an R&D tax credit, a manufacturing tax credit for renewable energy equipment, and loan guarantees[187]. Combined with incentives offered by states, government assistance has considerably narrowed the cost gap between building a photovoltaic plant in the U.S. and China that had been created by Chinese incentives.[188]

Federal funding for Department of Energy solar-energy programs has risen sharply in recent years. In 2008, the budget of the U.S. Department of Energy's Solar Program was doubled, to around $160 million a year, from levels of 2001 through 2007. The program received another $100 million boost in 2009 under the American Recovery and Reinvestment Act, with around half of that amount targeted at photovoltaic technologies.[189] The federal government has awarded $6.4 billion in grants in lieu of tax credits to renewable energy projects, with $593 million, or 9 percent, of that money going to solar energy.[190]

The government also has expanded the breadth of its assistance to the industry, not only funding research and demonstration projects but also helping finance manufacturing projects from the prototype phase to full-scale production. The DoE has awarded $1.1 billion in manufacturing tax credits to the solar industry, with $601 million going to plants for polysilicon cells and $264 million to thin-film.[191] In addition, the DoE has committed $12 billion in loan guarantees to 15 solar projects as of mid-2011 that have enabled companies to raise $35 billion in private investment.[192] For instance, in June 2011 the agency announced a $150 million loan guarantee to 1366 Technologies, a company based in Massachusetts that developed a method for casting 200-micron wafers rather than slicing them from a block, a breakthrough that the company says could reduce the manufacturing cost of a solar cell by 40 percent. The head of the DoE's loan guarantee program said the loan illustrates the agency's strategy to "develop a cradle-to-market innovation strategy that helps

---

[186] Asmus and Wheelock, op. cit.
[187] The loan guarantee program established pursuant to the 2009 stimulus legislation expired in 2011.
[188] John Lushetsky, manager of the Department of Energy's Solar Energy Technology Program, estimated that Chinese incentives had made the cost of building a photovoltaic plant there $131 million less expensive in the U.S. Incentives offered in the U.S. have closed that gap by approximately $96 million. From remarks in Nov. 1, 2010, "Meeting Global Challenges: U.S.-German Innovation Policy," op. cit.
[189] Lushetsky presentation in *The Future of Photovoltaic Manufacturing*, op. cit.
[190] U.S. Department of Treasury data cited in Minh Lee presentation in May 24-25, 2011, symposium "Meeting Global Challenges: German-U.S. Innovation Policy" in Berlin, op. cit.
[191] Ibid.
[192] The loan guarantees were awarded through the 1705 Loan Guarantee program established under the American Recovery and Reinvestment Act of 2009. The program is to end on Sept. 30, 2011. Data on private investment from Solar Energy Industries Association.

identify transformative technologies early in the process, and makes it possible for them to grow and mature rapidly, and leapfrog many of the steps along the way."[193] Conditional loans included $1.2 billion to SunPower, and $967 million to AguaCaliete. Finalized loans included $1.45 billion to Abengoa, $1.37 billion to BrightSource Energy, $535 million to Solyndra, and $400 million to Abound Solar.[194] Several of these loans have generated political controversy, however, especially after the bankruptcy of Solyndra in September 2011.[195] The loan guarantee program expired in 2011.

The DoE also has launched a Photovoltaic Technology Incubator program to accelerate commercialization of solar technologies. The incubator program has provided $59 million in support to 31 small businesses working on a range of promising solar technologies. These companies have in turn raised $1.3 billion in private capital and created 1,200 jobs.[196] The DoE continues to run the PV Manufacturing R&D project, started in 1991 with federal funds matched by an equal amount of private-sector money, and the Technology Pathway Partnerships program, which supports early-stage collaborations between universities and industry and also is funded by both the federal government and private sector.[197]

A new DoE initiative, SunShot, focuses on accelerating cost reduction in solar energy so that it is comparable to other sources of electricity on utility power grids. The target is to lower solar power costs to 6 cents per watt of installed capacity. Achieving that target would require a 75 percent reduction of the cost of systems compared to 2010. In 2008, the systems price for solar power came to $8 per watt of installed capacity. By 2010, system cost had dropped to $3.80 per watt, which included $1.70 for the photovoltaic module, 22 cents for power electronics, and $1.88 for the "balance of systems," which includes installation and permitting costs. The $1 per watt target envisions module prices dropping to 50 cents per watt through a combination of efficient improvements and manufacturing-cost reductions and 40 cents for "balance of system" costs, with the cost of power electronics dropping to 10 cents.[198] As part of the initiative, the DoE has awarded $27 million in grants "to encourage cities and counties to compete to streamline and digitize permitting processes, such as

---

[193] Matthew L. Wald, "Maker of Silicon Wafers Wins Millions in U.S. Loan Support," *New York Times*, June 17, 2011.
[194] See presentation by Kevin Hurst of the Office of Science and Technology Policy in *The Future of Photovoltaic Manufacturing* for explanation of DoE programs.
[195] See Eric Lipton and John M. Broder, "IN Rush to Assist Solar Company, U.S. Missed Signs," *New York* Times, Sept. 22, 2011, and Melissa C. Lott, "Solyndra —Illuminating Energy Funding Flaws?" *Scientific* American, September 27, 2011.
[196] Minh Lee presentation, op. cit.
[197] For an explanation of these DoE programs, see Margolis presentation, op. cit.
[198] DoE SunShot press release, "DOE Announces $27 Million to Reduce Costs of Solar Energy Projects, Streamline Permitting and Installation," June 1, 2011,
(http://www1.eere.energy.gov/solar/sunshot/about.html).

through information technology and streamlined local zoning and building codes."[199]

The Challenges Ahead

To sustain this positive momentum and enable solar energy to attain grid parity with fossil fuels will require sustained federal support and expanded public-private collaboration, especially given the intensifying competition for global leadership. Following are some major challenges confronting the industry.

Attaining Scale
Wide-scale deployment of solar power in a region lowers both the production and installation costs of photovoltaic modules. A rule of thumb used in the photovoltaic industry is that each doubling of production capacity leads to an average 17 percent drop in manufacturing costs.[200] Production scale, therefore, has a major influence on which companies have competitive advantage. Dramatic increases in production have helped drive First Solar's production cost for thin-film downs down from $1.40 per watt in 2006 to 77 cents in 2010,[201] for example. Because it usually is more cost-efficient to manufacture modules close to where they are installed, wide deployment also helps determine which nations or regions have comparative advantage in manufacturing.

But the very recent build-up in capacity in Asia, primarily China, has led to a divergence between where photovoltaics are produced and consumed. This has resulted in a large increase in trade flows in solar cells and modules. In 2010, for example, most new capacity was installed in Europe, principally Germany, while most supply was from Asia, primarily China. [See Figure 6.7]

Because the U.S. accounts for less than 6 percent of the world's installed solar capacity, the U.S. photovoltaic industry is at a competitive disadvantage against other several nations and regions in Europe and Asia. In 1997, U.S. manufacturers supplied 42 percent of the world market for photovoltaic models. In 2010, the U.S. produced only 6 percent while China and Taiwan accounted for 54 percent of the world market and Europe 15 percent.[202]

---

[199] DoE SunShot Initiative Web site. See http://www1.eere.energy.gov/solar/sunshot/news_detail.html?news_id=17408.
[200] Source: National Renewable Energy Laboratory cited in "Partnering for Photovoltaics Manufacturing in the United States," overview chapter in *The Future of Photovoltaic Manufacturing*.
[201] First Solar, "First Solar Overview," 2011, on company Web site at Web site, http://www.firstsolar.com/Downloads/pdf/FastFacts_PHX_NA.pdf.
[202] Paula Mints, Navigant Consulting, "Reality Check: The Changing World of PV Manufacturing," *Electro IQ*, October 3, 2011.

[See Figure 6.5] As of 2009, Chinese companies accounted for half of Applied Materials' order book for wafer-making equipment and 35 percent of equipment to produce photovoltaic cells, compared to just 5 percent by U.S.-based companies. What's more, some of the capacity by U.S. companies was being built in China.[203] The size of plants being built offshore also is larger than those being constructed in the United States. A number of manufacturers in China and India are adding production lines that will bring their capacity to 1 gigawatts to 2 gigawatts.[204] A one-gigawatt thin-film plant consumes enough glass to cover seven and a half football fields and can reduce production costs by around 20 percent, noted Mark Pinto of Applied Materials. Of new facilities capable of building solar panels the size of garage doors, Mr. Pinto added, China accounted for three in 2009, Taiwan one, India one, Abu Dhabi one, and Europe the rest. None were being built in California.[205]

Installation costs also drop with scale. A comparison with Germany illustrates the point. Germany had 7,408 megawatts of installed capacity as of mid-2011 compared to only 878 megawatts in the United States. Because Germany also has a smaller territory, capacity is more geographically concentrated. There are 53,728 watts of solar-generation capacity per million square feet in Germany and 90 watts per capita. In the U.S., there are 248 watts per million square feet of photovoltaic panels and only 2.8 watts per capita. As a result, it costs $3.83 to add one watt of capacity in Germany, compared to $6.50 in the U.S., mainly due to greater efficiencies in construction and permitting.[206]

The large scale of solar programs and well-established bureaucratic environment in other nations also makes installing solar systems much less expensive than in the United States. The non-module cost has dropped from around $7 per watt of capacity in Italy to around $2.50 since 1999. In Germany, non-module costs are around $2 per watt. In the United States, by contrast, installation costs have risen in the past three years, to nearly $5 per watt.[207]

The smaller scale of the U.S. industry also makes it harder for domestic manufacturers to compete with inexpensive modules flooding in from large plants in China, where rapid expansion of the photovoltaic industry has led to a supply glut. DeutscheBank projected serious global oversupply in 2011 due to a 53 percent increase in shipments compared to only 3 percent growth in demand.

---

[203] Presentation by Mark Pinto of Applied Materials in *The Future of Photovoltaic Manufacturing*.
[204] Sandra Enkhardt, "Small Island with Big Prospects," *PV Magazine*, December 2010 (http://download.taipeitradeshows.com.tw/2010/pv/news/201012_PV_Magazine.pdf).
[204] Industrial Technology Research Institute Web site.
[205] Ibid.
[206] Minh Lee presentation, op. cit.
[207] From presentation by Karsten Neuhoff of the DIW Berlin Climate Policy Initiative at the May 24-25, 2011, symposium "Meeting Global Challenges in Berlin.

NATIONAL SUPPORT FOR EMERGING INDUSTRIES 369

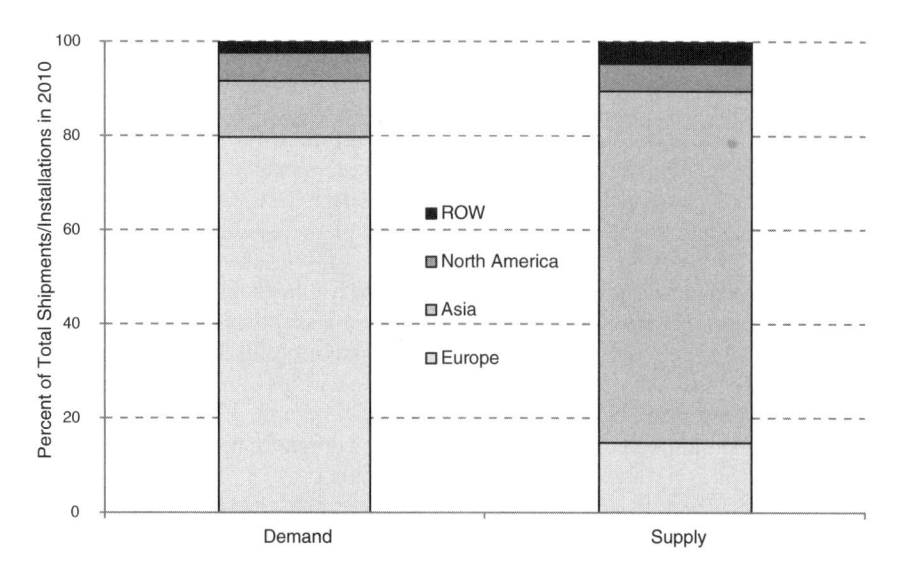

FIGURE 6.7 Photovoltaic demand is concentrated in Europe but supply is concentrated in Asia – 2010.
SOURCE: Demand: EPIA, *Global Market Outlook for Photovoltaics Until 2015*, April 2011; Supply: Paula Mints, "Reality Check: The Changing World of PV Manufacturing," *Electro IQ*, October 3, 2011 and SunPower and First Solar annual reports.

Average prices for crystalline silicon modules are expected to drop to $1.50 per kilowatt with the "potential to go much lower, and quickly."[208] While that makes solar power systems less expensive, it is even harder for U.S.-based manufacturers deploying next-generation technologies to compete with low-priced imports.[209] U.S. imports of Chinese-made photovoltaic modules surged by more than 300 percent from 2008 to 2010, when the U.S. imported more than $1.4 billion worth. In the first eight months alone, Chinese module exports to the U.S. passed $1.6 billion.[210] This surge, which helped push module prices down

---

[208] Peter Kin and Hari Polavarapu, "Solar Photovoltaic Industry 2011 Outlook—FIT Cuts in Key Markets Point to Over-Supply," Deutsche Bank, January 5, 2011 (http://www.strategicsiliconservices.com/wp-content/uploads/2010/07/2011solarpvindustryoutlook.pdf).
[209] For a discussion of the impact of excess Chinese capacity on the world market, see National Foreign Trade Council, "China's Promotion of the Renewable Electric Power Equipment Industry: Hydro, Wind, Solar, Biomass," prepared by Dewey & LeBoeuf LLP, March 2010 (http://www.nftc.org/default/Press%20Release/2010/China%20Renewable%20Energy.pdf).
[210] The Coalition for American Solar Manufacturing, "U.S. Manufacturers of Solar Cells File Dumping and Subsidy Petitions Against China," press release, October 19, 2011.

by 40 percent in 2011, prompted a group of U.S. crystalline silicon cell and module makers to file a dumping suit with the U.S. Department of Commerce and the International Trade Commission.[211]

Shipping glass panels from China, however, is costly and presents logistical risks. Companies, therefore, are likely to continue building capacity in nations where they are installed. The best way for the U.S. to regain leadership role in photovoltaic manufacturing is to become a market leader in installations, according to several industry experts. "Manufacturing will occur in the U.S. once we have adequate markets," said Ken Zweibel of the George Washington Solar Institute."[212] First solar CEO Michael J. Ahearn said the main reason that companies like his put most of their production offshore is because other countries have built a large market, while the U.S. market is "fragmented and sporadic."[213] Said Eric Peeters of Dow Corning: "It is going to be impossible to create a U.S.-based domestic industry if there is no domestic demand. This must be stimulated at every level, from residential to utility scale."[214]

If other nations and regions race too far ahead of the United States in establishing large-scale photovoltaic manufacturing industries, several industry experts warn, it may be difficult for the U.S. to regain competitiveness. Bob Street of the Palo Alto Research Center drew a parallel with the flat-panel display industry. The U.S. pioneered many of the early technologies for liquid-crystal displays, but the industry ended up being dominated by Japanese, South Korean, and Taiwanese companies. As U.S. plants closed, the substantial ecosystem of local equipment manufacturers, materials suppliers, and technology developers went with them, Mr. Street explained. Because flexible photovoltaic technology also requires large, capital-intensive plants and similar clean-room production expertise used in new displays, Mr. Street warned that well-capitalized Asian companies are in position to take over the industry when the market is ripe.[215]

Intense Global Competition

The competition over 21st century leadership in photovoltaic technology and manufacturing is intense. Established players such as Germany are investing to become leaders in innovation and to broaden their value chains. Relative newcomers such as China, Taiwan, India, and South Korea are investing aggressively to expand their global market share in crystalline silicon cells and modules and to catch up with Western companies in new thin-film technologies. They also are rapidly expanding deployment of solar power. Both

---

[211] Ibid.
[212] Zweibel presentation, op. cit.
[213] Ahearn presentation, op. cit.
[214] Presentation by Eric Peeters of Dow Corning in *The Future of Photovoltaic Manufacturing*.
[215] From presentation by Bob Street of Palo Alto Research center in *The Future of Photovoltaic Manufacturing*.

India and China, for example, have announced goals of having 20 gigawatts of installed capacity by 2020, three times more than the entire capacity in the world in 2009.[216]

Government financial incentives have played a big role in promoting the rapid growth of manufacturing and installation of photovoltaic systems in Europe and Asia.[217] Because solar-generated power is more expensive than electricity produced by coal, oil, or natural gas, most governments subsidize solar energy to make up all or part of the cost difference. Also, installing solar-power systems entails high up-front costs with a long-term payoff for consumers and businesses. Therefore, many governments offer assistance to assure that financing is available at affordable interest rates. At least 64 nations have some type of policy to promote renewable energy generation.[218] This has resulted in a rapid acceleration of solar power installations in the last decade. Cumulative installed photovoltaic capacity increased from 1.5 GW in 2000 to 6.9 GW in 2006, a compound average annual growth rate (CAGR) of almost 30 percent. Yet growth was even faster in more recent years as the CAGR accelerated to 54.2 percent from 2006 to 2010. [See Figure 6.8] The 16.6 GW of photovoltaic capacity added worldwide in 2010 equaled almost three-quarters of all the capacity added prior to that year.

Germany has set the global pace. The country has invested more than €2 billion in public-private photovoltaic R&D and €650 million in support for manufacturing, for example.[219] Germany also has been a leader in subsidizing installation of solar-power systems, starting with low-interest loans from the state-owned German Development Bank through the 1,000 Solar Roofs Initiative in 1991 to introduction of some of the world's most generous feed-in tariffs in 2000 (see discussion of feed-in tariffs below). Indeed, the U.S. Department of Energy estimates that Germany spends €4.6 billion on support for all kinds of renewable energies a year, equal to 0.2 percent of GDP. If the U.S. government devoted a similar share of GDP to renewable energy, it would invest $29 billion a year.[220]

Germany has been nurturing regional photovoltaic industrial clusters for the past two decades. In 2007, Germany also established a federally funded research consortium called SolarFocus that includes 12 universities and research institutions and 12 industrial partners. Foreign companies may participate as long as they manufacture domestically.[221]

---

[216] See Vikas Bajaj, "India to Spend $900 Million on Solar," *The New York Times*, November 20, 2009, and Steven Mufson, "Asian Nations Could Outpace U.S. in Developing Clean Energy." *Washington Post*, July 16, 2009.
[217] Zweibel, op. cit.
[218] Ahearn, op. cit.
[219] Grau, Huo, and Neuhoff, op. cit.
[220] Lee Minh presentation, op. cit.
[221] Coggeshall and Margolis, op. cit.

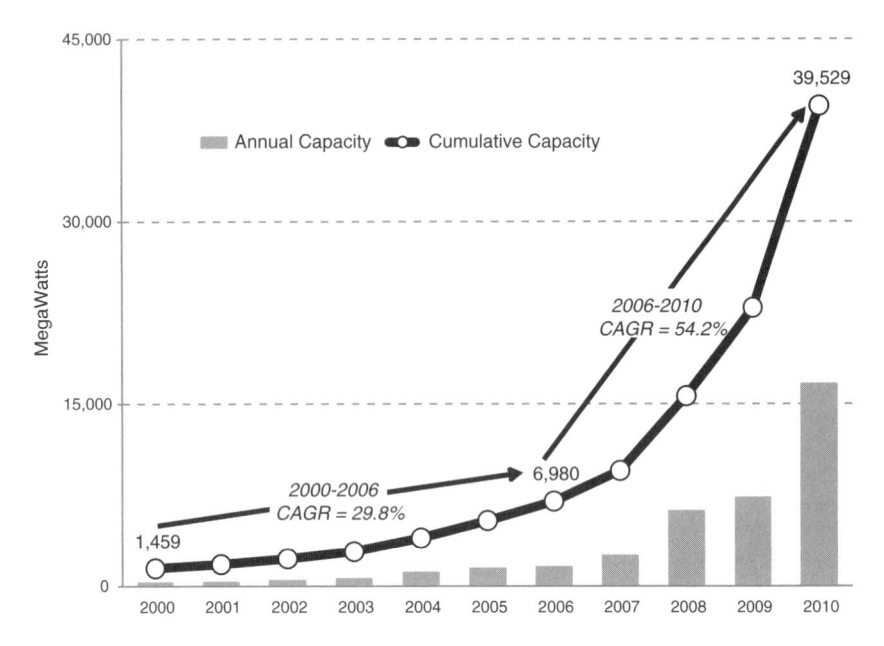

FIGURE 6.8 Worldwide annual and cumulative installed photovoltaic capacity, 2000 to 2010.
SOURCE: EPIA, *Global Market Outlook for Photovoltaics Until 2015*, April 2011.

China already is the world's biggest exporter of crystalline silicon cells and modules. Now, it is determined to become a leading market as well. In 2009 alone, , China invested $34.6 billion in renewable energy industries—more than any other nation and nearly twice as much as the United States,[222] – with solar power commanding greater attention.

Under the Golden Sun program, China is investing some $7.4 billion to install more than 600 megawatts of photovoltaic capacity, with at least 20 megawatts in each province. Through the Ministry of Finance, Ministry of Science and Technology, and the National Energy Administration, projects receive cash subsidies to cover 50 percent of investment in commercial buildings, 50 percent for large-scale photovoltaic systems connected to the power grid, and 70 percent of costs for remote rural residential buildings. In

---

[222] Pew Charitable Trusts, "Who's Winning the Clean Energy Race? Growth, Competition and Opportunity in the World's Largest Economies," 2010. This report can be accessed at http://www.pewtrusts.org/uploadedFiles/wwwpewtrustsorg/Reports/Global_warming/G-20%20Report.pdf.

addition, the cost of power is subsidized.[223] The Golden Sun subsidy "is so large that it is virtually certain to increase the demand for solar power generation equipment," according to a National Foreign Trade Council analysis.[224] As of mid-2011, 294 projects had been approved. In addition to these subsidies, feed-in tariff programs have been implemented in districts of Shanghai, Inner Mongolia and Gansu Province.[225]

China also offers many forms of support to photovoltaic manufacturers. For example, producers can access cash grants of between ¥200,000 and ¥300,000 ($30,900 to $46,300) available to high-tech startups that are less than three years old with no more than 3,000 employees. Large "demonstration projects" by manufacturers get grants of up to ¥1 million. The China Development Bank, meanwhile, offers low-interest loans of several billion dollars for major production plants. The bank reportedly provided $30 billion in low-cost loans to photovoltaic manufacturers in 2010.[226] A number of Chinese provinces offer further incentives, including refunds for interest on loans and electricity costs, 10-year tax holidays, loan guarantees, and refunds of value-added taxes.[227] To open its production plant in China, Massachusetts-based Evergreen Solar was reported to have received $21 million in cash grants, a $15 million property tax break, a subsidized lease worth $2.7 million, and $13 million worth of infrastructure such as roads.[228]

Such subsidies have spurred massive expansion of production capacity. By the first half of 2009, some 50 Chinese companies were constructing, expanding or preparing polycrystalline silicon production lines. Capacity for 2010 was forecast rise from 60,000 tons to more than 140,000 tons, even though much of that capacity is not being utilized.[229]

China's domestic photovoltaic industry has another major advantage in that government procurement rules require that products required for "government investment projects" be purchased from domestic sources unless they are unavailable. Purchases of imported equipment require government

---

[223] Grau, Huo, and Neuhoff, etc.
[224] National Foreign Trade Council, "China's Promotion of the Renewable Electric Power Equipment Industry: Hydro, Wind, Solar, Biomass," prepared by Dewey & LeBoeuf LLP, March 2010 (http://www.nftc.org/default/Press%20Release/2010/China%20Renewable%20Energy.pdf).
[225] For details of Chinese subsidies to photovoltaic plants, see Grau, etc.
[226] Stephen Lacey, "How China Dominates Solar Power: Huge Loans from the Chinese Development Bank are Helping Chinese Solar Companies Push American Solar Firms Out of the Market," Guardian Environment Network, guardian.co.uk, September 12, 2011.
[227] Ibid.
[228] Presentation by Doug Guthrie of George Washington School of Business, April 26, 2011, George Washington University Solar Institute conference.
[229] Jiao Ming, "Photovoltaic Bubble Shattered in China," China Development Gateway, August 28, 2009, *chinagate.cn* (http://en.chinagate.cn/features/earth/2009-08/28/content_18419484.htm).

approval.[230] China is requiring that at least 80 percent of the equipment for its solar power plants be domestically produced.[231] China's policies are the subject of trade friction. The U. S. is investigating a comprehensive trade case filed by United Steel Workers, for example, alleging that China's subsidies of renewable energies constitute unfair trade practices.[232]

Taiwan is leveraging its advantage as a leader in both semiconductor and flat-panel display manufacturing, which use similar production processes to those used in making both crystalline silicon and thin-film cells, to rival China as a photovoltaic exporter. Taiwan ranks behind only China in crystalline crystalline silicon cells, with some 230 companies across the entire supply chain,[233] and is projected to add around 13 gigawatts of capacity by the end of 2012. Three companies, Gintech, Motech, and Solar Power, each are building 1.2 gigawatts to 2.2 gigawatts in new production lines.[234] Industry consortia organized through Taiwan's Industrial Technology Research Institute are developing a range of processes for thin-film cells and printable photovoltaic cells,[235] technologies that also are being developed by Taiwanese producers of digital displays and solid-state lighting devices. Government incentives for manufacturers include a five-year tax holiday, credits that cover 35 percent of R&D and training, accelerated depreciation for facilities, and low-interest loans.[236]

Taiwan also offers an array of subsidies to accelerate domestic deployment of solar power, targeting 10 gigawatts of capacity. The government funds 100 percent of some photovoltaic projects in remote areas, as well as several "solar city" and "solar campus" demonstration projects.[237] Under the recently passed Renewable Energy Development Act, Taiwan implemented a feed-in tariff.

South Korea has recently joined the race to become a global photovoltaic leader. Solar power plays a big role in plans announced in 2009 to

---

[230] "Opinions on the Implementation of Decisions on Expanding Domestic Demand and Promoting Economic Growth and Further Strengthening Supervision of Tendering and Bidding Projects," Circular 1361, May 27, 2009.
[231] See Keith Bradsher, "China Builds High Wall to Guard Energy Industry." *International Herald Tribune,* July 13, 2009.
[232] Sewell Chang and Keith Bradsher, "U.S. to Investigate China's Clean Energy Aid," *New York Times*, October 15, 2010.
[233] Joeng Shein Chen, "Taiwan PV Roadmap: Strategies for PV Industry and Market Growth," Taiwan Photovoltaic Industry Association, November. 17, 2009, (http://www.mbipv.net.my/dload/NPVC%202009/Dr.%20Joeng-Shein%20Chen.pdf).
[234] Enkhardt, op. cit.
[235] Industrial Technology Research Institute Web site.
[236] Ministry of Economic Affairs, "Taiwan Photovoltaic Industry Analysis & Investment Opportunities," Department of Investment Services, (http://investtaiwan.nat.gov.tw/doc/industry/05Photovoltaic_Industry_eng.pdf).
[237] Chen, op. cit.

invest $84.5 billion, or 2 percent of GDP annually, over five years in environment-related and renewable energy industries.

South Korea also is rapidly expanding domestic photovoltaic production, targeting 5 percent of the world market.[238] Hyundai Heavy Industries is building a $200 million plant to make thin-film cells using copper, indium, gallium, selenide materials with France's Saint-Gobain.[239] In all, South Korea wants to capture 10 percent of global green technology market by 2020.[240] The government will require companies to source 10 percent of their electricity from renewable sources by 2022.[241]

The Feed-in Tariff Debate

The development of photovoltaic power requires policy measures to address the fact that it is more expensive than electricity generated through conventional means[242]. The most common measure is the feed-in tariff, a subsidy scheme under which utilities are compelled to purchase power generated by solar installations at a specified rate. The added costs generally are passed on to rate-payers or absorbed by the government. The United States introduced the first feed-in tariffs for renewable energy in 1978.[243] Germany introduced feed-in tariffs in 1990. While these early experiments led to some installation of wind turbines, they were not very successful in advancing solar power. The big boost came when Germany revised its feed-in tariffs in 2000 under the Renewable Energy Sources Act.[244] The prices utilities paid were based on the cost of

---

[238] *PV Magazine*, "Korea Expected to Pick up the PV Pace," January 12, 2011.

[239] *PV Magazine*, "Korea: Construction of 100 MW Thin Film Plant Underway," April 18, 2011.

[240] Jane Burgenmeister, "South Korea Taps Germany to Help Grow its Solar Industry," *Renewable Energy World.com*, April 29, 2009.

[241] For a comprehensive explanation of South Korea's renewable energy strategy, wee United Nations Environment Programme, *Overview of the Republic of Korea's National Strategy for Green Growth*, April 2010 (http://www.scribd.com/doc/30498024/UNEP-Report-on-Korea-s-Green-Growth).

[242] In order for photovoltaics to increase penetration of the electric power generation market financing, power-purchase arrangements and tariffs must be structured in a way that solar-generated power is cost competitive with other firms of power generation from the perspective of utilities. Various schemes have been employed to address the fact that PV electricity is much more expensive than electricity generated by conventional means. These usually involve some combination of subsidies/incentives and favorable feed-in tariffs based on the assumption that PV electricity will become less costly relative to conventional electricity over a long time horizon (e.g. 20 years). See generally Steve O'Rourke, "Financing Photovoltaics in the United States," in National Research Council, The Future of Photovoltaics Manufacturing in the United State (Washington, DC: The National Academies Press, 2011) pp. 88-93.

[243] The Public Utilities Regulatory Policy Act of 1978 required utilities to purchase power from independent power producers at rates designed to reflect the cost a utility would incur to provide the same electricity generation. The tariffs led to some wind installations but few solar-power systems and fell out of favor when oil prices dropped.

[244] The Renewable Energy Sources Act (EEG), passed in 2000 and renewed twice, offers feed-in tariffs for all kinds of renewable energy sources, including wind, water, biomass, biogas, geothermal

generating power for each renewable source, depending on the size of the project, plus a profit margin. Purchase guarantees were good for 20 years. Utilities were allowed to generate their own renewable energy. Italy, France, Spain, the Czech Republic, Japan, the United Kingdom, Greece, and the Canadian province of Ontario followed with their own feed-in tariffs.

Feed-in tariffs are popular with the financial community because the rate of return on a solar power system is guaranteed. This largely explains why Europe accounted for two-thirds of installed capacity in 2009, while the U.S., which lacks federal feed-in tariffs, had a small share, explained First Solar CEO Ahearn.[245] Another advantage of the German feed-in tariffs system, noted Lee Minh of the Department of Energy, is that the purchase-agreement process and incentive structure are far simpler and more stable than in the U.S., which has a mix of subsidies that vary from state to state.[246]

High electricity prices cannot be borne indefinitely by industry and consumers, however. Therefore, the ideal feed-in tariff program triggers rapid expansion of supply, enabling manufacturers and installers to attain economies of scale and to lower prices. Tariff rates must be adjusted frequently as the prices of photovoltaic modules and installation decline. If subsidies—and investor profits—are too high, then investors rush to build as much capacity as possible, straining government budgets. Also, high tariffs can reduce motivation to find innovative ways to lower costs.[247]

Germany's experience with feed-in tariffs illustrates the benefits and risks of the system. Between 2003 and 2009, Germany spent €4.26 billion for feed-in tariffs.[248] The program was so popular that it triggered rapid expansion in the global industry, causing prices to drop sharply as manufacturers added scale. The program also helped establish a globally competitive manufacturing industry. Germany has 70 manufacturers of silicon, wafers, solar cells, and modules that registered more than €9.5 billion in sales in 2008. Germany also has 100 photovoltaic equipment manufacturers with €2.4 billion in 2008 sales. The photovoltaic manufacturing sector employs more than 57,000 people.

One downside of German feed-in tariffs is that consumers pay much more for electricity than in the United States. German households paid an average of around 35.5 cents per kilowatt-hour for electricity as of January 2011, nearly twice as much as British households[249] and an average of just 11.2

---

and solar. But it grants the highest feed-in tariffs to electricity produced by photovoltaic devices. Tariffs are paid for 20 years.
[245] Ahearn presentation, op. cit.
[246] Minh presentation, op. cit.
[247] From Nov. 1, 2010, presentation by Bernard Milow, director of energy program at the German Aerospace Center, at "Meeting Global Challenges" symposium.
[248] Grau, Huo, and Neuhoff, op. cit.
[249] Data from *Europe's Energy Portal* (http://www.energy.eu/).

cents in the U.S.[250] Also, the photovoltaic industry tends to go from boom to bust. When tariffs are high compared to the cost of building capacity, developers race to build solar power systems. Tariffs for solar power ranged from 41 cents to 51 cents per kilowatt-hour in 2009. In 2010 alone, a record 7.1 gigawatts of capacity was installed. When tariffs drop, however, so does investment.[251] Germany is reducing tariff rates sharply. Nations such as Spain, Italy, France, and the Czech Republic also reduced feed-in tariffs, enacted moratoriums on new connections, or set limits on new capacity. As a result, global growth in the industry slowed dramatically in flagship nations like Germany and Spain in 2011.[252]

While the U.S. federal government does not offer feed-in tariffs, many states do. Such tariffs have been enacted in California, Maine, and New Hampshire, and have been proposed in Washington, Minnesota, Wisconsin, Michigan, Florida, New York, Indiana, and Illinois. In addition, 29 U.S. states have set renewable portfolio standards, with 16 of them requiring solar.[253] Such state programs are lowering costs of installing solar power systems, but one downside is that the wide variety of federal and state incentives makes investment processes very complex.[254]

Technological Challenges

Some industry experts maintain that another technological leap in materials and processes is required before solar power can become cost-competitive with carbon-emitting energy. John Kelly of IBM contended that incremental improvements, such as better production equipment or modules built from larger sheets of glass, won't boost energy output of photovoltaic panels fast enough to meet current implementation targets for solar power. The cost gap "has to be closed by leaps of technology," says Mr. Kelly. Nor can the U.S. remain competitive in manufacturing just by investing in more automation. "You have to innovate faster than anyone else," he said.[255]

Dramatically improving thin-film photovoltaic technology presents particularly hard challenges. In addition to inventing new substrates, for example, thin-film panels require a flexible, durable, protective front that keeps out moisture as effectively as glass. "From a polymer perspective, this is essentially unheard of," explained Steven C. Freilich of E. I du Pont de Nemours Co. Freilich of du Pont. Breakthroughs can only be achieved through substantial

---

[250] U.S. Energy Information Administration data as of March 11, 2011.
[251] See Lux Research, op. cit., and Neuhoff presentation, op. cit.
[252] Kim and Polavarapu, op. cit. See also James Montgomery, "Europe's 2011-2012 PV Installs: Two Tales of Growth," Renewable World.com, February 1, 2012.
[253] DESIREUSA.org data.
[254] Lushetsky comments at Nov. 1, 2010, "Meeting Global Challenges" symposium.
[255] Kelly remarks in *The Future of Photovoltaic Manufacturing*, op. cit.

investments and cooperative research in "radical new materials and processes," he said.[256]

Research to develop new materials that then can be produced in mass volume is expensive and risky, however. The challenge is made even more difficult by that fact that there are few widely accepted standards for materials and production processes. Unlike integrated circuits, most of which have been based on complementary metal-oxide-semiconductor (CMOS) technology for decades and are made from silicon wafers with defined parameters, the photovoltaic market is not yet well defined. John Lushetsky of the DoE compared the two industries in this way: "Put simply, the IC industry is one materials set with an infinite number of circuits; the PV industry is one circuit with an infinite number of materials."[257] There is a mix of large and small photovoltaic companies operating in different markets with different manufacturing targets.[258] Photovoltaic cells also are used in a wide range of formats. Nor does the industry have a well-defined technology roadmap delineating engineering benchmarks well into the future.

The inability to predict the technological direction of photovoltaic cells and the lack of widely accepted standards hampers efficiency in the industry and drives up cost and can result in uneven quality, according to Eric Peeters of BP Solar.[259] It also makes it difficult for materials companies to decide where to make expensive long-term R&D bets, Mr. Freilich of DuPont said. Among other materials used in the industry, DuPont makes polymers for coatings for roll-to-roll processing of thin-sheet modules only 20 nm thick "From a material supplier's standpoint, there can be a disincentive to do truly revolutionary work when you see this rapid change in markets and technologies," Mr. Freilich said. "We can do it, but the investment is so great, and the rate of return so dependent on the longevity of the technologies, that you're not going to see the kind of innovation you need."

Photovoltaic Policy Questions for the United States

The United States has an opportunity to reassert global leadership in the photovoltaic industry. It will require considerable national investment and public-private collaboration. The following are some of the major policy options facing the U.S.[260]

---

[256] Freilich presentation, op. cit.
[257] Lushetsky presentation, op. cit.
[258] See comments by Bettina Weiss of Semiconductor Equipment and Materials International (SEMI) in *The Future of Photovoltaic Manufacturing*.
[259] Peeters presentation in *The Future of Photovoltaic Manufacturing*, op. cit.
[260] See Powell, Buonassisi, et al, Crystalline Silicon Photovoltaics, a Cost Analysis Framework, Energy and Environmental Science (Feb 2012), 5, 5874,

Stimulating Demand

One of the most urgent decisions facing the U.S. is whether to extend tax credits for grid-connected solar installations, which currently are set to expire at the end of 2016. There was wide agreement among STEP Board symposium participants that federal incentives are necessary to promote the industry until the time when the costs of solar-power systems drop to the point where they can compete on their own with electrical generation from fossil fuels. Although technological advances are needed to bring down costs, so is greater domestic scale. Public commitment to continuing the expansion of solar power also is important to assure companies that are making long-term investments in research, new materials, and manufacturing capacity. "Government incentives that build market size and industry support can help industry make the right decision about programs on one side or another of that very gray line,"[261] Mr. Freilich of DuPont said.

There also is considerable agreement in the photovoltaic industry that, like most other industrial and industrializing nations, the U.S. should consider requiring utilities to purchase solar power and other renewable energy. A number of states, such as California, are pushing ahead with feed-in tariff requirements, meaning that the incentive structure varies from state to state. The question is whether there will be sufficient political support for a German-style feed-in tariff requiring utilities to buy solar power at premium prices if that leads to substantially higher electricity rates for businesses and consumers. The more likely option is that the U.S. continues to stimulate the solar industry's expansion through a combination of tax incentives, loan guarantees, and other measures.

Promoting Manufacturing

Although a number of industry experts stress that the best way to promote a domestic photovoltaic manufacturing industry is to stimulate domestic demand, public incentives also are regarded as necessary given the intense global competition for large-scale production capacity. The tax difference along between the U.S. and Asia is such that were a company to move 20 percent of its photovoltaic production to the U.S. its profitability would drop by 14 percent, estimated Steve O'Rourke of Deutsche Bank Securities.[262] In nations such as China and Germany, manufacturers receive tax credits rather than pay taxes, he noted. Malaysia, which has an ambitious goal to become the

---

http://pubs.rsc.org/en/journals/journalissues/ee, which is discussed in, Kevin Bullis, Technology Review: http://www.technologyreview.com/energy/39771/, 2012.
[261] Freilich presentation, op. cit.
[262] Presentation by Steve O'Rourke of Deutsche Bank Securities in *The Future of Photovoltaic Manufacturing*.

second largest solar producer in the world by 2020, provides a 15-year tax holiday for solar manufacturing profits.[263]

It will be difficult, and probably unnecessary, for the U.S. to match the kinds of generous concessions to manufacturers offered in nations such as China. The U.S. can close the competitiveness gap, however, with a combination of federal and state support. Steven O'Rourke of DeutscheBank Securities estimated that a modest drop in U.S. tax rates, a 27-cent-per-watt manufacturing credit for equipment produced in the U.S., and a subsidy for capital spending, such as offered by Germany, would essentially close the profitability gap.[264] Accelerated depreciation and state incentives also can make a difference, he said.

Another issue is financing where major unmet needs exist. The normal timeframe for venture capital investments of 5 to 7 years is not applicable to complex and capital intensive energy technologies subject to a long regulatory approval process. Although the U.S. accounts for most of the world's venture capital and private-equity investment in the photovoltaic sector, it is much harder for such companies to borrow funds. SBIR loan levels are completely inadequate. Of the $44 billion in debt financing provided to the solar industry around the world in 2010, the U.S. accounted for only 9 percent.[265] Mark Pinto of Applied Materials suggested that the U.S. create a clean-energy bank that offers low interest rates.[266] First Solar CEO Ahearn said it would be preferable to making loans available to all photovoltaic manufacturers rather than have the Department of Energy decide which applicants receive loan guarantees, a process that he said had little visibility. "I think we'd be much better off if the government simply enabled all banks to make loans that the market would direct to the right place," Mr. Ahearn said.[267]

Several U.S. regions are working to develop strong photovoltaic manufacturing clusters. Northern Ohio, for example, has long been an important center of innovation in solar cells and panels. The University of Toledo has had a strong basic research program and spun off several important startups, including thin-film cadmium pioneers Glasstech and Solar Cells Inc.,[268] which

---

[263] "Reasons Behind Malaysia's Surprising Success in Solar Industry Beating Larger Rivals USA and Japan," *Green World Investor*, October 26, 2010 at http://www.greenworldinvestor.com/2010/10/26/reasons-behind-malaysias-surprising-success-in-solar-industry-beating-larger-rivals-usa-and-japan/. The tax holiday has been cited by First Solar as a factor in its expanding manufacturing presence there. See, e.g., First Solar, Inc., "First Solar Announces 100MW Manufacturing Plant Expansion in Malaysia," *News Release*, January 25, 2007.
[264] Ibid.
[265] Department of Energy estimate based on Bloomberg NEF data. See Minh Lee presentation, op. cit.
[266] Mark Pinto remarks, op. cit.
[267] Ahearn remarks in *The Future of Photovoltaic Manufacturing*.
[268] Both Glasstech and Solar Cells were launched by Harold McMaster (1916-2003), a physicist who was regarded as the king of the Toledo glass industry.

later became First Solar. Manufacturing, however, has tended to move to other U.S. regions or overseas. Northern Ohio's tradition as a leader of the U.S. glass and polymer industries also meant that the region was rich in expertise in materials and developing panels.[269] In 2007, the state government awarded $18.6 million to establish the Wright Center for Photovoltaics Innovation and Commercialization at the University of Toledo. The state also mandated that 25 percent of Ohio's electricity come from renewable sources by 2025, formed a public-private partnership aimed at commercialization, and built a demonstration plant for new solar technologies at a military base.[270]

Collaborative Research

Given the expense, high risks, and long-term payoff of photovoltaic R&D, a number of industry experts said that public-private collaboration is required. While lacking a comprehensive research consortium and technology roadmap, the U.S. has many smaller research consortia supported by federal, state, and industry funding that focus on photovoltaic R&D as well as manufacturing and testing.

Universities lead several of consortia in addition to Ohio's Wright Center. The Silicon Solar Consortium, for example, combines the research efforts of four universities—North Carolina State, Georgia Tech, Lehigh, and Texas Tech—with several national laboratories and 15 companies. The consortium, which aims to reduce costs and boost performance of silicon photovoltaic materials, cells, and modules, is one of several dozen interdisciplinary Industry-University Collaborate Research Centers that receive seed funding from the National Science Foundation.[271] The Center for Revolutionary Solar Photoconversion conducts basic and applied research for third-generation photon conversion. Several Colorado universities and the National Renewable Energy Laboratory, based in Boulder, Colorado, lead the consortium. Corporate members include Applied Materials, DuPont, Lockheed Martin, Sharp, and Motech. The Energy and Environmental Technology Application Center, based at the University of Albany, has 50 corporate partners that include IBM, Applied Materials, SEMATECH, Global Foundries, and Tokyo Electron.[272] DOE National Laboratories such as Sandia and Oak Ridge

---

[269] See presentation by Norman Johnson of Ohio Advanced Energy in *The Future of Photovoltaic Manufacturing*.

[270] For an overview of Ohio's photovoltaic cluster activities, see remarks by U.S. Rep. Marcy Kaptur (D-OH) and Norman Johnson of Ohio Advanced Energy Association in National Research Council, *The Future of Photovoltaic Manufacturing in the United States, Summary of Two Symposia*, C. Wessner, Rapporteur, Washington, DC: The National Academies Press, 2011.

[271] Presentation by Thomas Peterson of the National Science foundation Directorate of Engineering in *The Future of Photovoltaic Manufacturing*.

[272] Profiles of these photovoltaic consortia are found in Coggeshall and Margolis, op. cit.

also have extensive photovoltaic programs and collaborate with industry and academia.

Photovoltaic research consortia in the U.S. have several limitations, however. Because the industry is still young and highly fragmented, there are many competing technologies and a lack of manufacturing standards. "There are dozens of groups and subgroups within the PV industry," according to a report by the National Renewable Energy Laboratory. "This diversity makes the development of any industry-wide consensus, such as manufacturing standards, extremely difficult." Because there are so many evolving technologies, there is good reason for companies to be protective of their intellectual property, the report added. "As a result, companies are less likely to participate in forums that could expose their proprietary information."[273] As a result, research collaborations tend to be on narrowly focused topics that meet the interests of companies funding the research.

Several industry executives questioned whether a SEMATECH-like research consortium would work for the photovoltaic industry. Doug Rose of SunPower noted that because CMOS already had become standard at the time SEMATECH was created, semiconductor manufacturers could share intellectual property that accelerated development of manufacturing processes on a predictable schedule and instead differentiate themselves on the basis of chip design. "There's no analog to that in PV," he said.[274] Mr. Pinto of Applied Materials agreed that the lack of an established common technology make such collaboration problematic.[275]

Still, industry experts say there are many other opportunities for pre-competitive research collaboration among manufacturers. Photovoltaic companies could share work on processes such as modeling, simulation, reliability, and characterization, for example.[276] Consortia also could accelerate solutions to technical issues such as metrology, material handling and deposition handling and in developing low-cost installation methods.[277]

Others in the industry believe such consortia could help. A technology roadmap similar to the one created by the semiconductor industry through SEMATECH[278] in the 1980s for lithography would help companies choose among the many options for major R&D investments, Mr. Freilich of DuPont said.[279] Governments in the European Union, China, India, Australia, and other

---

[273] Ibid.
[274] Comments by Doug Rose of SunPower in *The Future of Photovoltaic Manufacturing*.
[275] Mark Pinto comments in *The Future of Photovoltaic Manufacturing*.
[276] Ibid.
[277] Lushetsky, op. cit.
[278] For an explanation of how the SEMATECH experience may be applicable to the photovoltaic industry, see presentations by Eric Lin of the National Institute of Standards and Technology in *The Future of Photovoltaic Manufacturing*.
[279] Freilich, op. cit.

nations, meanwhile, are organizing efforts to define industry standards. Eric Daniels of BP Solar described standards as "critical in building consumer confidence."[280]

Conclusion

Having ceded the once-dominant position it held in the 1980s and 1990s in the photovoltaic industry to countries in Europe and Asia, the United States has an opportunity to regain global leadership. As European nations reduce feed-in tariffs, the U.S. has become one of the strongest growth markets for new solar capacity. The U.S. also remains at or near the forefront in photovoltaic research, and therefore could be the source of game-changing breakthroughs that lower the cost of solar power and dramatically improve efficiency.

Maintaining this momentum, however, will require consistent and substantial public financial support at a time of intense budget pressure. Expanding the U.S. market for solar power is essential to achieving the economies of scale needed to reduce production and installation costs of photovoltaic systems and to assure that the U.S. has a competitive manufacturing base in the face of intensifying international competition. Because solar power is not yet cost-competitive with electricity generated from fossil fuels, continued subsidies for solar-power installations are required. Given the scale, complexity, and long time frames needed for innovation solar technologies to come to market, public assistance, such as loan guarantees, early-stage capital, and R&D and manufacturing tax credits also will be required to enable fledgling U.S. photovoltaic companies to bring their products to market and establish domestic production at a time when Asian and European governments are increasing their aid to domestic manufacturers.

In this regard, public-private research collaboration can help accelerate the pace of photovoltaic innovation and reduce the costs and risks of developing the materials and production processes needed to make possible the widespread deployment of solar power.

ADVANCED BATTERIES

American researchers have long been at the technological forefront of lithium-ion batteries,[281] which produce electrical charges by lithium ions that

---

[280] Daniels presentation, op. cit.
[281] Development of the first commercially viable lithium-ion battery is generally credited to M. Stanley Whittingham of the State University of New York at Binghamton while working for Exxon Research & Engineering Co. in the 1970s. Other important breakthroughs were achieved by Bell Labs and teams led by University of Texas at Austin physicist John B. Goodenough. See J. B.

flow inside a liquid electrolyte mixture between anode and a cathode plates. But Sony Corp. was the first to market lithium-ion batteries in 1991. Japan has targeted lithium-ion batteries for vehicles since 1992, when the Agency of Industrial Science and Technology and the Ministry of International Trade and Industry established the New Sunshine Program.[282] Unable to compete, many U.S. battery makers and start-ups failed in the 1990s, including Duracell, Polystor, Motorola, MoliCell, Electro Energy, and Firefly.[283]

Until just a few years ago, the United States faced the prospect of entering the age of electrified transportation without a domestic advanced battery manufacturing industry. Virtually all lithium-ion cells and battery packs—projected to be a nearly $8 billion industry by 2015[284] and the dominant technology for electrified cars and trucks of the future—were manufactured in Asia[285]. There were many promising U.S. start-ups with innovative lithium-ion battery technology for cars, utility storage, and other uses, but few could raise funds to build capacity in America.

That situation began to change dramatically in 2009. The federal government awarded $2.4 billion in grants under the American Recovery and Reinvestment Act to dozens of makers of lithium-ion cells, battery packs, and materials.[286] A host of other state and federal financial incentives, such as manufacturing tax credits and research grants, provided further assistance. The federal government also boosted the U.S. market for advanced batteries with incentives for consumers who bought electrified cars, subsidies for solar and wind-power projects, and the $25 billion in debt capital made available under

---

Goodenough and M.S. Whittingham, *Solid State Chemistry of Energy Conversion and Storage,* American Chemical Society Symposium Series #163, 1977.

[282] Japan's New Sunshine Program established a 10-year research program for lithium-ion batteries that set very ambitious targets for the time for power output, battery density, and cycle life. See Rikio Ishikawa, "Current Status of Lithium-Ion Production in Japan," Central Research Institute of Electric Power Industry, Tokyo (http://www.cheric.org/PDF/Symposium/S-J3-0003.pdf).

[283] Presentation by Mohamed Alamgir of Compact Power at the National Research Council conference on *Building the U.S. Battery Industry for Electric-Drive Vehicles,* op. cit.

[284] John Gartner and Clint Wheelock, "Lithium Ion Batteries for Plug-in Hybrid and Battery Electric Vehicles: Market Analysis and Forecasts," executive summary, Pike Research, 2009.

[285] Although some question whether trucks, given their weight, are appropriate subjects for electrification, Taiwan is developing extended-range electric busses and electric commercial vehicles. "Taiwan Unveils First Electric Smart Commercial Vehicle," Asia Pulse (September 28, 2010); "Taiwan Alliance, Set Up to Develop Extended-Range Electric Buses," Taiwan Economic News (May 4, 2011).

[286] The American Recovery and Reinvestment Act of 2009 (P. L. 115-5) is a $787 billion economic stimulus packaged signed by President Barack Obama on Feb. 17, 2009. See Department of Energy, "The Recovery Act: Transforming America's Transportation Sector—Batteries and Electric Vehicles," July 14, 2010 (http://www.whitehouse.gov/files/documents/Battery-and-Electric-Vehicle-Report-FINAL.pdf).

the Advanced Technology Vehicles Manufacturing (ATVM) Loan Program to help automakers produce more energy-efficient cars.[287]

From less than two battery-pack plants before 2009, 30 now have been built or are under construction by the end of 2010. If all of these facilities are built as planned, the U.S. is on track to have 40 percent of global capacity to produce lithium-ion batteries for automobiles and utility storage by 2015.[288] As of mid-2010, some 16 battery-related factories that are expected to create 62,000 jobs in five years were being built just in Michigan, which aggressively targeted the industry with $1 billion in grants and tax credits.[289]

A major issue now is whether there will be enough demand for hybrid and electric vehicles for manufacturers to operate this capacity profitably.[290] Under most current projections of U.S. sales of hybrid and plug-in electric cars, the American battery industry will experience considerable overcapacity for several years.[291] Globally, significant excess capacity is expected to persist through 2015, resulting in significant consolidation. According to one projection, five producers will control 80 percent of the automotive lithium-ion battery market by 2015: AESC (a joint venture between Renault-Nissan and NEC), LG Chem, Panasonic, A123, and SB LiMotive (a joint venture between

---

[287] The Advanced Technology Vehicles Manufacturing (ATVM) Loan Program was authorized under Section 136 of the Energy Independence and Security Act of 2007. It makes available $25 billion to provide debt capital to the U.S. automotive industry for projects that help vehicles manufactured in the U.S. meet higher millage requirements and lessens U.S. dependence on foreign oil.

[288] U.S. Department of Energy, "The Recovery Act: Transforming America's Transportation Sector—Batteries and Electric Vehicles", July 14, 2010. Also see Rod Loach, Dan Galves, Patrick Nolan, "Electric Cars: Plugged In. Batteries Must be Included," Deutsche Bank Securities Inc., June 9, 2008.

[289] Data from remarks by Michigan Economic Development Corp. CEO Greg Main in National Research Council, *Building the U.S. Battery Industry for Electric-Drive Vehicles: Progress, Challenges, and Opportunities*, a symposium convened by the NRC STEP board in Livonia, MI on July 26-27, 2010, in cooperation with the Michigan Economic Development Corp. and the U.S. Department of Energy.

[290] As the *Wall Street Journal* has noted, the short term "mismatch between production and market demand" has led to lags in predicted job creation and production goals. In addition, it reports that "Ener1 Inc., a battery maker that built a plant in Indianapolis with $54.9 million of a $118 million government grant, sought bankruptcy protection earlier this year." See *Wall Street Journal*, "Car Battery Start-ups Fizzle," May 31, 2012.

[291] The Boston Consulting Group projects overcapacity in the U.S. industry from 2012 through 2015. See Boston Consulting Group, "Batteries for Electric Cars: Challenges, Opportunities, and the Outlook to 2020," accessible at http://www.bcg.com/documents/file36615.pdf. Battery industry consultant Menachem Anderman also contends the U.S. battery sector will face enormous overcapacity. See Anderman comments in *St. Petersburg Times PolitiFact.com*, "David Axelrod says U.S. will have 40 percent of global market for advanced batteries by 2015," St. Petersburg Times PolitiFact.com http://www.politifact.com/truth-o-meter/statements/2010/jul/15/david-axelrod/david-axelrod-says-us-will-have-40-percent-global-/.

Samsung and Bosch).[292] Another question is whether the U.S. industry will be able to compete in high-volume manufacturing with bigger, well-funded Asian battery producers who by some estimates have a 10-year lead.[293] The nascent U.S. advanced battery industry is at its "most critical stage of development," according to A123 Systems executive James M. Forcier.[294]

The advanced battery industry is regarded as strategic because it addresses several critical national needs, such as reducing greenhouse-gas emissions and dependence on imported oil. Advanced batteries are the enabling technology for electrified vehicles. The transportation sector accounts for two-thirds of U.S. petroleum consumption. The 240 million vehicles on U.S. roads, in turn, consume two-thirds of fuel used for transportation.[295] Utilities also require advanced batteries for storing energy generated by solar farms and wind turbines.

The U.S. military regards advanced batteries as important as well. Lightweight, rechargeable batteries could greatly extend the range of combat vehicles, support the ever-growing energy needs of modern weapons and surveillance systems, and ease the logistical challenges of hauling fuel to battle zones on long convoys of trucks. Such batteries would considerably lighten the heavy loads of equipment carried by soldiers in the field.[296] The U.S., which has one of the world's largest military vehicle fleets, has committed to cutting its fuel consumption by 20 percent in the next 10 to 15 years.

A domestic advanced battery industry also is strategically important for the future global competitiveness of America's automotive industry. Battery cells and packs are regarded as "the new power trains" of electrified automobiles, just as internal-combustion engine designs and technology are core to gasoline-powered cars, noted Eric Shreffler of the Michigan Economic Development Corp.[297] Reliance on foreign battery technology and products, some fear, could put competitiveness of the U.S. auto industry at risk. U.S. Senator Debbie Stabenow (D-MI) stated that "building the next generation of

---

[292] Roland Berger Strategy Consultants, "Global Study on the Development of the Automotive Li-ion Battery Market," *Press Release*, September 6, 2011.
[293] Estimate from battery industry analyst Menachem Anderman, ibid.
[294] Presentation by James M. Forcier of A123 Systems at the National Research Council conference on *Building the U.S. Battery Industry for Electric-Drive Vehicles,* op. cit.
[295] Data cited in presentation by Patrick Davis of the U.S. Department of Energy in *Building the U.S. Battery Industry.*
[296] Presentations by Grace Bochenek and Sonya Zanardelli of the U.S. Army Tank and Automotive Research, Development, and Engineering Center at the National Research Council conference on *Building the U.S. Battery Industry for Electric-Drive Vehicles,* op. cit.
[297] Presentation by Eric Shreffler of the Michigan Economic Development Corp. at the National Research Council conference on *Building the U.S. Battery Industry for Electric-Drive Vehicles,* op. cit.

NATIONAL SUPPORT FOR EMERGING INDUSTRIES					387

energy-efficient vehicles is do-or-die for all of the automakers, for the state of Michigan, and for America."[298]

Decades of experience in mass-producing rechargeable lithium-ion batteries for consumer electronic products such as cell phones and portable computers, however, have given Japanese, South Korean, and now Chinese companies a formidable edge.[299] While reliable estimates of production are difficult to come by, in part because of the lack of standard definitions and measurement techniques, the consulting firm GBI Research has estimated that only about 2 percent of advanced batteries were produced outside of Japan, South Korea, and China in 2009.[300] [See Figure 6.9] The United States produced only an estimated 1 percent of lithium-ion batteries.[301]

Large Asian producers also are more vertically integrated and better capitalized than most U.S. competitors. For example, South Korea's LG Chem, the world's third-largest producer of rechargeable lithium-ion batteries, is backed by the $113 billion LG Group. Having such deep pockets is important "to survive in this industry," explained Mohamed Alamgir, CEO of Compact Power, a U.S. unit of LG Chem. The Korean company plans to invest $1 billion over five years in battery R&D. Due to the LG Group's chemical businesses, LG Chem also has proprietary materials and processes. LG Chem supplies lithium-ion cells to both Ford and GM is and is building a $151 million complex in Michigan that will produce enough cells to make 50,000 vehicle batteries.[302]

Perhaps more importantly, demand for electrified vehicles has been stronger outside of the United States. Higher fuel prices in Europe and Japan, for example, make hybrids and plug-ins more affordable alternatives. Other nations have moved more aggressively to develop their domestic markets for electrified vehicles with subsidies, government vehicle purchases, and investments in public battery-charging infrastructure. Because of such factors, Pike predicts Asia will account for 53 percent of global demand in 2015—more than the U.S. and Europe combined.[303]

---

[298] Remarks by U.S. Sen. Debbie Stabenow at the National Research Council conference on *Building the U.S. Battery Industry for Electric-Drive Vehicles,* op. cit.
[299] See, for example, Ralph J. Brodd, "Factors Affecting U.S. Production Decisions: Why Are There No Volume Lithium-Ion Battery Manufacturers in the United States?" ATP Working Paper 05-01, June 2005.
[300] GBI Research, *Future of Global Advanced Batteries Market Outlook to 2020: Opportunity Analysis in Electronics and Transportation,* January 2010. Only 8 percent of the production of advanced batteries in 2009 was estimated to be for hybrid electric vehicles, with the rest destined for mobile phones, laptop computers, tablets and other electronic devices.
[301] Data cited in presentation by Patrick Davis , at the National Research Council conference on *Building the U.S. Battery Industry for Electric-Drive Vehicles,* op. cit. Davis estimated that Japan accounts for 46 percent, South Korea for 27 percent, and China for 25 percent of world production. This compares to the estimates of GBI Research for 2009 of 55 percent for Japan, 25 percent for China and 18 percent for Korea.
[302] Ibid.
[303] Forcier presentation, op. cit.

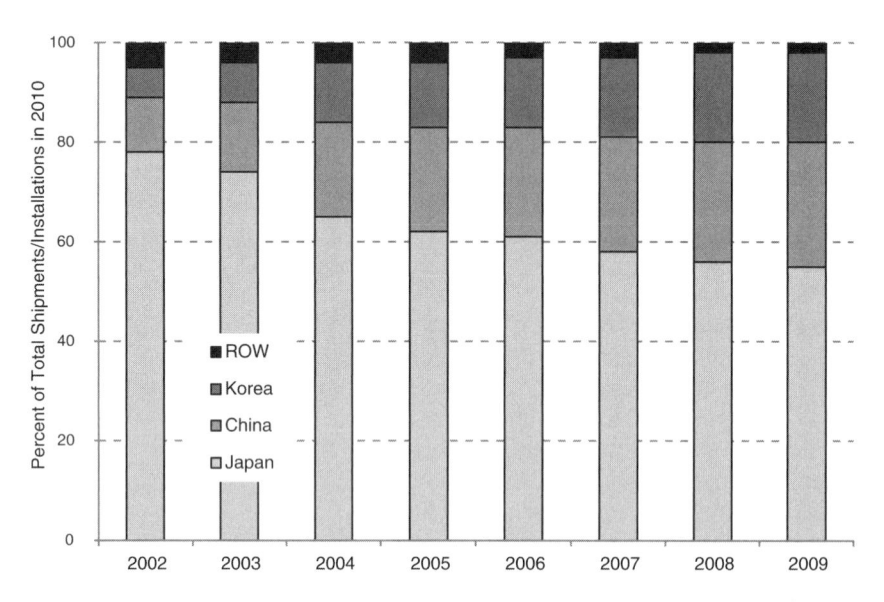

FIGURE 6.9 Advanced battery production by country, 2002 to 2009.
SOURCE: GBI Research, *Future of Global Advanced Batteries Market Outlook to 2020: Opportunity Analysis in Electronics and Transportation*, January 2010.

The global competition will only grow more intense. Governments around the world are funding aggressive plans to expand their national battery industries and domestic markets for electrified vehicles. For example—

- South Korea has announced that its government and companies will invest $12.5 billion over 10 years in a bid to become the world's dominant advanced battery producer. The Battery 2020 Project envisions Samsung and LG Chem boosting their combined share of the world lithium-ion battery market, which still is dominated by consumer electronics, to 50 percent. These two companies have aggressively entered the global lithium-ion market for cars. The national plan also calls for adding 1,000 engineers and technicians to R&D efforts to make the country's supply chain more self-reliant. Currently, South Korea produces less than 20 of the parts and materials used in batteries. The goal is to boost that to 75 percent for domestically made batteries, and to create up to 10 globally competitive battery makers in a decade.[304] The government of South Korea has budgeted $345 million for the research and development of high performance lithium batteries during the period 2011-13[305]. A Korean company, LG Chem, was selected by General Motors in 2009 to supply advanced batteries for the Chevrolet Volt plug-in vehicle[306] LG Chem's Ochang factory in Korea, opened in 2011, is the world's largest lithium-ion battery plant for electric vehicles.[307]

- Japan has launched a number of initiatives to shore up its share of the overall global lithium-ion battery market, which has declined from around 65 percent to 51 percent in the past five years.[308] Japan remains the world's biggest producer of lithium-ion cells for vehicles as well as materials such as cathodes, anodes, electrolytes, and separators.[309] Panasonic Corp.,[310] the industry leader and supplier to Toyota, is investing aggressively, as are Mitsubishi, Hitachi, Toyota, GS Yuasa, Fuji, and Toshiba.

Japan's New Energy and Industrial Technology Department Organization (NEDO) has developed an ambitious roadmap that sees lithium-ion as the dominant battery technology until 2030.[311] The Ministry of Economy, Trade, and Industry has a roadmap for the automotive industry that calls for up to 50 percent of cars to be "next-generation" electrified vehicles and up to 70 percent by 2030.[312] Under these roadmaps, the performance of advanced batteries is to increase 1.5 fold by 2015 while costs will drop to one-seventh current levels. By 2030, "innovative batteries" are to offer a seven-fold increase in performance and cost one-fortieth of current models. The roadmap also envisions up to 2 million regular chargers and 5,000 rapid chargers

---

[304] *Yonhap News Agency*, "S. Korea Aims to Become Dominant Producer of Rechargeable Batteries in 2020," July 11, 2010. Also see Foresight Science & Technology, "Regional Overviews: Asian Industry Overview (China, Japan, Korea)," July 27, 2010.
[305] United States and South Korea Invest in Lithium Battery Technology," The Street (December 22, 2011).
[306] "GM taps LG Chem, Compact Power to Supply Batteries for Volt Plug-in," Mlive.com (January 12, 2009).
[307] "LG Chem Builds World's Largest Electric-Car Battery Plant," Thai Press Reports (April 7, 2011).
[308] New Energy and Industrial Technology Department Organization data.
[309] Andy Bae, "Lithium-Ion Battery Materials: Japan Dominates in the EV Era," Pike Research, Feb. 4, 2011.
[310] Panasonic's battery unit was named Matsushita Electric Industrial Co. until 2009.
[311] New Energy and Industrial Technology Department Organization, "2008 Roadmap for the Development of Next Generation Automotive Battery Technology," Ministry of Economy, Trade, and Industry.
[312] Ministry of Economy, Trade and Industry, "Next-Generation Vehicle Plan 2010 (Outline)," (http://www.meti.go.jp/english/press/data/pdf/N-G-V2.pdf).

deployed across the country to "pave the way for full-scale diffusion."[313] The government's Fiscal Year 2010 budget includes ¥3 billion for collaborate R&D by the government, industry, and academia for innovative batteries.

- Taiwan seeks to become one of the top three lithium battery producers in the world. This effort is spearheaded by ITRI, which has developed STOBA technology, the first materials technology to enhance the safety of lithium-ion batteries[314]. STOBA was selected in 2009 for an "R&D 100 Award" by U.S.-based R&D magazine[315]. In 2008, ITRI collaborated with Taiwan's Welldone Co. to set-up a joint venture, High-Tech Energy Co., for the production of lithium batteries[316]. An ITRI battery expert left the institute to head up battery research at the new company[317]. ITRI formed the High Safety Lithium Battery STOBA consortium of Taiwanese companies to promote the development and diffusion of STOBA-based battery technology. As of 2011, four Taiwanese companies had entered into production of STOBA lithium batteries and the local industry was projected to invest $1.7 billion in 2012.[318]

- China's Ministry of Industry and Information Technology has pledged to invest around ¥100 billion ($15.2 billion) by 2020 in subsidies and incentives over 10 years to support new-energy vehicle production. The government set a target of selling 1 million electric vehicles a year by 2015 and 100 million by 2020.[319] Currently, the government offers a $9,036 subsidy to buyers of electric cars and subsidizes fleet operations in 25 cities.

The National Development and Reform Commission identifies lithium-ion cells and batteries as strategic industries, and several government programs subsidize China's industry through investment and tax credits, loans, and research grants. Argonne National Laboratories

---

[313] Ministry of Economy, Trade and Industry, "The Industrial Structure Vision 2010," June 2010 (http://www.meti.go.jp/english/policy/economy/pdf/Vision_Outline.pdf).
[314] STOBA stands for self-terminated oligomers with hyper-branched architecture. STOBA is designed to prevent battery explosions. "MOEA to invest more in safe lithium-ion Battery Development," Central News Agency (January 24, 2010).
[315] "ITRI's Battery Technology Wins Oscar of Invention," Taipei Times (October 17, 2009).
[316] "Welldone Ventures into Production of Lithium Battery," Taiwan Economic News (July 15, 2008).
[317] "Taiwan Spearheads Lithium-Battery Module Effort," Taipei Times (May 15, 2008).
[318] The companies are AMITA Technologies Inc., Ltd., E-one Moli Energy Corp., Synergy Science Tech Corp., and Lion-Tech Co., Ltd. "Taiwan's Investment in Lithium Batteries to Exceed NT$50 B. in 2012," Taiwan Economic News (March 22, 2011).
[319] *People's Daily*, "China to Sell 1 Million New-Energy Cars Annually by 2015," Nov. 223, 2010. English translation viewable at http://english.peopledaily.com.cn/90001/90778/90860/7207607.html.

estimated China had 60 lithium-ion battery makers as of 2008, including BYD, Tianjin Lishen, CITIC Guoan MGL, and Shenzhen BAK.[320] The government's goal is for Chinese companies to produce enough batteries to supply 150,000 electric vehicles in 2011.[321] To give its domestic industry an extra edge, the government essentially requires foreign battery companies to manufacture in China if they wish to sell there.[322]

- The French Atomic Energy Commission and the French Strategic Investment Fund have formed a joint venture with Renault and Nissan to manufacture lithium-ion batteries. The first plant, a €600 million investment, is to produce up to 100,000 batteries a year by mid-2012 in Flins, France. The venture also is building plants in Portugal, Great Britain, and Tennessee.[323] The French company Saft supplies lithium-ion batteries to Mercedes, BMW, and Ford.

The French government has set a target of having 2 million electric vehicles on the road by 2020. Government-linked companies such as Electricité de France, SNCG, Air France, France Telecom, and La Poste have committed to buying electric vehicles. In addition, the government is investing €1.5 billion to support up to 1 million public charging stations.[324]

Opportunities to Catch Up

Even though U.S. battery manufacturing is behind Asia, there is considerable confidence that the American industry has the potential to catch up and become a powerful force. American companies, universities, and national laboratories remain leading innovators of new lithium-ion chemistries, and the coatings and materials used in cathodes and anodes, which are more suitable to the demanding needs of automakers and the military. The U.S. supply chain is growing; a Duke University study identified at least 50 U.S.-based firms that manufacture or conduct R&D at 119 locations in 27 states, including 21 lithium-ion battery pack makers relevant to the auto industry.[325] Many are increasing the

---

[320] Pandit G. Patil, "Developments in Lithium-Ion Battery Technology in the Peoples Republic of China," Argonne National Laboratories, ANL/ES/08-1, January 2008.
[321] Comments by Minister of Science and Technology Wan Gang cited in *Reuters*, "China Electric Vehicles to Hit 1 Million by 2020: Report," October 16, 2010.
[322] Forcier presentation, op. cit.
[323] Details on Renault Web site at http://www.renault.com/en/capeco2/vehicule-electrique/pages/sites-de-production.aspx.
[324] David Pearson, "France Backs Battery-Charging Network for Cars," *Wall Street Journal*, Oct. 1, 2009.
[325] See Marcy Lowe, Saori Tokuoka, Tali Trigg, and Gary Gereffi, "Lithium-ion Batteries for Electric Vehicles: The U.S. Value Chain," The Center for Globalization Governance and

capabilities to manufacture cells domestically. Dow, A123, and EnerDel acquired or formed strategic partnerships with South Korean manufacturers. While U.S. producers still must import most cathodes and anodes, there are several large American suppliers of electrolytes, separators, and lithium. Companies such as 3M, DuPont, and Dow Kokam, meanwhile, have created divisions to domestically produce anodes and cathodes.

Another cause for optimism is that the advanced battery industry for vehicles is still young and technological standards have not yet been established—leaving room for new entrants. Most analysts predict it will be at least five years before the costs and performance of battery-powered cars reach levels at which they will attain widespread consumer appeal. Pike Research, for instance, predicts hybrid and plug-in electrics will account for only 2 to 3 percent of the U.S. market in 2015 and 5 percent in 2020.[326] Although Ford Motor plans to offer a full portfolio of hybrid and plug-in cars and trucks, it projects it will take at least 15 years for electrified vehicles to account for 25 percent of sales.[327]

Industry experts also believe lithium-ion batteries will have to go through several more generations of technology and manufacturing improvements before they are affordable, efficient, and light enough to win wide consumer acceptance for electric cars. The cost of a 25 kilowatt hybrid battery pack has dropped by more than two-thirds since 1997. Densities and life cycles have more than doubled.[328] However, rechargeable auto batteries remain very expensive. Current lithium-ion batteries for cars cost an average of $800 per kilowatt-hour, which translates to more than $20,000 for a battery for an all-electric car such as the Ford Focus and $10,000 to $12,000 for a battery to power a typical hybrid. A general industry assumption is that those costs should drop nearly two-thirds to make such cars affordable enough to convince consumers to abandon gasoline-powered cars. The DOE roadmap calls for cutting costs to $300 per kilowatt-hour by 2014 for plug-in hybrids. Some analysts believe reaching that target will be difficult.[329] What's more, the battery for a Focus weighs 500 pounds, too large to make them easily replaceable, explained Nancy Gioia, Ford Motor's director of global electrification. Ms.

Competitiveness, Duke University, Oct. 5, 2010. http://www.cggc.duke.edu/pdfs/Lithium-Ion_Batteries_10-5-10.pdf.

[326] Pike Research predicts the penetration rate of hybrid and plug-in vehicles will be 2.41 percent in 2015.

[327] Presentation by Nancy Gioia of Ford Motor at the National Research Council conference on *Building the U.S. Battery Industry for Electric-Drive Vehicles,* op. cit.

[328] Data cited by David Howell of the DOE in his presentation at the National Research Council conference on *Building the U.S. Battery Industry for Electric-Drive Vehicles,* op. cit.

[329] Ford's goal is for hybrid battery packs to cost $250 per kilowatt hour by 2020. See Gioia presentation, op. cit.

Gioia said it would be five or six years before batteries weighing a more manageable 250 pounds are mass-produced.[330]

The Growing Federal Role

The U.S. government has long supported basic battery research programs. Federal programs now address the full value chain, from accelerating development of commercial products and manufacturing to workforce training and charging infrastructure for electrified vehicles. The Department of Energy's Vehicle Technologies Program[331] has made lithium-ion battery research and development a high priority since 2000.[332] The DOE also leads a government-industry partnership called the U.S. Advanced Battery Consortium,[333] which funds projects aimed at commercializing new battery technologies and sets cost and performance development targets. The Duke University study counted 59 battery-technology development projects underway at U.S. universities and national laboratories such as Lawrence Berkeley, Argonne, Sandia, Oak Ridge, and the National Renewable Energy Laboratory.[334]

In terms of battery-related R&D, the U.S. has increased spending at every level. The DOE's Basic Energy Sciences program has expanded research into fundamental materials and electrochemical processes. The DOE funds 60 energy storage R&D projects at 10 national laboratories and 12 universities, as well as projects with companies such as A123, Johnson Controls, and EnerDel. Five of the agency's 46 Energy Frontier Research Centers are involved with batteries and vehicle technology. The DOE also has awarded a number of research grants to companies and partnerships working on advanced anode, cathode, electrolyte, and lithium materials and processing technologies.[335] The DOE's Advanced Research Projects Agency (ARPA-E) is providing $100 million for "transformational" advanced-storage research, including projects in lithium-air batteries at the Missouri University of Science & Technology, an all-

---

[330] Gioia, ibid.
[331] The Vehicle Technologies Program is administered by the Energy Efficiency and Renewable Energy Office of the Department of Energy. It funds projects aimed at developing "leap frog" technologies that will lead to more energy-efficient and environmentally friendly transportation.
[332] Presentation by David Howell of the Department of Energy's Vehicle Technologies Program at the National Research Council conference on *Building the U.S. Battery Industry for Electric-Drive Vehicles,* op. cit.
[333] The United States Advanced Battery Consortium is a collaborative effort between the Department of Energy and the United States Council for Automotive Research, whose members consist of General Motors, Ford, and Chrysler. The group's stated mission is "to develop electrochemical energy storage technologies that support commercialization of fuel cell, hybrid, and electric vehicles."
[334] Lowe et al, op. cit.
[335] Howell presentation, op. cit.

electron battery at Stanford, and high-performance and ultra-low-cost rechargeable batteries at MIT.[336]

The 2009 Recovery Act grants to 48 cell, pack, and materials production projects marked the federal government's biggest move to directly support domestic battery manufacturing and to create jobs. The Advanced Technology Vehicle Manufacturing program also supports battery-manufacturing projects.[337] The Advanced Energy Manufacturing Tax Credit program provides credits that cover 30 percent of investments in new, expanded, or refurbished manufacturing plants producing renewable-energy equipment.[338] The Obama Administration has expanded the advanced manufacturing tax credit program to $7 billion. Other incentives include the DOE's 1703 and 1705 loan guarantee programs[339] and the 1603 program that gives cash grants in lieu of tax credits for renewable-energy projects,[340] many of which use advanced batteries.

To promote market acceptance of electrified cars, the U.S. government has offered $7,500 tax credits to purchasers of plug-in hybrid cars. The DOE is funding projects that will deploy 10,000 electric-drive vehicles, ranging from light-duty trucks to passenger busses, as well as home and public-access chargers across the nation. The DOE's Clean Cities program works with 86 coalitions in 45 states to introduce electrified vehicles and charging stations.

Tougher federal and state environmental standards further boost the industry. The Obama Administration has set a target of reducing greenhouse-gas emissions by at least 30 percent by 2016.[341] California has more aggressive emission targets. The state also is raising requirements on automakers to sell a

---

[336] Ibid.
[337] Davis presentation, op. cit.
[338] The Advanced Energy Manufacturing Tax Credit was authorized in Section 1302 of the American Recovery and Reinvestment Act and also is known as Section 48C of the Internal Revenue Code. It authorizes the Department of Treasury to award $2.3 billion in tax credits to cover 30 percent of investments in advanced energy projects, to support new, expanded, or re-equipped domestic manufacturing facilities.
[339] Section 1703 of Title XVII of the Energy Policy Act of 2005 ("EP Act 2005") authorizes the DOE to issue loan guarantees to acceleration commercialization of technologies that "avoid, reduce, or sequester air pollutants or anthropogenic emission of greenhouse gases." Section 1705 of the EP Act is a temporary program set up under the American Recovery and Reinvestment Act authorizing the DOE to make loan guarantees to renewable energy systems, electric transmission systems and leading-edge bio-fuels projects that commence construction no later than September 30, 2011.
[340] Section 1603 of the American Recovery and Reinvestment Act created a program administered by the U.S. Department of Treasury that extends grants covering between 10 percent and 30 percent of the cost of certain renewable-energy property.
[341] The U.S. Environmental Protection Agency and the Department of Transportation's National Highway Traffic Safety Administration (NHTSA) are finalizing greenhouse gas-emission standards for model years 2012 to 2016 under the Energy Policy and Conservation Act. For details, see http://www.epa.gov/oms/climate/regulations/420f10014.htm.

certain number of zero-emission vehicles and wants the carbon-intensity of all fuels to be cut by 10 percent.[342]

The Military's Electrification Drive

The Defense Department is another major driver of advanced-battery development. The U.S. Army's Tank-Automotive Command Research, Development, and Engineering Center (TARDEC) and the Army Research Laboratory collaborate with the DOE and industry on several battery, new material, and electrical system R&D projects. TARDEC, based in the Detroit area, oversees maintenance of the Army's 400,000-vehicle fleet and development of next-generation vehicle capabilities. TARDEC has 60 battery-related research projects underway. These projects encompass basic research, applications, manufacturing processes, battery management, and safety.[343]

The Army has ambitious plans to introduce electrified vehicles into its fleet that require lighter, longer-lasting, more powerful batteries that will not fail in extreme climates and are safe when under heavy fire. Achieving greater energy independence for tactical units is a top priority. The Army wants to boost fuel-efficiency of future light tactical vehicles by nearly 50 percent, to 61 ton-miles per gallon. The Army also wants tanks that can operate two or three days without refueling and Stryker armored cars with cruising ranges of up to 360 miles.[344]

Dramatic improvements in batteries also are required to meet the ever-rising power requirements of combat vehicles, weapons, and other equipment. The Army consumes about 20 gallons of gasoline per day to support one soldier in the field. Half of that generates electricity for jammers, remote sensing devices, and other equipment. A high Army priority is to fit combat vehicles with Silent Watch capability, enabling them to operate essential systems while stationary without running the engine. Future light tactical vehicles will require 40 kilowatts of power, compared to 10 kilowatts now, according to Grace Bochenek of TARDEC. Future ground combat systems will need nearly 50 kilowatts. Cost reduction also is critical. Although current lithium-ion battery packs for light tactical vehicles weigh one-third as much as advanced lead-acid batteries and produce 50 percent more power, they cost nearly 20 times as much—around $10,000 each.[345]

---

[342] Presentation by Daniel Sperling of the University of California at Davis in *Building* at the National Research Council conference on *Building the U.S. Battery Industry for Electric-Drive Vehicles,* op. cit.
[343] Presentation by Sonya Zanardelli of TARDEC at the National Research Council conference on *Building the U.S. Battery Industry for Electric-Drive Vehicles,* op. cit.
[344] Bochenek presentation, op. cit.
[345] Ibid.

Other branches of the military also have an interest in advanced batteries. The Air Force is developing hybrid systems for unmanned aerial vehicles that operate 40 to 50 hours and need thousands of watts of power, for example. The U.S. Navy is looking to use hybrids for unmanned underwater vehicles, shallow-water combat submersibles, submarine distributed power systems, and surface ship fuel economy. In each scenario for reducing energy use, "batteries run rampant throughout them in almost every capacity," explained John Pellegrino of the Army Research Laboratory.[346]

Collaboration with industry and academia is essential if the Army is to achieve its targets. Sharp cost reductions of domestically produced batteries can only be achieved with high-volume production, noted Dr. Pellegrino. Therefore, the Army is forming major partnerships with the private sector and collaborating earlier with industry to make sure new devices are manufacturable. "We don't want each of those vehicles to cost $1 billion," Dr. Pellegrino said.

Future Policy Priorities

Now that the U.S. has established a manufacturing base for advanced batteries, policymakers face a new set of challenges to make this nascent industry sustainable and globally competitive. If U.S.-based manufacturers cannot survive, the implications for American competitiveness in next-generation vehicles may be severe. "While the risk of overcapacity is very real for U.S. firms," warns the Duke University study, "it may actually pale in comparison to the opposite risk: that of not being prepared to lead this new industry, with serious implications for the U.S. edge in the global automotive sector."[347]

The following are several of the key policy issues identified by industry experts—

Accelerate R&D: Not all analysts agree on what kind of battery performance will be required for electrified vehicles to win wide consumer acceptance. Research by the Institute of Transportation Studies at the University of California at Davis, for example, suggests that drivers of plug-in hybrids adapt to limited driving ranges and battery recharging needs the more they drive their cars.[348] Other analysts, however, contend that battery cost and performance are not improving fast enough.[349]

---

[346] Pellegrino presentation, op. cit.
[347] Lowe, et al, op. cit.
[348] Presentation by Daniel Sperling of the Institute of Transportation Studies at the National Research Council conference on *Building the U.S. Battery Industry for Electric-Drive Vehicles,* op. cit.
[349] Boston Consulting Group, for example, concludes that a "major breakthrough in battery chemistry" that leads to much higher energy densities without increasing costs of either battery materials or manufacturing processes is essential. See Boston Consulting Group, op. cit.

There is general agreement in the industry that the federal government should increase battery R&D through public-private partnerships in order to accelerate advances that will make electrified vehicles viable alternatives to gas-powered cars for the mass market in the near future. Experts also stressed that it is important to increase R&D funding for research into technologies beyond lithium-ion (such as hydrogen fuel cells) that can yield breakthroughs. GM estimates alternative technologies could be commercially viable by 2016.[350]

Government Purchases: Large-scale production is the surest way of bringing down the costs of advanced battery cells and packs. Government purchases of advanced batteries from U.S. based manufacturers can help the domestic industry to attain economies of scale. A number of industry executives and experts recommend the U.S. government to take stronger action to help stimulate demand enough to launch the industry, as are governments in Asia and Europe. The federal policy priority should shift to "demand-driven stimulation rather than stimulating manufacturing and research," said Les Alexander, A123's general manager for government solutions. "We can create the best battery in the world, but without vehicles to put them in, this industry will go back overseas and we will have stimulated another country's industries," he said.[351] It is important to note that increased sales of electric vehicles in the U.S. will not necessarily result in increased sales of U.S. made advanced batteries.

Government purchases of electrified vehicles are one policy option. Several experts noted that the federal government could help create a substantial market by purchasing U.S.-made hybrid and plug-in electric cars and trucks to replenish the fleet of some 700,000 vehicles owned by the military and agencies such as the U.S. Postal Service. The Advanced Vehicle and Power Initiative, a program backed by TARDEC, calls for replacing 8 percent of the government truck fleet annually with electrified vehicles.[352]

The General Services Administration recently announced a goal to buy more than 40,000 alternative-fuel and fuel-efficient vehicles to replace aging and less-efficient sedans, trucks, tankers, and wreckers across federal agencies.[353] Such programs may need to be increased and implemented over a longer term. Michael E. Reed of Magna E-Car Systems noted that manufacturers make investments based on a five- to seven-year time horizon.[354]

---

[350] Smyth presentation, op. cit.
[351] Presentation by Les Alexander of A123 at the National Research Council conference on *Building the U.S. Battery Industry for Electric-Drive Vehicles,* op. cit.
[352] See presentation by Bill Van Amburg of CALSTART in *Building the U.S. Battery* Industry. The Advanced Vehicle and Power Initiative is an effort facilitated by TARDEC to advance collaboration among manufacturers, academia, and government to accelerate deployment of advanced vehicle technologies. A May 25, 2010, draft of AVPI's policy white paper is available on the CALSTART Website (www.calstart.org/Libraries/HTUF_Documents/AVPI.sflb.ashx).
[353] Department of Energy press release, January 26, 2011 (http://www.energy.gov/10034.htm).
[354] Presentation by Michael Reed of Magna E-Car Systems at the National Research Council conference on *Building the U.S. Battery Industry for Electric-Drive Vehicles,* op. cit.

Improve Incentives: Federal incentives such as consumer tax credits for purchases of electrified vehicles will likely be required for several more years before the U.S. market is large enough to support fledgling advanced-battery manufacturers.[355] Several industry experts suggested the life of such programs be extended.

Current incentive programs also can be improved. Under the current incentive program, for examples, buyers of plug-in hybrids receive a $7,500 credit that they can apply the following year on their income tax return. Such an incentive program would have a more immediate impact if consumers could receive the credit at the time they complete the purchase of the car.[356] Some experts also recommend extending more federal incentives to purchasers of hybrid and plug-in commercial trucks.[357]

Public Charging Infrastructure. There is general agreement among experts that some level of public charging infrastructure is needed to ease so-called "range anxiety" by drivers who fear they will be stranded if their electric car batteries run out of power. Several countries have made nationwide public-charging networks a top priority.

There is disagreement over how extensive such charging infrastructure must be. Research by the Institute of Transportation Studies, for instance, suggests that most drivers of electric cars do not use public charging stations, and rarely use them at work. Instead, they prefer to charge their cars overnight at home.[358] Executives from GM and Ford agreed that public infrastructure is a lower priority than providing affordable battery-charging systems for homes, with charging stations at work sites a next priority. Home chargers for small, basic plug-in hybrids can be installed for less than $200, but units for all-battery electric cars cost around $2,000. Workplace or public stations can cost $50,000 each.[359] Several experts suggested that policy should focus on R&D aimed at bringing down the costs of home and workplace charging units.

Developing the Value Chain: Developing a more extensive domestic supply base for advanced batteries also is required to make the U.S. industry globally competitive. Although some companies are investing in U.S. plants to make materials and key components, several industry executives describe the efforts as inadequate. In addition to cathodes, anodes, and separators, according to Johnson Controls executive Mr. Watson, most software and mechanical components must be imported.[360] Nor is there a sufficient domestic supply base for key components used in electrified cars. Mr. Reed of Magna E-Car explained

---

[355] Mr. Forcier of A123 estimates that consumer incentives will be required for at least another five years. Forcier presentation, op. cit.
[356] Rep. Stabenow presentation, op. cit.
[357] Van Amberg presentation, op. cit.
[358] Sperling presentation, op. cit.
[359] Gioia presentation, op. cit.
[360] Watson presentation, op. cit.

that the complexity of the supply chain "adds significant cost" to U.S.-based manufacturing and is time-consuming.[361] The impressive investment in North American lithium-ion cell production since 2008 has not "been balanced by necessary investment in the supply chain itself," he observed.

Conclusion

The emerging U.S. advanced battery industry represents a bold experiment by the federal government in direct financial support of private companies to establish a domestic manufacturing industry. Prior to 2008, the U.S. had a number of lithium-ion battery start-ups but virtually no production plants. It now has dozens of battery-related factories, thanks in part to $2.4 billion in aid under the American Recovery and Reinvestment Act, making it an active competitor in the advanced vehicle battery industry.

The major question is whether the U.S. will be able to sustain the policy support needed for the nascent advanced battery industry through what are expended to be challenging years ahead. Most experts predict it will be at least five years before a combination of technological improvements and higher production volumes will bring costs of lithium-ion car batteries—and therefore the prices of hybrid and electric vehicles—to the point that they can compete with gas-powered vehicles in the market. There also is some doubt whether demand will be big enough to justify the battery-manufacturing capacity that is coming online.

Continued federal incentives to promote consumer purchases of hybrid and electric cars and perhaps greater public procurement—as other nations are doing--will likely be essential to make the U.S. battery industry viable. Continued federal and state support for public-private research collaborations also will be required to accelerate the advances in technology and manufacturing processes needed to bring the cost of rechargeable car batteries down to the point where electrified vehicles can compete on their own with gasoline-powered vehicles.

PHARMACEUTICALS AND BIOPHARMACEUTICALS

The U.S. pharmaceutical industry produces medicines through chemical synthesis and biological processes (biopharmaceuticals). Historically the industry has lagged behind the European pharmaceutical sector in innovation, but the phenomenal growth of the U.S. biotechnology industry after the late 1970s catapulted the U.S. into a clear leadership position in the development of new and innovative drugs. The ascent of the U.S. industry has been driven by massive federal support for life sciences R&D, primarily by the

---

[361] Reed presentation, op. cit.

National Institutes of Health (NIH). In addition, the implementation of federal legislation and regulatory policies designed to foster innovation in the 1980s — most notably the enactment of the Bayh-Dole Act of 1980, which enabled universities to own intellectual property rights for technologies they developed through federal funding — enabled a burst of innovative entrepreneurial activity in biotechnology. During the decade of 2001, U.S. firms developed 57 "new chemical entities" (NCEs) compared with 33 by European firms and nine by Japanese firms, erasing the European lead which existed in prior decades.[362]

Despite the spectacular successes of the past two decades, the U.S. pharmaceutical industry's future prospects are uncertain. Many of the blockbuster drugs that drove the industry's success have gone off patent or will do so soon, including first-generation biotechnology drugs, and branded producers face growing competitive pressure from generic drug makers. The costs and risks of developing new drugs and bringing them to market are rising, while the productivity of the industry's R&D appears to be declining. Current research spending by the industry and by NIH is stagnant. The industry faces numerous growing risks, including various kinds of high stakes litigation, regulatory pressure, counterfeit drugs, and the hazards of operating in sometimes disorderly emerging markets, where the industry sees its best growth prospects.

Strategic importance

The pharmaceutical industry produces medicines and vaccines that are essential to the well-being of the U.S. population. Pharmaceuticals ward off epidemics, treat cancer, and cure diseases. The availability of medicine to treat wounded military personnel drastically reduces the death rate from infections in wartime. The history of conflict in the Twentieth Century demonstrates the vulnerability of countries that lack the capability to produce medicine.[363] The ability of U.S. forces to wage war in the Pacific during World War II was strengthened by the development of synthetic alternatives to the anti-malarial quinine, the principal sources of which were controlled by Japan.[364]

---

[362] According to the U.S. Food and Drug Administration, an NCE is a drug that does not contain any active moiety that has already been approved by the FDA pursuant to an application submitted under Section 505(b) of the Federal Food, Drug and Cosmetic Act. NCEs are molecules a company has developed in the drug discovery phase which — assuming it passes clinical trials — could become a drug used to cure diseases and address chronic ailments such as arthritis.

[363] When World War I broke out, Britain was entirely dependent on a hostile power, Germany, for the supply of aspirin, one of the most important painkillers then available. Britain tried but did not succeed in securing adequate alternative sources in Sweden and Switzerland, and "had no alternative but to continue importing German drugs via neutral countries." Corelli Barrett, *The Collapse of British Power* (Marrow, 1922).

[364] Malaria could have a devastating effect on combat effectiveness. During an Australian battalion's retreat from Rabaul, New Britain, in January 1942, 50 out of 252 men died of cerebral malaria because of the lack of quinine. The remaining infected men experienced a long period of

Evolution of the Industry

The modern pharmaceutical industry is descended from small apothecary shops, principally in Germany, which began systematic production of drugs in the mid and late Nineteenth Century.[365] In the United States, during the same era the principal drug companies were wholesaler/producers offering a full line of drugs, many of which were imported from Germany. The embargo of German goods during World War I compelled these companies to enhance their own technical ability to make refinements on existing drug technologies and to develop new drugs.[366] While some sophisticated technological centers devoted to pharmaceutical science arose in the U.S., the European industry led in new drug development through most of the Twentieth Century, and as recently as 1980 eight of the top ten drugs were discovered in Europe.[367]

Economic historian Alfred Chandler observes that during the first quarter of the Twentieth Century a relatively small number of European and American companies developed internal organizational structures permitting the use of science for the systematic discovery, manufacture and commercialization of new drugs. Most current industry leaders trace their origins to these companies, including Pfizer, Merck, Eli Lilly, Squibb and Abbott Laboratories.[368] During the 1940s the so-called "therapeutic revolution," driven initially by the U.S. government's emergency programs to develop antibiotics and synthetic antimalarial drugs, saw a "cascade of discoveries" including antibiotics, antihistamines, tranquilizers, steroids, and new prescription drugs for heart and lung disease, cancer, diabetes and ulcers.[369] Between 1939 and 1957, total U.S. drug sales volume increased by seven fold.[370]

Federal support. Federal support for the development of medicines began in the latter part of the Nineteenth Century. In 1887, a laboratory was created within the Marine Hospital Service (MHS) to pursue bacteriological research into deadly diseases such as cholera and diphtheria. In 1902 Congress

---

recuperation before recovering their fitness. Roy N. MacLeod, *Science and the Pacific War: Science and Survival in the Pacific 1939-45* (Dordrecht: Kluwer Academic Publishers, 2000), p. 54.

[365] These included serum antitoxins and vaccines drawing on the discoveries of Louis Pasteur and Robert Koch in microbiology and immunology, as well as synthetic organic drugs from coal tar, including aspirin, vernal, phenacetin, and Salvarsan, the first cure for syphilis.

[366] Alfred Chandler, *Shaping the Industrial Century: The Remarkable Story of the Evolution of the Modern Chemical and Pharmaceutical Industries.* (Cambridge and Lander: Harvard University Press, 2005) pp. 177-179.

[367] Ross DeVol, Armen Bedsoussian, and BejaminYeo, *The Global Biomedical Industry: Preserving U.S. Leadership* (Milken Institute, 2011). DuPont established an experimental research station near Wilmington early in the 20th Century which evolved into one of the world's first industrial medicine centers. Ibid.

[368] Brian D. Smith, *The Future of Pharma: Evolutionary Threats and Opportunities* (Farakaan, U.K.: Gower Publishing Limited, 2011).

[369] Chandler (2005) op. cit. p. 179.

[370] Smith (2011) p. 34.

renamed MHS the Hygienic Laboratory and delegated to it the authority to regulate the safety of biologics (technologies such as vaccines produced in animals), oversight which the laboratory continued until 1972. In 1930 the Hygienic Laboratory was renamed the National Institute of Health (NIH)[371]. During World War II the institute expanded dramatically and Congress enacted legislation converting existing divisions within the NIH into institutes and centers with topic-specific research and training missions. The National Cancer Institute became part of NIH in 1944. From the 1940s to the 1960s NIH budgets grew substantially, enabling the institute to increase research grants to academic institutions, to construct research infrastructure, and to expand training.[372] The NIH budget also doubled between 1998 and 2003.

In addition to funding research and training in the life sciences, the federal government has contributed to the growth of the pharmaceutical industry through several landmark pieces of legislation. The 1962 Kefauver-Harris Amendment to Food, Drug and Cosmetic Act established the FDA drug approval process in its current form and this well-defined process has become the world "gold standard" for assessing the safety and effectiveness of new drugs, giving U.S.-based companies a major global competitive advantage. The Drug Price Competition and Patent Term Restoration Act of 1984 (the Hatch-Waxman Act) amended U.S. patent laws with respect to drugs to take into account the long time required to bring a new product to market by giving firms a larger period of protection in which to recoup their investments.[373] The Prescription Drug User Fee Act (PDUFA), which became effective in 1992, authorized the FDA to collect fees from drug makers to expedite the drug review process, a measure which substantially reduced the average review time for new medicines.

Emergence of biotechnology. In the early 1970s researchers at Stanford University developed techniques for creating "recombinant DNA" — DNA sequences which combine genetic material from multiple sources.[374] This discovery, coupled with advances in biochemistry, microbiology and enzymology, enabled the emergence of a radical new discipline, molecular biology, addressing the molecular basis of biological processes.[375] Advances in

---

[371] "Institute" became plural after additional institutes were formed and added to NIH.
[372] Annual increases in NIH's budget of 40 percent or more occurred between 1957 and 1963. National Research Council, *Research Training in the Biomedical, Behavioral and Clinical Research Sciences*, Washington, DC.; The National Academies Press, 2011.
[373] The Hatch-Waxman Act protects drug patents for either 20 years from the date of a patent's first filing or 17 years from the patent issued date.
[374] P. Lobban and A. Kaiser, "Enzymatic End-to-End Joining of DNA Molecules," *Journal of Molecular Biology* 78(3) (1973).
[375] Biochemistry is the study of the chemical elements and processes which occur in living organisms. Microbiology is the study of microscopic organisms, which are one cell, a cell cluster or no-cell (acellular) organisms. Enzymology is the study of proteins that increase the rates of chemical reactions.

microbiological understanding, coupled with the development of techniques of genetic engineering, led to the formation and rapid growth of the biotechnology industry. In 1976, Genetic Engineering Technology, Inc. (Genentech) was established to take advantage of advances in large molecule drug development and to commercialize drugs developed with recombinant DNA technology. Its first product was synthetic human insulin (1978). Thereafter the industry grew rapidly, reflecting the fact that biomedicines could address clinical areas that were not reachable with conventional therapeutics, such as oncology and treatment of HIV and autoimmune disorders.

The advent of new learning in the biological sciences was paralleled by the concept of "discovery by design" which emerged from advances in the information industry. Traditional drug development relied on screening large numbers of chemical variants to find one that acted against disease agents. In the 1970s, researchers began applying computational technology, x-ray crystallography and nuclear magnetic resonance to develop hypothetical molecules that could interfere with biochemical sequences in disease agents. These "ideal" molecules were then given to chemists to search for real molecules whose structures most closely matched those of the ideal ones.[376]

Fostering innovation. The rapid advances being made in the biological sciences in the 1970s were paralleled by a public policy debate in the U.S. arising out of a slowdown in U.S. economic and productivity growth. An influential group of economists at the University of Chicago encouraged a reappraisal of the U.S. patent system due to a perceived "anti-patent" bias in the legal system and a "general concern about industrial stagflation and a lack of significant technological innovations."[377] Reflecting changing attitudes in government and the courts, a series of policy measures followed which established the institutional basis for the explosive growth of the U.S. biotechnology industry—

- ***The Bayh-Dole Act of 1980*** provided that universities conducting federally-funded research could own the patents for technologies they developed, opening the door for the commercialization of university-based R&D.

---

[376] Chandler (2005) op. cit. p. 181.
[377] Federal Trade Commission, *To Promote Innovation: the Proper Balance of Competition and Patent Law and Policy* (October 2003). This view was supported by an advisory body established by President Carter to study U.S. innovation policy which found that "diminished patent incentive" was contributing to economic stagnation. Advisory Committee on Industrial Innovation, Industrial Subcommittee for Patent and Information Policy, *Report on Patent Policy* (1979). David M. Hurt, "Antitrust and Technological Innovation," *Issues in Science and Technology* (Winter 1998); William C. Kovacic and Carl Shapiro, "Antitrust Policy: A Century o Economic Thinking," 14 *Journal of Economic Perspectives* (200); Richard A. Posner, "The Chicago School of Antitrust Analysis," 127 *University of Pennsylvania Law Review* (1979).

- A key Supreme Court decision, **Diamond v. Chakabarty**, expanded the scope of patentable technologies to include living organisms, after which the biotechnology industry "virtually exploded."[378]
- The U.S. competition agencies adopted **new antitrust guidelines** which were less hostile to patent monopolies and which substantially broadened the exclusive rights of innovators to exploit their inventions.[379]
- In 1982, Congress created the **Court of Appeals for the Federal Circuit**, giving it exclusive jurisdiction over all federal district court appeals of patent-related decisions, creating an institution which has upheld patent validity with more consistency than previously occurred. Stanford University took advantage of the new Bayh-Dole rules and the new legal environment after the Chakrabarty decision to secure a patent on process technology for genetic engineering that had been developed by Drs. Stanley Cohen and Herbert Boyer, launching a new era of university-industry collaboration in biotechnology. The patent was granted in 1980 and between that year and its expiration in 1997, it was licensed on a non-exclusive basis at relatively low fee levels to 468 companies — many of them fledgling biotech firms — and had been utilized to develop 2,442 new products. This massive transfer of basic enabling genetic engineering process technology "was the real technological foundation for the commercial biotechnology industry."[380]

The advent of innovative biotechnology firms caused some established branded drug producers to seek to sidestep costly early stage drug development through acquisition of biotechnology firms and/or in-licensing of technology from such firms.[381] Biologics like Avastin, Rituxan and Enbrel have

---

[378] "Patenting of a Living Organism," *Patent Home* (February 11, 2012); In Diamond v. Chakrabarty et. Al, 447 U.S. 303 (1980) the Court upheld a patent in a new form of bacterium developed by a microbiologist.
[379] In 1981, the Department of Justice Antitrust Division renounced the so-called "nine No-nos," which set forth fee arrangements and contractual restraints that could not be incorporated in technology licensing arrangements. Beginning in 1988, the Department of Justice issued Antitrust Enforcement Guidelines which commit the competition agencies to apply the rule of reason extensively in intellectual property rights cases, ensuring that antitrust challenges to patents will be subject to extensive antitrust analysis.
[380] Rachel Schurman and Dennis Kelso, *Engineering Trouble: Biotechnology and its Discontents* (Berkeley and Los Angeles: University of California Press, 2003).
[381] "How Big Pharma's New Direction Might Help Little Research Firms," *Chemical Business Newsbase* (November 24, 2009); "Big Drug Groups Urged to Buy in Test Products," *Financial Times* (January 31, 2010).

TABLE 6.2 Branded Pharmaceutical Firms Biotech Acquisitions

Year	Acquiring Firm	Acquired
2009	Roche	Genentech
2009	Johnson & Johnson	Elan Corp
2008	Eli Lilly	ImClone Systems
2009	Johnson & Johnson	Cougar Biotechnology
2009	Sanofi-Aventis	BiPar Sciences
2009	Bristol-Myers Squibb	Medarex
2009	Sanofi-Pasteur	Shantha Biotechnics
2009	Sanofi-Pasteur	Acambis
2008	Johnson & Johnson	Omrix Biopharmaceuticals

SOURCE: "Roche Wins Fight for Genentech," The Express (March 13, 2009); "Johnson & Johnson Completes Deal with Elan, Acquiring its Alzheimers Assets," M2 Equitybytes (September 21, 2009); "Johnson & Johnson Completes Acquisition of Cougar Biotechnology," Datamonitor (July 14, 2009); "Eli Lilly Completes $6 Billion Acquisition of ImClone Systems," Financialwire (November 25, 2008); "Pharma Japan: Sanofi-Aventis too Acquire BiPar Sciences," Chemical Business NewsBase (April 27, 2009); "Bristol-Meyers Squibb Completes Acquisition of Medarex, Inc.", Chemical Business NewsBase (September 1, 2009); "Sanofi Pasteur's Shantha a Shot in the Arm for Indian Pharma," Financial Express (August 2, 2009); "Sanofi-Pasteur Acquires Acambis for GBP 285 Million," Datamonitor (September 26, 2008); "Johnson & Johnson Completes Acquisition of Omrix Biopharmaceuticals Inc.," Chemical Business NewsBase (December 30, 2008).

demonstrated "blockbuster" potential and fueled a mergers-and-acquisitions wave. Perhaps the most dramatic acquisition has been Roche Holding AG's acquisition of Genentech, the first biotechnology start-up in the United States, in 2009. Roche was previously a pharmaceutical company, and now, following the purchase of Genentech, it looks more like a biopharmaceutical company."[382] Eli Lilly's 2008 acquisition of biopharmaceutical producer InClone Systems, a maker of oncology drugs, in 2008 transformed Lilly into the world's fifth largest biotech firm, with biologic drugs comprising over half its pipeline.[383]

The U.S. achieves global leadership. As early as 1980, eight out of the world's top ten drugs had been discovered in Europe. But advances in U.S. science, the rapid growth of the U.S. biotechnology industry and the sweeping changes in U.S. legal and regulatory structures which began in the late 1970s

---

[382] "Pharmaceutical Companies Seek Biotech Acquisitions to Boost Drug Pipelines," *ICIS.com* (February 12, 2010).
[383] IMAP, *Pharmaceuticals & Biotech Industry Global Report — 2011* p. 10.

TABLE 6.3

Decade	Percent Total NCES by Headquarter of Inventing Firm		
	U.S.	Europe	Japan
1971-80	31	54	15
1981-90	32	40	29
1991-2000	42	49	9
2001-2010	57	33	9

SOURCE: Ross C. DeVol, Armen Bedroussian and Benjamin Yeo, "The Global Biomedical Industry: Preserving U.S. Leadership" (Milken Institute, 2011).

gave rise to an environment that was more conducive to innovation than was the case in Europe. Small firms and universities in Europe experienced difficulty in commercializing new drugs, at the same time that policy reforms was opening up opportunities for such entities in the United States. By the decade of 2001-10 the United States had reversed a prior European lead in the production of wholly-innovative "new chemical entities."[384]

Current impact of federal policies. The federal government has played a major role in the development of the U.S. pharmaceutical industry. During World War II government scientists developed techniques for producing penicillin efficiently and transferred technology to a handful of small companies which arguably enabled them to emerge as major, research-intensive manufacturers in the 1940s.[385] Since the war, federal institutions have conducted or funded much of basic research necessary for the development of new drugs, with the private sector then conducting the applied research and development necessary to bring new medicines to the market. A recent study by researchers from the National Institutes of Health (NIH) and Boston University estimated that between 1990 and 2007, public sector research institutions (PSRIs) – most of which are themselves federal entities or federally funded – contributed up to 21.2 percent of all products in new drug applications, including "virtually all of the important, innovative vaccines that have been introduced in the past 25 years." These organizations also tended to "discover drugs that are expected to have disproportionately important clinical effect."[386]

By far the most important federal research organization is the National Institutes of Health (NIH), a part of the U.S. Department of Health and Human Services. Roughly 10 percent of its budget supports research in NIH's own

---

[384] DeVol, et. Al., *Global Biomedical Industry* (2011) op. cit. p. 19.
[385] These firms included Roche, Abbott Laboratories, Merck, Squibb, Pfizer, Parke David, Eli Lilly, Lederle, Winthrop and Upjohn. Peter Yourkin, "Making the Market: Howe the American Pharmaceutical Industry Transformed itself During the 1940s," (University of California at Berkeley, November 2008).
[386] "U.S. Public Research 'Responsible for Many Major New Drugs,'" *Pharma Times* (February 18, 2011).

laboratories, which are staffed by around 6,000 scientists supporting the NIH Intramural Research Program (IRP). The IRP is the largest medical research organization in the world. 80 percent of the research budget is awarded via the NIH Extramural Research Program in the form of about 50,000 competitive grants to over 300,000 researchers at universities, medical schools and research organizations in the U.S. and abroad. NIH awards research grants to small pharmaceutical businesses for the development of new drugs through the Small Business Innovation Research (SBIR) program.[387]

In addition to research for the explicit purpose of developing new pharmaceuticals, much of NIH's research spending supports basic research on the mechanisms of disease and augments the private sectors' own research efforts. NIH funding supports graduate students and postdoctoral researchers at U.S. universities, and helps train researchers who are eventually hired by drug companies.[388]

In the 1970s the NIH Deputy Director for Science, Dr. DeWitt Stetter, chaired a national committee of scientists to develop guidelines for the emerging research in recombinant DNA, which transformed the manner in which scientists study diseases. On the basis of new legislation NIH spearheaded major research efforts against cancer and heart disease.[389] In the late 1980s NIH launched the Human Genome Project to map and sequence the entire set of human genes. To date over 80 Nobel Prizes have been awarded for NIH-sponsored research, which has led to cures for some forms of cancer, a substantial reduction in the occurrence of heart attacks and strokes, and the development of drugs targeting proteins involved in some disease processes.[390]

---

[387] "Winston Pharmaceuticals, Inc. Receives SBIR Grant from the NIH to Investigate Treatment for Postherpetic Neuralgia of the Trigeminal Nerve," *Business Wire* (April 9, 2012); "Achillion Pharmaceuticals Receives Phase I SBIR Grant from NIH," *Datamonitor* (March 19, 2010). See also National Research Council. VENTURE FUNDING AND THE NIH SBIR PROGRAM. Washington, DC: The National Academies Press, 2009.

[388] In a 2006 study the Congressional Budget Office warned that the sheer scale of the federal investment in life sciences research could "crowd out private sector investment." CBO noted that federal spending tended to be directed toward basic research while the private sector concentrated on applied research and development. However, "the distinction between basic and applied research is not well defined, and the division of labor between the two has become less pronounced as the potential commercial value of basic life sciences research has become more widely recognized." Federal crowding out in research could also occur via competition between the federal government and the private sector with respect to the supply of labor, which could drive costs upward. CBO, "Research and Development in the Pharmaceutical Industry," October 2006, p. 4. CBO, "Research and Development in the Pharmaceutical Industry," October 2006, p. 3.

[389] The legislation was National Cancer Act of 1971 and National Heart, Blood Vessel, Lung and Blood Act. The Cancer Act created 15 specialized research, training and demonstration centers. The heart legislation mandated expanded research directed at heart disease, including high blood pressure, stroke, high cholesterol levels, and blood diseases such as sickle cell anemia. NIH, A Short History of the National Institutes of Health, http://history.nih.gov/exhibits/history/docs/page_09.html>.

[390] Ibid.

NIH's annual budgets have been flat since 2009, although it received $10 billion in one-off funding for short term stimulus in 2009-10.[391] Its 2012 budget of $30.9 billion was only slightly more than the 2009 level of $30.5 billion.[392] Critics charge that the lack of growth in NIH funding will inhibit innovation. The president of the nonprofit group Research America said in 2012 that "we strongly believe a frozen budget for the NIH will flat line medical breakthroughs in the coming years and stifle the business and job creation that begins with R&D. . . . Researchers will leave the field, potential breakthroughs will be shelved and new business opportunities grounded in medical discovery will evaporate as research institutions grapple with learner budgets."[393]

The Human Genome Project. In 1990, the U.S. government launched a $3 billion dollar project to identify and map the genes of the human genome and determine the sequence of chemical base pairs that comprise DNA. The Human Genome Project (HGP) was jointly administered by the Department of Energy and NIH.[394] This effort was paralleled by a privately funded project undertaken by the company Celera Genomics, which sought to patent several hundred genes.[395] The competition between the public and private efforts drove the effort forward more rapidly than anticipated in the original timetable and resulted in midcourse adjustments in strategy by both sides. In 2001, the HGP and Celera published drafts of their results, including analysis of sequences covering around 83 percent of the human genome.[396] The data generated by HGP was deposited in the GenBank sequence database, which is managed by the National Center for Biotechnology Information, which is part of NIH. This data is available to any biomedical scientist in the world.[397] The data generated

---

[391] The American Recovery and Reinvestment Act (ARRA), enacted in 2009, made available to NIH $10.4 billion in funding for use in 2009 and 2010. Of this, $8.2 billion was used to support scientific research priorities and most of the remainder was used to upgrade infrastructure. NIH, "NIH's role in the American Recovery and Reinvestment Act (ARRA), www.nih.gov/about/director/02252009statement_arra.htm.
[392] NIH, History of Congressional Appropriations, Fiscal Years 2000-2012.
[393] "Pharma Blasts Obama Budget, FDA to Get $4.49 Billion," *Pharma Times*, February 14, 2012.
[394] DOE's National Laboratories concentrated on developing technologies for mapping, sequencing and informatics. Seven NIH centers were involved in the project, and scores of smaller research projects were funded by NIH to undertake gene mapping and sequencing research directed at single-disease-associated genes. National Research Council, *Large-Scale Biomedical Science* (Washington, DC: The National Academies Press, 2005), p. 35.
[395] In 2000, President Clinton indicated that the human genome sequence could not be patented.
[396] "Initial Sequencing and Analysis of the Human Genome," *Nature* (February 15, 2001); "The Sequence of the Human Genome," *Science* (February 16, 2001).
[397] GenBank is part of an international effort to pool and share data on the human genome, the International Nucleotide Sequence Database (INSDC) which includes GenBank, the DNA Data Bank of Japan, and the European Molecular Biology Laboratory. New data on nucleotide sequences contributed from laboratories around the world are incorporated in the databases in a coordinated manner on a daily basis.

by the genome projects is expected to produce major benefits in medicine and biotechnology.[398]

The genome projects have given rise to the field of "omics," involving application of the new knowledge about genes, proteins and other molecular characteristics of living organisms to detect disease, predict how individuals will react to drugs, and eventually to develop treatments. The patentability of various forms of DNA remains murky.[399] However, a decade after the HGP and Celera published their drafts, few if any new medicines have been developed based on genomic knowledge, or "pharmagenomics."[400] Premature use of "omics"-based clinical tests at Duke University, and alleged improper alteration of data, has led the Institute of Medicine of the National Academies to establish a committee to develop recommendations for strengthening omics-based research.[401]

Translational research. NIH devotes substantial effort toward translational research, that is, the translation of scientific ideas into practical application at the clinical level. In 2006, NIH established the Clinical and Translational Science Awards (CTSA) Consortium, which has grown to 60 linked medical research institutions dedicated to developing the discipline of clinical and translational science. This program aims to develop "a cadre of well-trained multi-and inter-disciplinary" research teams and investigators, to create an incubator for innovative research tools and information technologies, and to combine multi-disciplinary and inter-disciplinary knowledge and techniques for application in a clinical context. CTSA-sponsored programs create teams which may include biologists, basic scientists, pharmacists, geneticists, biomedical engineers and other specialists in "bench-to-bedside"

---

[398] Because of DNA's key role in cellular processes, the detailed information generated by the genome projects is expected to foster major advances in medicine in areas such as cancer and Alzheimer's disease.
[399] Notwithstanding the Supreme Court's 1980 decision in Diamond v. Chakrabarty that new life forms could be patented, a federal district court in New York ruled in 2010 that isolated DNA gene sequences were not patentable. The Department of Justice has taken the position that isolated but otherwise unaltered genomic DNA is not patentable subject matter. "Gene Sequence Patents are Being Questioned," *Michigan Lawyers Weekly* (June 27, 2011).
[400] "Cancer, Diabetes, Dementia and Cystic Fibrosis: Having the Genome Has Not Meant an End to These Afflictions," *Irish Times* (February 25, 2011). Cytrix Pharmaceutics, Inc. is reportedly raising venture capital to pursue novel drug-based cancer therapeutics based on data from the Human Genome Project which revealed that various forms of extra-hepatic cytochrome P450s are "over-expressed" during the malignant progression of most cancer. "Cyterix Pharmaceuticals Raises $9.2M in a Series A Venture Financing," *Chemical Business Newsbase* (June 7, 2011).
[401] Institute of Medicine, "Evolution of Translational Omics: Lessons Learned and the Path Forward," (Report brief, March 2012); "Panel Calls for Closer Oversight of Biomarker____ Tests," *Science Insider* (March 23, 2012).

developmental efforts which include designating technologies for licensing and commercialization.[402]

In December 2011, Congress created the National Center for Advancing Translational Sciences (NCATS) under the supervision of NIH to accelerate the development of new medicines. NCAT's a 2012 budget is $574.7 million. Its intended role has been compared with that of a "home seller who spruces up properties to attract buyers in a down market." The idea behind the center is for NIH researchers to evaluate novel drugs and to develop leads with respect to promising compounds — work traditionally done by the private sector.[403] NCAT will perform "as much research as it needs to do so that it can attract drug company investment." NCAT will screen chemicals for potential use in medicines, perform animal tests and conduct some human tests – activities which have "traditionally been done by drug companies, not the government." Existing translational research programs under way at other NIH organizations will be transferred to NCATS.[404] The creation of NCATS reflects governmental frustration over industry's reluctance to "follow the latest genetic advances with expensive clinical trials."[405]

Development of orphan drugs. In 1983 Congress enacted the Orphan Drug Act to promote the development and commercialization of "orphan drugs" – medicines to treat rare diseases. The Act relaxes certain requirements in the regulatory development path for new drugs, provides for enhanced patent protection, and authorizes tax incentives and subsidies. Most importantly the Act provides for seven years of market exclusivity that is independent of the drug's patent status and which does not begin to run until FDA approval is granted. The result was an exponential increase in the development of orphan drugs which, while having a relatively small market, could be sold at high

---

[402] "US $23 Million Grant Makes Cincinnati University CTSA Members" *Pharma Times* (April 8, 2009).
[403] "New $1b NIH Center Will Tackle Early-Stage Drug Development to Ease Industry Risk of Failure," *Centerwatch* (February 7, 2011). "Increased Funding for NIH: A Biomedical Science Perspective," Life Sciences Forum.
[404] R&D programs will be transferred from NIH's National Human Gerome Research Institute, National Center for Research Resources, and the NIH Director's Common Fund "US Govt Drug Research Agency 'To Start Work in October,'" *Pharma Times* (January 25, 2011).
[405] "Federal Research Center Will Help Develop Medicines," *New York Times* (January 22, 2011).

prices.[406] A few orphan drugs have become blockbuster drugs.[407] As of early 2011, pharmaceutical companies had 460 orphan drugs under development.[408]

In 2003 NIH launched the Rare Diseases clinical Research Network (RDCRN) to promote research on rare diseases. In 2009 NIH awarded $117 million over a five year period to 19 research consortia and a data management center to fund research into the natural history, epidemiology, diagnosis and treatment of over 95 rare diseases (defined as affecting less than 200,000 people in the U.S.). RDCRN has enrolled thousands of patents for clinical studies and established an extensive data management system.[409] In 2009 NIH invested $24 million in the Therapeutics for Rare and Neglected Diseases (TRND) program, which develops research collaborations with universities working on rare diseases. The NIH Director observed that—

> *The federal government may be the only institution that can take the financial risks needed to jump-start the development of treatments for these diseases, and NIH clearly has the capability to do the work.*[410]

Stem cell research restrictions. Some federal policies have hampered critical research. In 1995, the so-called Dickey-Wicker Amendment was attached as a rider to an appropriations bill that was passed by Congress, prohibiting the use of appropriated funds for the creation of human embryos for research purposes or research in which embryos are destroyed.[411] In 2001, President George W. Bush issued an executive order that prohibited NIH from funding research on embryonic stem cells beyond using the 60 cell lines which then existed. He subsequently vetoed a number of bills which would reduce limitations on federally-funded research on embryonic stem cells. In 2009, President Barak Obama signed an executive order lifting the ban and a memorandum establishing more independence for federal science program. The President commented that "in recent years, when it comes to stem cell research,

---

[406] In the 1970s, prior to enactment of the Orphan Drug Act, fewer than 10 orphan drugs were approved by the FDA. Between the effective date of the Act and the beginning of 2011, the FDA approved over 350 orphan drugs. "U.S. Pharma's 'Record' 460 Drugs in Development for Rare Diseases," *Pharma Times* (February 28, 2011).

[407] Vioxx, Botox, Cialis, Provigil and Abilify are orphan drugs. Olivier Wellman-Laback and Youwen Zhou, "The U.S. Orphan Drug Act: Rare Disease Research Stimulator or Commercial Opportunity?" *Health Policy* (May 2010).

[408] "U.S. Pharma's 'Record' 460 Drugs in Development for Rare Diseases," *Pharma Times* (February 28 ,2011).

[409] "NIH Award US $117 Million to Rare Disease Consortia," *Pharma Times* (October 7, 2009).

[410] "NIH to Create Development Pipeline for Rare, Neglected Diseases," *Pharma Times* (May 25, 2009).

[411] The Dickey Amendment language was added to subsequent appropriations on a yearly basis. The Dickey Amendment was an impediment to researchers seeking to create their own stem cell lines.

rather than furthering discovery, our government has forced what I believe is a false choice between sound science and moral values."[412] The Dickey-Wicker Amendment remained in force, however, and provided the basis for an unsuccessful legal challenge to NIH guidelines which permitted federal funding of research projects using embryonic stem cells but not for the destruction of embryos.[413] Dismissal of the case in 2011 was seen as a decisive victory for NIH, but as one stem cell researcher at Harvard Medical School put it, "I hope we're done for now, but nothing surprises me anymore."[414]

The politicization of stem cell research has hampered the development of stem cell-based therapies in the U.S. Other countries which encourage stem cell research have captured R&D activity that otherwise probably would have taken place in the United States.[415] Korea, not the U.S., introduced the world's first stem cell-based medication, a drug developed by a domestic bio-venture company, FCB-Pharmcell, to help regenerate damaged coronary arteries.[416]

Challenges

While the U.S. pharmaceutical sector currently leads the world in innovation, the industry faces daunting challenges in maintaining its position. The costs and risks of developing new drugs are increasing, and have become so substantial that many major, traditionally innovative companies are cutting or not increasing their R&D spending. While the innovation crisis is the most serious problem confronting the industry, it faces other significant risk factors, including high stakes civil and criminal litigation, compulsory licensing by foreign governments, counterfeiting, and an increasingly complex and hazardous global supply chain.

The innovation crisis. The pharmaceutical industry "has plunged ever deeper into a crisis that threatens to turn off the tap of all new medicines."[417] The traditional innovation model of the large U.S. pharmaceutical firms

---

[412] "Obama Overturns Bush Policy on Stem Cells," *CNN Politics* (March 9, 2009).
[413] "Obama's Stem Cell Policy Hasn't Reversed Legislative Restrictions," *Fox News* (March 14, 2009).
[414] Stem Court Ruling a Decisive Win for NIH," *Science* (July 27, 2011).
[415] In 2006, Singapore established a $45 million consortium for stem cell research headed by an American scientist Roger Pederson.
[416] This development would represent a recovery from a 2005 incident in which an eminent scientist, Huang Wo-suk, was found by a review board to have manipulated key stem cell research data. Two more stem cell-based medications were approved in Korea in 2012: Cartistem, which uses stem cells to regenerate knee cartilage, and Cupistem, which uses stem cells to treat anal fistula occurring as a result of Crohn's disease, an inflammatory bowel disease. "Major Stem Cell Medication Given Green Light," *Chosun Ilbo* (January 20, 2012). "Korea Set to Approve World's First Stem Cell Drug," *The Korea Herald* (June 24, 2011).
[417] "Why the Gene Revolution Has Been Postponed — It Costs $1bn to Develop a New Drug, so Don't Expect Personalized Treatments, But the Genome Project is Still Worthwhile," *The Times* (London, August 25, 2011).

emphasizes the pursuit of proprietary "blockbuster" drugs which generate $500 million to $1 billion or more in annual revenues. Patents on these drugs last 20 years, and given that the time frame from patent filing to market is seven to ten years, the patent holder typically enjoys a legal monopoly on the drug for 10 to 13 years, which can result in huge profits during the protected period. However, upon expiration of the patent, the drugs come under intense competitive pressure from generic drug makers, eroding if not eliminating the profit margins achieved under patent. An obvious response to the expiration of drug patents is innovation — develop new blockbuster medicines that can be patented and offset the effect of the drugs going off patent. However, the development of new drugs by the pharmaceutical industry appears to be slowing down, and the looming expiration of patents will leave some drug companies with no clear replacement with equivalent profit potential. Between 2012 and 2014, pharmaceutical firms will lose patent exclusivity on over 110 products in the United States.[418] In some cases generic versions of proprietary drugs appear on the market on "day one of patent expiry."[419] A 2009 study by the Congressional Budget Office observed that—

> *[T]he patents for many top-selling drugs have expired, subjecting them to competition from cheaper generic compounds. The resulting decline in spending on those drugs has not been fully offset by added spending or new brand-name drugs because, at the same time, the rate at which new drugs are being introduced has slowed substantially.*[420]

The FDA approval process. Historically FDA approval process has worked to the advantage of U.S. firms, reflecting its comparative efficiency relative to regulatory regimes in Europe and elsewhere. But this edge is eroding, with the approval process becoming more protracted, uncertain, and costly for drug developers.

Development of new drugs entails a long time frame between initial discovery and actual commercialization, and most new drugs never become products. During the pre-discovery phase, of every 5-10,000 compounds tested, roughly 250-500 are identified as promising. These are subjected to preclinical testing to identify a "lead molecule" capable of altering the course of a disease or condition, the compounds are tested for safety and effectiveness, and redesigned variations are pursued. The combined pre-discovery and preclinical

---

[418] "Drug R&D Spending Fell in 2010, and Heading Lower," *Reuters* (June 26, 2011).
[419] "Teva Fast Out of the Blocks to Sell Generic Seroquel," *Pharma Times* (March 27, 2012).
[420] CBO, "Pharmaceutical R&D and the Evolving Market for Prescription Drugs," October 26, 2009.

phases take 3-6 years.[421] Before a new drug is approved, it must undergo clinical trials in which its potential benefits and risks are assessed based on tests using human volunteers. Of every 250-500 compounds subject to pre-clinical testing, about 5-10 are ultimately submitted to clinical trials. Clinical trials typically take 6-7 years to complete and involve thousands of people in three research phases—

- *Phase I trials*. Phase I trials are usually performed with healthy volunteers and are intended to ascertain whether a new drug is safe (20-100 volunteers).
- *Phase II trials*. Phase II trials involve assessing a drug's effectiveness using volunteers who actually have the condition or illness the drug is intended to address and identifying common short-term side effects (100 to 500 volunteers).
- *Phase III trials*. The final and largest phase of trials involved testing the drug on a large population to generate data with respect to safety and effectiveness of the drug (1,000 to 5,000 volunteers often at multiple sites).

When clinical trials are completed, the drug is reviewed by the Food and Drug Administration and either approved or disapproved. With approval, the pharmaceutical company can begin investment in production capability. On average, of every 5,000 compounds that are examined in the preclinical phase, only one becomes an FDA-approved commercial product.

In recent years, the FDA has established more stringent requirements for clinical trials, making them more time-consuming and costly. In 2008, Congress heard testimony that the FDA was "barely hanging on by its fingertips" and that it suffered from a shortage of scientists who understood the newest technologies, inability to speed the development of new drugs, and an information technology infrastructure that was a pervasive source of risk.[422] The fact that the FDA is underfunded "is a consensus held by patient organizations, consumer and research groups, the professional community, and all the industries regulated by the FDA."[423]

Clinical trials are becoming more complex and the failure rate for drugs entering clinical trials is growing rapidly. Between 2000-2003 and 2004-2007, the median number of procedures per trial increased by 49 percent and the work

---

[421] Pharmaceutical Research and Manufacturers of America, www.innovation.org/index.cfm/InsideDrugDiscovery/Inside_Drug_Discovery> visited April 24, 2012.
[422] "U.S. Congress Warned of 'Gatering Storm' at FDA," *Pharma Times* (February 8, 2008).
[423] "FDA Has Critical Budget Shortfall," *AJC* (February 16, 2010). The FDA's Science Committee recently reported that the FDA had to bring back retired computer experts to repair its computers, which so obsolete that younger repairmen did not know how to fix them. Ibid.

burden per protocol increased by 54 percent.[424] Increasing complexity results in stricter eligibility criteria for volunteers, which has translated into declining volunteer enrollment and retention rates. Over 50 Phase III trials were terminated in 2010 and the number of drugs entering Phase III fell by 55% from the prior year. Phase I and Phase II trials also fell by 47% and 53%, respectively, in 2010.[425]

U.S. clinical trials' lengthening time frame and complexity is a factor underlying U.S. firms' increasing resort to clinical trials in Asia and Central and Eastern Europe, which enable them to reduce costs and accelerate time-to-market.[426] South Korea, for example, is emerging as a global center for clinical trials for new drugs, reflecting the government's sustained efforts to establish an excellent infrastructure for clinical tests in the nation's hospitals. In 2004, multinational corporations sponsored 61 clinical trials in Korea, a figure which surged to 216 in 2008. The President of Bayer Korea, Friedrich Gause, commented in 2010 that—

> *These clinical trials provide enormous benefits to Korea. They benefit Korean patients, Korean medical institutions and clinical experts, and the Korea economy in general ... [T]hey offer immediate access to innovative treatments to patients involved in Phase II and III global trial program.*[427]

Stagnant R&D spending. The average cost of researching and developing a successful drug is estimated by the U.S. pharmaceutical industry at $800 million to $1 billion. These figures include the costs associated with thousands of failures.[428] Because the industry does not make detailed R&D investment data available, these figures cannot be independently assessed, and some estimates place the average development cost of a new drug at much lower levels. However, even at reduced levels the cost of R&D is substantial.[429]

The U.S. pharmaceutical industry's trade association, PhRMA, takes the position that U.S. biopharmaceutical company R&D "remains strong."[430]

---

[424] PhRMA, *2011 Profile*, p. 13.
[425] Stephanie Sutton, "The Status of Pharma R&D," *BioPharm* (July 5, 2011).
[426] "Looking Abroad: Clinical Drug Trials," *Food and Drug Law Journal* (2008), p. 673; "Novartis Stays Ahead with New Ideas: Country Head Says Dedication," *The Korea Herald* (March 31, 2004).
[427] "Korea Emerging as Global Trial Hub," *The Korea Herald Online* (May 26, 2010).
[428] The source of these figures is the Pharmaceutical Research and Manufacturers of America (PhRMA) the trade association representing U.S. research-based pharmaceutical and biotechnology companies.
[429] A study published in BioSocieties journal in 2011 calculated the cost of R&D for a new drug at "a controversially low $75m." "Cost of New Drug Development Remains High," *Evaluate Pharma* (March 10, 2011).
[430] PhRMA, *2011 Profile*, p. 11.

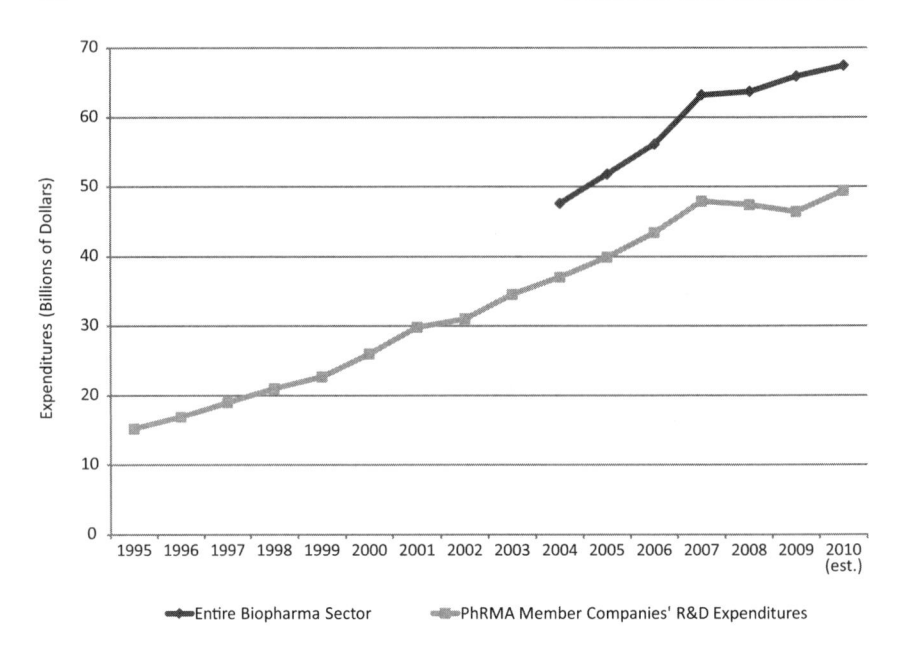

FIGURE 6.10 Total biopharmaceutical company R&D and PhRMA member R&D: 1995-2010.
SOURCE: PhRMA, *2011 Profile*, p. 11

But PhRMA's own figures indicate that R&D spending by its members has been stagnant since 2007 and actually declined in absolute dollars in 2008 and 2009.

The rate of introduction of "priority drugs" – defined by the FDA as drugs that represent a "significant therapeutic or public health advance" – has dropped from an average rate of over 13 per year in the 1990s to about 10 a year in the 2000s.[431] According to a number of analysts, the stagnation in pharmaceutical R&D spending reflects disillusionment with the shrinking returns on R&D investment.[432] Evaluate Pharma, a London-based research firm, calculated in 2011 that the pharmaceutical industry was spending $57 billion per year more on R&D than the value of the new products it was launching, and concluded that "the industry as a whole is not yet generating a return on R&D investment."[433] Pfizer, the world's largest pharmaceutical firm, will cut its R&D

---

[431] CBO, "Pharmaceutical R&D and the Evolving Market for Prescription Drugs," October 26, 2009.
[432] "Drug R&D Spending Fell in 2010, and Heading Lower," *Reuters* (June 26, 2011); Stephanie Sutton, "The Status of Pharma R&D," *BioPharm* (July 5, 2011); Ben Hirschler, "Analysis: Big Pharma Strips Down Broken R&D Engine," *Reuters* (May 11 2011).
[433] "R&D Spending Soars Above Value of New Drugs," *Indianapolis Business Journal* (July 5, 2011).

budget by about 25 percent during the period 2011-2013.[434] Eli Lilly CEO John Lechleiter commented in 2011 that "Our industry [R&D] is taking too long, we're spending too much, and we're producing far too little."[435] Chris Viebacher, CEO of Sanofi, observed in 2011 that—

> *Five years ago people would say the more I spend on R&D, the more shots in the goal I will have, the more successful I will be. Now you have got some investors out there who believe that what we do in R&D is actually value destroying.*[436]

Patent litigation. The profit sanctuary represented by proprietary drugs is undergoing pressure in the courts. Generic drug manufacturers have been challenging branded pharmaceutical companies' patents aggressively since 2000, when Barr Laboratories broke Eli Lilly's patent on Prozac. By the end of 2008 Lilly was engaged in litigation to protect drugs which collectively represented half of its revenues.[437] In 2009 Johnson & Johnson won a $1.67 billion award against Abbott Laboratories based on rival claims for the two firms' rheumatoid arthritis drugs.[438] Under such circumstances a pharmaceutical firm's very survival is linked to the outcome of patent litigation.

Antitrust risk. What is widely perceived as the high cost of proprietary medicines, coupled with the monopoly associated with patent rights, gives rise to an abiding risk of antitrust action, both formal and informal, against proprietary drug makers in the U.S. and a number of other key countries. In 2009, the U.S. Federal Trade Commission scuttled a proposed $3.1 billion acquisition by Australia's CSL of U.S.-based Talecris Biotherapeutics on the grounds that the acquisition would "hasten the market's path toward cartelization."[439]

A particular area of antitrust vulnerability is the so-called reverse payments or "pay for delay" agreements pursuant to which branded drug makers pay generics manufacturers to delay their market entry upon expiration of the branded firms' patents, giving the latter an additional interval of comparatively high-priced sales. In the European Union, a competition policy authorities have initiated investigations against branded pharmaceutical makers such as Johnson

---

[434] Pfizer announced plans in February 2011 to close an R&D facility in the United Kingdom employing 2,400 people. "Pfizer to Close UK Research Site," *BBC News* (February 1, 2011).
[435] "The World's Biggest R&D Spenders," *Fierce Biotech* (March 8, 2011).
[436] "Analysis: Big Pharma Strips Down Broken R&D Engine," *Reuters* (May 11, 2011).
[437] "Generic Meds Don't Come Cheap," *Indianapolis Business Journal* (December 15, 2008); "Patent Battles Could Savage Drug Giant," *The Independent on Sunday* (March 18, 2007).
[438] "J&J Wins $1.67 Billion Lawsuit Against Abbott," *Modern Healthcare* (June 30, 2009).
[439] "Obama Administration Plans to Take More Regulatory Approach on Healthcare Mergers," *Modern Healthcare* (June 24, 2009).

& Johnson, Pfizer, GlaxoSmithKline and AstraZenica on suspicion of conspiracy to maintain the prices of their drugs after the patents expired.[440] In the U.S. the FTC has denounced pay-for-delay deals for over a decade, although its challenges to such agreements in the courts have thus far proven unsuccessful.[441] The Obama Administration has proposed legislation banning pay-for-delay agreements.[442] Senator Kohl and Grassley are backing bipartisan legislation to prohibit pay-for-delay deals.[443]

White collar prosecutions. The marketing of pharmaceutical products entails substantial legal risks. A manufacturer which touts the therapeutic benefits of a drug may be severely penalized for "fraud."[444] The massive outlays of public money for health care, combined with complicated and opaque payment systems creates opportunities and motivation for individuals employed by drug companies to engage in kickback schemes and other practices which are prohibited by law. These realities expose pharmaceutical companies to penalties which can be staggering.[445] If anything the risks to industry are increasing, given the Obama administration's stated intention of attacking health care fraud through more aggressive prosecutions and deployment of advanced monitoring technology.[446]

---

[440] In October, 2011 EU competition authorities disclosed that they had opened an investigation into pay-for-delay arrangements between Johnson & Johnson and the generic branches of the Swiss-based company Novartis. EU Competition Commissioner Joaquin Alumunia said that "paying a competitor to stay out of the market is a restriction of competition that the Commission will not tolerate." He said that with respect to this issue, the Commission "has been firmly on the sector's back for the last couple of years." "EU Antitrust Authorities Probe Johnson & Johnson, Novartis," *Agence France-Presse* (October 21, 2011). "Drug Companies Trigger European Ire for Holding Back Supplies of Cheap Medicine," *The Times* (London, November 29, 2008); "EU Says Investigation Raid Pharma Giants," *Agence France-Presse* (October 6, 2009).

[441] "FTC Loses Bid to Block Pay-for-Delay Drug Settlements," *Thomson Reuters News & Insight* (April 25, 2012); FTC, *Pay-for-Delay: How Drug Company Pay-Offs Cost Consumers Billions* (FTC Staff Study, January 2010).

[442] "Obama Seeks $135B Drug Price Cuts Over 10 Years," *Pharma Times* (September 23, 2011).

[443] "U.S. CBO Doubles Estimated Savings from Pay-for-Delay Ban," *Pharma Times* (November 13, 2011).

[444] In April 2012 Johnson and Johnson and a subsidiary were ordered to pay over $1.2 billion in fines after an Arkansas jury concluded that they had minimized or concealed damages associated with the antipsychotic drug Risperdal when marketing it. "J&J Fined $1.2 Billion in Drug Case," *New York Times* (April 11, 2012).

[445] Pfizer was fined $2.3 billion in September 2009 in the settlement of charges that it had promoted use of its drugs for purposes not approved by the FDA, and for entertaining doctors as an inducement to prescribe the drugs. Eli Lilly agreed to a $1.4 billion fine to settle federal criminal and civil charges to the effect that it had illegally promoted the sale of Zyprexa, an antipsychotic medication. "Officials: Pfizer to Pay Record $2.3 B Penalty," *Forbes* (September 3, 2009); "Eli Lilly Owes $1.4B Over Off Label Use," *CBS News* (February 11, 2009).

[446] "Making Them Pay," *Modern Healthcare* (October 12, 2009); "Attorney General Holder and HHS Secretary Sebelius Announce New Interagency Health Care Fraud Prevention and Enforcement Action Team," Department of Justice Press Release (May 21, 2009).

TABLE 6.4 Class Action Lawsuits Against Life Sciences Firms on Behalf of Consumers Claiming Injury

Defendant	Product	Allegation
Advanced Medical Optics, AMO Canada Company	COMPLETE contact lens solution	Solution caused serious eye infections, ancanthamoeba keratitis
Baxter International	Heparin	Tainted drug caused several deaths
Pfizer	Trovan, Rocephin	Eleven Nigerian children died after being given these drugs in a human trial
Novartis	Zelnorm	Increase in cardiovascular events by users of the drug
Merck	Vioxx	Heart attacks attributable to drug use
Hoffman LaRoche	Accutane	Several, chronic stomach injuries
Pfizer	Bextra, Celebex	Increase in cardiovascular events by users of the drug

SOURCE: "Advanced Medical Optics Sued Over Lens Solution," OCRegister.com (June 5, 2007); "Baxter Loses First Heparin Lawsuit," Pharmalot (June 10, 2011); "Pfizer to Pay $75 Million to Settle Nigerian Trovan Drug-Testing Suit," Washington Post (July 31, 2009); "Novartis in Tentative Pact to Settle Zelnorm Lawsuits" Drug-Injury.com (July 15, 2010); "Jury: Merck Negligent," CNN (August 22, 2005); "First Acculane Verdict Yields $2.6 M in NJ Superior Court," Lawyers USA (July 2, 2007); "Pfizer Reaches Massive Settlement in Celebex, Bextra Lawsuit," Huffingtonpost.com (October 17, 2008).

Class action lawsuits. Consumers who believe that they have been injured by a pharmaceutical product can bring a product liability lawsuit against the manufacturer for damages. Many of these suits are class actions involving massive damage claims. While these product liability lawsuits can make a valuable contribution to protecting consumer interests, they also represent a significant business risk for pharmaceutical manufactures. The relationship of product liability lawsuits to both drug innovation and drug safety/effectiveness involves complex matters that the committee has not had an opportunity to examine in detail.

Stock prices of life sciences firms are frequently volatile and can be affected by disclosure (whether or not authorized) of the results of clinical trials and FDA proceedings associated with approval of a promising new drug. Allegedly misleading disclosure or nondisclosure of problems can result in volatility in the share prices of a company's stock. The collapse of share prices under such circumstances commonly gives rise to costly class action lawsuits.[447] Companies may also face enforcement proceedings by the Securities and Exchange Commission.[448]

Compulsory Licensing. Pharmaceutical companies with proprietary drugs are under price pressure from many governments outside the United States, and one powerful legal tool that is sometimes utilized is the compulsory licensing of patented drugs.[449] The World Trade Organization (WTO) Agreement on the Trade-Related Aspects of Intellectual Property Rights (TRIPS) permits governments under certain conditions to compel a patent holder to allow the subject of the patent to be used by others.[450] This clause has been invoked by several countries.[451] A number of governments have used a threat of

---

[447] The experience of Sequenom illustrates this phenomenon. In June 2008, Sequenom disclosed that a non-invasive prenatal test which it had developed to screen maternal blood for Downs syndrome was effective in all samples, sending its shares up 21.8 percent on eight times average volume. However, on the eve of the product launch, Sequenom revealed that the introduction of the test would be delayed "due to the discovery by company officials of employee mishandling of R&D test data and results," and that the company's board had launched an independent internal investigation. The special committee charged with conducting the investigation concluded that Sequenom "failed to provide adequate protocols and controls" of results of the prenatal test. The company's CFO and other executives resigned. The company fired its CEO and head of research and development. Share prices collapsed, and numerous class action suits were brought on behalf of shareholders who bought Sequnom shares after the 2008 disclosure of a promising new drug. The complaints alleged that the company made "materially false and misleading statements regarding the clinical performance of the Company's developmental Down syndrome test." "Sequenom Announces Additional Positive Tests Results for Down Syndrome Test at Analyst Briefing. "*Chemical Business NewsBase* (September 23, 2008); "Sequenom Raises Bar in Prenatal Test Field," *Investor's Business Daily* (December 16, 2008); "Sequenom Readies Tests for Market," *Business Review Western Michigan* (March 26, 2009); "Sequenom Announces Delay in Launch of SEQureDx Trisomy 21 Test," *Business Wire* (April 29, 2009); "Sequenom: Bloodied and Unbowed," *Barron's* (September 29, 2009).

[448] In 2008, the SEC filed a civil fraud action against Biopure Corporation, alleging that the company materially misled the investment community by failing to disclose — or by framing as positive developments — certain negative information from the FDA regarding the approval prospects of its synthetic blood product, Hemopure." "Increased Scrutiny of Investor Communications by Federla Regulators," *Food and Drug Law Institute* (January/February 2006).

[449] "Big Pharma Learns to Live With Generics," *Bangkok Post* (August 15, 2009).

[450] Compulsory licensing can be used in "a national emergency as other circumstances of extreme urgency." TRIPS Article 31.

[451] In 2007, Brazil issued a compulsory license for Merck's anti-AIDS drug Efavirenz. In 2006, Thailand issued compulsory license for two anti-AIDS drugs made by Merck and Abbott Laboratories, and a compulsory license for the anti-cancer drug Docetaxel, patented by the French

TABLE 6.5 Shareholder Class Action Lawsuits Against Life Sciences Firms

Year	Defendant	Product	Allegation
2009	Pozen Inc	Treximet	False or misleading statements about migraine drug candidate, Treximet
2009	Caraco Pharmaceutical Laboratories	various tablets	Failure to disclose material information re FDA warning letter on drug manufacturing.
2009	Rigel Pharmaceuticals	R788	False and misleading statements with respect to clinical trial of a drug, R788 for treatment of rheumatoid arthritis
2008	KV Pharmaceutical Co	Makena	Failure to disclosure compliance problems with FDA requirements
2009	Immucor	Blood reagents and related equipment	Failure to disclose compliance problems with FDA requirements

SOURCE: Brian Johnson et al v. Pozen Inc. et al, U.S. District Court, Middle District of North Carolina (2009)1 ; Wilkof v. Caraco Pharmaceutical Laboratories, Ltd., U.S. District Court, Eastern District of Michigan (2009); Immucor, Inc. Form 10-K for the fiscal year ended May 31, 2011, p. 16; "KV Pharmaceutical Company Hit by Investor Class Action Over Alleged Securities Law Violations, " Shareholders Foundation (October 19, 2011).

compulsory licensing to pressure foreign pharmaceutical firms into reducing drug prices.[452] TRIPS requires that compulsory license "shall be authorized predominantly for the supply of the domestic market," but in 2009 the WTO

---

firm Sanofi-Aventis. "Compulsory Thai Licensing of AIDS Drug Sets Precedent," *Deutsche Press Agentur* (July 29, 2008); "Commerce Ministry Asks Council of State for Opinion on Legality of Compulsory Licensing of Cancer Drug," *Thai Press Reports* (August 22, 2008).

[452] In 2009, Korea threatened Roche with compulsory licensing in negotiations over the supply of Tamiflu to Korea. The government of Brazil has applied similar pressure to multinational drug makers, particularly with respect to the supply of anti-retroviral drugs to treat HIV/AIDS. "Tamiflu Generics Protection Planned," *Korea Times* (September 9, 2009); "GSK and Fiocruz to Develop and Product Vaccines," *Economist Intelligence Unit* (September 14, 2009).

ruled that Pakistan could grant compulsory licenses on patented drugs for export to third countries that lacked their own manufacturing capacity.[453] In March 2012, the Controller of Patents, Mumbai, granted Natco Pharma, an Indian company, a compulsory license for manufacture of a generic version of sorafenib toyslate, a drug developed by Bayer to treat liver and kidney cancer, stating that the drug was "exorbitantly priced."[454]

Supply chain vulnerabilities. Governments in western countries are pressing pharmaceutical firms to reduce the cost of their products, and one way in which the industry is responding is to move the manufacture of drugs to lower cost countries and to source ingredients from those countries. Roughly 80 percent of the active ingredients used in U.S. prescription drugs originate outside the U.S.[455] "[W]hether locally made generics, or patented drugs produced by either a multinational or a contract-manufacturing organization, Chinese-made prescription drugs will soon become unavoidable." Imports from China and India accounted for about 20 percent of the generic and over-the-counter drugs sold in the U.S. in 2008.[456] A number of scandals have occurred in which U.S. consumers have been harmed through use of drugs with adulterated ingredients derived from unregulated or under-regulated companies in China.[457] Recently the Chinese government has taken steps to strengthen supervision of companies which comprise the pharmaceutical supply chain, but a recent incident in which large numbers of commonly used capsule drugs were found to contain high levels of toxic chromium indicates that significant risks still exist.[458]

---

[453] "WTO Allows Pakistan to Grant License," *Business Recorder* (October 3, 2009).

[454] "India Uses Arm-Twist Rule for Cancer Drug," *The Telegraph Online* (March 13, 2012).

[455] "Counterfeit Avastin Seized in the US," *Pharma Times* (February 16, 2012).

[456] "Clamping Down on Fakes," *Chemical Business NewsBase* (September 8, 2008).

[457] "Chinese Chemicals Flow Unchecked Onto World Drug Market,:" *The New York Times* (October 31, 2007). In 2008 Baxter International suspended sales of the anti-coagulant heparin produced at an uncertified plant in China which was not inspected by the government after four U.S. users died and 350 suffered complications. "China Didn't Check Drug Suppliers, Files Show," *The New York Times* (February 16, 2008). "Will US Inspections Help Improve the Safety of Chinese Drugs?" *Economist Intelligence Unit* (April 15, 2008).

[458] The capsules were made of industrial gelatin, and the chromium could cause digestive disorders and internal organ failure. An advisory expert at the State Food and Drug Administration commented that "drug quality control has been quite strict on end products. We examine all the quantities and qualities of medical substances inside the capsules. But somehow we have left out instrumental materials like the capsules themselves. That's a loophole, and we certainly need to address it." The government shut down two of the capsule plants and took four plant owners into police custody. "Capsule Scandal Exposes Loopholes in Drug Quality Control," *China Radio International Online* (April 17, 2012). In China, Good Manufacturing Practices (GMP) standards were introduced in the late 1970s but were phased in very slowly. The State Food and Drug Administration (SFDA) issued revised GMP standards in 1999, requiring all pharmaceutical manufacturers to meet GMP standards and secure GMP certification by June 30, 2004. New and more stringent GMP rules governing pharmaceutical production took effect October 1, 2010, requiring producers to apply for supplementary registration if the new standards were not met.

Supply chain vulnerabilities arise out of the increasing use of lower-cost bulk active pharmaceutical ingredients (APIs) as ingredients in manufactured drugs. In some major countries makers of APIs can sidestep regulatory scrutiny by not disclosing that their chemicals will be used in pharmaceutical products.[459] Bulk APIs are now sold over the Internet, which is also a global platform for marketing and sale of counterfeit drugs. Some contaminated substances find their way into the U.S. healthcare system.[460]

Counterfeiting and mislabeling. Counterfeit and mislabeled medicines are a growing global concern both for legitimate pharmaceutical manufacturers and consumers. According to the World Health Organization, fake drugs account for under one percent in developed countries but from 10 to 30 percent of drug sales in emerging markets.[461] Counterfeit medicines "are often produced in unsanitary conditions by people without any medical or scientific background."[462] Spuriously/falsely-labeled/falsified/counterfeit (SFFC) medicines can result in treatment failure and death. In 2012 the FDA sent out letters to 19 medical practices warning that counterfeit versions of Avastin, made by Roche and Greentech, had been detected in the U.S. and "may have left patients without their therapy."[463] The World Health Organization cites a number of other examples of known SFFC incidents.

Counterfeit drugs increase the business risks of legitimate pharmaceutical manufacturers. Branded firms may find themselves targeted by lawsuits based on consumer use of worthless or toxic counterfeit medicine bearing the company's brand. U.S. and European pharmaceutical firms which have Chinese operations or incorporate Chinese APIs in their manufacturing processes risk legal actions by consumers. Historically legitimate

---

Many Chinese manufacturers are finding compliance with GMP standards to be financially burdensome. Some companies reportedly received GMP certification despite their deviation from GMP requirements, and "one factor causing this poor state of GMP implementation is believed to be a lack of transparency in the drug administration system" (Royan Gai, et al, "GMP Implementation in China: A Double-Edged Sword for the Pharmaceutical Industry", Drug Discoveries and Therapeutics (January 2007))

[459] Chinese regulators do not supervise the production of raw materials used in pharmaceutical manufacture, so-called "intermediates" which are used to make APIs. The lack of oversight has contributed to tragedies such as the death and disability of 128 Panamanians who used cold medicine manufactured in China which contained diethylene glycol, a toxic substance normally used as engine coolant but sometimes utilized as a substitute for glycerine. "Chemicals Flow Unchecked from China to Drug Market," *Kyodo* (November 1, 2007).

[460] In 2007 University Health Care System, based in Augusta, Georgia was warned by one of its suppliers that some of the oral care kits used by the hospital might contain toothpaste made in China containing toxic diethylene glycol. "This Problem Made in China," *Modern Healthcare* (October 22, 2007).

[461] "Just How Big is the Counterfeit-Drug Problem?" *FiercePharma* (September 13, 2010).

[462] "Pfizer Steps Up Campaign in Fight Against Counterfeit Drugs," *Pharma Times* (September 30, 2011).

[463] "Counterfeit Avastin Seized in the US," *Pharma Times* (February 6, 2012).

TABLE 6.6 Examples of SFFC Medicines

SFFC medicine	Country/Year	Report
Anti-diabetic traditional medicine (used to lower blood sugar)	China, 2009	Contained six times the normal dose of glibenclamide (two people died, nine people hospitalized)
Metakelfin (antimalarial)	United Republic of Tanzania, 2009	Discovered in 40 pharmacies: lacked sufficient active ingredient
Viagra & Cialis (for erectile dysfunction)	Thailand, 2008	Smuggled into Thailand from an unknown source in an unknown country
Xenical (for fighting obesity)	United States of America, 2007	Contained no active ingredient and sold via Internet sites operated outside the USA
Zyprexa (for treating bipolar disorder and schizophrenia)	United Kingdom, 2007	Detected in the legal supply chain: lacked sufficient active ingredient
Lipitor (for lowering cholesterol)	United Kingdom, 2006	Detected in the legal supply chain: lacked sufficient active ingredient

SOURCE: WHO Fact Sheet No. 275 (January 2010) www.who.int/mediacentre/factsheets/fs275/en/.

pharmaceutical companies have been reluctant to complain publicly about fake drugs because it could damage their business.[464]

---

[464] Robert Cockburn, Paul Newton, Kyermateng Agyarko, Dora Akunyii and Nicholas White, "The Global Threat of Counterfeit Drugs: Why Industry and Government Must Communicate the Dangers," *Plos Medicine* (March 2005).

Looking Ahead

The U.S. pharmaceutical industry continues to pursue growth strategies despite the numerous challenges it confronts. Major branded pharmaceutical companies will seek to offset declining R&D productivity through partnerships with innovative biotechnology firms, a strategy which also may help to counter competitive pressure from generics makers. U.S. pharmaceutical firms will increase investments in R&D in emerging markets, where demand for medicines is growing at a far more rapid rate than in developed country markets. And the industry will pursue niche strategies in areas such as biosimilars and orphan drugs.

Strategic combinations. Pharmaceutical and biotechnology firms are increasingly entering into complex strategic alliances with other companies, including licensing and cross-licensing of patents, joint ventures, joint development and trials, and distribution alliances. Such combinations mitigate the costs and risks associated with development of new drugs and enable companies to enter new product and geographic markets. Development of biopharmaceuticals may also help branded pharmaceutical firms to counter competition from generic drug makers. The high cost of developing biologics such as monoclonal antibodies serves as a partial competitive foil to generics makers. On industry analyst observed in 2010 that—

> *It's not going to be that easy for generic players to be very successful in the biotech area. They are not easy to copy and not easy to manufacture.*[465]

In 2009 the CEO of Johnson & Johnson, William Weldon, said that J&J would acquire minority shareholding and develop alliances with its competitors in order to share costs and risks.

> *[Weldon's] remarks reflect a trend even by large, cash, generative pharmaceuticals companies to fund new ways to share the potential costs as well as the profits in proving the safety and efficacy of new drugs to regulators and winning agreement by health care systems to reimburse them.*[466]

---

[465] Rajith Gopinathan, analyst with industry market research firm Frost & Sullivan, in "Pharmaceutical Companies Seek Biotech Acquisitions to Boost Drug Pipelines," *ICIS.com* (February 12, 2010).
[466] J&J Wants Deals with Rivals to Share Risk," *Financial Times* (October 25, 2009).

TABLE 6.7 Strategic Alliances in Pharmaceuticals

Year	Companies	Activity
2008	Sequenom, MetaMorphix	Apply Sequenom genotyping to enhance livestock DNA screening
2009	PRA International, LSK Global Pharma Services, Mediscience Planning	Joint management of clinical trials in Asia
2009	Illumina, Agilent	Scalable solution for researchers conducting targeted sequencing studies
2009	Eli Lilly, Cadila Heath care	Development of cardiovascular drugs
2009	Johnson & Johnson, Elan	J&J acquires rights to Elan Alzheimer immunotherapy program, 18 percent stake in Elan, and links to Elan partners Biogen Idec and Wyeth (Pfizer)
2009	Johnson & Johnson, Crucell N.V.	Develop monoclonal antibodies for prevention/treatment of influenza
2009	Johnson & Johnson, Gilead	Use joint trials to develop a once-daily HIV therapy
2009	GlaxoSmithKlein, Pfizer	Combine experimental and existing HIV medicines with joint venture
2009	AstraZeneca, Bristol-Meyers Squibb	Joint development of diabetes treatment drugs

SOURCE: "Johnson & Johnson Completes Deal with Elan, Acquiring its Alzheimers Assets, " Business Wire (October 14, 2009); "Johnson & Johnson and Crucell form Drug Discovery Collaboration," Datamonitor (September 30, 2009); "MetaMorphix and Sequenom Agree to Build on Success," Business Wire (January 9, 2008); "PRA International, LSK Global Pharma Services and Mediscience Form Partnership," Datamonitor (January 15, 2009); "Illumina and Agilent Sign Co-Marketing Agreement," Datamonitor (April 20, 2009); "PharmaChem, Cadila, Eli Lilly in Drug Development Deal," Chemical Business NewsBase (March 31, 2009); "j&J, Gilead HIV Drug Wins FDA Approval," Blomberg (August 10, 2011); "GaxoSmithKline, Pfizer Inc. HIV Venture Plans Russian Manufacturing," Chemical Business NewsBase (November 3, 2011); "Onglyza Study by Bristol-Meyers Squibb and Astrazenica,"Asia Pulse (June 29, 2010).

Emerging markets. Pharmaceutical markets are growing far more rapidly in emerging economies than in mature markets in the United States, Europe and Japan.[467]

The pharmaceutical industry will necessarily pursue growth by increasing its presence in emerging markets, particularly countries with large populations and rising standards of living.[468]

China. China is now the world's third largest pharmaceuticals market, is reportedly growing at a rate of over 25 percent per year, and is forecast to overtake Japan as the world's second largest market in 2016. In 2011, the government announced its intention to boost healthcare spending by 16.3 percent to about $26 billion. At present over 90 percent of China's population is covered by some form of insurance, making modern medicine more affordable. Demand is particularly strong for drugs to treat chronic illnesses, which account for 80 percent of deaths in China.[469] In 2011, Merck indicated its R&D spending in China would reach $1.5 billion over the next five years, and that it would construct a 600-person R&D headquarters in Beijing.[470] U.S. pharmaceuticals companies investing in China face a number of challenges, including government intervention in drug pricing, competition from locally-produced generics, and infringement of intellectual property.

Major foreign pharmaceutical makers have made significant commitments in China.[471] Novartis announced in 2009 that it would invest $1 billion in R&D in China over the next five years, augmented by acquisition of an 85 percent stake in one of the largest private makers of vaccines in the country, Zhejiang Tianyuan Bio-Pharmaceutical Co. Ltd. Eli Lilly opened an R&D center in Shanghai in 2008 and has entered into a venture capital initiative to launch new products in collaboration with Chinese institutes and companies.[472]

South Korea. Major U.S. pharmaceuticals firms are establishing a presence in South Korea, a country with a strong university and science infrastructure, a large pool of skilled manpower, and the ability to conduct

---

[467] "A 2010 study by Thomson Reuters Pharma observed that demand for pharmaceuticals was growing at an annual rate of 25-27 percent in China and 15-17 percent in markets such as Brazil, India, Poland and Russia. Western European markets were growing at an annual rate of 1-3 percent and the United States 3-5 percent." Thomson Reuters Pharma, "The Ones to Watch: A Pharma Matters Report," (July-September 2010).

[468] Merck has reportedly embraced an aggressive growth plan for emerging markets which would up its 18 percent growth rate in 2012 to 25 percent in 2013, focusing R&D in each country on products that are important for that country. "Merck and Company Firms Up Plan for Emerging Markets," *The Economic Times* (Mumbai, February 17, 2012).

[469] "Alliances Form in Growing Pharmaceutical Market," *Business Daily Update* (August 3, 2011).

[470] "Merck Play R&D Centre in China," *Chemical Business Newsbase* (December 12, 2011).

[471] "Foreign Giants Dominate China Pharmaceutical Market," *SinoCast* (November 5, 2010).

[472] "Eli Lilly Opens China R&D Headquarters in Shanghai,"*SinoCast* (October 17, 2008); "Eli Lilly Asia VC Fund Settles in Shanghai," *SinoCast* (November 16, 2007).

clinical trials in an extremely efficient manner.[473] Pfizer announced in 2007 that it would make Korea a "key research bank for its new medicine development" and invest $300 million over a five year period.[474] In 2007, VGX Pharmaceutical Inc., a U.S. firm that specializes in hepatitis and HIV treatments, announced it would invest $200 million to establish its Asian headquarters in Korea.[475] Johnson & Johnson manufactures drugs in Korea through a subsidiary, Janssen Korea, which functions as J&J's production base for the entire Asian market.[476] Foreign pharmaceutical firms operating in Korea face significant challenges, including pressure by healthcare providers to give suppliers rebates,[477] lack of transparency with respect to Korea's pricing and reimbursement of drugs,[478] and government pressure on the intellectual property of branded drug firms.[479]

Biosimilars. The first generation of biotechnology drugs is going off-patent, giving rise to a promising new market for "follow-on biological," also known as biosimilars. A number of the major branded pharmaceutical producers are entering the biosimilars markets, including Merck, Eli Lilly, and AstraZenica. In contrast to small molecule drugs formed through chemical synthesis, biologics are molecularly complex and potentially sensitive to changes in manufacturing processes, raising the prospect that they might not have the same effects in human beings as the original drug.[480] As a result, biosimilars face an uncertain regulatory path to approval which is still evolving

---

[473] Korea has a unique advantage in the form of large hospitals in a dense area; with so many patients, clinical trials can be done quickly. In addition, Korean hospitals have strong links with university R&D organizations. "Novartis Stays Ahead with New Ideas: Country Head Says Dedication," *The Korea Herald* (March 31, 2004).
[474] "Pfizer Pharmaceutical Company to Invest 300m Dollars in South Korea by 2012," *Yonhap* (June 14, 2007).
[475] "US Drug Maker to Have Headquarters in Korea," *Korea Times* (July 9, 2007).
[476] "Pharmaceutical Giant to Expand Korea Operations," *Dong-A Ibo* (February 18, 2008).
[477] Since 2007, a significant number of manufacturers, including Eli Lilly, Pfizer and GlaxoSmithCline have been fined by the Korea Fair Trade Commission (KFTC) for illegal payment of rebates to hospitals, doctors and pharmacists. The U.S. government has noted concerns expressed by U.S. companies targeted by the KFTC that they have not been accorded a significant opportunity to review and respond to the evidence against them, including an opportunity to cross-examine witnesses at KFTC hearings. "10 Pharmaceutical Firms Face Heavy Fines for Rebates," *Korea Times* (October 25, 2007); "War Declared on Drug Makers' Rebates to Doctors," *Dong-A Ilbo* (July 31, 2009); "Cleanup Drive to Sweep Pharm Industry," *Korea Times* (March 31, 2009). Office of the U.S. Trade Representative, *2009 National Trade Estimate on Foreign Trade Business* (2009) p. 316.
[478] Imported pharmaceuticals are subject to multiple price reduction mechanisms under the Korean Drug Expenditure Rationalization Plan (DERP) cost containment measures, enacted in 2006, which affects not only drugs entering the market since DERP was adopted, but retroactively affects drugs approved for reimbursement in the pre-DERP era. Office of the U.S. Trade Representative, *2009 National Trade Estimates Report on Foreign Trade Barriers* (2009) p. 317.
[479] "ROK Firms Plan Tamiflu Generics Production," *Korea Times* (September 9, 2009).
[480] The makers of follow on biologic drugs do not have access to the originating company's active drug substances, cell bank, molecular clone or fermentation and purification processes.

in the U.S. and Europe.[481] The Patient Protection and Affordable Cure Act, enacted in 2010, establishes a 12 year period of data exclusivity for new biological drugs between the date of FDA approval and the filing date for biosimilar approval based on the innovator's original data, a measure which may inhibit the introduction of biosimilars.

IN CLOSING

The global competitive environment is being shaped to an important degree by the national policies of our competitors. This chapter has explored the major policy issues affecting the competitiveness of the semiconductor, photovoltaic products, advanced batteries, and pharmaceuticals industries. Each of these industries can be regarded as strategic to the United States. While many nations in Europe and Asia use the full force of government to attain commercial competitive advantage in industries they regarded as strategic, the idea of proactive government help for private industry in the name of economic development has sometimes raised concerns in the United States about distorting market forces and the wisdom of letting public servants "pick winners." In reality, the U.S. federal government has long played an integral role in the early development of numerous strategic industries, not only by funding research and development but also through financial support for new companies and government procurement.

Each of the four industries studied face unique circumstances and challenges. At the same time, they illustrate the important role that national investments have played in supporting their development and the need for public policies to ensure that the nation captures the benefits of these investments in terms of economic growth and high value employment.

---

[481] In the U.S. the Biologics Price Competition and Innovation Act of 2009 was enacted in 2010 to create a shortened path to regulatory approval for biosimilars. The FDA is currently developing guidelines for the approval process for biosimilars. As of March 2012 it had not yet received its first biosimilars application. "Fitch Looks at Implications of FDA Biosimilar Guidance," *Pharma Times* (February 13, 2012).

Chapter 7

Clusters and Regional Initiatives

Clusters foster the collaboration needed to develop new ideas and bring them to market. In this way, successful clusters significantly improve the return on public investments in R&D and provide global leadership in key technologies. Recognizing this impact, both advanced and emerging economies are making investments and promulgating polices to encourage cluster development.

This chapter explores several ways in which U.S. regions are rising to the challenge, focusing on Regional Innovation Cluster initiatives and new types of science and research parks. In this chapter, we explore a sample of some of the more interesting regional innovation cluster initiatives underway in the United States. The second part of this chapter assesses new strategies for developing research parks, both in the United States and abroad.

THE INNOVATION CHALLENGE

U.S. regional economies face mounting global competitive challenges. No longer do U.S. states and cities primarily compete among themselves for talent, investment, and entrepreneurs in technology-intensive industries. They also compete against national and regional governments that are executing comprehensive strategies that seek to create innovation clusters in many of the same important, emerging industries. National and regional governments in Europe, Asia, and Latin America are backing up these strategies with heavy investment in universities, public-private research collaborations, workforce training, early-stage capital funds, and modern science parks.[1] They are further

---

[1] Francisco Grando, Brazil's Secretary of Innovation, and Alberto Duque Portugal, State Secretary for Science, Technology and Higher Education of the Brazilian state of Minas Gerais presented a review of initiatives underway in Brazil at the National Academies conference on *Clustering for 21st Century Prosperity*, Washington, DC, February 25, 2010.

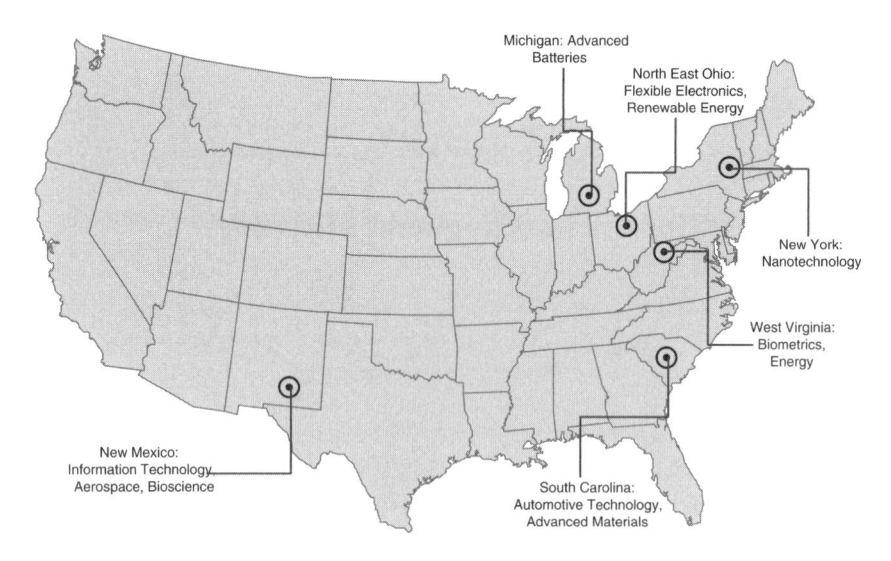

FIGURE 7.1 U.S. regional innovation clusters discussed in Chapter 7.

Map labels:
- Michigan: Advanced Batteries
- North East Ohio: Flexible Electronics, Renewable Energy
- New York: Nanotechnology
- West Virginia: Biometrics, Energy
- New Mexico: Information Technology, Aerospace, Bioscience
- South Carolina: Automotive Technology, Advanced Materials

reinforced by strong policy focus from top leaders. National and regional governments also can offer investors financial incentives that state governments cannot, such as exemption from all corporate taxes.

In a number of Asian clusters, most notably Taiwan's Hsinchu Science Park, research and manufacturing functions are tightly linked, an entire industry chain is present within the cluster to manufacture and commercialize the technologies emerging from the laboratories. While this phenomenon is observable in many U.S. clusters, a number of the clusters featured in this study, have seen U.S. developed technologies[2] manufactured outside the United States because so much of the value chain is located there.

John A. Matthews, an Australian academic who has extensively studied the cluster phenomenon in Asia recently noted the actual and prospective advent

---

[2] In nanotechnology, a specialty of a number of U.S. clusters, a number of U.S. firms that have originated promising new technologies have outsourced the manufacturing to Asia. In 2011, U.S.-based Nova Centrix entered into an agreement with Japan's Showa Denko pursuant to which the latter would manufacture and sell nanoparticle inks developed by Nova Centrix. An industry journal commented as follows: "Nova Centrix is one of several nanomaterials suppliers working with Japanese and other Asian partners to support production and commercialization of their technology. Experience of industrialized production methods can be leveraged as these technology developers try to commercialize their technologies, and much of the world's display and electronics manufacturing occurs in Asia." "Nanomaterials firms turn to Asia for Commercial Opportunities" Plastic Electronics (April 15, 2011).

of research and industrial clusters in China and India and commented that '"the success of these emerging industrial giants of the 21st century cannot be understood without reference to the industrial cluster phenomenon that is embedded within them, housed within such institutional settings as *Special Economic Zones* and science-based industry parks. All the intellectual machinery developed to understand the rise of clusters in the advanced world is now going to have to be applied in order to make sense of this same phenomenon in the developing world, but in a new context defined by globalization and the emergence of global production networks and global value chains".[3]

POLICIES TO FOSTER INNOVATION

The new competitive landscape is prompting state and regional authorities around the U.S. to take creative, comprehensive, and proactive approaches to developing innovation-led economies. Indeed, just as foreign governments have absorbed lessons from successful U.S. innovation zones such as Silicon Valley and Research Triangle Park, U.S. economic-development officials have studied the strategies and experiences of other nations. The growing global challenges also have prompted fresh discussion of and experimentation with closer collaboration between federal agencies and state and local bodies to improve innovation capacity and boost industrial competitiveness.

REGIONAL INNOVATION CLUSTERS

Communities across the world have long tried to mimic the success of innovation hot spots such as Silicon Valley and Boston's Route 128. Only in the past decade or two, however, have innovation clusters become a matter of serious public policy in the United States. Today, a growing number of state and regional governments are developing comprehensive strategies to nurture new concentrations of growth industries.

No longer is regional economic development merely a competition among states for corporate investment on the basis of tax breaks and subsidized land and labor. State development officials also know that it takes more than funding for university research and building science parks for high concentrations of innovative companies to take root in a given region.[4] It

---

[3] John A. Matthews. "The Hsinchu Model: Collective Efficiency, Increasing Returns and Higher-Order Capabilities in the Hsinchu Science-Based Industry Park, Taiwan". Keynote Address, Chinese Society for Management of Technology, 20th Anniversary Conference, Tsinghua University, Hsinchu, Taiwan, December 10, 2010.

[4] For the perspectives of state economic development officials from Ohio, Pennsylvania, Virginia, Kansas and Washington state, see National Research Council, *Growing Innovation Clusters for*

requires an entire ecosystem in which high densities of talented people—researchers, entrepreneurs, and investors—collaborate to develop and launch new products and companies.[5] As Michael Porter observed, to secure competitive advantage against other regions, communities must be able to fully exploit knowledge, relationships, and motivation that "distant rivals cannot match."[6]

Early U.S. innovation clusters such as Silicon Valley and Greater Boston emerged from the interaction between the private sector and major universities that received substantial federal research funding,[7] but with little government design. By contrast, Research Triangle in North Carolina is the result of early, substantial, and patient public and private support. In recent decades, however, economic development agencies across the U.S. and around the world have devised policy strategies to stimulate the rapid development of regional innovation clusters.[8] Governments are investing in universities, public-private research partnerships, skilled workforce training, shared prototyping facilities, and early-stage capital funds for entrepreneurs. Innovation America President Richard Bendis describes the conceptual shift of the past decade as going from "technology-based development" toward "innovation-based economic development."[9] Egils Milbergs of the Washington Economic Development Commission contends that "a new model of economic development for states" has come to the fore, one that focuses on talent, infrastructure, productivity growth, open innovation systems, and global connections.[10]

American Prosperity, Summary of a Symposium, C. Wessner, Rapporteur, Washington, DC: The National Academies Press, 2011.
[5] See Robert E. Lucas, Jr., "On the Mechanics of Economic Development," *Journal of Monetary Economics* 22, 1988, pp. 38-39. Richard Florida has popularized the characteristics and economic advantages of innovative clusters. See Richard Florida, *The Rise of the Creative Class*, New York: Basic Books, 2002.
[6] Michael E. Porter, "Clusters and the New Economics of Competition," *Harvard Business Review*, 76(6), pp. 77-90, 1998.
[7] See AnnaLee Saxenian, *Regional Advantage: Culture and Competition in Silicon Valley and Route 128*, Cambridge, MA: Harvard University Press, 1994, p. 161. Also see Martin Kenney, ed., *Understanding Silicon Valley: The Anatomy of an Entrepreneurial Region*, Stanford: Stanford University Press, 2000.
[8] Regional cluster development policies are proliferating so fast that rigorous assessments of their effectiveness are lagging. As one researcher has summed it up: "Cluster policy has not only surged ahead of cluster potential, it has also outpaced our theoretical and empirical understanding of the cluster phenomenon." Matthias Kiese, "Cluster Approaches to Local Economic Development," in Uwe Blien and Gunther Maier, eds., *The Economics of Regional Clusters: Networks, Technology and Policy*, Cheltenham: Edward Elgar Publishing, 2008, p. 290.
[9] Presentation by Richard Bendis of Innovation America in National Research Council, *Growing Innovation Clusters for American Prosperity: Summary of a Symposium,* op. cit.
[10] Presentation by Egils Milbergs of the Washington Economic Development Commission in National Research Council, *Growing Innovation Clusters for American Prosperity,* ibid.

CLUSTERS AND REGIONAL INITIATIVES   435

Until very recently, U.S. federal agencies have done little to support state and regional innovation cluster initiatives. This is not the case abroad. Clusters have been embraced globally as effective vehicles for mobilizing and coordinating public and private activities to spur economic growth. The growing movement among governments around the world to shift from outright subsidies to companies and poor regions to investing in public goods that enable industry, universities, and communities to compete represents "a new paradigm in regional policy," according to Mario Pezzini of the Organization for Economic Co-operation and Development.[11]

Andrew Reamer of the Brookings Institution noted in 2009 that 26 of 31 European Union nations have cluster development programs at the national level and the EU even operates a European Cluster Observatory that maps clusters across the European continent[12]. A number of Asian and Latin American nations and regions have also promulgated cluster strategies. A few examples—

- **Brazil:** Minas Gerais, a Brazilian state with 20 million people and a territory roughly the size of France, is investing $300 million in emerging clusters in micro-electronics, bio-fuels, and software. Minas Gerais also has identified hundreds of "poles of excellence" in traditional industries scattered across the state that it hopes to develop further. Sistema Mineiro de Inovação (SIMI), the agency coordinating the campaign, is promoting development of science parks, incubators, and training programs.[13]
- **Hong Kong:** Realizing that it needed to diversify its industry base after the 1997 Asian financial crisis, the Hong Kong government launched an initiative to develop innovation clusters in fields that leverage its technology strengths, its reputation as a world-class business environment, and its strategic location on the doorstep of mainland China. Some 250 companies in electronics, green technology, information and communication technology, precision engineering, and

---

[11] Presentation by Mario Pezzini of the Organization for Economic Co-operation and Development at the National Academies conference on *Clustering for 21st Century Prosperity*, Washington, DC, February 25, 2010. See also "National Innovation Systems," OECD, 1997 http://www.oecd.org/dataoecd/35/56/2101733.pdf.
[12] The OECD examined 26 cluster programs in 14 countries. Notably, the programs examined for the United States were state programs – the Georgia Research Alliance and the Oregon Cluster Network. OECD, *Competitive Regional Clusters: National Policy Approaches*, Paris: OECD, 2007.
[13] Presentation by Alberto Duque Portugal of the Minas Gerais Secretariat for Science, Technology, and Higher Education, op. cit. SIMI also is encouraging research organizations and entrepreneurs to consolidate their activities into hubs in locations strong in particular fields so that they can achieve greater scale and draw more foreign investment.

biotechnology clusters are based in the new Hong Kong Science and Technology Park (see below).[14]

- **Canada:** As part of its goal of developing at least 10 internationally recognized technology clusters,[15] Canada has established a network of 17 Centers of Excellence since 2008 in fields such as brain research, optics, and theoretical physics[16]
- **Singapore:** Singapore is investing billions of dollars in comprehensive strategies to expand innovation clusters in biomedicine, digital media, and high value-added manufacturing, including microelectronics and new materials (see chapter 3).[17]
- **France:** The Grenoble region is rising fast as one of Europe's premier hubs for micro-electronics and nanotechnology companies, and is a showpiece of the French government's *pôles de croissance* initiative to develop globally competitive innovation clusters.[18]
- **Taiwan:** Taiwan's Industrial Technology Research Institute (ITRI) already has helped establish some of the world's most successful clusters in notebook PCs, digital displays, and semiconductors. Now ITRI and other government agencies are working with industry to develop promising clusters of manufacturers in solid-state lighting,

---

[14] Now Hong Kong is focusing on developing innovation clusters in areas like thin-film photovoltaic cells, environmental engineering, and energy management for buildings. Presentation by Nicholas Brooke of Hong Kong Science and Technology Parks Corp. in National Research Council, *Understanding Research, Science and Technology Parks: Global Best Practice: Report of a Symposium,* Charles W. Wessner, editor, Washington, DC: National Academy Press, 2009.

[15] Each center is based at a university and receives a mix of government and private industry funding for collaborative research and commercialization programs. The centers are credited with creating more than 100 spin-off companies, training 36,000 personnel, and attracting $71 million in private investment. Industry Canada, *Achieving Excellence: Investing in People, Knowledge and Opportunity—Canada's Innovation Strategy, 2001.* (http://dsp-psd.pwgsc.gc.ca/Collection/C2-596-2001E.pdf.

[16] Networks of Centers of Excellence, "About the Networks of Centres of Excellence," accessible on the Web at http://www.nce-rce.gc.ca/About-APropos/Index_eng.asp.

[17] The initiative, led by the Agency for Science, Technology, and Research (A*STAR), includes development of several multibillion-dollar science parks, recruitment of top international scientists, a training program for 1,000 Singaporean science and engineering Ph. Ds, revamped university curriculum, and a $275 million program to support technology entrepreneurs with start-up capital and incubators. National Research Foundation, "National Framework for Innovation and Enterprise," Prime Minister's Office, Republic of Singapore, 2008, (http://www.nrf.gov.sg/nrf/otherProgrammes.aspx?id=1206).

[18] Gilles Duranton, Philippe Martin, Thierry Mayer, and Florian Mayneris *The economics of clusters. Lessons from the French experience.* Oxford: Oxford University Press, 2010. The cluster is centered around MINATEC, a 3,000-student campus that represents a €3.35 billion investment by the national and local government (see Science Park chapter). Minatec has brought together public-private research collaborations involving four universities and has spawned start-ups in optoelectronics, biotechnology, circuit design, motion sensing, and other fields. From presentation by David Holden of MINATEC in *Understanding Research, Science, and Technology Parks*, op. cit.

flexible displays, thin-film photovoltaic cells, medical devices (see chapter 3).[19]

Cluster Dynamics

Industrial clusters have been the subject of study since the pioneering study of Sheffield's cluster by the British economist Alfred Marshall in the late 19th century.[20] He identified three basic advantages of clusters which are still acknowledged and have come to be known as "Marshall's trinity". They are: 1) a pool of skilled labor; 2) knowledge spillovers; and 3) inter-firm linkages. These factors are widely recognized to convey benefits to enterprises located in a cluster, but the benefits have proven difficult to quantify.[21] In addition to the traditional sources of cluster advantages cited by Marshall, a number of contemporary analysts, notably Michael Porter, have argued that highly clusters localities in which intense competition for ideas occurs are more conducive to innovation.[22]

---

[19] Presentation by John Chen, Industrial Technology and Research Institute of Taiwan at the National Academies Conference on Flexible Electronics for Security, Manufacturing, and Growth in the United States, September 24, 2010 in Washington, DC.

[20] Alfred Marshall, *Principles of Economics*, London: Macmillan, 1920. The first edition of Marshall's classic textbook appeared in 1890. While the analysis of the spatial concentration of economic activity goes back to Marshall's analysis of the localization of industry it was given more recent attention by Paul Krugman, *Geography and Trade*, Cambridge: The MIT Press, 1991 See also W. Brian Arthur, "Industry Location Pattern and the Importance of History," in W. Brian Arthur, *Increasing Returns and Path Dependence in the Economy*, Ann Arbor: The University of Michigan Press, 1994. Arthur examines the relationship between two different theories of spatial concentration, agglomeration economies and the historical accident/path dependence viewpoint. Recent empirical work by Delgado, Porter and Stern find significant evidence for cluster-driven agglomeration. Mercedes Delgado, Michael E. Porter and Scott Stern, "Clusters, Convergence, and Economic Performance," March 11, 2011, submitted for publication, accessible at http://www.isc.hbs.edu/econ-clusters.htm.

[21] Paul Krugman, who popularized Marshall's thinking in the late 20th century, observed that "technological spillovers leave no paper trail." Stephern Klepper, "Nano-economics, Spinoffs, and the Wealth of Regions", Small Business Economics (2011) 37: 141-154.

[22] Michael Porter, "Location, Competition, and Economic Development: Local Clusters in a Global Economy," Economic Development Quarterly (2000); Eric Y Cho and Hideki Yamawaki, "Clusters, Productivity, and Experts in Taiwanese Manufacturing Industries". (University of Michigan Quantitative Analysis of Newly Evolving Patterns of Japanese, U.S. and International Trade: Fragmentation; Off-shoring of activities; and vertical intra-industry trade, October 16th, 2009). See also the empirical analysis by Walter Powell et al. of the emergence of life sciences clusters. The authors point out that "necessary conditions are a diversity of for-profit, nonprofit, and public organizations, a local anchor tenant, and a dense web of local relationships. These features make possible cross-network transposition, whereby experience, status, and legitimacy in one domain are converted into 'fresh' action in another. The argument does not hinge on specific types of organizations or ingredients; indeed, it is general enough to accommodate multiple pathways." Walter W. Powell, Kelley A. Packalen, and Kjersten Bunker Whittington, "Organizational and Institutional Genesis: The Emergence of High-Tech Clusters in the Life Sciences." In John Padgett,

John Matthews, who has extensively studied Taiwan's Hsinchu technology cluster, cited data from the Hsinchu Science Park to the effect that firms located in the park were 66 percent more productive than firms located outside of the park.[23] He attributed that fact in substantial part to the existence of "inter-firm linkages", cited by Marshall, which facilitated the establishment of highly efficient industry chains based on specialization by individual companies.[24] If Matthews' productivity estimate is anywhere near accurate, the implication is that companies' presence in a successful cluster gives them a major cost advantage relative to other companies, regions, and countries. A further implication is that current trends, with see U.S.-originated designs bring manufactured in Asia, could be at least partially offset through the establishment of local manufacturing industry chains in U.S. technology clusters.

The Taiwanese production chains which operate in and around the Hsinchu cluster include many of the companies which originated as spin-offs and start-ups. The fact that venture capital was available to such companies in their initial stages was an important aspect of their subsequent success[25]. In U.S. Innovation clusters, the creation of comparable spin-offs and start-ups will depend in significant part, to the availability of early stage funding.

An Emerging U.S. Cluster Strategy

Compared to the national cluster-development initiatives of other nations, U.S. federal programs have tended to be "siloed" and "uncoordinated."[26] Ginger Lew of the White House National Economic Council

---

Walter W. Powell, eds., *The Emergence Of Organization And Markets*, Princeton: Princeton University Press, 2012. Chapter 13.
[23] Matthews (2010). Op. cit. p. ii.
[24] "Firms that form part of a network have access to many more resources than would be available to them individually and such firms can contract with third parties to accomplish many more activities than would otherwise be under their control [and] the scope for specialization and intermediation grows. Matthews (2010) op. cit p. ii. Ding Yuan Yang, founder of Winland Electronic Corporation, located in Hsinchu Park, described this dynamic as follows: "Taiwanese companies may not coordinate well enough, but each company clearly defines its own focus. And [they] break down the PC industry into parts. Each company does what it does best. Some do the keyboards, some do the monitors, some do the motherboards, and some do the casing. That is what I call the ability to innovate." Interview with Ding-Yuan Yang, recorded February 23, 2011 (Computer History Museum, 2011).
[25] The government has contributed directly and indirectly to making Taiwan one of the world's largest sources of venture capital. "Taiwan—A Growing Model for Startup Companies" Central News Agency (November 27, 2011); "Fund to Invest in Venture Capital Firms" Taipei Times (March 19th, 2009); "Cabinet Inks Deal with Israeli Fund" Taipei Times (October 19, 2004).
[26] See presentation by Andrew Reamer of The Brookings Institution in *Growing Innovation Clusters for American Prosperity,* op. cit. Stockinger, Sternberg and Kiese examine differences between the "liberal market economy" approach of the United States and the "coordinated market economy" approach of Germany. Dennis Stockinger, Rolf Sternberg and Matthias Kiese, "Cluster Policy in

agreed that state and regional efforts have been "occurring on an ad-hoc basis without a formal U.S. policy."[27]

The federal government has become far more engaged in the past few years. Concerns that the U.S. is ceding global leadership in technology and innovation competitiveness in the wake of the National Academies' *Gathering Storm* report have prompted Congress to address clusters in legislation such as the America COMPETES Act.[28] Cluster building took on greater urgency in the wake of the financial crisis of 2008 and deep recession that followed. The departments of Energy, Commerce, Defense, Agriculture, Labor, and Education now all have programs devoted to regional innovation clusters.

Congress allocated substantial financial support for clusters such as advanced batteries through the American Recovery and Reinvestment Act of 2009, and the Obama Administration's budget for fiscal year 2011 included more than $300 million in new funding for federal agencies to assist regional innovation cluster initiatives. The Administration also developed a strategy to coordinate programs of various federal agencies to support "holistic, integrated solutions to building regional economies," according to Ms. Lew of the National Economic Council.[29] New federal programs include—

- The Energy Regional Innovation Clusters (ERIC) program, in which the DOE is leading six other federal agencies to help U.S. regions develop innovation zones. Regions compete for funds.[30]
- The Energy Innovation Hubs program, also led by the DOE, provides funds for multidisciplinary teams to deploy new clean-energy technologies at scale.
- The Economic Development Agency of the Commerce Department received $50 million under the Recovery Act to map cluster activities

---

Co-Ordinated vs. Liberal Market Economies: A Tale of Two High-Tech States," paper presented at Copenhagen Business School Summer Conference 2009, Denmark June 17-19, 2009.

[27] Presentation by Ginger Lew of the White House National Economic Council at the National Academies conference on *Clustering for 21st Century Prosperity*, Washington, DC, February 25, 2010.

[28] Sec. 603 of The America Creating Opportunities to Meaningfully Promote Excellence in Technology, Education and Science Reauthorization Act of 2010 (P. L. 111-358), known as the America COMPETES Act, provides for the Department of Commerce to provide competitive grants to regional innovation clusters and create a research and information program on regional innovation strategies.

[29] Lew presentation, op. cit. The Taskforce for the Advancement of Regional Innovation Clusters (TARIC), under the auspices of the National Economic Council, is overseeing the development and implementation of interagency clusters efforts. The TARIC was chaired by Ginger Lew until her retirement in June 2011.

[30] A public-private consortium led by Pennsylvania State University won the first grant of up to $130 million to form an innovation hub focusing on energy-efficient building technologies. For an explanation of the Energy Regional Innovation Clusters program, see Lew presentation, op. cit. Details on the announcement to fund the Energy Innovation Hub in Philadelphia can be found in the DOE press release of Aug. 24, 2010 at http://www.energy.gov/news/9380.htm.

across the country, develop evaluation metrics, and spread best practices.[31]
- The Small Business Administration is supporting efforts to develop robotics clusters in Michigan, Virginia, and Hawai'i with the help of state agencies and the Department of Defense.[32]
- The Department of Agriculture proposes a Regional Innovation Initiative in its FY 2011 budget. The agency would set aside 5 percent of the funding from around 20 programs, or about $280 million, would be granted on a competitive basis to pilot projects for regional planning in rural areas to create new industries.[33]
- The i6 Challenge program, announced by the Department of Commerce in May 2010, announced a $12 million partnership with the National Institutes of Health and the National Science Foundation to award grants to six teams around the country with the most innovative ideas to drive technology commercialization and entrepreneurship.[34]
- The Department of Labor proposes to use part of its FY 2011 budget request for a Workforce Innovation Fund pursuant to which states and regions would compete for funds by demonstrating a commitment to transforming their workforcesa program which will support cluster initiatives such as ERIC.
- The National Science Foundation plans to invest $12 million to promote "NSF Innovation Ecosystems" that support regional innovation clusters by helping faculty and students to commercialize innovations, form industry alliances, and launch start-ups.

Most of these new federal cluster initiatives are too new to assess.

Provided they are funded, however, and taken together with growing activity at the state and regional level, they mark a clear new direction for U.S. economic and innovation policy.

---

[31] The EDA is requesting $75 million to continue such activities.EDA, along with the Institute for Strategy and Competitiveness at Harvard Business School, has launched www.clustermapping.us the U.S. Cluster Mapping Web site. EDA sees this website, which creates a national database of cluster initiatives and other economic development organizations, as "a new tool that can assist innovators and small business in creating jobs and spurring regional economic growth." See EDA Update, October 6, 2011, "U.S. EDA Announces Registry to Connect Industry Clusters Across the Country."

[32] The SBA also proposes to use $11 million to train and advise small businesses on how to participate in clusters For explanation of Small Business Administration cluster activities, see the summary of remarks by SBA Administrator Karen Mills in National Research Council, Growing Innovation Clusters for American Prosperity, C. Wessner, rapporteur, Washington, DC: The National Academies Press, 2011.

[33] U.S. Department of Agriculture Fiscal Year 2011 Budget Summary and Annual Performance Plan http://www.obpa.usda.gov/budsum/FY11budsum.pdf).

[34] National Science Foundation press release, May 3, 2010.

"Regional innovation clusters have a proven track record of getting good ideas more quickly into the marketplace," Commerce Secretary Gary Locke explained at an NAS symposium. "The burning question becomes, 'How do we create more of them?'"[35] The best ways to create sustainable clusters and the appropriate role of public policy remain subjects of extensive debate. Perhaps what experts do agree on is that there are no standard recipes to develop new clusters. Strategies and public policies that are successful in some U.S. regions may not be appropriate in others. "If you attempt to replicate what was done in Silicon Valley, it just will not work," said Arizona State University President Michael Crow.[36] "You need to learn from them, draw on their lessons, and then work out your own solution." Andrew Reamer of the Brookings Institution warns that too many states have attempted to launch clusters in the same industries, such as biotechnology, regardless of whether they have any compelling competitive advantage. Economic development agencies also tend to jump onto fads. "Today, clusters have that danger," he said. "They're the next magic bullet."[37] Wholesale attempts to transport successful Asian strategies, where governments often dictate where clusters are to be located, also would be problematic in American regions, not least because clusters are "complex, self-organized, and composed of a broad patchwork of people and institutions," noted Maryann Feldman of the University of North Carolina. The role of government in the U.S., she said, is to provide incentives.[38]

To assess the wide range of experimentation at the state, regional, and federal level, the National Academies STEP Board has hosted wide-ranging dialogues over the past few years on how to stimulate innovation clusters. These symposia explored the role that clusters play in promoting economic growth, the role of government and universities in stimulating clusters, and specific strategies in place around America and abroad. The aim was to identify institutions and programs that can be leveraged to grow and sustain clusters.

Several common themes emerged from this extensive dialogue regarding guidance and best practices for state, federal, and regional policymakers. To maximize chances of success, regional innovation clusters need to—

- **Leverage local strengths:** Regional innovation cluster initiatives should be built upon existing knowledge clusters and comparative strengths of a geographic region. Government should promote proven

---

[35] Presentation by Commerce Secretary Gary Locke at the National Academies conference on *Clustering for 21st Century Prosperity*, Washington, DC, February 25, 2010.
[36] From remarks by Arizona State University President Michael Crow in *Growing Innovation Clusters for American Prosperity*, op. cit.
[37] Reamer presentation, op. cit.
[38] From presentation by Maryann Feldman of the University of North Carolina at Chapel Hill in *Growing Clusters for American Prosperity*, op. cit.

methods and practices, and federal support should leverage existing institutions and programs rather than create new ones.

- **Encourage self-organization:** Clusters should be developed from the ground up rather than designed and driven from afar. Private businesses and local education institutions and economic-development agencies are in the best position to identify opportunities, gauge competitive strengths, and mobilize wide community support for regional cluster initiatives. These initiatives should then compete for federal funds. Federal agencies can, however, make valuable contributions by spreading best practices and facilitating collaborations.

- **Pool resources:** Cluster initiatives can maximize their impact with limited funds if federal and state agencies, corporate leaders, higher education, charitable foundations, and nonprofits coordinate and pool their resources and organize their programs within the framework of comprehensive, overarching strategies.

- **Share risks:** The public and private sectors should share risks. Government investments in research and development infrastructure often are essential to kick-start innovation clusters and secure "buy-in" from the private sector. Many of the more successful initiatives require that corporations and private donors match or exceed public funds at the outset and through subsequent rounds of expansion.

- **Grow a trained workforce:** Attracting R&D centers and factories isn't enough to build a sustainable cluster. The entire ecosystem and supply chain must be considered. Programs should be in place to provide for workforce training, infrastructure, materials and component suppliers, shared prototyping and early-production facilities, and assistance for start-ups.

- **Connect clusters with local universities and labs:** Government-funded research in universities and national labs should be coordinated with nearby regional innovation clusters. Historically, federally funded R&D has not been connected to state and regional industrial development. Bridging that gap can create the local talent and technology base needed to convert these U.S. investments into domestic companies, industries, and jobs.

- **Provide long-term commitment:** Given the long-time horizon of serious R&D programs, corporations must know that federal and state incentive schemes and support for research infrastructure will be consistent, predictable and sustained. Steady public commitment is critical to give the private sector confidence to invest.

- **Provide incentives:** Public incentives often are necessary. Given the increasingly intense global competition in key industries, government seed grants, loan guarantees, tax credits, and other financial incentives can influence corporate decisions on where to locate corporate R&D and manufacturing investments. Such incentives must be carefully

designed to spur sound private investment rather than merely distort the market.
- **Monitor and measure**: Performance must be monitored and measured. Systems should be in place to evaluate the effectiveness of public investment in regional innovation cluster programs. Measuring performance is important to gauge which public policy tools work, make a compelling case for continued public support, and keep a focus on results.

Why Clusters are Relevant Now

One might ask why clusters are relevant now. Industries have congregated in certain geographic areas throughout history, and economists such as Andrew Marshall began studying such concentrations in England more than a century ago.[39] More recently, a number of European and U.S. academics such as Michael Porter of Harvard Business School have developed theoretical frameworks to explain how industrial clusters enhance regional development.[40] Governments in states such as New York, South Carolina, Ohio, New Mexico, and Michigan began to develop comprehensive cluster-development strategies over the past decade in an attempt to create new sources of high-paying jobs. Cluster strategies attracted more national attention in the wake of the 2008 economic crisis.[41]

Experts offer several reasons why regional innovation clusters have suddenly gained prominence. Maryann Feldman of the University of North Carolina suggests that there has been a shift in development thinking toward the notion that "all growth is local and grounded in place." There also is a greater

---

[39] See Alfred Marshall, *Principles of Economics*, London: Macmillan, 1920. The first edition of Marshall's classic textbook appeared in 1890. While the analysis of the spatial concentration of economic activity goes back to Marshall's analysis of the localization of industry it was given more recent attention by Paul Krugman, *Geography and Trade*, Cambridge: The MIT Press, 1991.

[40] Michael Porter, *The Competitive Advantage of Nations*, New York: The Free Press, 1990. Also see AnnaLee Saxenian, *Regional Advantage: Culture and Competition in Silicon Valley and Route 128*, Cambridge, Mass.: Harvard University Press, 1994. Other influential early works on global policies to promote innovation include Charles Freeman, *Theory of Innovation and Interactive Learning*, London: Pinter, 1987 and Bengt-Åke Lundvall, ed., *National Innovation Systems: Towards a Theory of Innovation and Interactive Learning*, London: Pinter, 1992. For an analysis of the historical evolution of the clusters for automobiles in Detroit, tires in Akron, Ohio, semiconductors in Silicon Valley, cotton garments in Bangladesh, see Steven Klepper, "Nano-Economics, Spinoffs, and the Wealth of Regions," *Small Business Economics*, 2011, vol. 36, issue 2, pp. 141-154. See also Christos Pitelis, Roger Sugden, and James R. Wilson, eds., *Clusters and Globalisation: The Development of Urban and Regional Economies*, Cheltenham: Edward Elgar Publishing, 2006.

[41] In his presentation at the National Academies conference on *Clustering for 21st Century Prosperity*, (Washington, DC, February 25, 2010) Assistant Secretary of Commerce for Economic Development John Fernandez observed that the deep recession "in many ways may have been an opportunity for a bit of a wake-up call across the board, not only for the federal government but also for the private sector and in public agencies across the country."

appreciation that innovation is a "cognitive and contextual process" that is based on face-to-face interactions, serendipity, and chance encouragers and their outcomes.[42] Mark Muro and Bruce Katz of the Brookings Institution offer three reasons why clusters have recently gained the attention of U.S. policy makers. First, new research confirms that strong clusters foster higher employment and wages, economic growth, and opportunities for innovation.[43] Second, clusters help provide a more grounded focus on the dynamics of the real economy as opposed to abstract macroeconomic management. Third, clusters provide a conceptual "framework for rethinking and refocusing economic policy" that help policymakers set priorities and get maximum impact out of limited resources.[44]

A number of think tanks and non-government organizations, meanwhile, recently have begun urging the federal government to more actively support regional clusters. Rather than call for massive new funding and new national institutions, however, several cluster advocates have urged federal agencies to make more effective and efficient use of resources they already deploy[45]. Michael Porter has said that federal programs are "appropriately criticized as often fragmented, duplicative, and inefficient."[46] An influential paper by Karen Mills, Elizabeth Reynolds, and Andrew Reamer tallied some 250 often-overlapping federal programs budgeted at $77 billion aimed at

---

[42] Feldman presentation, op. cit.
[43] For example, see Mercedes Delgado, Michael E. Porter and Scott Stern, "Clusters, Convergence, and Economic Performance," March 11, 2011, submitted for publication, accessible at http://www.isc.hbs.edu/econ-clusters.htm. Also see Karl Wennberg and Gören Lindqvist, "How Do Entrepreneurs in Clusters Contribute to Economic Growth?" SSE/EFI Working Paper Series in Business Administration No 2008:3 (http://swoba.hhs.se/hastba/papers/hastba2008_003.pdf).
[44] Mark Muro and Bruce Katz, "The New 'Cluster Moment': How Regional Innovation Clusters Can Foster the Next Economy," Brookings Institution Metropolitan Policy Program, September 2010.
[45] The Small Business Administration, Department of Energy, Department of Labor, the National Institute of Standards and Technology, the Department of Defense, and the National Institutes of Health, to name a few, all have programs aimed at promoting economic development. But rarely have these programs been coordinated with those of local development agencies, educational institutions, or non-government organizations pursuing similar aims. Inside the Department of Commerce alone, the Economic Development Administration, Technology Innovation Program, Manufacturing Extension Partnership, International Trade Administration, and the National Telecommunications and Information Administration all engage in activities that can be coordinated to promote regional clusters. See Jonathan Sallet, "The Geography of Innovation: The Federal Government and the Growth of Regional Innovation Clusters," in National Research Council, *Growing Innovation Clusters for American Prosperity, Summary of a Symposium*, C. Wessner, ed., Washington, DC: The National Academies Press, 2011.
[46] See Michael Porter, "Clusters and Economic Policy: Aligning Public Policy with the New Economics of Competition," ISC White Paper, Harvard Business School, November 2007 (http://www.isc.hbs.edu/pdf/Clusters_and_Economic_Policy_White_Paper.pdf).

assisting regional economy policy. The authors called on agencies to "link, leverage, and align" their resources with regional innovation cluster initiatives.[47]

State and Regional Case Studies

Michigan's New Battery Cluster

The steep decline in Michigan's auto manufacturing industry, which led to the loss of 800,000 jobs over the past decade, prompted state economic development officials to launch an intensive drive to develop new industrial clusters. The goal was to both diversify the state's industrial base and to expand on its existing strengths in automotive technologies and advanced manufacturing. Some 80 percent of U.S. automotive R&D is done within a 50-mile radius of downtown Detroit.[48]

After an extensive analysis, the Michigan Economic Development Corp. (MEDC) in 2005 targeted six industries: advanced energy storage, solar power, wind turbine manufacturing, bio-energy, advanced materials, and defense. The campaign to nurture a cluster in advanced batteries—a manufacturing industry that at the time was based almost entirely in Asia—was launched. Of the $2.4 billion allocated by the Department of Energy to advanced battery manufacturing projects under the American Reinvestment and Recovery Act of 2009, $1.3 billion went to Michigan-based factories.[49] At a National Academies symposium on Michigan's battery initiative, then Michigan Governor Jennifer Granholm declared that the state "is well on its way to becoming the advanced battery capital of the world."[50] (See Table 7.1 for a list of advanced battery and energy storage investments in Michigan.)

Michigan's approach is characterized by a comprehensive strategy that included investments in R&D, generous tax incentives, extensive training programs for engineers and skilled production workers, and public-private

---

[47] Karen G. Mills, Elisabeth B. Reynolds, and Andrew Reamer, "Clusters and Competitiveness: A New Federal Role for Stimulating Regional Economies," Metropolitan Policy Program at Brookings, April 2008.
[48] Southeast Michigan also has more than 2,500 parts suppliers, some 65,000 engineers, and tens of thousands of mechanical engineers, skilled machinists and veteran factory managers who can quickly turn conceptual prototypes into workable products that can be mass produced. Michigan Economic Development Corp. data.
[49] The factories included facilities by A123, Johnson Controls-Saft, Dow Kokam, and Compact Power, a unit of South Korea's LG Chem. The 16 battery-related plants being built in the state as of mid-2010 represent nearly $6 billion in private investment and are expected to create 62,000 jobs in five years. Ibid.
[50] Remarks by then-Gov. Jennifer Granholm at the symposium "Building the U.S. Battery Industry for Electric-Drive Vehicles: Progress, Challenges, and Opportunities" in Livonia, Mich., on July 26-27, 2010. Presentations from this symposium will be summarized in the forthcoming volume National Research Council, *Building the U.S. Battery Industry for Electric-Drive Vehicles: Progress, Challenges, and Opportunities*, Charles W. Wessner, rapporteur, Washington, DC: The National Academies Press.

partnerships that brought together universities, industry, government agencies, and the U.S. Army—a large potential customer for high-performance, energy-saving rechargeable batteries. What's more, the MEDC knew Michigan needed more than battery assembly plants and front-end R&D to build a sustainable industry and to compete with Asia. The state also needed an entire supply chain of materials and core components, most of which currently must be imported (see the advanced battery case study in this chapter).

Michigan targeted advanced batteries well before federal aid was available. The MEDC believed the state's base in car manufacturing and engineering gave it a clear advantage in an industry expected to surge as automakers boosted production of hybrid and electric vehicles.[51] The MEDC viewed advanced batteries as strategically important because they represent the core technology of future automobiles.[52] "Michigan did not want to stand by and cede leadership in power-train development to other states and countries," explained Eric Shreffler, who leads the MEDC's advanced energy storage program.[53]

The MEDC began by recruiting battery pack manufacturing and vehicle electrification programs. Michigan launched the Centers for Energy Excellence, which granted $13 million to lithium-ion battery developers Sakti3 and A123 on condition they secure federal funds and establish university partnerships. The agency also introduced the Michigan Advanced Battery Tax Credits program.[54] Industry response was so strong that the legislature tripled funding, to $1.02 billion. Under the scheme, Michigan refunds up to $100 million of a company's capital investment. Battery pack manufacturers receive a credit for each pack they assemble in Michigan.[55] The $1.3 billion in Recovery Act grants went to many of the same companies that received state aid.

---

[51] Presentation by Greg Main, then of the Michigan Economic Development Corp., at the National Academies conference on *Clustering for 21st Century Prosperity*, Washington, DC, February 25, 2010.

[52] A recent study out of the Center on Globalization, Governance & Competitiveness at Duke University concluded "If the United States is to compete in the future auto industry, it will need to be a major player in lithium-ion batteries." Marcy Lowe, Saori Tokuoka, Tali Trigg and Gary Gereffi, *Lithium-ion Batteries for Electric Vehicles: The U.S. Value Chain*, Center on Globalization, Governance & Competitiveness, Duke University, October 5, 2010.

[53] Presentation by Eric Shreffler of MEDC in *Building the U.S. Battery Industry for Electric Drive Vehicles, op. cit.* Another advantage is that the Detroit area is home to the U.S. Army's Tank Automotive Research, Development and Engineering Center (TARDEC), which leads Army development programs for fuel-efficient vehicles.

[54] Michigan's Advanced Battery Tax Credits initiative was created through an amendment to the Michigan Business Tax Act, Public Act 36 of 2007, to allow the Michigan Economic Development Authority to tax credits for battery pack engineering and assembly, vehicle engineering, advanced battery technology development, and battery cell manufacturing.

[55] The state of Michigan has since scaled back its tax credit program for manufacturers under a policy of new Governor Rick Snyder, who instead eliminated business income taxes. Instead, Gov. Snyder has said that future business incentives will be handled as appropriates. Previously

Michigan also invested in skilled-worker training and research programs for electrified vehicle technologies. It established the Center of Energy Excellence for advanced batteries under a program in which state funds for research projects are matched by corporations, universities, and national laboratories.[56] Michigan also is upgrading its workforce for the demanding needs of the electrified vehicle industry. The No Worker Left Behind program, which granted up to $10,000 for two years of college tuition to any person laid off or about to be laid off, enabled 135,000 residents to complete associate degrees or complete bachelor's or master's degrees. Michigan developed a special program for the electric-vehicle sector based on input from General Motors, Ford, Chrysler, Japanese automakers, and universities. Wayne State University and Michigan Technological University and the Michigan Academy for Green Mobility have trained hundreds of engineers. State agencies also formed a "skills alliance" that works with small tool-and-die suppliers that must diversify.[57] Wayne State University in Detroit has established a comprehensive degree program in electric-drive and battery technologies. The program's advisory board includes Ford, TARDEC, and Compact Power.[58]

Going forward, the MEDC is focusing on building out the advanced-battery supply chain in Michigan.[59] Broadening the state's advanced manufacturing base beyond automobiles to such industries as renewable energy equipment, aerospace, and defense is another goal.[60]

---

committed tax credits will be honored through 2013. See Amy Lane, "Snyder Budget: The Era of the Tax Credit is Over," *Crain's Detroit Business*, February 18, 2011.

[56] Michigan's Centers of Energy Excellence Program was established under Senate Bill 1380, Public Act 175. State contributions come from the Michigan Strategic Fund Board. For-profit companies receiving grants must secure matching federal funds and financial backing. Public Act 144 of 2009 allowed a second phase of the COEE program. These research programs also seek federal dollars. Partners in the advanced battery center include A123, Mascoma, Volvo, Mistra, and Smurfit Kappa. Another center of excellence involving Dow Corning and Oak Ridge National Laboratories focuses on low-cost carbon-fiber materials.

[57] From presentation by Andy Levin, former acting director of the state's Department of Energy Labor, and Economic Growth in *Building the U.S. Battery Industry for Electric Drive Vehicles*.

[58] See presentation by Simon Ng of Wayne State University in *Building the U.S. Battery Industry for Electric Drive Vehicles*, op. cit.

[59] Commitments so far include a cathode material plant by Toda America, electric motor component production by Magna, battery-testing facilities by AVL and A&D Technology, and an electric-drive testing operation by Eaton. MEDC currently lists 31 investments in Michigan's advanced battery and energy storage cluster. And more investments are planned. Johnson Controls is persuading Asian suppliers of materials to Michigan to supply its big lithium-ion battery joint venture in Holland, Mich., with France's Saft Advanced Power Solutions. Shreffler presentation, op. cit.

[60] For detailed information on non-auto manufacturing industries in Michigan, see the Michigan Economic Development Corp. Web site called "Michigan Advantage," (http://www.michiganadvantage.org). A sizeable cluster in solar power equipment is taking root. Michigan's Photovoltaic Tax Credit, which rebates up to 25 percent of a company's investments in manufacturing facilities, helped entice companies such as Dow, Uni-Solar, Hemlock Semiconductor, and Solar Ovanic to build or expand major production facilities.[60] Michigan's photovoltaic tax credit plan also has been scaled back.

TABLE 7.1 Advanced Battery and Energy Storage Investments in Michigan

	Company	City
1	AVL	Ann Arbor
2	TSC Michigan	Northville
3	Ricardo	Van Buren Twp
4	Magna	Kalamazoo
5	FEV	Auburn Hills
6	Detroit Testing Laboratory	Warren
7	Compact Power	Troy
8	Cobasys	Orion Twp
9	Battery Solutions	Howell
10	A&D Technology	Ann Arbor
11	Dow Kokam	Midland
12	fortu PowerCell	Muskegon
13	Johnson Controls - Saft	Holland
14	LG Chem	Holland
15	Eaton	Galesburg
16	Toda America	Battle Creek
17	Magna Electronics	Grand Blanc
18	Sakti3	Ann Arbor
19	ALTe	Auburn Hills
20	Xtreme Power	Wixom
21	Techno SemiChem	Northville
22	Azure Dynamics	Oak Park
23	Ford Wayne Assembly	Wayne
24	Chrysler LLC	Auburn Hills
25	GM	Warren
26	A123 Systems	Romulus
27	A123 Systems	Livonia
28	Magna Holdings of America	Troy
29	A123 Systems	Ann Arbor
30	Bright Automotive	Auburn Hills
31	Piston Group	Detroit

SOURCE: Michigan Economic Development Corporation, 2010.

The MEDC is forging deeper partnerships between state agencies, federal agencies such as the DOE and the DOD, and national laboratories. [61]

---

[61] One example of such state and federal collaboration is a new $27 million, three-year joint program involving Michigan, Oak Ridge National Laboratories, and TARDEC to commercialize advanced-storage and lightweight material research in DOE labs and adapt the technologies for military use.

New York State's Nano Initiative

Once a thriving center of advanced manufacturing, upstate New York fell on hard economic times as companies such as General Electric, IBM, Eastman Kodak, and Xerox began shifting production in the 1970s to other states and then overseas. The state government's decision in the early 1990s to invest heavily in nanotechnology research was part of a bold campaign to restore the region's industrial dynamism. The "main mantra" from the outset was public-private partnership involving government, industry, and academia, explained Pradeep Haldar of the Energy and Environmental Technology Applications Center in Albany.

New York's nano initiative began in the early 1990s, when then-Governor George Pataki gathered a diverse group of stakeholders to develop a strategy to revive the Upstate economy. The group decided to start with nanotechnology and concluded the region needed a plan that integrated R&D, education, and business. The vision was to bring a complete value chain to the region, including manufacturers, end users, suppliers, and construction firms specializing in clean rooms.[62]

The effort began modestly in 1993, when the state allocated $10 million over 10 years to a small research center for thin-film technology at the University at Albany-SUNY, run by Professor Alain Kaloyeros. Eight years later, the state named the university a center of excellence in nanotechnology. The state contributed $50 million and IBM $100 million to the center. Around the same time, IBM announced it would build its wafer plant in the Albany area. Then International Sematech, a consortium of 12 major chip manufacturers, picked the Albany campus as the site of a new 300mm computer chip R&D facility. Sematech invested $193 million and the state provided $160 million.[63] Semiconductor-equipment maker Tokyo Electron and lithography leader ASML also announced major R&D centers on the campus.

The region still lacked a sufficient pool of scientists, engineers, and highly skilled workers, however. New York's next major move was to establish the College of Nanoscale Science and Engineering in 2004. The NanoCollege, led by Dr. Kaloyeros, was the first of its kind in the U.S. The new college drew R&D investment from Applied Materials, Micron, AMD, Infineon, and a partnership between NIST and the U.S. Army, among others.

The new NanoCollege is designed to encourage collaboration with industry. Because it is was built from scratch, there were no long-standing silos to break down. Rather than organize faculty and students in rigid departments,

---

By demonstrating that such collaborations work, the MEDC hopes to secure further funding for "dual use" projects that can fuel new innovation clusters. Shreffler presentation, op. cit.

[62] Haldar, op. cit.

[63] For a concise history of the SUNY-Albany nanotechnology program, see Saul Spigel, "University of Albany Nantechnology Program, OLR Research Report, 2005-R-0146, February 9, 2005 (http://www.cga.ct.gov/2005/rpt/2005-R-0146.htm).

the focus has been on constellations of engineering and business people who can communicate easily.[64]

The campus also is home to the Institute for Nanoelectronics Discovery and Exploration (INDEX), a $500 million collaboration among 11 top U.S. universities, the National Science Foundation, NIST, and companies including Intel, IBM, Advanced Micron Devices, and Texas Instruments.[65]

In addition to performing research, the nanotechnology center fills important gaps in the industrial value chain for advanced manufacturing.[66] Indeed, one of the projects stated objectives was "to bring together in a single cluster the entire value chain of the nanotechnology industry. This includes not only manufacturers and end users, but also suppliers and construction firms." [67]

The progress has been dramatic. Anchored by the new College of Nanoscale Science and Engineering, the campus of the State University of New York at Albany has quickly emerged as one of the world's most important research bases for nano-scale materials, the building blocks for everything from tomorrow's computer chips and renewable energy devices to consumer electronics and medical devices.

Seven years after the launch of the NanoCollege, as it is called, boasts some of the best public-sector research facilities in the world.[68]

---

[64] "Since we built from the ground up, 70 percent to 80 percent of the people we hired came from industry, so they know what industry needs," explained Dr. Haldar. The college does not even have a technology-transfer office, which it regards as a barrier to commercializing intellectual property. Instead, the college gets its money from companies that pay it to perform research. Dr. Haldar suggested such arrangements are a model for the future. "Universities are being forced to deliver for companies in exchange for support," he said Haldar, op. cit.

[65] An overview of the Institute for Nanoelectronics Discovery and Exploration can be found on the Semiconductor Research Corp. Web site at http://www.src.org/program/nri/index/ INDEX is developing materials to replace complementary metal-oxide semiconductor technology (CMOS). Processes aren't expected to be introduced commercially for another decade. Interview with Lee Ji Ung, CNSE professor for Nanoscale Engineering, 2010.

[66] The center has a small business incubator, state-of-the-art prototyping labs, and testing facilities. To help develop a broad high-skills base needed for manufacturing, the nanotech consortium works with community colleges and high schools to train engineers, equipment and material suppliers, and clean-room construction professionals.

[67] Pradeep Haldar, "New York State's NANO Initiative," in National Research Council, Growing Innovation Clusters for American Prosperity, C. Wessner, rapporteur, Washington, DC: The National Academies Press, 2011.

[68] The complex includes one of the most advanced public-sector research prototyping facilities for 300 mm silicon wafers and four other "nano fabs" with clean rooms. The campus employs 2,600 and has 50 faculty, 29 masters and 126 doctoral students. The state's commitment has been rewarded with more than $5 billion in private investment. The 300 corporate partners include IBM, Applied Materials, and Tokyo Electron, which all have major labs at the 800,000-square-foot complex. Another 500,000 square feet in facilities are being added. Data are from College of Nanoscale Science and Engineering at the University of New York, "CNSE Quick Facts," accessible at http://cnse.albany.edu/AboutUs/CNSEQuickFacts.aspx.

Just as important for the state of New York, the nano-science compound is starting to make the region a magnet for high-tech manufacturing.[69]

The biggest industrial investment so far is a $4.5 billion silicon wafer plant being built on once-barren brush land north of Albany by GlobalFoundries, a joint venture between Advanced Micron Devices and an investment vehicle of the Abu Dhabi government. GlobalFoundries plans to become a new power in so-called chip foundries, which fabricate semiconductors on a contract basis.[70]

New York's nano initiative is branching far beyond semiconductors. In 2010, the NanoCollege announced it will develop degree programs with SUNY's Downstate Medical Center in Brooklyn to train a "new hybrid generation of research physicians" in nano-scale medical applications.[71] The NanoCollege took over management of a state-funded Smart System Technology & Commercialization center in Canandaigua in New York's Finger Lakes region. The facility had once been owned by Rochester-based Eastman Kodak—a company that had developed the first organic light-emitting diodes (OLEDs) and had sharply reduced its workforce in the region since the 1980s. The new center's biggest project is a $20 million collaboration between India-based Moser Baer Technologies, Universal Display Corp., and the NanoCollege to begin what is billed as the world's first pilot production of lighting devices using ODEDs.[72] If successful, Moser Baer plans to manufacture the devices on adjacent land earmarked as a future industrial estate.[73]

While SUNY's nanotechnology activities most attention, "some of the most pioneering innovation to nanosicence are taking place nearby at the

---

[69] Establishment of a state-of-the-art 300 mm research fab was a factor in IBM's decision to build and then expand a new multibillion-dollar wafer fab in East Fishkill, N Y, along with a generous state investment package.[69] Vistec Lithography moved to the campus from Cambridge, England, and now is shipping electron-beam lithography systems from a plant in nearby Watervliet, NY.[69] General Electric has announced plans for a $100 million advanced-battery plant nearby. Valerie Bauman, "IBM Will Invest $1.5B to Expand NY Operations," *Associated Press*, July 15, 2008; See Jack Lyne, "IBM's Cutting-Edge $2.5 billion Fab Reaps $500 Million in NY Incentives," *Site Selection* (http://www.siteselection.com/ssinsider/incentive/ti0011.htm); College of Nanoscale Science & Engineering press release, July 1, 2009 (http://cnse.albany.edu/Newsroom/NewsReleases/Details/09-07-01/Advanced_electron_beam_lithography_shipment_from_Vistec.aspx).
[70] Taiwanese companies dominate this industry (see semiconductor industry case study in this chapter). The state of New York contributed $1.2 billion in grants and tax credits to cover construction costs. Larry Rulison, "GlobalFoundries Board Approves Malta Fab Go-Ahead," *Albany Times Union,* March 20, 2009.
[71] College of Nanoscale Science & Engineering press release, March 1, 2011.
[72] College of Nanoscale Science & Engineering press release, October 23, 2010.
[73] Interview with Moser Baer CEO Gopalan Rajeswaran. In April 2011, the school received a $57.5 million Department of Energy grant to become the base of the U.S. Photovoltaic Manufacturing Consortium, a partnership that includes SEMATECH and the University of Central Florida. College of Nanoscale Science and Engineering news release, April 5, 2011, (http://www.albany.edu/news/12770.php).

Rensselaer Polytechnic Institute (RPI) in Troy, New York".[74] RPI operates an NSF Nanoscale Science and Engineering Center on campus which is pursuing research in areas such as carbon nanotubes and nanotube fabrication, graphenes, and liquids embedded with nanoparticles. In 2007, RPI opened the Computational Center for Nanotechnology Innovation (CCNI) in collaboration with IBM and New York State in North Greenbush, N.Y. to apply massive supercomputing power to the development of nanotechnology, and shrinking electronic device dimensions[75]. In 2010, researchers at RPI developed a new technique for mass producing graphenes, nanostructures which are "considered…potential heir[s]to copper and silicon as fundamental building blocks of nanoelectronics."[76]

New Industries from Old in West Virginia

The city of Morgantown, West Virginia, has become the hub of rapidly growing innovation clusters in biometrics and energy technologies, helping transform a regional economy long dominated by the exploitation of natural resources such as coal, natural gas, and timber. A key element in this biometrics initiative has been the FBI's relocation of its fingerprint center from the Washington, DC area to Clarksburg, WV, as encouraged by Senator Robert C. Byrd. West Virginia University serves as the catalyst for these clusters by using a variety of methods to leverage its research activities to promote businesses in the region, explained WVU President James Clements.[77]

One successful approach has been to "target and create" a cluster in an emerging technology niche in which West Virginia has competitive advantage. The cluster in biometrics[78] is an example. The region's advantages are WVU's 40-year history of research in technologies used to identify individuals through distinguishing biological traits [79] and its proximity to Washington, DC. In the late 1990s, WVU become the main partner to the Federal Bureau of Investigation's center of excellence in biometric identification, which is especially active in border-security technology.

---

[74] "Top Ten Regions for Nanotech Start-ups" Nanotechnology Law and Business (September 2006) p. 383.
[75] "Rensselaer Polytechnic Institute Appoints Cyberinfrastructure Expert James Myers to Lead the Computational Center for Nanotechnology Innovations," M2 Presswire (August 30, 2010).
[76] "Researchers at Rensselaer Polytechnic Institute Develop New Method for Mass Producing Graphene," *Nanotechnology Now* June 23, 2010.
[77] Presentation by West Virginia University President James Clements at the National Academies conference on *Clustering for 21st Century Prosperity*, Washington, DC, February 25, 2010.
[78] Biometrics is the use of science and technology to measure and statistically analyze biological data.
[79] For a concise history of the development of West Virginia's biometrics cluster, see Kim Harbour, "WV Biometrics: Fertile Ground for Innovation," on the West Virginia Department of Commerce Web site (www.wvcommerce.org/business/industries/biometrics/fertileground.aspx).

Interest in the field by law enforcement agencies and industry surged after the 2001 terrorist attacks on New York and Washington. WVU added one of the nation's first degree-granting programs in biometrics. Morgantown then became home to CITeR,[80] a National Science Foundation center that serves as a hub for identification technology research conducted around the country. The critical mass in R&D brought more private investment. Twenty corporations now have operations close to the center, including Booz Allen Hamilton, Northrup Grumman, Lockheed Martin, and Raytheon.

Now that West Virginia's biometrics cluster has critical mass, "more companies are coming in, and more people want to connect with our researchers and students," Dr. Clements said. The university is working with the Department of Defense to develop algorithms to measure the iris, for example, and on biometric fusion algorithms. These programs are generating spin-off companies.[81]

West Virginia is using a more regional approach to develop its energy innovation cluster, which capitalizes on the state's endowments of fossil fuels and timber. Morgantown is a hub because it is home to the National Energy Technology Laboratory. More than 100 faculty researchers in West Virginia work on advanced energy projects in areas such as liquefied coal for transportation fuel, environmentally safe access to natural gas reserves, and carbon sequestration. WVU also takes advantage of its proximity to two major Pennsylvania research universities, Carnegie Mellon University and the University of Pittsburgh.

One key to building a cluster is to coordinate all research activities, Dr. Clements said. "An ad-hoc series of projects is good, but when not properly coordinated you don't get to leverage them." West Virginia University, Carnegie Mellon University, the University of Pittsburgh, and the National Energy Technology Laboratory launched an applied-research collaboration aimed at commercializing the institutions' energy technologies.[82]

---

[80] CITeR stands for the Center for Identification Technology Research. It is an Industry/University Cooperative Research Center funded by the National Science Foundation. The center was founded by West Virginia University and is the I/UCRC's lead site for biometrics research and related identification technologies. CITeR also works with such agencies as the FBI, Department of Homeland Security, the Federal Aviation Administration, and the National Security Agency. CITeR established a second site for credibility assessment at the University of Arizona. A third is planned at Clarkson University in Potsdam, N.Y.

[81] Clements presentation, op. cit.

[82] The state also has organized the Advanced Energy Initiative, which is building public-private R&D research partnerships in new energy areas. To build the region's talent base, the state created a trust fund known as Bucks for Brains that allows WVU and Marshall University to recruit scientists who want to commercialize their research in energy and other fields. Bucks for Brains, officially known as, The West Virginia Research Trust Fund is a $50 million endowment established in 2008 by Senate Bill 287 that is to be matched by private contributions. West Virginia University and Marshall are to use the funds to recruit research scientists that intend to commercialize their work.

WVU also uses the traditional "linear model" of cluster building, Dr. Clements explained, in which research faculty help convert inventions into local businesses that then spawn other businesses. Protea Biosciences, a developer of technologies to discover new proteins in human blood and tissue samples, is an example. The fast-growing company began as a WVU research project, moved to a campus incubator, and then opened its own facility in Morgantown, where it continues to collaborate with university researchers on new products. Protea now is fostering its own spin-off companies, Dr. Clements said.[83]

Cluster-Building in Ohio

Like Michigan, Ohio is trying to diversify an economy whose manufacturing base has been battered by recession and offshore outsourcing. Economic development officials are designing road maps to nurture clusters in sectors such as energy storage, photovoltaic cells, smart-grid technology, electric transportation, and conversion of biomass and waste into energy. The leading universities in northeastern Ohio have long been at the cutting edge of important technologies—but not always good at translating them into local industries.[84]

Concerted efforts are underway to assure that the next round of innovations translate into regional industries. The Northeast Ohio Technology Coalition (NorTech), a nonprofit economic development organization funded by business associations and foundations, is spearheading efforts to create new clusters in technologies such as flexible electronics and renewable energy in a 21-county region that contains 42 percent of Ohio's population, including the cities of Cleveland, Akron, and Youngstown.[85] A group called Ohio Advanced Energy is trying to advance the region's small but growing cluster in photovoltaic cells and modules. Another group, PolymerOhio, is working to expand Ohio's strong bases in polymers and plastics, which includes 2,800 companies and research institutions employing 140,000 skilled workers.[86]

---

[83] Clements, op. cit.
[84] Kent State University has a Liquid Crystal Institute that helped pioneer that technology and patented the first LCD wristwatch in 1971, for example. Yet Japanese, Korean, and Taiwanese companies have dominated the vast LCD display industry for decades. Likewise, the University of Toledo has been at the forefront in thin-film photovoltaic technology. Yet little manufacturing of solar cells and modules has been based Northern Ohio. See presentation by Norman Johnston of Solar Fields, Calyxo, and Ohio Advanced Energy in National Research Council, *The Future of Photovoltaic Manufacturing in the United States: Summary of Two Symposia*, Charles W. Wessner, editor, Washington, DC: The National Academies Press, 2011. Most of the manufacturing capacity of industry leader First Solar, which originated as a University of Toledo spinoff, is in Germany and Malaysia. First Solar, "First Solar Corporate Overview Q2 2011," accessible on the company's Web site at Web site http://files.shareholder.com/downloads/FSLR/1301877449x0x477649/205c17cb-c816-4045-949f-700e7c1a109f/FSLR_CorpOverview.pdf.
[85] Presentation by Rebecca Bagley of NorTech. at the National Academies conference on "Building the Ohio Innovation Economy," Cleveland OH, April 26, 2011.
[86] PolymerOne data

Drawing on the scale and reputation of the Cleveland Clinic, a biomedical cluster in Greater Cleveland also is becoming well established, with more than 600 companies, including imaging giants such as Philips, General Electric, Siemens, Hitachi, and Toshiba. In 2008, the cluster attracted $395 million in venture capital as wells as National Institutes of Health funding.[87] The state of Ohio is financially backing these initiatives.[88]

According to NorTech President Rebecca Bagley, NorTech acts as a "quarterback" for regional cluster initiatives.[89]

Flexible Electronics

Flexible electronics is a top NorTech priority.[90] Ms. Bagley noted that northeast Ohio has a unique capability in liquid-crystal display technologies and electronics that can be printed on flexible substrates. Ohio has 11 core companies in flexible electronics, including start-ups such as Kent Displays, Alpha Micro, and Hana.[91] Among northern Ohio's chief assets are five universities that are leaders in new materials.[92] In the past, Ms. Bagley noted, Kent State produced technology breakthroughs, but the resulting manufacturing activity migrated elsewhere in the world. "How do we not make that mistake again?" she asked. The challenge in developing a roadmap is determining "what

---

[87] Muro and Katz, op. cit.

[88] Under the Ohio Third Frontier program, the state is investing $2.3 billion to support applied research, commercialization, entrepreneurial assistance, early-stage capital, and worker training to create an "innovation ecosystem" for a number of clusters. Since its launch in 2002, Third Frontier is credited with creating 55,000 direct and indirect jobs as of 2009; creating, capitalizing, or attracting more than 600 companies; and generating $6.6 billion in economic impact—nine times more than the state has invested. In 2010, Ohio taxpayers approved a $700 million funding boost so that Third Frontier can continue its activities through 2015. The availability of early-stage investment doubled from 2004 to 2008 to $445.6 million, much higher than the average U.S. growth rate. SRI International, *Making an Impact: Assessing the Benefits of Ohio's Investment in Technology-Based Economic Development Programs,* September 2009,
(http://development.ohio.gov/ohiothirdfrontier/documents/recentpublications/OH_impact_rep_sri_final.pdf). Details on the Third Frontier program can be found at http://thirdfrontier.com/History.htm M. Camp, K. Parekh, and T. Grywalski, *2007 Ohio Venture Capital Report*, Fisher College of Business, Ohio State University.

[89] Nortech identifies opportunities, maps the region's value chains, and coordinates resources and programs among a wide range of stakeholders. Partners include private companies, government agencies, and universities. Non-profit allies include JumpStart Inc., which helps develop early-stage business, and the Manufacturing Advocacy and Growth Network, which helps manufacturers adopt best practices and new technologies. Bagley presentation, op. cit.

[90] For a more detailed discussion on flexible electronics, see chapter on Industry Case Studies.

[91] SRI International, op. cit.

[92] The University of Akron is a global research power in polymers, for example, and Kent State's Liquid Crystal Institute remains at the top of its field. Case Western University has a strong program in new materials, Ohio State University is a leader in manufacturing technologies and nanotechnology, and the University of Cincinnati is strong in nano-scale sensors. NorTech's FlexMatters program has 10 staff, has raised $2 million, and is developing a roadmap for flexible electronics. In addition to seeding start-ups, the goal is to keep manufacturing of flexible electronics technologies invented in Ohio anchored in the region.

can we keep here, how do we build a research capacity, and how do we keep manufacturing processes that make sense for northeastern Ohio?" she said.[93]

Advanced Energy

One of the region's biggest cluster efforts is the Advanced Energy Initiative. Northeast Ohio has more than 400 companies in the advanced-energy space, according to Ms. Bagley. The organization believes the region is strong in 10 energy areas, including solar power, bio-fuels, and technologies for electric vehicles. The state has 49 companies involved in fuel cells.[94] Specific projects include an advanced-energy incubator in Warren, Ohio, a city hit especially hard by the loss of auto-related manufacturing jobs.

The photovoltaic industry is particularly promising. The Toledo area has been an early pioneer. The city was a hub of the glass industry for more than a century.[95] Industrialist Harold McMaster and a group of colleagues founded Glasstech Solar in 1984 and invested in manufacturing and basic research at the University of Toledo and other institutions.[96] An early start-up, Solar Fields LLC, was founded in a business incubator at the University of Toledo in 2003 but production was moved to Germany.[97] Norman Johnston, founder of Solar Fields and now head of Ohio Advanced Energy, a trade association promoting the renewable energy technology industries, said the dream is to convert Toledo from "glass city" to "solar city." Concerted efforts to build a regional photovoltaic cluster began in 1993, but progress was slow.[98]

Photovoltaic manufacturing investment is starting to grow, but still far below the levels of some countries in Asia or Germany.[99]

---

[93] Bagley presentation, op. cit.

[94] SRI International, op. cit.

[95] Pioneering Toledo firms included Edward Ford Plate Glass Company (1899-1930), Toledo Glass Company (1895-1931), and Libbey-Owens Glass Company (1916-1933).

[96] Harold McMaster (1916-2003 was once called "The Glass Genius" by Fortune magazine. In 1939 he became the first research physicist ever employed by Libbey Owens Ford Glass in Toledo and went on to found four glass companies. These included Glasstech Solar, in 1984, and Solar Cells, Inc., formed to develop thin-film cadmium telluride technology. Solar Cells was later bought and renamed First Solar, currently a world leader in thin-film PV.

[97] Solar Fields used cadmium telluride thin-film molecules, which were first demonstrated at a lab at the University of Toledo. After beginning small-scale production in Ohio, however, Solar Fields licensed its technology to Germany's Q Cells in a joint venture, Calyxo. Production shifted to Germany. After production was shifted to Germany, the company evolved into First Solar. Johnston presentation, op. cit.

[98] In 1997, the Ohio Department of Development awarded $18.6 million to Ohio Advanced Energy to establish the Wright Center for Photovoltaics Innovation and Commercialization, which has research operations at the University of Toledo, Ohio State University, and Bowling Green State University. Matching funds from federal agencies, universities, and industrial partners boosted that amount to $50 million. The state legislature also has supported the industry by mandating that at least 25 percent of Ohio's electricity come from clean and renewable sources by 2025. Ibid.

[99] First Solar recently expanded its production lines in Perrysburg, Ohio. Xunlight Corp., a Toledo start-up that is developing roll-to-roll thin film modules, will keep some of its production in the area.

Polymers

Efforts to broaden Ohio's polymer cluster also are underway. Akron has been a global center for the industry, due to its legacy as the rubber tire capital of America. Polymers have a wide variety of uses in industries such as automobiles, construction, medical equipment, and consumer electronics. The Ohio polymer ecosystem includes more than 250 mold builders and 1,600 plastics and polymer processors. It also includes makers of rubber, components, inks, fibers, and machinery. In addition to such companies as DuPont, 39 foreign-owned companies have subsidiaries and joint ventures in Ohio.[100]

To enhance Ohio's global competitiveness in the polymer industry, the Ohio Department of Development funded an Edison Technology Center—one of seven around the state that provide technical services to industries.[101]

The University of Akron is playing a major role in expanding the polymer industry. Over the past decade, the university has worked to become a national leader in research commercialization in general, ranking seventh in the nation in licensing revenue among universities without a medical school.[102] By 2009, Akron had 450 active and pending patents, had generated 30 start-ups, and was hosting more than 100 active industry-sponsored research projects.[103]

South Carolina's Innovation Cluster Push

South Carolina has long used incentives to attract industrial investment. In 1992, for instance, it outbid Nebraska to land a BMW auto assembly plant by offering $150 million in subsidies. But only in the past decade has the state seriously begun efforts to develop innovation clusters rather than compete

---

Another startup, Willard & Kelsey Solar Group, plans to begin production in Perrysburg in late 2009. Dr. Johnson said northern Ohio has more cadmium telluride and glass expertise than any other region in the world. Another startup, inverter company Nextronics in Toledo, has made the area's supply chain more complete. Dr. Johnson said that with 830 acres of abandoned but usable industrial space in Toledo alone, there is plenty of room for more capacity and for solar farms. Ibid.
[100] PolymerOhio, "Strength of Workforce," Sept. 23, 2008, accessible on Web site at http://www.polymerohio.org/index.php?option=com_content&view=article&id=70&Itemid=87.
[101] Called PolymerOhio, the center is a networking group linking companies, academic institutions, and service providers. Among other things, PolymerOhio set up a "polymer portal" to help small and midsized businesses obtain productivity-improving software with a grant from the NIST Manufacturing Extension Partnership. The center also supports training programs for middle-skill jobs needed in the polymer industry and is working with companies to develop for-credit and continuing education programs. Another PolymerOhio program promotes "re-shoring." It helps polymer companies maintain operations in Ohio or repatriate production from Asia. Details of Ohio's Edison Technology Centers can be found at http://www.development.ohio.gov/Technology/edison/tiedc.htm.
[102] Association of University Technology Managers (AUTM) data, February 2009.
[103] Akron's new Bioinnovation Institution leverages the university's expertise in polymers by working with three major hospitals and a medical school in the area to develop biomaterials. The aim is to build top biomedical and orthopedic research program in the world, according to University of Akron President Luis Proenza. From presentation by Luis M. Proenza of the University of Akron in *Growing Innovation Clusters for American Prosperity*.

mainly on its low cost advantage.[104]

The state began in 2002 by upgrading research programs at South Carolina's universities. The state legislature funded an endowed chair program to attract high-quality academic researchers, provide facilities and equipment for academic research, and establish the International Center for Automotive Research (CU-ICAR). In 2005, the South Carolina government hired Michael Porter and Monitor Group to develop a strategic plan to develop innovation clusters.[105] The legislature also passed the 2005 Innovation Centers Act and created SC Launch, a program managed by the South Carolina Research Authority that provides seed funding, guidance, networking, and commercialization services to South Carolina start-ups.[106]

South Carolina's cluster strategy focuses on five areas in which the state was deemed to have strengths and that had good commercial potential: automotive technology, advanced materials and fibers, alternative energy, life sciences, and related information technology. The South Carolina Research Authority has the mandate to build innovation systems to commercialize knowledge produced at the state's three research universities—Clemson, the University of South Carolina, and the Medical University of South Carolina.[107]

Clemson's CU-ICA has made a particularly strong impact in moving South Carolina's automotive industry beyond assembly work and into design. In 2004, the center consisted of 250 acres of undeveloped land with no funds or master plan. Today CU-ICAR includes a 90,000-square-foot graduate engineering center with world-class faculty holding well-funded endowed chairs.[108]

The automotive industry was seen to present a good opportunity for an innovation cluster because there were more assembly and parts makers within a 500-mile radius of Clemson than there are within a similar distance from

---

[104] See presentation by David McNamara of South Carolina Research Authority in *Growing Innovation Clusters for American Prosperity.*
[105] Michael E. Porter and Monitor Group, *South Carolina Competitiveness Initiative: A Strategic Plan for South Carolina,* South Carolina Council on Competitiveness, 2005, (http://www.isc.hbs.edu/pdf/200504_SouthCarolina_report.pdf). For an analysis of Michael Porter's impact on South Carolina economic development policy, also see Douglas Woodward, "Porter's Cluster Strategy Versus Industrial Targeting," University of South Carolina, presentation at ICIT Workshop, July 1, 2005,
(http://nercrd.psu.edu/Industry_Targeting/ResearchPapersandSlides/IndCluster.Woodward.pdf).
[106] Information on SCLaunch can be found on the organization's Web site, http://www.sclaunch.org/.
[107] McNamara presentation, op. cit.
[108] The center offers Master's and Doctoral programs in automotive engineering. BMW and Timken have R&D facilities, and the center has new partnerships with Michelin, IBM, Dale Earnhardt Inc., Sun Microsystems, the Society of Automotive Engineers, and the Richard Petty Driving Experience. In its first four years, CU-ICAR generated more than $220 million in public and private investment and created more than 500 new jobs with an average salary of $72,000. From presentation by Clemson University President James Barker in National Research Council, *Understanding Research, Science and Technology Parks: Global Best Practice: Report of a Symposium,* Charles W. Wessner, editor, Washington, DC: National Academy Press, 2009.

Detroit, according to Clemson President James Barker[109]. The Clemson faculty has been regarded as pioneers in vehicle-related R&D since the 1970s, first in rail systems and then in auto modeling and engineering.

Private companies and other contributors matched state grants. BMW, for instance, contributed $25 million for construction of the graduate engineering center of CU-ICAR and $15 million for an IT facility. In 2004, the Research Universities' Infrastructure Act offered $210 million for facilities and equipment that also was matched from other sources. Clemson received another $38 million in state funds matched by private sources to build CU-ICAR's physical plant and infrastructure.

A recent study estimated that the automotive cluster in South Carolina supported 84,935 full-time equivalent jobs in the state in 2008,[110] with 314 manufacturing companies and four non-manufacturing establishments engaged in research, logistics and wholesaling. [See Table 7.2]

TABLE 7.2 Automotive Cluster Economic Impact in South Carolina

	Value Added (Billions of Dollars)	Output (Billions of Dollars)	Employment
Direct	9.47	19.44	29,844
Indirect	3.82	4.47	26,774
Induced	2.57	3.14	28,317
Total	15.86	27.05	84,935

SOURCE: Douglas P. Woodward, Joseph C. Von Nessen and Veronica Watson, "The Economic Impact of South Carolina's Automotive Cluster," Darla Moore School of Business, University of South Carolina, January 2011, study prepared for South Carolina Automotive Council.

---

[109] South Carolina has assembly plants by six companies and more than 1,800 auto-related factories and companies. The automotive sector also was one of South Carolina's biggest sources of job growth between 1998 and 2008, adding around 10,000 jobs at a time when tens of thousands of jobs were lost in industries like textiles, apparel, chemical products, and furniture. At the time CU-CAR was launched, BMW was planning a $400 million expansion. The region also is in the middle of the Charlotte-to-Atlanta I-85 corridor, which not only ranks as the world's eighth-largest regional economy but also is the base of two-thirds of U.S. auto-racing teams. Michael E. Porter, "South Carolina Competitiveness: State and Cluster Economic Performance," Harvard Business School, prepared for Governor Nikki Haley, February 26, 2011, (http://www.isc.hbs.edu/nga/NGA_SouthCarolina.pdf).

[110] Douglas P. Woodward, Joseph C. Von Nessen and Veronica Watson, "The Economic Impact of South Carolina's Automotive Cluster," Darla Moore School of Business, University of South Carolina, January 2011, study prepared for South Carolina Automotive Council.

Clemson wants CU-ICAR to make an industrial impact far beyond car design. Because automobiles integrate so many complex parts and advanced technologies, Clemson regards the industry as "a platform for innovation that can be translated to countless other products and manufacturing processes," Present Barker explained. The vision is to create a workforce of systems engineers, people "who understand and improve how extremely complex systems interact with each other and apply these principles to a broad spectrum of applications."

The SC Launch program, meanwhile, is making progress in fostering South Carolina start-ups. Founded with a budget of only about $6 million, the program has helped start about 130 companies in its first three years as of 2009, according to David McNamara of South Carolina Research Authority. One-third of companies are in life sciences, with most of the rest in engineering, chemicals, and information technology.[111]

Creating Clusters in New Mexico

New Mexico's predicament a decade ago was that the state had one of the highest concentrations of scientists in the United States and received $6 billion in federal research dollars a year that went to its universities and major national laboratories, including Sandia National Laboratories, Los Alamos National Laboratory, and the Air Force Research Laboratory.[112] Yet New Mexico had produced few technology start-ups and high-tech industries. It also had one of the nation's highest unemployment rates.

Ambitious initiatives since then have spurred growth in several innovation clusters. In 1999, Sandia inaugurated a science park in Albuquerque for commercial offshoots, the first of its kind for a U.S. national laboratory (see science park section below).[113] Several years later, the state government developed a technology and economic-development roadmap.[114] The plan called

---

[111] "We have to use a lot of leverage," Mr. McNamara explained. "The good news about being small is that we can get all the legislators and economic development people we need in one room when a company wants to come to town." Mr. McNamara estimated that SC Launch had brought to the state about $65 million in follow-on funding secured by launch companies, and that the salaries at companies it works with average $77,000. In 2008, SC Launch received a national award for "Achievement in Building Knowledge-Based Economies" from the State Science & Technology Institute (SSTI). While SC Launch was not charged explicitly with the mission of forming clusters, "they seem to be forming on their own," Mr. McNamara said. McNamara presentation, op. cit.
[112] Presentation by Thomas Bowles, science advisor to then-New Mexico Governor Bill Richardson, at National Academies Technology Innovation Program Symposium, Washington, DC, April 24, 2008.
[113] Presentation by J. Stephen Rottler of Sandia National Laboratories, "Sandia National Laboratories as a Catalyst for Regional Growth," at http://sites.nationalacademies.org/PGA/step/PGA_056081.
[114] New Mexico's strategy is explained in *Technology 21: Innovation and Technology in the 21st Century Creating Better Jobs for New Mexico*, New Mexico Economic Development Department and Office of New Mexico Governor Bill Richardson, January 2009, (http://www.edd.state.nm.us/publications/Technology21.pdf).

for developing clusters in energy and environmental technologies, aerospace, film production, bioscience, information technology, and nanotechnology. The goal was to create new industries and "bridge the gap between federally funded basic R&D and the commercial sector," explained Thomas Bowles, science advisor to then-Gov. Bill Richardson.[115]

New Mexico invested in infrastructure, such as a supercomputer center and a $250 million "space port" in southern New Mexico that would serve as a base for a future commercial space industry. It greatly expanded science and technology education at the K-12 level and at universities[116]. The state even made direct-equity investments in private films and in a small jet manufacturer, Eclipse Aviation.[117]

The New Mexico Computing Application Center illustrates the state's use of public-private partnerships to build infrastructure for a 21st century knowledge economy. The center's 172-teraflop super computer, called Encanto, is billed as the fastest public-use computer in the world and is a collaborative effort by the state, Sandia, Los Alamos, the University of New Mexico, New Mexico State University, and the New Mexico Institute of Mining and Technology. Los Alamos's advanced simulation and computing program, which created a new hybrid supercomputer called Roadrunner that can perform 1,000 trillion calculations a sector, is a major contributor.[118]

Economic development is the super computer's express mission, explained Mr. Bowles. Encanto provides R&D support to New Mexico businesses and is an asset for attracting large corporations to the state.[119]

While not all of New Mexico's investments have paid off,[120] a number of these initiatives have changed the dynamics in the state economy. Germany's

---

[115] Bowles presentation, op. cit.
[116] New Mexico tapped a multibillion-dollar trust fund that manages royalties on oil, gas, and minerals extracted from public lands to set aside $500 million for early-stage investments in start-ups to be managed by venture capital firms that establish offices in the state. New Mexico also offered some of the most generous financial incentives to companies shooting films in the state and building high-tech manufacturing or R&D facilities.
[117] Many of these investments by New Mexico are described in Pete Engardio, "State Capitalism," *BusinessWeek*, February 9, 2009.
[118] Sandia offers its expertise in massively parallel computing and has its own 40-teraflop supercomputer, Red Storm. Encanto is based at Intel's new Energy Research Center in Rio Rancho. The state committed $42 million over five years, while other partners contributed $60 million.
[119] Dreamworks Animation is among the high-profile clients. All colleges and universities are to be equipped with "gateways" to Encanto. Gateways are large, high-definition displays with high-speed connections to the super computer through a secure network. So far, 10 of a planned 38 gateways have gone into operation. Businesses, community groups, and public schools all have access to the gateways, which provide services such as 3-D visualization theaters and distance learning. Eventually, the network will connect health centers, schools, libraries, museums, and homes. Information about the New Mexico Computing Applications Center is available on the Web site, http://nmcac.net/.
[120] Eclipse Aviation filed for bankruptcy in 2008. Production of planes has not yet resumed under new management. Virgin Galactic's plans to begin commercial space flights at the Space Port have

Schott Solar has opened a module plant in Albuquerque. Intel announced a $2.5 billion upgrade to a plant to make 32nm chips. Fidelity and Hewlett Packard announced 1,000-worker financial- and technical-support service centers. More than 155 major movie and television productions have been filmed in New Mexico since 2003, including *Terminator Salvation, True Grit, Stargate, The Book of Eli,* and *Cowboys and Aliens,* contributing nearly $700 million to the economy in 2010.[121] Until activity was slowed by the recession, venture capital investment in New Mexico had surged from just $6.6 million in 2003 to a peak of $128 million in 2007.[122] Forty companies that had received $370 million from the state as of 2009 went on to raise an additional $1.7 billion.[123]

The science park next to Sandia, meanwhile, has filled up with 30 companies. The jobs pay salaries that are twice as high as the Albuquerque average. "For a state such as New Mexico, which still tends to rank at the bottom of most national statistics, this is something that the city, the county, the state, and our laboratory are quite proud of," Dr. Rottler said. The park has a "very aggressive" goal to account for 6,000 jobs in another decade.[124]

Policy Lessons for U.S. Innovation Clusters

Intensifying international competition for leadership in next-generation industries means that U.S. state and regional governments no longer compete only against each other for investment. They also must compete against regions around the world with comprehensive and increasingly well-funded strategies to develop world-class innovation clusters that have absorbed many lessons from the United States. For U.S. regional innovation clusters to remain globally competitive, therefore, their policies and strategies should be benchmarked against those of rival clusters in Europe, Asia, and Latin America.

The wide range of initiatives across the U.S. and around the world shows there are many methods of leveraging regional technology advantages into a cluster of innovative companies. There are, however, several common features of successful clusters studied by the STEP Board. Each had research universities or national laboratories at their core, and in some cases, both. They sprang from pre-existing regional industries and R&D strengths. Successful clusters also feature strong collaborations among academia, industry, economic-development agencies, and nonprofits and can require investment in the entire innovation ecosystem—public-private research programs, work-force

---

been postponed. See Dan Frosch, "New Mexico's Bet on Space Tourism Hits a Snag," *New York Times*, Febraury 23, 2011.
[121] New Mexico Film Office, "Film/Media Production Statistics FY2003-FY2011."
[122] National Venture Capital Association data, fastest growth in country.
[123] Adam Bluestein and Amy Barrett, "How States Can Attract Venture Capital," *Inc. Magazine,* July 1, 2010.
[124] Rottler presentation, op. cit.

development, entrepreneurial training, shared infrastructure such as incubators and prototyping facilities, and access to early-stage capital. Some cluster initiatives entail substantial new public investments. In other cases, progress was achieved largely because a range of stakeholders aligned existing programs around common goals.

National governments in some nations take the lead in targeting and developing innovation clusters. The approach that has proved most effective in the United States is for cluster initiatives to emerge from the ground up, from companies, research institutions, and public agencies that identify unique opportunities based on their strengths and share a common interest in economic development. Federal agencies can play a valuable support role, however. Federally funded university and national laboratory basic and applied research programs can be oriented toward the activities of local industrial clusters and university research programs. Government agencies such as the departments of Energy, Defense, Commerce, Labor, and Agriculture can align a wide range of existing programs intended to accelerate development of strategic technologies or to promote economic development with regional cluster initiatives that are deemed to have the best chances of success. Federal agencies also can contribute by sharing best practices with regional agencies and by facilitating networking with researchers, investors, and support organizations across the United States.

TWENTY-FIRST CENTURY RESEARCH AND INDUSTRY PARKS

At the heart of many major innovation clusters are found dynamic research parks that integrate scientists, engineers, and entrepreneurs from universities, government research institutes, and the private sector.[125] Such innovation zones have been a distinct U.S. advantage since the first was established in 1948 in Menlo Park, California, soon followed by research parks affiliated with universities such as Stanford, the Massachusetts Institute of Technology, and those surrounding North Carolina's Research Triangle.[126]

New research parks are now appearing in the rest of the world. Nations and regions such as Singapore, Taiwan, Germany, France, China, Mexico, and Spain see science and technology parks as critical infrastructure to spur innovation, create new industries, and generate tens of thousands of high-paying

---

[125] Dr. Albert N. Link defines a university research park as "a cluster of technology-based organizations that locate on or near a university campus in order to benefit from the university's knowledge base and ongoing research." See Albert N. Link, "Research, Science, and Technology Parks: An Overview of the Academic Literature," paper published in National Research Council, *Understanding Research, Science and Technology Parks: Global Best Practice: Report of a Symposium,* Charles W. Wessner, editor, Washington, DC: National Academy Press, 2009.
[126] See Rachelle Levitt, ed., *The University/Real Estate Connection: Research Parks and Other Ventures*, Washington, DC: Urban Land Institute, 1987. See also Roger Miller and Marcel Cote, *Growing the Next Silicon Valley: A Guide for Successful Regional Planning*, Toronto: DC Heath and Company, 1987.

jobs.[127] There are more than 700 research, science, and technology parks of various stages of development around the world, according to the Association of University Research Parks.[128] Research parks help create clusters of knowledge among researchers, academic institutions, companies, and government agencies.[129] They incubate and spin off innovation-based companies.[130] They provide value-added services and high-quality space and lab facilities that cannot be found on universities campuses.[131] Science and research also are valuable for universities and national laboratories that seek to make a broader economic and societal impact.[132] As former University of Maryland president C. D. Mote explained, a research park is an "essential tool for institutions with an entrepreneurial and innovative culture that hope to benefit from complicated partnerships on a global scale."[133]

Some of the newer parks erected around the world greatly exceed the size, scope, and pubic commitment of those in the U.S.[134] The average major research park in China, for example, covers more than 10,000 acres, compared to an average of 358 for those in the U.S. science parks.[135] Unlike their American counterparts, many of these parks also are parts of comprehensive national strategies aimed at improving competitiveness and accelerating the transition to 21st century knowledge economies. Whether they are known as research parks, science parks, or technopoles, such districts now are found in 60 countries at all stages of development. Some of the largest—

---

[127] Many of the findings in this chapter are from a March 13, 2008, symposium at the National Academy of Sciences in Washington, DC, convened by the National Academies' Board on Science, Technology, and Economic Policy (STEP) in partnership with the Association of University Research Parks (AURP). The proceedings are summarized in National Research Council, *Understanding Research, Science and Technology Parks: Global Best Practice: Report of a Symposium,* Charles W. Wessner, editor, Washington, DC: National Academy Press, 2009.
[128] Data on Association of University Research Parks Web site at http://www.aurp.net/history-of-aurp.
[129] See, for example, presentation by Pradeep Haldar of the Energy and Environmental Technology Applications Center at the University of New York in Albany in National Research Council, *Growing Innovation Clusters for American Prosperity: Summary of a Symposium,* Charles W. Wessner, rapporteur, Washington, DC: The National Academies Press, 2011.
[130] For example, see presentation by David Holden of France's MINATEC in *Understanding Research, Science, and Technology Parks,* op. cit.
[131] See presentation by C. D. Mote, former president of the University of Maryland, in *Understanding Research, Science and Technology Parks,* op. cit.
[132] See presentation by Richard Stulen of Sandia National Laboratories in *Understanding Research, Science and Technology Parks,* op. cit.
[133] Mote, op. cit.
[134] See remarks by U.S. Senator Jeff Bingaman in *Understanding Research, Science and Technology Parks,* op. cit.
[135] "Average North American Research Park" data are from "Characteristics and Trends in North American Research Parks: 21st Century Directions," commissioned by AURP and prepared by Battelle, October 2007; "Average IASP Member Park" data are from the International Association of Science Parks annual survey, published in the 2005-2006 International Association of Science Parks directory.

- Zhangjiang High-Tech Park in Shanghai's Pudong district sits on what was farmland in 1992. Now, more than 6,000 companies and 160,000 workers cover 20 square kilometers, with plenty of room to expand. The more established Zhongguancun Science Park in Beijing hosts more than 20,000 enterprises and 950,000 employees, and produced $110 billion worth of income as of 2009.[136]
- Singapore is pouring some $10 billion into a network of research parks in a 500-acre urban district called One North. Developments include Biopolis, a 4.5 million-square-foot campus housing 5,000 life science researchers from universities, hospitals, and multinationals such as Eli Lilly and Novartis. Another is Fusionopolis, a futuristic 24-story tower filled with media, communications, and information-technology companies.[137]
- Barcelona is transforming an old industrial district into a 100-block zone called 22@Barcelona. The project involves transforming 115 blocks in Barcelona's historic cotton district into an international hub for more than 1,000 media, information technology, and medical technology companies; research institutes; and university labs that could employ 150,000 in 15 years.[138]

Because they are relatively new, many big research parks outside the U.S. have the benefit of learning from the experiences of American parks. They also can be designed to take advantage of the modern, fast-evolving demands of 21st century global competition. New parks are building bridges between academia and industry, paying for top international talent in multiple disciplines, building state-of-the-art labs, and establishing programs to train entrepreneurs and incubate new companies. Many offshore parks also offer financial incentives that many U.S. parks cannot match, such as tax holidays, research grants, low-cost rent and lab space. As at most U.S. parks, they also have programs, such as incubators, entrepreneurial coaching, and prototyping facilities, aimed at launching new companies. Unlike in other nations, parks in the U.S. are supported by state and local governments, with limited federal support.[139]

With so many alluring options now available worldwide to industry and entrepreneurs, American research parks must rise to meet the tougher global

---

[136] Data from Shanghai Zhangjiang Group and from presentation by Zhu Shen of BioForesight in *Understanding Research, Science and Technology Parks,* op. cit.
[137] Singapore Economic Development Board data cited in Pete Engardio, "Innovation Goes Downtown," *BusinessWeek*, Nov. 19, 2009. Also see presentation by Yena Lim, Singapore Agency for Science, Technology and Research, in *Understanding Research, Science and Technology Parks.*
[138] Pete Engardio, "Barcelona's Big Bet on Innovation," *BusinessWeek Online,* June 8, 2009.
[139] See remarks by Phillip H. Phan of Rensselaer Polytechnic Institute in *Understanding Research, Science and Technology Parks.*

competition. Presentations at STEP symposia suggests that U.S. science and technology parks can indeed remain globally competitive with strong public-private partnerships, proper investment, and consistent policy support.

American parks still possess several important advantages. These advantages include well-established ecosystems for nurturing new high-tech companies and a strong network of research universities and national laboratories. Although American global dominance of science and engineering has waned, 15 of the world's 30 top-rated engineering schools and 29 of the top 100 are still based in the U.S., according to *U.S. News and World Report* magazine's 2010 rankings. The U.S. also boasts 12 of the top 30 universities in natural science and physics and 14 in life sciences.[140] America's national labs, devoted primarily to research for national defense, energy, and medicine, are among the world's greatest depositories of scientific and engineering knowledge.

The first part of this chapter describes a sampling of science park initiatives in Singapore, China, Germany, France, India, Hong Kong, and Mexico, as well as some of the key lessons they offer. [See Figure 7.2] The next section of this chapter explains how a sampling of U.S. science and technology parks—some new, some old—are addressing the challenges of intensifying global competition by fostering innovation and creating new companies, industries, and high-paying jobs.

Research Parks Around the World

Singapore's One North Masterplan

Singapore illustrates the ways in which nations are using science parks as focal points for developing 21st century knowledge economies. Having already established itself as a leading global R&D hub for multinationals, Singapore now wants to evolve into a leading base of innovation. Singapore's advantages include a highly educated workforce proficient in math and science[141] and a government that has long been willing to invest in world-class infrastructure for next-generation industries.

Singapore is building a network of science parks in a 500-acre urban district called One North, located close to the National University of Singapore, National University Hospital, and Singapore Polytechnic. The goal is to create "an ecosystem designed to nurture new ideas and push them quickly to reality," explained Yena Lim of the Singapore Agency for Science, Technology and Research at a STEP symposium.[142]

---

[140] *U.S. News and World Report* World's Best University Rankings based on QS World University Rankings, Sept. 21, 2010.
[141] See comments by Phillip Phan of Rensselaer in *Understanding Research Parks*.
[142] See presentation by Yena Lim of the Singapore Agency for Science, Technology, and Research in *Understanding Research, Science and Technology Parks*.

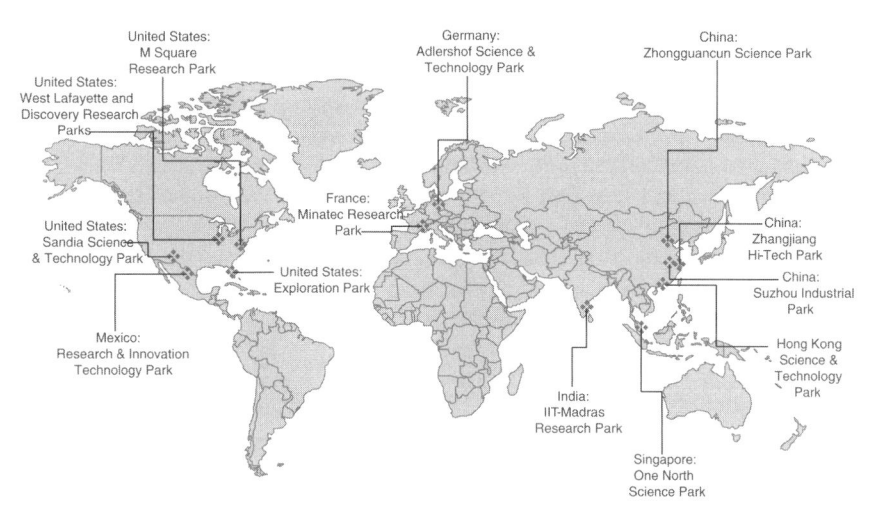

FIGURE 7.2 Global research parks discussed in Chapter 7.

The Biopolis project in One North is the furthest along. This city within a city is central to the government's plan to make Singapore "the biomedical hub of Asia," Ms. Lim said, by attracting scientists, researchers and entrepreneurs from around the world. Unlike traditional research parks that are located in suburbs, Biopolis is in the heart of Singapore. The campus is designed to encourage scientists and researchers in disciplines as diverse as proteomics, X-ray crystallography, and DNA sequencing to intermingle and collaborate on new projects. Lab buildings are situated intentionally close together and include amenities such as convenience stores, a gym, child care, restaurants, and a pub.[143] Singapore officials expect that, within a few years, Biopolis will have 5,000 researchers, making it bigger than any U.S. biomedical cluster aside from San Diego.[144]

Fusionopolis, a development that opened in 2008, is nearby. It serves as a one-stop science and R&D haven mixing companies and research labs in new energy technologies, aerospace, nanotechnology, sensors and sensor networks,

---

[143] Two Biopolis phases have opened since ground was broken in 2001. Buildings house seven research institutes, including the Genome Institute of Singapore, the Institute of Bioengineering and Nanotechnology, the Institute of Molecular and Cell Biology, and labs of 20 companies, including Novartis, Eli Lilly, and GlaxoSmithKline. The largest private tenant will be Procter & Gamble, which announced it is building a Singapore $250 million (US$195 million) global innovation hub that will cover 34,000 square feet when it opens in 2013. Linette Lim, "P&G Invests $250 million in Innovation Centre," *The Business Times*, January 27, 2011.

[144] Singapore Economic Development Board Executive Director Yeoh Keat Chuan quoted in Pete Engardio, "Singapore's One North," *BusinessWeek,* June 1, 2009,
(http://www.businessweek.com/innovate/content/jun2009/id2009061_019963.htm).

cognitive science, and devices for wired homes. Fusionopolis is housed in a 24-story building designed by renowned Japanese architect Kisho Kurokawa that includes service apartments, experimental theater space, hotels, and a shopping mall featuring smart-shopping technologies. A*STAR, which manages the complex, also helps integrate the work of research institutes with multinationals, small enterprises, and start-ups, as well as with agencies such as the Economic Development Board.[145]

China's Mega Parks

Sprawling research and science parks are the most visible manifestations of China's big push in innovation and science-based development.[146] Susan Wolcott identifies three basic types of Chinese research parks—"multinational development zones" such as those in Shenzhen and Suzhou designed to attract foreign companies as growth engines, "multinational learning zones," and "local innovation zones" catering mainly to domestically generated technology with some interactions with foreign companies.[147]

The scale of China's leading science parks surpasses that of Research Triangle Park, by far America's largest.[148] The Chinese government invested $1.4 billion in Suzhou Industrial Park, for example, home to operations of 113 of the Fortune 500 companies.[149] The more established Zhongguancun Science Park in Beijing hosts more than 20,000 enterprises and 950,000 employees, and has produced $110 billion worth of income as of 2009.[150]

Strong government policy and financial support at the national, regional, and local level therefore is important in China, said Zhu Shen, CEO of the San Diego pharmaceutical consulting firm BioInsight.[151] Some parks offer tax waivers, free rent, and financing to attract multinationals and "sea turtles," as overseas Chinese who return to the mainland are known.[152] Good ones also

---

[145] Details on Fusionopolis and Biopolis can be found on the A*STAR Web site at ttp://www.a-star.edu.sg/?tabid=860.

[146] Kazuyuki Motohashi and Xiao Yun, "China's innovation system reform and growing industry and science linkages." *Research Policy*, 36, pp. 1251-1260, 2007.

[147] For a review of China's science and technology industrial parks, see Susan M. Walcott, *Chinese Science and Technology Industrial Parks*, Aldershot: Ashgate, 2003. Also see Kazuyuki Motohashi and Xiao Yun, "China's innovation system reform and growing industry and science linkages." *Research Policy*, 36, pp. 1251-1260, 2007.

[148] Research Triangle Park is about 28 square kilometers in size. Beijing's Zhongguancun Science Park is about 280 square kilometers, or larger by a factor of ten. "Zhongguancun Going Ahead", www.sing.com.cn (June 26, 2002).

[149] The Suzhou Industrial Park is undergoing a transformation, however, because industries were not developing as planned, with many companies producing low value-added goods and foreign producers relying on markets and supply-chains outside of China. Zhou Furong and Zhang Zhao, "Suzhou Industrial Park Faces Challenges on Path to Change," *China Daily*, March 16, 2010.

[150] From Zhu Shen presentation, op. cit.

[151] Ibid.

[152] *China Daily*, "China Luring 'Sea Turtles Home." December 18, 2008. The recent U.S. financial crisis appears to be accelerating the trend of repatriating Chinese professionals and scholars.

offer business resources, such as one-stop services for accounting, intellectual property advice, and counseling.

Some of the most prominent Chinese science parks are not single industry clusters; they are instead characterized by industrial diversity and a high concentration of R&D facilities of universities, corporations, and government research institutes. They are major centers of innovation efforts for industries such as pharmaceuticals, information technology, and high-tech electronics.[153]

The Zhangjiang High-Tech Park in Shanghai's Pudong district also illustrates the breadth and scale of modern Chinese science parks. Built on what was farmland in 1992, it has more than 6,000 companies—2,500 of them from overseas—and covers 20 square kilometers.[154] The park is expanding at the astounding rate of two kilometers a year and has another 58 square kilometers of undeveloped land.[155] Zhangjiang's workforce has grown from 5,000 in 2000 to 160,000 in 2011.[156]

The Zhangjiang park has become Shanghai's premier innovation zone. More than 30 government research institutes and more than 100 multinational R&D centers have located in the district in industries as diverse medical equipment, life sciences, new energy, information technology, semiconductors, and multimedia gaming.

For companies, the advantages of being inside the park include low taxes and land costs that are much lower than in other areas of Shanghai, one of

---

[153] "The idea is that these are places where a lot of the top talents from different fields are clustered—this then is what attracts private enterprises," she said. Zhu Shen presentation, op. cit. Beijing's Zhongguancun Science Park, for example, features companies in information technology, new energy, biomedicine, advanced manufacturing, and new materials. The life science district alone has 100 companies, around 80 percent of them Chinese start-ups. "Sea turtles" founded or run many of these companies For information on some of the most prominent "sea turtles" in the Chinese pharmaceutical research industry, see the slide show, Pete Engardio, "Who's Who in Chinese Sea Turtles," *Bloomberg BusinessWeek*, at http://images.businessweek.com/ss/08/09/0904_chinese/index.htmhttp://images.businessweek.com/ss/08/09/0904_chinese/index.htm. A major attraction of the park is affordable land close to the life-sciences research programs of Tsinghua University, Peking University, and the Chinese Academy of Sciences, according to Jin Guowei, vice general manager of Beijing Zhonguancun Life Science Park Development Co. Interview with Vice General Manager Jin Guowei and Chairman Yuan Shugang of Beijing Zhongguancun Life Science Park Development Co. in Beijing; The park claims that companies on campus have 40 to 50 drugs that are in the first phase of clinical trials. Beijing Zhonguancun Life Science Park Development, a state-run company that manages the park, offers tenants technical support services, such as molecular analysis, and helps them apply for national research funds. The administration also organizes seminars to explain government programs. A second phase is under construction. Interview with Jin and Chairman Yuan Shugang of Beijing Zhongguancun Life Science Park Development Co. in Beijing.
[154] Zhangjiang High-Tech Park data.
[155] Interview with Yin Hong of Shanghai Zhangjiang Group in Shanghai.
[156] Zhangjiang High-Tech Park data.

China's most expensive real estate markets[157]. Companies can draw from some 9,000 researchers, scientists, and workers from several nearby universities. The elite Fudan University has moved its research institutes for software, integrated circuits, and pharmaceuticals to Zhangjiang. Just as importantly, Shanghai is one of China's biggest magnets for international talent, especially Taiwanese. More than 10,000 non-Chinese nationals work in Zhangjiang. The ability to hire engineers, scientists, and managers that have lived and worked abroad is one of the main features that draw multinationals to Zhangjiang, according to Yin Hong, vice general manager of Shanghai Zhangjiang Group In Beijing, by contrast, multinationals mainly recruit from local universities.[158] Shanghai Zhangjiang Group, the company that manages the park, serves as a one-stop shop for handling red tape.[159]

Zhangjiang has become the core of several of Shanghai's most promising emerging innovation clusters. In life sciences alone, Zhangjiang includes R&D centers by Roche, Eli Lily, Pfizer, Novartis, GE, and AstraZeneca, all of which have announced major expansion plans. There also are 60 small-molecule drug-development companies, 35 medical device and diagnostic firms, and 15 traditional Chinese medicine companies. Small and midsized companies get financial help via grants from the National Technology Innovation Fund and some $2.5 billion in venture funding set aside for the park.[160] Zhangjiang also includes two of China's most important drug-research companies—Wuxi PharmaTech and Hutchison MediPharma—that assist multinationals in early-stage discovery.[161]

---

[157] The Zhangjiang High-Tech Park also offers affordable apartments to staff of companies based there. Location is another selling point. Zhangjiang is in the center of Pudong, a district across the Huangpu River from downtown Shanghai that is a major industrial zone and is home of Shanghai's financial district. Zhangjiang is within 50 minutes of both Shanghai airports. Three ring roads pass through or alongside Zhangjiang, and the park has three subway stops, making it within commuting range of much of Shanghai.

[158] Yin Hong, op. cit.

[159] If all documents are ready, according to Vice General Manager Yin Hong, the company can approve an application to enter the park within 10 working days. Mr. Yin said the park concentrates on "intelligence-intensive" companies primarily engaged in research in five main clusters: semiconductor manufacturing and design, pharmaceutical research, renewable energy, information technology and gaming, and advanced manufacturing. Of the park's 160,000 workers, only around 10 percent are engaged in manufacturing. Two-thirds of those employees have at least a bachelor's degree. Mr. Yin said that the focus on research and development sets Zhangjiang apart from most other "research parks" in China, many of which lease out much of their space for manufacturing. Yin Hong, op. cit.

[160] Zhangjiang High-Tech Park data.

[161] For more information on China's role as a drug-research base for multinationals, see Pete Engardio, "Chinese Scientists Build Big Pharma Back Home," *BusinessWeek*, Sept. 15, 2008 (http://www.businessweek.com/magazine/content/08_37/b4099052479887.htm). Also see Vivek Wadhwa, Ben Rissing, Gary Gereffi, John Trumpbour, and Pete Engardio, "The Globalization of Innovation: Pharmaceuticals," Duke Pratt School of Engineering, Kauffman Foundation, Harvard Law School Labor and Worklife Program, June 2008, (http://www.kauffman.org/uploadedFiles/global_pharma_062008.pdf).

Tenants in other industries include Hewlett-Packard, Lenovo, Infineon, Intel, IBM, Citibank, Infosys, SAP, eBay, Dow, and DuPont.[162]

Like management companies at other research parks in China, Shanghai Zhangjiang has its own direct-investment fund, which it sees not only as a source of capital to seed start-ups but also as a money-making opportunity. The company's has ¥2 billion ($310 million) in capital.[163]

One of the park's top priorities is to deepen the area's talent pool. As labor gets more expensive, the Shanghai area will have to compete more on the basis of innovation. Shanghai Zhangjiang would like to attract a campus of a major Western university.[164] So far, however, the Chinese government has been slow to approve campuses by foreign universities.

Chinese science parks still must overcome many challenges. Some parks are run largely as real-estate projects, rather than as real innovation zones. Phillip H. Phan of Rensselaer Polytechnic Institute noted that China's weak protection of intellectual property and an academic culture that discourages scientists from thinking like entrepreneurs are other handicaps to developing world-class science parks.[165]

Rejuvenating Berlin's Adlershof

The ruins of old wind tunnels, engine-testing facilities, and military barracks that still stand on the six-square-kilometer campus of the Science and Technology Park Berlin Adlershof testify to the site's storied past as the birthplace of German motorized aviation, a development base of fighter aircraft for two world wars, and a science center of the former East Germany.

Today, Adlershof is one of Europe's largest and most established science parks. The campus includes 17 research institutes and operations of 866 companies. Employment within the park more than doubled between 1997 and 2010 to around 14,000.[166] The campus also includes nearly 8,000 students of Berlin's Humboldt University, which has moved its computer science, mathematics, chemistry, physics, geography, and psychology institutes to

---

[162] The R&D centers of most multinationals focus on localizing products and technology for China's domestic market or for products manufactured in China for export, Mr. Yin explained. He also estimated that around 90 percent of revenue by chip-design companies in the area are from the domestic market Yin interview, op. cit.

[163] Its biggest investment is a startup called MicroPort Scientific, a maker of medical devices such as cardiovascular stents and insulin pumps, with $113 million in 2010 sales. Mr. Yin said. Shanghai Zhangjiang also makes low-interest loans to small and midsized Chinese companies. Ibid.

[164] "We would love to have more and more global education resources in this area," Mr. Yin said. "It will help foster the talent pool. This also offers a good opportunity for these institutions' globalization strategies." Ibid.

[165] Phan presentation, op. cit.

[166] Data from *2010 Report on Adlershof.*

Adlershof. Several more institutes and business accelerators are under construction.[167]

A study by the German Institute for Economic Research (DIW) concluded that Adlershof directly contributed more than €1 billion in economic value-added in the area in 2010 and another €740 million in other parts of Berlin. The park also was responsible for 28,000 jobs in the city, and generated €340 million in tax revenue—more than half of which stayed in Berlin.[168]

Adlershof regards itself as a successful model of how public subsidies can stimulate sustainable development of private industry. Between 1991 and 2005, 80 percent of the €1.3 billion invested came from public sources with the remainder from companies. Of the €500 million invested between 2005 and 2011, 70 percent came from private investors.[169] Government funding accounts for only 6.4 percent of the park's budget.[170]

The park was founded in 1991 after the fall of the Berlin Wall, when the city's government faced a quandary over what to do with some 5,500 East German scientists and highly skilled staff, many of them at the forefronts of fields such as laser technology, space research, new materials, and chemistry.[171]

Adlershof was established in 1991 mainly as a means of providing employment in both research and industry for the remaining 4,100 East German scientists, who did not easily fit into West German scientific research

---

[167] Briefing by Peter Strunk of Wista-Management GMBH in Berlin.
[168] Wista-Management GMBH, "The Economic Significance of Adlershof: Impact on Added Value, Employment, and Tax Revenues in Berlin," study by the German Institute for Economic Research commissioned by Wista-Management, 2011.
[169] Ibid.
[170] *2010 Report on Adlershof.*
[171] The science park is located on what originally was the Johannisthal Air Field, which at the turn of the century became one of the world's first development bases for motorized aircraft. German companies such as Albatros, Fokker, and Rumpler all developed early flying machines on the grounds, as did the Wright brothers, who built 60 aircraft there. The German Research Center for Aviation was established in 1912, and 6,000 fighter planes used in World War I were built at Adlershof. After the war, hundreds of films—including Friedrich Murnau's *Nosferatu*--were shot in the unused hangars. When the Nazis came to power, Adlershof once again was used to develop ultra-fast warplanes. After Germany's defeat, Adlershof's aviation research laboratories were dismantled and shipped to the Soviet Union as war reparations. After Germany's partition, Adlershof was home to East German national television and a 12,000-strong regiment of the Ministry of State Security, or Stasi. The East German Academy of the Sciences made Adlershof its base for chemistry and physics. Many of the historical details are taken from Hardy Rudolph Schmitz, "100 Years of Innovation from Adlershof: Dawns, Damage, and Determination," Wista-Management GMBH, Sept. 9, 2009, (http://www.adlershof.de/fileadmin/web/ansprechpartner/netzwerke/internationales/events/Hardy_Schmitz_-_Adlershof_100_years_of_innovation_speech.pdf). Also see a brief history of Adlershof on the Adlershof Web site at http://www.adlershof.de/geschichte/?L=2 and on the Web site of the Gorman Aerospace Center (DLR) http://www.dlr.de/en/desktopdefault.aspx/tabid-2039/2510_read-3894/.

organizations.[172] Many of these former East German scientists now are entrepreneurs.[173]

Management of the science park is modeled after North Carolina's Research Triangle. Governments provided most of the funds for new buildings. But the budget of Wista, the organization that manages the park, comes from renting space to corporate tenants. Companies also pay to use lab space and other services. Tenants must be in industries related to each facility's specialty. There are five research centers that as of 2010 were 95 percent full. After a center has been in operation for 10 years, Wista-Management is allowed to sell the building and lease it back—using the proceeds to fund new construction. Centers devoted to microsystems and materials and information technology and media opened in 2011. A new center for photovoltaic technologies is under construction. Some German high-tech companies are opening manufacturing plants at Adlershof.[174] One key to Adlerhof's success, Dr. Strunk said, is that government funding agencies have maintained support over the long term but did not interfere with management. "A science park needs 10 to 15 years to reach a tipping point," Dr. Strunk said. "During that time, it must be free of political constraints."[175]

Minatec and France's Nanotech Push

France has long been known as a great place to do research and is seeking to become a more attractive place to start companies. The French government is investing in research parks to develop regional competitive clusters, or *pôles de croissance*.

Minatec, a campus of 3,000 students and researchers in Grenoble, is regarded as a model of what the government hopes to achieve.[176] The facility

---

[172] "The main idea was to prevent a social catastrophe," recalls Peter Stunk, executive manager of public relations for Wista-Management GMBH, which runs the park. "They lost everything." The Berlin government dismantled many of the aging buildings and built new ones to house reorganized research institutes and incubators for starting new business.
[173] Former scientists from the Academy of Sciences founded Röntec, a leading manufacturer of X-ray spectrometers that subsequently was acquired by the Nasdaq-listed Bruker Group. Other Adlershof spinoffs founded by East German scientists include FMB Feinwerk und Messtechnik GmbH, a world leader in vacuum systems and beamlines for infrared and soft X-radiation, and LLA Instruments, a maker of devices that can detect 20 different kinds of plastics that are used in recycling facilities. Descriptions of these start-ups are found in Berlin Adlershof, *2010 Report on Adlershof*, (http://www.adlershof.de/newsview/?no_cache=1&L=2&tx_ttnews%5Btt_news%5D=8888).
[174] Berlin-based Soltecture, a manufacturer of thin-film photovoltaic modules and solar-energy systems that has raised more than €104 million in venture and private-equity investment, is building a major production plant on the campus. One draw is a new "competence" center for cutting-edge research in thin-film and nanotechnology for photovoltaics that will be a joint venture between Helmholtz Center Berlin for Materials and Energy and Berlin Technical University.
[175] Strunk, op. cit.
[176] See presentation by David Holden of Minatec in *Understanding Research, Science, and Technology Parks*.

began as an extension of the national nuclear research institute[177] and the Laboratory of Electronics and Information Technologies (Leti), which spawned Thompson Semiconductor in 1973.[178] It then took on a broader mission of promoting public-private research partnerships and the region's industrial base. Over the past decade, Minatec has emerged as one of Europe's premier hubs for nano-technologies and micro-systems. Covering 20 hectares, the campus represents a €3.2 billion investment by the French government and €150 million by local government. As of 2009, the French government also has awarded 113 research projects to Minatec over two years worth about €1.2 billion.

In a nation where researchers tend to be scattered in small groups, Minatec brought together academic programs from four universities with 60,000 students, half of them studying sciences.[179]

The high concentration of R&D activity at Minatec has led to the creation of start-ups in fields such as optoelectronics, biotechnologies, components, circuit design, and motion sensing. Minatec also has attracted significant corporate investment and has forged major international research alliances.[180]

Minatec is the focal point of Nano 2012, described as France's biggest industrial project. The aim of the program, which involves nearly €4 billion in funding from the national, state, and local governments for R&D and equipment, is to make the Grenoble region the world center for development of 32nm and 22nm CMOS technologies.[181] Minatec is diversifying into biotechnology and clean-energy technologies to complement its strength in micro-systems.

India's Research Parks

Science parks are relatively new to India, where until the early 1990s the government discouraged partnerships between academia and business. A number of high-tech industrial estates have since been set up around the country to incubate technology ventures or attract research facilities by foreign and domestic companies. The city of Hyderabad has parks devoted to biotech and

---

[177] See Junko Yoshida, "Grenoble Lure: Un-French R&D," *EE Times*, June 12, 2006.
[178] Thompson Semiconductor merged with Italy's SGS Microelectronics in 1988 and became STS Thompson, one of the world's largest semiconductor companies.
[179] Minatec's state-of-the-art facilities include a 300mm silicon wafer center that operates around the clock, a 200mm micro-electro-mechanical systems (MEMS) prototyping line for fast development of new products, and one of Europe's best facilities for characterizing new nano-scale materials. The campus is home to 2,400 researchers and numerous technology-transfer experts. Researchers have filed nearly 300 patents and published more than 1,600 scholarly papers. Data from Minatec Web site and Dr. Holden presentation.
[180] Minatec's 200 industrial partners include Mitsubishi, Philips, Bic, and Total. Two-thirds of its annual €300 million annual budget comes from outside contracts.
[181] Among the initiative's partners are CEA Leti, IBM's Fishkill, N. Y., semiconductor production complex, ST Microelectronics, the University of New York at Albany, ASML Holdings of the Netherlands, and ST Mentor Graphics of Wilsonville, Oregon. Anne-Francoise Pele, "Mentor Joins 2012 R&D Alliance," *EE Times*, March 16, 2010.

information technology, for example, and Uttar Pradesh capital Lucknow has industrial and research parks for software and life sciences. Several of the nation's famed Indian Institutes of Technology (IIT) and Institutes of Science (IISc) have launched science and technology parks in cities such as Mumbai, Kanpur, Bangalore, and Madras. Most Indian research parks are very small by world standards, however, and focus on incubating start-ups. Other research parks set up by private investors tend to have weak links to universities.[182]

The new IIT-Madras Research Park is one of India's first modern research parks, aspiring to "create a knowledge and innovation ecosystem through collaboration between industry and academia."[183] Funded by an independent company and promoted by the university and local government, the park plans to build 1.2 million square feet of office space in three phases in Chennai, the city formerly known as Madras. It will be built at a cost of just 3 billion rupees (around $65 million), mainly with funds from government, bank loans, and alumni donations. The local government provided 11.5 acres of land and infrastructure.[184] The first tower opened in March 2010, and so far 27 companies have signed up.

The park intends to boost India's role as a "design house" for developing higher-quality products and intellectual property, according to IIT-Madras park director M. S. Ananth. The automotive industry is a major focus. The state of Tamil Nadu is the base of 25 percent of Indian auto assembly and 35 percent of the auto parts industry.[185]

Some 15 percent of the park's space will be reserved for start-ups and training facilities. The park also will feature facilities for prototyping and consulting services and help raise venture funding. The rest of the space is reserved for corporate R&D partnerships with the university.

One of the keys to making research parks work in India is to make sure corporations do not simply treat them as cheap industrial real estate, Mr. Ananth said. IIT-Madras has set up a system in which companies may stay in the park as long as they engage in a certain level of joint-research, consult faculty, sponsor students for advanced degrees, teach, or mentor or employ students.[186]

---

[182] From presentation by M. S. Ananth of the Indian Institute of Technology-Madras in *Understanding Research, Science and Technology Parks.*
[183] From IIT-Madras Research Park Web site, http://respark.iitm.ac.in/about_us.php.
[184] Ananth, op. cit.
[185] India Department of Scientific and Industrial Research data Nissan, BMW, Ford, Hyundai, Ashok Leyland, and Mitsubishi all have major assembly operations in the region. Caterpillar, Bridgestone, Michelin, BorgWarner, and Delphi are among the many component suppliers. See Eric Bellman, "A New Detroit Rises in India's South," *The Wall Street* Journal, July 8, 2008.
[186] Ananth, op. cit.

Leveraging Geography in Hong Kong

Hong Kong is using an impressive new science park to develop a range of new industrial clusters and to position itself as a corporate research hub for China and Southeast Asia.

The government has invested $1.5 billion so far to build the first two phases of the Hong Kong Science and Technology Park in the New Territories, close to the Chinese city of Shenzhen. The park has 250 companies, 80 percent of which are foreign, and employs 7,000 people. When a third phase is completed, the park is expected to have 450 companies and employ 15,000, according to Nicholas Brooke, who chairs the Hong Kong Science and Technology Parks Corp., the park's manager.[187]

The park's strategy is to pick clusters based on existing Hong Kong strengths, and capitalize on its position as a world-class business environment with strong legal protections just across the border from the Chinese city of Shenzhen. It selected electronics, green technology, information and communication technology, precision engineering, and biotechnology. Phase III facilities will focus on new clusters, such as thin-film photovoltaic panels, environmental engineering, and energy management for buildings. The park's laboratories, design center, and incubators focus on niche technologies within these broad areas, such as chips for wireless telecom devices, smart cards, and RFID applications, areas where Hong Kong already is strong. While Hong Kong is hardly on the cutting-edge of scientific research, as Mr. Brooke conceded, it serves as an important integration platform of technologies from around the world for markets in China and elsewhere in Asia.[188]

Mexico's First Modern Technology Park

The new Research & Innovation Technology Park (PIIT) in Monterrey has become a symbol of Mexico's ambition to move beyond *maquiladora* assembly manufacturing and develop a knowledge-based economy. Spread over 172 acres near the airport, PIIT will the first in Mexico to integrate the labs in an array of technologies by leading universities, foreign and domestic corporations, small-business incubators, and national laboratories at a single site.

PIIT's first $145 million phase includes major labs by companies as diverse as Motorola, PepsiCo, AMD, Bosch, and India's Infosys. The park also is building public R&D centers for electronics, biotechnology, mechatronics, advanced materials, the food industry, product design, IT, and water research. There are business incubators devoted biotechnology and nanotechnology companies. The University of Texas at Austin will run an IC2 business

---

[187] From presentation by Nicholas Brooke of the Hong Kong Science and Technology Park Corp. in *Understanding Research, Science and Technology Parks.*
[188] Ibid. Many of the 250 companies in the park conduct sensitive R&D in Hong Kong and manufacture their products in China. DuPont, Philips, Freescale, Xilinx, and Nvidia are among the multinationals using this "Hong Kong-Shenzhen model."

incubator.[189] Texas A&M and Arizona State University also are among the partners.[190]

The goal is to help Monterrey develop new, hybrid industries and innovative companies that will be pillars of the region's growth and "promote a new culture of innovation in Nuevo Leon society," according to PIIT Director Jaime Parada.[191]

Monterrey has several key ingredients for innovation clusters in industries from auto parts and appliances to information technology and life sciences. The state of Nuevo Leon has the highest education level in Mexico.[192] The Monterrey metropolitan area has several of Latin America's best universities, including Tecnológico de Monterrey and the University of Monterrey, as well as several major research hospitals. The city also has a dynamic industrial base that produces 11 percent of the nation's manufactured goods. It is home to such large Mexican companies as Cemex and the operations of 2,000 foreign companies, including United Technologies' Carrier unit, Ford, General Electric, Lenovo, and Whirlpool.

The PIIT campus received $250 million in investment from the federal government. The state of Nuevo Leon also established a $30 million seed fund with private backers. Mexico offers tax incentives covering 30 percent of annual R&D expenses for those who invest in research and development.[193]

A New Generation of U.S. Research Parks

American universities have long been in the vanguard of using research parks as conduits for disseminating technological know-how from universities to private industry. It is likely that research parks will therefore have to play an important role if America is to extract more economic value from the $100 billion the federal government invests each year in research at universities and national laboratories.

---

[189] The IC² Institute at the University of Texas at Austin includes the Austin Technology Incubator.
[190] Nuevo Leon Government, "Monterrey: International City of Knowledge," Power Point. This presentation can be accessed at
http://info.worldbank.org/etools/docs/library/244614/IC4Session4_Parada.pdf.
[191] See presentation by Jaime Parada of Research and Innovation Technology Park (PIIT) in *Understanding Research, Science and Technology Parks*. The park stems for an initiative called Monterrey International City of Knowledge, which aims to coordinate the public and private sectors to upgrade industry in the state of Nuevo Leon. The project is part of a larger goal to boost per-capita GDP in Nuevo Leon state from about $16,000 today to $35,000, the current level of industrialized nations, by 2030. The state of also wants to be regarded as among the world's top 25 locations according to international rankings and to have a "world class education, research and innovation system." Nuevo Leon Government, "Monterrey: International City of Knowledge," op. cit. http://info.worldbank.org/etools/docs/library/244614/IC4Session4_Parada.pdf.
[192] Ibid.
[193] Parada, op. cit.

At a time of intensifying global competition, there is considerable room for improvement in U.S. university research parks. Although American universities and national labs made great strides in commercializing research since passage of the Bayh-Dole Act[194] in 1980, not all are proficient at it. There also are signs that overall progress has slowed. While the number of start-ups spun out of elite research universities rose from 200 in 1994 to 600 in 2008, successful patent applications and new technology licenses have remained flat for a decade, according to the Association of University Technology Management.[195] Of 19,554 invention disclosures by universities in 2009, only 16 percent resulted in issued U.S. patents and 3 percent of those inventions led to the formation of start-up companies. Fifty-two percent of the 130 technology-transfer programs studied lose money for their universities. Only 16.2 percent reported that their programs are financially self-sustaining, meaning they do not depend on a university's budget to remain in operation.[196]

Among the reasons cited for this performance are underfunding of university technology-transfer offices[197] and federal rules that make it too difficult for principal investigators to commercialize federally funded innovations. Another explanation is that the system for allocating federal R&D funds and for rewarding faculty focus overwhelming on scientific discovery, rather than applied research or development of prototypes.[198]

Some U.S. university officials note that the walls between academia and the private sector remain high. Former MIT President Charles M. Vest, now president of the National Academy of Engineering, observes that the role of economic development and technology is necessarily secondary in importance to the university's prime missions of education and research.[199] At universities such as Johns Hopkins, only recently have administrations encouraged scientists to interact with business.[200] Even in universities with active technology transfer

---

[194] The Patent and Trademark Law Amendments Act (35 USC Sec. 200-212), known as the Bayh-dole Act, gave universities control over intellectual property that results from publicly funded research.
[195] Presentation by Ashley J. Stevens of the Association of University Technology Management in National Research Council, at the National Academies conference on *Clustering for 21st Century Prosperity*, Washington, DC, February 25, 2010.
[196] Ibid. With respect to barriers to innovations, see Box 1.2 in Chapter 1.
[197] Ibid.
[198] Remarks by Brian Darmody of the Association of University Research Parks at the National Academies conference on *Clustering for 21st Century Prosperity*, Washington, DC, February 25, 2010i.Also see Association of University Research Parks, "The Power of Place 2.0: The Power of innovation—10 Steps for Creating Jobs, Improving Technology Commercialization and Building Communities of Innovation," March 5, 2010, (http://www.matr.net/article-38349.html).
[199] See remarks by National Academy of Engineering President Charles M. Vest in National Research Council, *Building the 21st Century: U.S. - China Cooperation in Science, Technology, and Innovation*, Charles. W. Wessner, editor, Washington DC: The National Academies Press, 2011.
[200] Presentation by Johns Hopkins University technology-transfer director Aris Melissaratos at the National Academies conference on *Clustering for 21st Century Prosperity*, Washington, DC, February 25, 2010.

offices, bureaucracy can move too slowly or make obtaining technology too costly for entrepreneurs to seize rapidly evolving market opportunities.[201] In other cases, commercialization is stymied by inadequate investment in physical infrastructure and a lack of capital to back promising start-ups and see them through the Valley of Death.

Across America, however, new 21st century research parks affiliated with universities and national labs are being established that have been designed after studies of contemporary best practices and the demands of a knowledge economy. Following are some examples featured in NAS STEP board symposia.

Using Research Parks to Expand Maryland's Mission

The University of Maryland at College Park not only has been a leader at promoting entrepreneurialism among its faculty and students and engaging with private industry.[202] It also has been a pioneer at forging research and economic-development collaborations across Maryland, the nation, and around the world. This high level of engagement with industry has enabled the university to generate $20 billion in economic activity over the past quarter century at a total cost to the state of approximately $88 million, according to former University of Maryland-College Park president C. D. Mote.[203]

Now the university is using research parks in innovative ways to expand these missions by serving as multi-purpose structures where partners from different sectors can interact and innovate. Such parks help "by adding dimension to (the university's) partnership opportunities with industry and government on a global scale that cannot be fulfilled in any other manner that we have discovered," Dr. Mote said.

The new M Square research park, adjacent to the University of Maryland-College Park, is the focal point of research and business clusters in homeland and national security, environmental and earth sciences, and food safety and security. The park benefits from the campus's proximity to

---

[201] For example, see Robert E. Litan, Lesa Mitchell, E. J. Reedy, "Commercializing University Innovations: Alternative Approaches," National Bureau of Economic Research working paper JEL No. O18, M13, O33, O34, O38 (http://papers.ssrn.com/sol3/papers.cfm?abstract_id=976005).
[202] For example, the University of Maryland-College Park has a special dormitory for student entrepreneurs, an award competition for student business proposals, a center for entrepreneurship, a technology enterprise institute run by the engineering school, Maryland's oldest business incubator, a "venture accelerator" to help faculty and student businesses develop commercial products, "boot camps" for technology engineers, and programs to train engineers for industry jobs. See presentation University of Maryland-College Park President C. D. Mote in National Research Council, *Building the 21st Century: U.S. - China Cooperation in Science, Technology, and Innovation*, Charles. W. Wessner, rapporteur, Washington, DC: The National Academies Press, 2011.
[203] See presentation by C. D. Mote in *Understanding Research, Science, and Technology Parks*, op. cit. Also see Dr. Mote's presentation in National Research Council, *Building the 21st Century: U.S. - China Cooperation in Science, Technology, and Innovation*, Charles. W. Wessner, editor, Washington, DC: The National Academies Press, 2011.

Washington, DC and important nearby research institutions such as the American Center of Physics.

M Square will cover 138 acres and have more than 2 million square feet of space when fully built out. It has attracted some $500 million in private investment and is expected to employ 6,500 people.[204]

The research park has helped the University of Maryland, which already had important institutes for physics and telecommunication sciences and the Center for Advanced Study of Languages, land a cluster of research centers tied to federal national security organizations[205]. The University of Maryland has set up other innovative research parks to extend its global reach. The UM-China Research Park, established in 2002, hosts 10 Chinese companies and offers services provided by the university's engineering and business schools. Another 11 Chinese companies have set up operations at the university's special international incubator, Dr. Mote noted, including a developer of software for the construction industry that raised $2 billion in a stock offering valued at $20 billion within six months. Another "international research park" affiliated with the university serves as a "foothold" for foreign companies in Maryland. The park "shows what universities can do on an international scale to build enterprises," he said.

Purdue's Regional Approach

A science park initiative need not be limited to one area. The Purdue Research Parks is a network of parks launched a decade ago by the Purdue Research Foundation, a nonprofit set up in the mid-1990s. In addition to the 725-acre park near Purdue's main West Lafayette, Indiana, campus, there are campuses in Indianapolis, Merrillville, and New Albany, all aimed at helping students and faculty commercialize technology to enhance the Indiana economy. In all, Purdue is a partner in at least 10 Indiana science parks, a half dozen of which are doing quite well, according to Victor L. Lechtenberg, Purdue's vice provost for engagement.[206]

---

[204] A new 38-acre mixed-use University Town Center under development will allow researchers and entrepreneurs to both live and work at the park. Key tenants for climate and weather research include the National Oceanic and Atmospheric Administration, which will occupy 10 acres, employ 800, and partner with the university and the NASA/Goddard Space Flight Center. A climate change institute run jointly by the university and the Pacific Northwest National Laboratory and a $25 million center for earth systems modeling also are at M Square. Data from M Square Web site at http://www.msquare.umd.edu/about/um-research-park.

[205] They include the Intelligence Advanced Research Project Activity, a new government program that consolidates high-level, forward-looking intelligence research. Tenants relating to food research include facilities for the U.S. Department of Agriculture, the Food and Drug Administration, and the university's own Center for Food, Nutrition, and Agriculture policy. M Square also houses a number of start-ups incubated at the university, ranging from developers of medical devices and Internet security software to nutritional products.

[206] From presentation by Victor Lechtenberg of Purdue in *Understanding Research, Science, and Technology Parks*.

The West Lafayette park has nearly 100 high-tech businesses and entities and the nation's largest incubation program, covering 259,000 square feet and housing 57 start-ups. Some $121 million in venture capital has been invested in businesses. The park's 2,800 employees earn an average of $58,400 each.[207] Combined, the research parks have 214 companies in fields such as life sciences, information technology, advanced manufacturing, digital imaging, agri-science, and engineering.[208]

A second Purdue initiative is Discovery Park. Founded in 2001, Discovery Park is a network of integrated research centers at the Purdue campus, each dedicated to large-scale, interdisciplinary research in topics such as biosciences, nanotechnology, advanced manufacturing, energy, oncology, and healthcare engineering. Discovery Park also includes a $25 million Hall for Discovery and Learning Research. Discovery Park has 113,000 square feet of laboratory space, has raised nearly $150 million in research funding as of mid-2010, and recruited 300 faculty.[209] The Lilly Endowment is a major contributor.

Discovery Park's mission is to help "redefine" the academic culture for research and discovery."[210] The park has a number of project-based centers sponsored by different funders and that are affiliated with the core research centers. One major project is developing systems to predict the reliability of micro-electromechanical systems (MEMS) used in security, defense, and space applications. Scientists at Purdue's Bindley Bioscience Center work closely with engineers at the Birck Nanotechnology Center to pioneer new cancer treatments using tiny micro-sensors implanted into tumors to allow doctors to monitor radiation. Other R&D projects study processes to convert biomass into energy and develop low-cost diagnostic tools to detect the AIDS virus. The park is also home to the George E. Brown, Jr. Network for Earthquake Engineering Simulation (NEES), which is housed in Purdue University's Discovery Learning Research Center in Discovery Park. Led by Purdue University, NEES connects research equipment sites and the earthquake engineering community from universities and research centers across the country. NEES is supported by a $105 million grant from the National Science Foundation.[211]

Seeding and nurturing start-ups by blending different disciplines is a major objective of Discovery Park. It has helped launch 30 companies so far, as well as six start-ups initiated by students. Discovery Park also has helped Purdue

---

[207] Details on Purdue Research Parks from Lechtenberg presentation, ibid.
[208] A current list of companies in the parks are found on the Purdue Research Web site, http://www.purdueresearchpark.com/companies/index.asp.
[209] Data from Purdue University Discovery Park Web site, http://www.purdue.edu/discoverypark/.
[210] Discovery Park Web site, op. cit.
[211] See Peter Folger, Earthquakes, Risk, Detection, Warning and Research, Congressional Research Service, September 2, 2011. Access this CRS report at http://www.fas.org/sgp/crs/misc/RL33861.pdf

raise more research funds, which surged to $342 million in the 2008-2009 fiscal year.

Spurring Entrepreneurialism at Sandia

As at many national laboratories engaged in weapons research, the technologies developed at Sandia National Laboratories in Albuquerque, N.M., did little through most of its history to stimulate civilian industries in the surrounding area. That began to change in a big way in 1999, when Sandia became the first national laboratory to open an industrial park next to its compound. The motive was not only to spur development of science-based companies in New Mexico, but also to enable the laboratories to share costs and expertise with industry so that their scientists can keep pace with the latest technological innovations.[212]

Today, the Sandia Science and Technology Park has 18 buildings housing 29 companies and more than 2,000 employees, who earn an average annual salary of around $70,000.[213] The park also has plenty of room to expand: More than two-thirds of the 240 acres of land reserved for the park have yet to be developed.

Sandia was founded as part of the Manhattan Project in the late 1940s as a spinoff of Los Alamos National Laboratories, also located in New Mexico, in order to manufacture weapons for the U.S. nuclear stockpile. Owned by the DOE and managed by Lockheed Martin, Sandia subsequently broadened its mission to meeting other national needs, such as technology for homeland security and renewable energy. Sandia's core technological strengths include computer science, micro systems, materials, engineering sciences, and biosciences.[214] Sandia also is strong in fields such as solar power, a legacy of its decades of work developing power systems for spacecraft.

The park is located alongside what Sandia Chief Technology Officer Richard Stulen describes as an "innovation corridor." Major facilities include the Microsystems and Engineering Sciences Applications (MESA) complex, a $$516 million investment by the Department of Energy, the National Nuclear

---

[212] For an early analysis of the Sandia science park, see National Research Council, *Industry-Laboratory Partnerships: A Review of the Sandia Science and Technology Park Initiative*, Charles W. Wessner, editor, Washington, DC: National Academy Press, 1999.
[213] Sandia Science and Technology Park, "Facts and Figures," on Web site at http://www.sstp.org/Pages/FactsFiguresPage.html Partners in the park include the DOE, Lockheed Martin, New Mexico's Economic Development Administration, and local governments. The park claims to have created more than 5,400 indirect jobs in the Albuquerque area and that the $68 million in public investment as of 2009 brought in $243 million in private investment. Data are from 2009 report by the Mid-Region Council of Governments. See Sandia news release, "Report: Sandia Science & Technology Park Fuels Economy With Jobs, Tax Revenue, Spending," Aug 3., 2010 (https://share.sandia.gov/news/resources/news_releases/report-sandia-science-technology-park-fuels-economy-with-jobs-tax-revenue-spending/).
[214] Presentation by J. Stephen Rottler of Sandia National Laboratories at the National Academies conference on *Clustering for 21st Century Prosperity*, Washington, DC, February 25, 2010.

Security Agency, and Sandia.[215] MESA is used for both classified and non-classified research and is a state-of-the-art microelectronics fabrication facility that can integrate single chips using different materials in ways not normally available to industry.[216] Some of the park's tenants are sizeable. EMCORE, a developer of fiber-optic transmission equipment and solar cells used in spacecraft and terrestrial systems, employs 500 and has invested $104 million in Albuquerque. The company licensed key laser, solar cell, and transponder technology from Sandia. KTech, a local company that provides technicians for the Sandia Pulsed Power Facility, also employs 500 and has invested $34 million.[217] Other tenants include radar-imaging developer Microwave Imaging Systems, the spacecraft electronics operations of Moog Inc., and Applied Technology Associated, a small maker of devices such as sensors and testing instruments.[218]

Among the biggest challenges for the Sandia science park is "keeping the federal government engaged" and maintaining interest by government agencies in maintaining incentives to lure small businesses, explained Dr. Stulen. "Parks don't just happen. They require energy, devotion, and passion from leaders – not only of the institution but also of the region." He said that Sandia and other national laboratories need to improve the ways in which they collaborate with private companies, such as by reducing red tape involved with licensing intellectual property and meeting government regulations. "We need more speed in working with industries, to be able to work at their pace," he said.

Kennedy Space Center: A New Mission
The final voyage of Space Shuttle Atlantis, which landed for the last time on July 21, 2011, in Cape Canaveral, Fla., not only marked the end an era of American manned space travel. It also marked the beginning of a major economic challenge for the area surrounding the Kennedy Space Center. The end of the program was estimated to have cost up to 25,000 jobs.[219]

Regional officials hope that a new science park just outside the security gates of Kennedy Space Center will help rebuild the region's economy and

---

[215] "DOE/NNSA to Dedicate Half Billion Dollar Microsystems Engineering Sciences Complex at Sandia," News Release, National Nuclear Security Administration, August 20, 2007.
[216] See presentation by Sandia Chief Technology Officer Richard Stulen in *Understanding Research, Science and Technology Parks*. Sandia also is home to the Red Storm, one of the world's most powerful supercomputers, and a Joint Computation and Engineering Lab used by corporations such as Goodyear and Procter & Gamble to simulate complex industrial designs. Sandia is a partner in a new Center for Integrated Nanotechnologies, a federally funded public-private research partnership. Sandia has moved some research facilities "outside the fence," from the highly secured laboratory compound and into the science and technology park itself. They include the Computer Science Research Institute.
[217] Ibid.
[218] Sandia Science and Technology Park Web, "Tenants."
[219] Donna Leinwand Leger, "End of Shuttle Program Slams Space Coast Economy, *USA Today*, July 5, 2011.

reposition it for the future. Dubbed Exploration Park, the campus will support the emerging commercial space industry and new companies spun off from the center's research projects. Construction began in March 2011 on the first 60-acre, nine-building phase of the 139-acre site.[220] The project is a public-private partnership involving Space Florida—the state's aerospace economic development organization—and real estate developer The Pizzuti Companies.

When it opens, 5,000 technicians, engineers, and administrative support staff will transfer to the park, guaranteeing what NASA Kennedy Space Center Director and former astronaut Robert Cabana described as "a really high-quality workforce that will be transitioning from the end of the shuttle program to the future."[221] The science park also is adjacent to the University of Central Florida, which has an excellent engineering program and the third-largest enrollment of any U.S. university. "If we can capitalize on universities, industry, and government partnerships with the state of Florida, it is amazing what I think we can accomplish," Mr. Cabana said.

The federal government is providing considerable assistance for the transition. The Obama Administration set up a $40 million transition fund and appointed a presidential task force to promote worker retraining and economic development on the Space Coast. The Administration also announced it intends to invest $6 billion over five years in new NASA space initiatives that will stimulate the space industry and that should provide new economic opportunities in the region.[222] The federal government also is allocating funds to use the Space Shuttle as a national laboratory for experiments conducted in space and for various technology-demonstration projects. "The question is, 'How do we tie all of this together, to where we can bring industry in and really make this beneficial to everyone?'" Mr. Cabana said.[223]

The Kennedy Space Center began focusing more on commercializing technology several years ago. It already opened what is to be the new park's anchor facility--the Space Life Sciences Lab. Currently the building is located within the gates of Kennedy Space Center, but will move to a new building in the park, where it will be easier for civilians to enter. The facility, built by the state of Florida, has 25 fully equipped scientific laboratories for life-sciences research and administrative offices. NASA will lease the space. The space center has a number of public-private partnerships that focus on applied research and commercialization that can create spin-offs based in the park.[224]

---

[220] Space Florida press release, March 10, 2011.
[221] From presentation by Robert Cabana of NASA Kennedy Space Center in *Understanding Research, Science and Technology Parks*.
[222] See Presidential Task Force on Space Industry Workforce & Economic Development, "Report to President," August 15, 2010,
(http://www.explorationpark.com/feeds/Space_Industry_Report_to_the_President.pdf).
[223] Cabana presentation, op. cit.
[224] The NASA Innovative Partnerships Program provides bridge funding to help start-ups and launch projects. Innovation Partnerships provided around $400,000 to help initiate a program called Lunar

In the field of lighting, the Kennedy Space Center is developing light-emitting diode (LED) technology to help plants grow in controlled environments such as space. It also is developing LEDs in different frequencies and colors that have a direct influence on human performance. The technology has applications on earth as well as space, Mr. Cabana said. For example, it could be used to adjust office lighting during certain times of the day to help people work more efficiently.[225]

Other space center collaborations with industry with terrestrial applications include a "self-healing wire" developed for the Space Shuttle with ASRC Aerospace Corp. The wire can detect breaks and release polymers to repair the damage. A collaboration with PPG Industries is developing "micro-encapsulated"[226] materials that inhibit corrosion in paint, while a joint project with Louisiana Tech University is developing biological instruments that detect radiation damage to DNA during space travel. A partnership with Florida Power & Light is installing a 10-megawatt solar-array system that Mr. Cabana says could help attract solar-array companies to Exploration Park. Yet another collaboration is with Starfighters Inc., a company that operates a fleet of F-104 jets[227] that now are used for training. The company is developing a system that can track and monitor rockets fired in test ranges, reducing the chance of human error.[228]

Mr. Cabana observed that Kennedy Space Center is a "critical resource for our future" and added that he wants to "make sure that it is maintained so that we have the ability to explore." With an extensive research commercialization program and construction of Exploration now underway, Mr. Cabana said, "we really are doing all the right things."

Observations on Factors in the Success of Research Parks

The proliferation of research parks, and the sheer scale of those being built abroad, highlights the need for U.S. policy makers to better understand the role of such parks in a nation's innovation system. The ways in which successful parks are structured, financed, and operated have important implications for the

---

Analog Field Demo of ISRU for lunar prospecting, for example. The program, a collaboration with the Goddard and Johnson space centers, Carnegie Mellon University, and the Pacific International Space Center for Exploration Systems run by the University of Hawai'i, uses a simulation of the lunar surface to find and develop natural resources on the moon. ISRU standards for In-Situ Resource Utilization. The program's goal is to develop ways to use resources already on the moon to establish lunar habitats and sustain human life.
(http://microgravity.grc.nasa.gov/Advanced/Capabilities/ISRU/).
[225] Cabana, op. cit.
[226] Micro-encapsulation is a process in which tiny particles are surrounded by a coating.
[227] The Lockheed F-104 Starfighter is a single-engine supersonic interceptor jet used by the U.S. Air Force from 1958 until 1967. NASA used F-104s for test flights until 1994.
[228] Examples in this paragraph cited in Cabana presentation.

competitiveness of the U.S. and other nations in a 21st century global economy. Yet despite the significant investments in such parks, there has been little rigorous study of which practices work best or to precisely quantify their economic impact. As a result, there is no systematic framework to understand the dynamic interactions among the various stakeholders and participants in research parks and the outcomes that result.[229]

To advance that understanding, the National Academies' Board on Science, Technology, and Economic Policy (STEP) made research parks a major area of focus in its study of comparative innovation policy. The major policy findings from the examination of research parks around the world are summarized below.

- Successful research parks tend to have a large research university or national laboratory at the core and support a critical mass of highly trained knowledge workers.
- Strong public-private partnerships among government, corporations, universities, and national laboratories are increasingly important to the success of research parks.
- There is ample evidence that public investment in research parks have a high "spillover" effect in terms of attracting corporate investment, creating jobs, and forming new companies, although more work must be done to measure such impact with precision.
- Public financial and policy support must be sustained over the long-term if research parks are to win support from corporate investors. Given the long-time horizons of major corporate research programs, public commitment must be viewed as reliable.
- Research parks must be viewed as much more than real estate projects if they are to be catalysts of regional innovation. Successful parks not only offer corporations access to first-rate public research institutions and talent, but also valuable services such as low-cost shared laboratory and prototyping facilities, small-business incubators, advice on intellectual property, and assistance in raising early-stage capital.
- Successful research parks outside of the U.S. tend to benefit from strong government-supported programs to promote applied research as well as basic research.
- There is substantial room for improvement in the flow of research from universities and national laboratories to the commercial sector. This is especially true in nations such as China, India, Japan, and some nations in Europe, where academic cultures traditionally have not encouraged

---

[229]See Phillip H. Phan, Donald S. Siegel, and Mike Wright, "Science Parks and Incubators: Observations, Synthesis and Future Research," *Journal of Business Venturing*, 20(2): 165-182, March 2005.

entrepreneurialism, but also in the United States. Greater incentives and reform of technology-transfer policies may be required.

IN CLOSING

As we have seen, both advanced and emerging economies are making significant investments and promulgating polices to encourage cluster development as a way to maximize their investments in research and development. This chapter has explored several ways in which U.S. states and regions as diverse as Michigan, New York, West Virginia, and South Carolina are rising to the challenge by developing regional innovation clusters and new types of science and research parks. In many cases, these regional initiatives leverage federal investments to achieve scale. In recent years, federal policies have also sought to develop a more integrated approach to supporting regional efforts. Given that innovation clusters typically coalesce over many years, a key issue is whether these initiatives will benefit from steady commitment over the long term.

APPENDIXES

Appendix A

List of Workshops and Symposia for the Study of

Comparative National Innovation Policies: Best Practice for the 21st Century

- September 19, 2011: (Washington, DC) *U.S.-China Policy for Science, Technology, and Innovation.*
- June 30, 2011: (Beijing) *China-U.S. Forum on Biomedical Innovation and Health Policy.*
- June 28, 2011: (Chinese Academy of Engineering, Beijing): *Comparative Innovation Systems: China and the U.S.*
- May 24-25, 2011: (Berlin) Meeting Global Challenges: German-U.S. Innovation Policy.
- November 1, 2010: (Washington, DC) *Meeting Global Challenges: U.S.-German Innovation Policy.*
- May 17-18, 2010: (Washington, DC) *Building the 21st Century: U.S.-China Cooperation on Science, Technology, and Innovation.*
- December 3-4, 2009: (Washington, DC) *Rebuilding the Transatlantic Bridge: U.S.-Polish Cooperation on Innovation.*
- October 8-9, 2008: (Washington, DC) *Opportunities & Challenges in the U.S. & Polish Innovation Systems.*
- March 12-13, 2008: (Washington, DC) *Understanding Research S&T Parks—Global Best Practice.*
- December 18, 2007: (New Delhi) *Growing Indian Innovation: Issues, Opportunities, and Solutions.*
- September 22, 2006: (Leuven) *Innovative Flanders: Synergies in Regional and National Innovation Policies.*

- June 16, 2006: (Washington, DC) *India's Changing Innovation System: Achievements, Challenges, and Opportunities for Cooperation.*
- January 11, 2006: (Tokyo) *Creating 21st Century Innovation Systems in Japan and the United States: Lessons from a Decade of Change.*
- January 6, 2006: (Taipei) *21st Century Innovation Systems for the United States and Taiwan.*
- April 15, 2005: (Washington, DC) *Innovation Policies for the 21st Century.*

Appendix B

Bibliography

Acs, Zoltan, and David Audretsch. 1990. *Innovation and Small Firms*. Cambridge, MA: The MIT Press.

Aerts, Kris, and Dirk Czarnitzki. 2005. "Using Innovation Survey Data to Evaluate R&D Policy: The Case of Flanders." K.U. Leuven: Department of Applied Economics and Steunpunt O&O Statistieken.

Aerts, Kris, and Dirk Czarnitski. 2006. "The Impact of Public R&D Funding in Flanders." *IWT Studies* 54.

Agence France-Presse. 2009. "EU Says Investigation Raid Pharma Giants." October 6.

Agence France-Presse. 2011. "EU Antitrust Authorities Probe Johnson & Johnson, Novartis." October 21.

Aghion, Phillipe, Robin Burgess, Stephen Redding, and Fabrizio Zilibotti. 2003. "The Unequal Effects of Liberalization: Theory and Evidence from India." Washington, DC: Center for Economic Policy Research.

Agrawal, A. and R. Henderson. 2002. "Putting Patents in Context: Exploring Knowledge Transfer from MIT." *Management Science* 48(1):44-60.

Ahluwalia, Montek Singh. 2001. "State Level Performance Under Economic Reforms in India." Stanford University. Stanford Institution for Economic Policy Research Working Paper No. 96. March.

Aizcorbe, A., K. Flamm, and A. Khurshid. 2002. "The Role of Semiconductor Inputs in IT Hardware Price Decline: Computers vs. Communications." Federal Reserve Finance and Economics Discussion Paper 2002-37. Washington, DC: The Federal Reserve Board of Governors. August 2002; revised 2004. November 2001; Revised: June 2002. JEL classification: L63, 030, 047.

Aizcorbe, A., S. Oliner, and D. Sichel. 2006. "Shifting Trends in Semiconductor Prices and the Pace of Technological Progress." Federal Reserve Board Finance and Economics Discussion Series No. 2006-44. September. Original version: July 2003; Current version: September 2006.

Alcacer, Juan and McKinsey & Co. 2009. "Mapping Innovation Clusters." McKinsey Digital, March 19.

Alic, John A., Lewis M. Branscomb, Harvey Brooks, Ashton B. Carter, and Gerald L. Epstein. 1992. Beyond Spin-off: Military and Commercial Technologies in a Changing World. Boston, MA: Harvard Business School Press.

Allen, Jonathan. 2011. "House continuing resolution would bar NASA from China ties." *Politico* February 12.

Allen, Stuart D., Albert N. Link, and Dan T. Rosenbaum. 2007. "Entrepreneurship and Human Capital: Evidence of Patenting Activity from the Academic Sector." *Entrepreneurship Theory and Practice* 31(6):937-951.

Allison J. and M. Lemley. 1998. "Empirical Evidence on the Validity of Litigated Patents." *Aipla Quarterly Journal* 26:185-277.

Altenburg, Tilman, Hubert Schmitz, and Andreas Stamm. 2008. "Breakthrough?: China's and India's Transition from Production to Innovation." *World Development* 36(2):325-344.

AmCham-China. 2010. "American Business in China: 2010 White Paper." May 22.

AmCham-China. 2010. "U.S. Export Competitiveness in China: Winning the World's Fastest-Growing Market." September.

American Association for the Advancement of Sciences. 2011. "R&D in the FY 2011 year-Long Continuing Resolution." May 2. Accessed at <*http://www.aaas.org/spp/rd/fy2011/ 03/07/2012, Posted April 12, 2011*>.

American Society of Civil Engineers, 2009 Report Card for America's Infrastructure, March 25. Accessed at <*https://apps.asce.org/reportcard/2009/grades.cfm on 3/19/2011*>.

Amos, Paul, Dick Bullock, and Jitendra Sondhi. 2010. High-Speed Rail: The Fast Track to Economic Development? World Bank. July.

Amparo San José, Juan Roure, and Rudy Aernoudt. 2005. "Business Angel Academies: Unleashing the Potential for Business Angel Investment." *Venture Capital* 7(2).

Amsden, Alice H. 2001. *The Rise of "the Rest": Challenges to the West from Late-industrializing Economies*. Oxford, UK: Oxford University Press.

Amsden, Alice H. and Wan-wen Chu. 2003. *Beyond Late Development: Taiwan's Upgrading Policies*. Cambridge, MA: The MIT Press.

Amsden, Alice H., Ted Tschang, and Akira Goto. 2001. "Do Foreign Companies Conduct R&D in Developing Countries? A New Approach to Analyzing the Level of R&D, with an Analysis of Singapore" Tokyo, Japan: ADB Institute Working Paper Series March 2001.

Andonian, Andre, Christoph Loos, and Luiz Pires. 2009. "Building an Innovation Nation." McKinsey & Co. March 4.

Aoki, Reiko, and Sadao Nagaoka. 2004. "The Consortium Standard and Patent Pools." *Institute of Economic Research*. Hitotsubashi University. Discussion Paper Series. No. 32. Accessed on 7/9/2012 at <http://hi-stat.ier.hit-u.ac.jp/research/discussion/2004/pdf/D04-32.pdf>.

Aoki, Reiko, and Sadao Nagaoka. 2005. "Coalition Formation for a Consortium Standard through a Standard Body and a Patent Pool: Theory and Evidence from MPEG2, DVD and 3G." IIR Working Paper WP#05-01. February.

Applied Research Institute, Inc. 2006. Survey of the Environment for Startups. Applied Research Institute. November.

Archibugi, Danielle, Jeremy Howells, and Jonathan Michie, eds. 1999. *Innovation Policy in a Global Economy*. Cambridge, UK: Cambridge University Press.

Argyres, N. S., and J. P. Liebeskind. 1998. "Privatizing the Intellectual Commons: Universities and the Commercialization of Biotechnology." *Journal of Economic Behavior & Organization* 35:427-454.

The Arizona Republic. 2011. "Arizona Board Approves Steep Tuition Hikes." April 8.

Arora, A., and R. P. Merges. 2004. "Specialized Supply Firms, Property Rights, and Firm Boundaries." *Industrial and Corporate Change* 13(3):451-476.

Arora, A., M. Ceccagnoli, and W. Cohen. 2001. "R&D and the Patent Premium." Carnegie-Mellon University and INSEAD: paper presented at the ASSA Annual Meetings. January 2002. Atlanta, Georgia. (See also same title by same authors in 2003 NBER working papers series <http://www.nber.org/papers/w9431>.

Arora, A., A. Fosfuri, and A. Gambardella. 2003. "Markets for Technology and Corporate Strategy." In O. Granstrand, ed., *Economics, Law, and Intellectual Property*. Boston, MA: Kluwer Academic Publishers.

Arora, Ashish, Lee G. Branstetter, and Matej Drev. 2010. "Going Soft: How the Rise of Software-Based Innovation Led to the Decline of Japan's IT Industry and the Resurgence of Silicon Valley." National Bureau of Economic Research. Working Paper 16156. July.

Arora, Ashish Ralph Landau, and Nathan Rosenberg, 1999. "Dynamics of Comparative Advantage in the Chemical Industry." in *Industrial Leadership, Studies of Seven Industries*. David C. Mowery and Richard R. Nelson, eds. Cambridge: Cambridge University Press.

Arrow, K. J. 2000. "Increasing Returns: Historiographic Issues and Path Dependence." *European Journal of the History of Economic Thought* 7(2):171-180.

Arrow, Kenneth. 1962. "Economic welfare and the allocation of resources for invention." In The Rate and Direction of Inventive Activity: Economic and Social Factors. Princeton, NJ: Princeton University Press. Pp. 609-626.

Arthur, W. Brian 1989. "Competing Technologies, Increasing Returns, and Lock-in by Historical Small Events." *Economic Journal* 99(394):116-131.

Arthur, W. Brian. 1994. "Industry Location Pattern and the Importance of History." In W. Brian Arthur. *Increasing Returns and Path Dependence in the Economy.* Ann Arbor: The University of Michigan Press.

Asheim, Bjorn T. et al., eds. 2003. *Regional Innovation Policy for Small-medium Enterprises.* Cheltenham, UK: Edward Elgar.

Asmus, Peter, and Clint Wheelock. 2011. "Clean Energy: Ten Trends to Watch in 2011 and Beyond." Pike Research.

Association of University Research Parks. 1998. "Worldwide Research & Science Park Directory 1998." New York: Coral Springs, FL: BPI Communications for Association of University Research Parks.

Association of University Research Parks. 2008 "The Power of Place: A National Strategy for Building America's Communities of Innovation." Tucson, AZ: Association of University Research Parks.

Association of Public Land-Grant Universities. 2010. "Ensuring Public Research Universities Remain Vital: *A Report to the Membership on the Research University Regional Deliberations.*" November.

Athreye, Suma S. 2000. "Technology Policy and Innovation: The Role of Competition Between Firms." In Pedro Conceicao et al., eds. *Science, Technology, and Innovation Policy: Opportunities and Challenges for the Knowledge Economy.* Westport, CT: Quorum Books.

Atkinson, Robert D. 2004. *The Past and Future of America's Economy-Long Waves of Innovation that Power Cycles of Growth.* Cheltenham, UK: Edward Elgar.

Atkinson, Robert, D. 2006. "Is the Next Economy Taking Shape?" *Issues in Science and Technology* 62(2): 62-65.

Atkinson, Robert D. 2007. "Expanding the R&E tax credit to drive innovation, competitiveness and prosperity." *Journal of Technology Transfer* 32:617-628.

Atkinson, Robert. 2009. "Effective Corporate Tax Reform in the Global Innovation Economy." Washington, DC: The Information Technology & Innovation Foundation. July.

Atkinson, Robert D. 2010. "Commentary on Gregory Tassey's 'Rationales and Mechanisms for Revitalizing U.S. Manufacturing R&D Strategies.'" *Journal of Technology Transfer* 35(3). DOI 10.1007/s10961-010-9164-9. (Available at SSRN: <*http://ssrn.com/abstract=1722875*>.

Atkinson, Robert D. 2012. "Worse than the Great Depression: What Experts Are Missing about American Manufacturing Decline." Washington, DC: The Information Technology & Innovation Foundation. March.

Atlanta Journal-Constitution. 2010. "FDA Has Critical Budget Shortfall." February 16.

Asia Pulse. 2010. "Taiwan Unveils First Electric Smart Commercial Vehicle." September 28.

Audretsch, D. B., ed. 1998. *Industrial Policy and Competitive Advantage.* Volumes 1 and 2. Cheltenham, UK: Edward Elgar.

Audretsch, D. B. 1998. "Agglomeration and the Location of Innovative Activity." *Oxford Review of Economic Policy* 14(2):18-29.
Audretsch, D. B. 2001. "The Prospects for a Technology Park at Ames: A New Economy Model for Industry-Government Partnership?" In National Research Council. *A Review of the New Initiatives at the NASA Ames Research Center.* Charles W. Wessner, ed. Washington, DC: National Academy Press.
Audretsch, D. B. 2007. *The Entrepreneurial Society.* Oxford, UK: Oxford University Press.
Audretsch, D. B., and M. P. Feldman. 1996. "R&D Spillovers and the Geography of Innovation and Production." *American Economic Review* 86(3):630-640.
Audretsch, D. B., and M. P. Feldman. 1999. "Innovation in Cities: Science-based Diversity, Specialization, and Localized Competition." *European Economic Review* 43(2):409-429.
Audretsch, D. B., B. Bozeman, K. L. Combs, M. P. Feldman, A. N. Link, D. S. Siegel, P. Stephan, G. Tassey, and C. Wessner. 2002. "The Economics of Science and Technology." *Journal of Technology Transfer* 27:155-203.
Audretsch, D. B., H. Grimm, and C. W. Wessner. 2005. *Local Heroes in the Global Village: Globalization and the New Entrepreneurship Policies.* New York: Springer.
Auerbach, Alan J. 2009. "Public Finance in Practice and Theory." Richard Musgrave Lecture. CESifo. Munich. May 25.
Auerswald, Philip E., Lewis M. Branscomb, Nicholas Demos, and Brian K. Min. 2005. *Understanding Private-Sector Decision Making for Early-Stage Technology Development: A "Between Invention and Innovation Project" Report.* NIST GCR 02-841A. Gaithersburg, MD: National Institute of Standards and Technology.
Augustine, Norman. 2007. *Is America Falling Off the Flat Earth?* Washington, DC: The National Academies Press.
Bae, Andy. 2011. "Lithium-Ion Battery Materials: Japan Dominates in the EV Era." Pike Research. February 4.
BBC News. 2011. "Pfizer to Close UK Research Site." February 1.
Bajaj, Vikas. 2009. "India to Spend $900 Million on Solar." *The New York Times* November 20. (Accessed at <http://green.blogs.nytimes.com/2009/11/20/india-to-invest-900-million-in-solar>).
Baker, Stephen. 2005. "New York's Big Hopes for Nano." Bloomberg BusinessWeek February 4. Accessed at: <http://www.businessweek.com/technology/content/feb2005/tc2005024_1576_tc024.htm>).
Bakouros, Y. L., D. C. Mardas, and N. C. Varsakelis. 2002. "Science Parks, a High-Tech Fantasy? An Analysis of the Science Parks of Greece." *Technovation* 22(2):123-128.

Baldwin, J. R., P. Hanel, and D. Sabourin. 2000. "Determinants of Innovative Activity in Canadian Manufacturing Firms: The Role of Intellectual Property Rights." Statistics Canada Working Paper No. 122. March 7.

Baldwin, John Russel, and Peter Hanel. 2003. *Innovation and Knowledge Creation in an Open Economy: Canadian Industry and International Implications.* Cambridge, UK: Cambridge University Press.

Balfour, Frederik. 2010. "IPad Assembler Foxconn Says it Has More Than 1 Million Employees in China." *Bloomberg.* December 10.

Balzat, Markus, and Andreas Pyka. 2006. "Mapping National Innovation Systems in the OECD Area." *International Journal of Technology and Globalisation* 2(1-2):158-176.

Bangkok Post. 2009. "Big Pharma Learns to Live With Generics." August 15.

Baptista, R. 1998. "Clusters, Innovation, and Growth: A Survey of the Literature." In G. M. P. Swann, M. Prevezer, and D. Stout, eds. *The Dynamics of Industrial Clustering.* Oxford, UK: Oxford University Press.

Barboza, David, Christopher Drew and Steve Lohr. 2011. "G.E. to Share Jet Technology with China in New Joint Venture." *New York Times* January 17. P. B1.

Bardham, Ashok Deo and Dwight M. Jaffeee. "Innovation, R&D, and Off-shoring." University of California at Berkeley: Fisher Center Research Reports.

Barron's. 2009. "Sequenom: Bloodied and Unbowed." September 29.

Barrett, Corelli. 1922. *The Collapse of British Power.* Marrow.

Barose, Galen Naïm Darghouth, Ryan Wiser, and Joachim Seel. 2011. "Tracking the Sun IV: An Historical Summary of the Installed Cost of Photovoltaics in the United States from 1998 to 2010." Lawrence Berkeley National Laboratory. September.

Bartzokas, Anthony, and Morris Teubal. 2002. "The Political Economy of Innovation Policy Implementation in Developing Countries." *Economics of Innovation and New Technology* 11(4-5): Pp. 271-274.

Battelle. 2008. 2009 Global R&D Funding forecast" R&D Magazine December. <http://www.battelle.org/news/pdfs/2009RDFundingfinalreport.pdf>.

Bauman, Valerie. 2008. "IBM Will Invest $1.5B to Expand NY Operations." Associated Press. July 15.

BBC. 2010. "China claims supercomputer crown." October 28.

Beinhocker, Eric D. 2007. *Origin of Wealth—Evolution, Complexity, and the Radical Remaking of Economics.* Cambridge, MA: Harvard Business School.

Belitz, Heike Marius Clemens, Martin Gornig, Florian Mölders, Alexander Schiersch, and Dieter Schumacher. 2011. "After the Crisis: German R&D-Intensive Industries in a Good Position." DIW Economic Bulletin 2.

Bellman, Eric. 2008. "A New Detroit Rises in India's South." *The Wall Street Journal* July 8.

Bennis, Warren, and Patricia Ward Biederman. 1997. *Organizing Genius*. New York: Basic Books.

Benoit, Bertrand. 2007. "German Skills Gap Costs €20 bn." *Financial Times* August 20.

Berglund Dan and Christopher Coburn. 1995. Partnerships: A Compendium of State and Federal Cooperative Programs. Columbus, OH: Battelle Press.

Berlin, Leslie. *The Man Behind the Microchip: Robert Noyce and the Invention of Silicon Valley*. 2005. New York: Oxford University Press.

Bernanke, Ben. 2011. "Promoting Research and Development: The Government's Role." *Issues in Science & Technology* XXVII(4):37-41.

Bessen, J., and M. J. Meurer. 2005. "The Patent Litigation Explosion." Research on Innovation and Boston University School of Law: Working Paper No. 05-18.

Bhidé, Amar. 2006. "Venturesome Consumption, Innovation and Globalization." Paper for a Joint Conference of CESifo and the Center on Capitalism and Society "Perspectives on the Performance of the Continent's Economies." Venice. July 21-22, 2006. Paper presented at the Centre on Capitalism & Society and CESifo Venice Summer Institute 2006. "Perspectives on the Performance of the Continent's Economies." July 21-22, 2006. Held at Venice International University. San Servolo, Italy.

Biegelbauer, Peter S., and Susana Borras, eds. 2003. *Innovation Policies in Europe and the U.S.: The New Agenda*. Aldershot, UK: Ashgate.

Bilstein, Roger E. 1984. *Flight in America: From the Wrights to the Astronauts*. Baltimore: Johns Hopkins University Press.

Birch, David. 1981. "Who Creates Jobs?" *The Public Interest* 65:3-14.

Birgeneau, Robert J. and Frank D. Yeary. 2009. "Rescuing Our Public Universities." Washington Post, Sept. 27. (Accessed at <http://www.washingtonpost.com/wp-dyn/content/article/2009/09/25/AR2009092502468.html>).

Birkler, John, et al. 2012. "Keeping a Competitive U.S. Military Aircraft Industry aloft." Santa Monica CA: RAND.

Black, Jeff. 2011. "Germany's Future Rising in East as Exports to China Eclipse U.S." *Bloomberg*. April 6.

Blanke, Jennifer and Thiery Geiger. The Lisbon Council & Allianz Dresdner Economic Research. 2008. "The Lisbon Review 2008: Measuring Europe's Progress in Reform." World Economic Forum.

Blanpied, William A. 1998. "Inventing U.S. Science Policy." *Physics Today* 51(2):34-40.

Block, Fred, and Matthew Keller. 2008. "Where Do Innovations Come From? Transformations in the U.S. National Innovation System, 1970-2006." Washington, DC: The Information Technology and Innovation Foundation. July.

Blockab, Joern and Philipp Sandnerc, 2009. "What is the effect of the financial crisis on venture capital financing? Empirical evidence from US Internet start-ups." *Venture Capital: An International Journal of Entrepreneurial Finance* 11(4).

Blockab, Joern, De Vries, Geertjan and Sandner, Philipp G. 2010. "Venture Capital and the Financial Crisis: An Empirical Study Across Industries and Countries." January 24.

Blomström, Magnus, Ari Kokko, and Fredrik Sjöholm. 2002. "Growth & Innovation Policies for a Knowledge Economy: Experiences from Finland, Sweden, & Singapore." EIJS Working Paper. Series No. 156.

Bluestein, Adam and Amy Barrett. 2010. "How States Can Attract Venture Capital." *Inc. Magazine* July 1.

Bonvillian, William B. 2006. "Power Play, The DARPA Model and U.S. Energy Policy." *The American Interest* II(2):39-48.

Bonvillian, William B. and Richard Van Atta. 2011. "ARPA-E and DARPA: Applying the DARPA Model to Energy Innovation." *Journal of Technology Transfer* 36:469-513.

Borras, Susana. 2003. *The Innovation Policy of the European Union: From Government to Governance*. Cheltenham, UK: Edward Elgar.

Borrus, Michael, and Jay Stowsky. 2000. "Technology Policy and Economic Growth." In Charles Edquist and Maureen McKelvey, eds. *Systems of Innovation: Growth, Competitiveness and Employment*, Volume 2. Cheltenham, UK: Edward Elgar.

Boston Consulting Group. 2010. "Batteries for Electric Cars: Challenges, Opportunities, and the Outlook to 2020." (Accessed at <http://gerpisa.org/en/system/files/file36615.pdf>, March 28, 2012).

Bosworth, Brian. 2007. "Lifelong Learning, New Strategies for the Education of Working Adults." Center for American Progress, December. (Access at <http://www.americanprogress.org/issues/2007/12/pdf/nes_lifelong_learning.pdf>).

Bradsher, Keith. 2009 "China Builds High Wall to Guard Energy Industry." *The New York Times* July 13.

Bradsher, Keith. 2009. "China-U.S. Trade Dispute Has Broad Implications." *The New York Times* September 14.

Bradsher, Keith. 2011. "Chasing Rare Earths, Foreign Companies Expand in China." *The New York Times* August 24.

Bradsher, Keith. 2011. "China Benefits as U.S. Solar Industry Withers." *The New York Times* September 1.

Bradsher, Keith. 2011. "Hybrid in a Trade Squeeze." *The New York Times* September 5.

Bradsher, Keith. 2011. "U.S. Solar Panel Makers File Case Accusing China of Violating Trade Rules." *The New York Times* October 19.

Brady, Tim ed. 2001. *The American Aviation Experience: A History*. Southern Illinois University Press.

Brandhorst, Jr., Henry W. "Photovoltaics—The Endless Spring." NASA Technical Memorandum 83684. (Accessed at <http://ntrs.nasa.gov/archive/nasa/casi.ntrs.nasa.gov/19840023712_198402 3712.pdf>).

Branscomb, L. M., and P. E. Auerswald. 2001. *Taking Technical Risks: How Innovators, Executives, and Investors Manage High-Tech Risks*. Boston, MA: The MIT Press.

Branscomb, Lewis M., and Philip E. Auerswald. 2002. *Between Invention and Innovation: An Analysis of Funding for Early-Stage Technology Development*. NIST GCR 02-841. Gathersburg, MD: National Institute of Standards and Technology. November.

Branstetter, Lee and Yoshiaki Nakamura. 2003. "Is Japan's Innovation Capacity in Decline?" National Bureau of Economic Research. Working Paper 9438. January.

Braudel, Fernand. 1973. *Capitalism and Material Life 1400-1800*. London, UK: Harper Colophon Books.

Breschi, S. and F. Lissoin. 2001. "Knowledge Spillovers and Local Innovation Systems: A Critical Survey." *Industrial and Corporate Change* 10(4):975-1005.

Breznitz, Dan. 2007. *Innovation in the State: Political Choice and Strategies for Growth in Israel, Taiwan, and Ireland*. New Haven, CT: Yale University Press.

Breznitz, D. and Murphree. 2011. *Run of the Red Queen; Government, Innovation, and Globalization and Economic Growth in China*. New Haven, CT: Yale University Press.

Brodd, Ralph. 2005. "Factors Affecting U.S. Production Decisions: Why are There No Volume Lithium-Ion Battery Manufacturers in the United States?" ATP Working Paper Series, Working Paper 05–01,

Broder, J. and J. Ansfield. 2009 "China and U.S. Seek a Truce on Greenhouse Gases." *The New York Times*. June 7

Brooks, Harvey. 1996. "The Evolution of U.S. Science Policy." Chap. 2 in Bruce L. R. Smith and Claude E. Barfield, editors. *Technology, R&D, and the Economy*. Washington, DC: The Brookings Institution and American Enterprise Institute.

Brown, Jeffrey R. 2010. "Why I Lost My Secretary: the Effect of Endowment Shocks on University Operations." NBER. May 29.

Browner, Carol. 2010. "White House Blog: 183 projects, 43 states, Tens of Thousands of High Quality Clean Energy Jobs." January 8.

Browning, L., and J. Shetler. 2000. *SEMATECH: Saving the U.S. Semiconductor Industry*. College Station, TX: Texas A&M University Press.

Bruche, Gert. 2011. "A new Geography of Innovation--China and India Rising" in Karl P. Sauvant et al. ,eds. *FDI Perspectives: Issues in International Investment*. New York: Vale Columbia Center on Sustainable International Investment. January.

Buday, Sarah K, Jayne E. Stake and Zoë D. Peterson. 2012. "Gender and the Choice of a Science Career: The Impact of Social Support and Possible Selves." *Sex Roles-Journal of Research* 66(3-4):197-209.

Burgenmeister, Jane. 2009. "South Korea Taps Germany to Help Grow its Solar Industry." Renewable Energy World.com. April 29.

Burrelli, Joan and Alan Rapoport. 2009. *Reasons for International Changes in the Ratio of Natural Science and Engineering Degrees to the College-Age Population*, National Science Foundation, Directorate for Social, Behavioral and Economic NAF 09-308, January.

Bush, Nathan. 2005. "Chinese Competition Policy, It Takes More than a Law." *China Business Review* May-June.

Bush, Vannevar. 1945. *Science: The Endless Frontier*. Washington, DC: U.S. Government Printing Office.

Business Daily Update. 2011. "Alliances Form in Growing Pharmaceutical Market." August 3.

Business Daily Update. 2012. "Caterpillar Expands China Research Center." January 10.

Business Recorder. 2009. "WTO Allows Pakistan to Grant License." October.

Business Review Western Michigan. 2009. "Sequenom Readies Tests for Market." March 26.

Business Software Alliance and IDC. 2009. 2008 Piracy Study. May.

Business Wire. 2009. "Sequenom Announces Delay in Launch of SEQureDx Trisomy 21 Test." April 29.

Business Wire. 2012. "Winston Pharmaceuticals, Inc. Receives SBIR Grant from the NIH to Investigate Treatment for Postherpetic Neuralgia of the Trigeminal Nerve." April 9.

Cai, Yong. 2012. "China's Demographic Reality and Future." *Asian Population Studies* 8(1). March.

Camp, M., K. Parekh, and T. Grywalski. 2007. Ohio Venture Capital Report. Fisher College of Business. Ohio State University.

Canada Council on Learning. 2010. Taking Stock of Lifelong Learning in Canada (2005-2010): Progress or Complacency? August 25. (Access at <*http://www.ccl-cca.ca/ccl/aboutccl/PresidentCEO/20100825TakingStockReport.html*>).

Canada Foundation for Innovation. Evaluation and Outcome Assessment Team 2008. 2008 Report on Results: An Analysis of Investments in Infrastructure (as of December 19, 2008).

Cao, Cong, Richard P. Suttmeier, and Denis Fred Simon. 2006. "China's 15-Year Science and Technology Plan." *Physics Today* 59(12):38-43. December.

Capron, Henri, and Michele Cincera. 2006. *Strengths and Weaknesses of the Flemish Innovation System: An External Viewpoint*. Brussels, Belgium: IWT.

Caracostas, Paraskevas, and Ugur Muldur. 2001. "The Emergence of the New European Union Research and Innovation Policy." In P. Laredo and P. Mustar, eds. *Research and Innovation Policies in the New Global Economy: An International Comparative Analysis.* Cheltenham, UK: Edward Elgar.

Carnevale, Anthony, Nicole Smith and Michelle Melton. 2011. *STEM.* Georgetown University Center on Education and the Workforce. October.

Carney, Richard W. and Loh Yi Zheng, "Institutional (Dis)Incentives to Innovate: An Explanation for Singapore's Innovation Gap." *Journal of East Asia Studie*s 9(2):291-319.

Carter, Jimmy. 1979. "Solar Energy Message to the Congress." June 20. (<*http://www.presidency.ucsb.edu/ws/index.php?pid=32503&st=foreign+oi l&st1=#axzz1OmYKbnIb*>).

Castells, M., and P. Hall. 1994. *Technopoles of the World: the making of twenty-first-century industrial complexes.* London, UK: Routledge.

CBS News. 2009. "Eli Lilly Owes $1.4B Over Off Label Use." February 11.

Cebrowski, Arthur, and John Garska. 1998. "Network Centric Warfare: Its Origin and Future." U.S. Naval Institute Proceedings. January.

Center for Economic Development and Business Research. 2008. "Kansas Aviation Manufacturing."Wichita State University: W. Frank Barton School of Business. September.

Centerwatch. 2011. "New $1b NIH Center Will Tackle Early-Stage Drug Development to Ease Industry Risk of Failure." February 7.

Central News Agency. 2010. "MOEA to invest more in safe lithium-ion Battery Development." January 24.

Central News Agency. 2011. "Taiwan—A Growing Model for Startup Companies" November 27.

Chambers, John, ed. 1999. *The Oxford Companion to American Military History.* Oxford, UK: Oxford University Press.

Chan, K. F., and Theresa Lau. 2005. "Assessing Technology Incubator Programs in the Science Park: The Good, the Bad and the Ugly." *Technovation* 25(10):1215-1228.

Chand, Satish, and Kunal Sen. 2002. "Trade Liberalization and Productivity Growth: Evidence from Indian Manufacturing." *Review of Development Economics* 6, February.

Chandler, Jr., Alfred. 1990. *Scale and Scope: The Dynamics of Industrial Capitalism.* Cambridge and London: Harvard University Press. pp. 181-193.

Chandler Jr., Alfred. 2005. *Shaping the Industrial Century: The Remarkable Story of the Evolution of the Modern Chemical and Pharmaceutical Industries.* Cambridge and Lander: Harvard University Press. pp. 177-179.

Chang, Connie, Stephanie Shipp, and Andrew Wang. 2002. "The Advanced Technology Program: A Public-Private Partnership for Early-stage Technology Development." *Venture Capital* 4(4): 363-370.

Chang, Sewell and Keith Bradsher. 2010. "U.S. to Investigate China's Clean Energy Aid." *The New York Times* October 15.

Chao, Loretta. 2011. "China Plans to Ease Rules That Irked Companies." *The Wall Street Journal* July 1. (Accessed at <http://online.wsj.com/article/SB10001424052702303763404576417621905338368.html>).

Chaturvedi, S. 2005. "Evolving a National System of Biotechnology Innovation, Some Evidence from Singapore." *Science Technology & Society.*

Chemical Business NewsBase. *2008.* "Sequenom Announces Additional Positive Tests Results for Down Syndrome Test at Analyst Briefing." September 23.

Chemical Business Newsbase. 2009. "How Big Pharma's New Direction Might Help Little Research Firms." November 24.

Chemical Business Newsbase. 2011. "Cyterix Pharmaceuticals Raises $9.2M in a Series A Venture Financing." June 7.

Chemical Business Newsbase. 2011. "Merck Play R&D Centre in China." December 12.

Chemical Business NewsBase. "Clamping Down on Fakes." September 8.

Chemical Week. 2007. "DuPont opens Tech Center in Russia." March 21.

Chen, Duanjie and Jack Mintz. 2010. "U.S. Effective Corporate Tax Rate on New Investments: Highest in the OECD." Tax & Budget Bulletin No. 62. Washington, DC: Cato Institute.

Chen, Joeng Shein. 2009. "Taiwan PV Roadmap: Strategies for PV Industry and Market Growth." Taiwan Photovoltaic Industry Association. November 17.

Cheng, Dawei. "China SMEs: Today's Problem and Future's Cooperation." School of Economics. Renmin University of China.

Chesbrough, Henry. 2003. Open *Innovation: The New Imperative for Creating and Profiting from Technology.* Cambridge, MA: Harvard Business School Press. April.

China Daily. 2008. "China Luring 'Sea Turtles' Home." December 18.

China Daily. 2010. "Boeing, Tsinghua Open Research Center" October 21.

China Daily. 2010. "China to invest 7t yuan for urban infrastructure in 2011-15." May 13.

China Daily. 2010. "Toyota rolls out wholly owned Research Center." November 22.

China Daily. 2011. "Corning Sets Up Research Center on the Mainland." June 29.

China Trade Extra. 2005. "China Agrees to Delay Software Procurement Rule While Talking with U.S." July 11.

Chinese Ministry of Finance. 2006. Opinions of the Ministry of Finance on Implementing Government Procurement Policies That Encourage Indigenous Innovation. Cai Ku [2006]. No. 47. June 13.

Chinese Ministry of Information Industry. 2006. "Outline of the 11th Five-Year Plan and Medium-and-Long-Term Plan for 2020 for Science and Technology Development in the Information Industry." Xin Bu Ke [2006]. No. 309.

China Radio International Online. "Capsule Scandal Exposes Loopholes in Drug Quality Control." April 17.

China Research and Intelligence. 2009. "Brief of the LED Lighting Program of 10,000 Lights in 10 Cities in China." July 23.

Chi-ping, Ho 2012. "Demography could threaten China's lead in manufacturing*." China Daily* April 25.

Cho, Eric Y and Hideki Yamawaki. 2010. "Clusters, Productivity, and Experts in Taiwanese Manufacturing Industries". *World Scientific Studies in International Economics*

Chonja Sinmun. 1998. "Taedok to Become Mecca for Venture Firms." April 10.

Chordà, I. M. 1996. "Towards the Maturity State: An Insight into the Performance of French Technopoles." *Technovation* 16(3):143-152.

Chosun Ilbo. 2012. "Major Stem Cell Medication Given Green Light." January 20.

Christensen, Clayton M. 1997. The Innovator's Dilemma: The Revolutionary Book that Will Change the Way You do Business, NY : Harper Business Essentials.

Ng, Y. C. and S. k. Li. 2009. "Efficiency and productivity growth in Chinese universities during the post-reform period." *China Economic Review:* 20 (2): 183-192.

Chuma, Hiroyuki. 2006. "Increasing Complexity and Limits of Organization in the Microlithography Industry: Implications for Science-based Industries." Research Policy 35:394-411.

Chuma, Hiroyuki, and Norikazu Hashimoto. 2007. "Moore's Law, Increasing Complexity and Limits of Organization: Modern Significance of Japanese DRAM ERA." NISTEP Discussion Paper No. 44. National Institute of Science and Technology Policy.

Chang, C. Y. and C. V. Trappey (2003). "The National Si-Soft Project." Applied Surface Science 216(1-4 SPEC.): 2

Chun-Yen, Chang and Wei Hwang, 2004."Development of National System-on-Chip (NSoC) Program in Taiwan." National Chiao Tung University. Proceedings. 2004 International Symposium on System-on-Chip.

Cimoli, Mario, and Marina della Giusta. 2000. "The Nature of Technological Change and its Main Implications on National and Local Systems of Innovation." IIASA Interim Report IR-98-029.

Cincera, Michele. 2006. Comparison of Regional Approaches to Foster Innovation in the European Union: The Case of Flanders. Brussels, Belgium: IWT.

Cincera, Michele. 2006. R&D Activities of Flemish Companies in the Private Sector: An Analysis for the Period 1998-2002. Brussels, Belgium: IWT.

Clark, B. 1995. Places of Inquiry. Berkeley, CA: University of California Press.

Clarke, P. 2004. "LETI, Crolles Alliance Open $350-million 32-nm Research Fab." EE Times April 24.

Clarke, P., M. LaPedus, and M. Santarini. 2005. "IBM-led Consortium to Build Fab in N.Y." EE Times January 5.

Cleantech. 2012. "Busting the Myth of the 'Clean-tech' Crash". February 15.
Cleantech. 2012. "There is No Cleantech Venture Bust, Sorry Wired." February 14.
Clemins, Patrick J. 2009. "Historical Trends in Federal R&D." in AAAS Report XXXVI: Research and Development FY 2010 Intersociety Working Group, American Association for the Advancement of Science, May.
Clemins, Patrick J. 2011. "R&D in the Federal Budget". AAAS. May 25 Accessed at http://www.aaas.org/spp/rd/presentations/aaasrd20110525.pdf.
Cliff, Roger, Chad J. R. Ohlandt, and David Yang. 2011. *Ready for Takeoff: China's Advancing Aerospace Industry*, RAND National Security Research Division for U.S.-China Economic and Security Review Commission.
Clinton, Hillary Rodham. 2011. "On Principles of Prosperity in the Asia Pacific, speech at Shangri-La Hotel, Hong Kong, July 25.
Clough, G. Wayne. 2007. "The Role of the Research University in Fostering Innovation." The Americas Competitiveness Forum. June 12.
CNN Politics. 2009. "Obama Overturns Bush Policy on Stem Cells." March 9.
Coakes, Elayne, and Peter Smith. 2007. "Developing Communities of Innovation by Identifying Innovation Champions." *The Learning Organization: An International Journal* 14(1):74-85.
The Coalition for American Solar Manufacturing. 2011. "U.S. Manufacturers of Solar Cells File Dumping and Subsidy Petitions Against China." press release, Oct. 19.
Cockburn, Robert, Paul Newton, Kyermateng Agyarko, Dora Akunyii and Nicholas White. 2005."The Global Threat of Counterfeit Drugs: Why Industry and Government Must Communicate the Dangers." Plos Medicine. March.
Coggeshall, Charlie and Rorbert M. Margolis. 2010. Consortia Focused on Photovoltaic R&D, Manufacturing, and Testing: A Review of Existing Models and Structures, National Renewable Energy Laboratory, Technical Report NREL/TP-6A2-47866, March.
Cohen, Linda R. and Roger G. Noll. 1991. *The Technology Pork Barrel*, Washington, DC: Brookings Institution Press, June.
Cohen, Linda R. and Roger G. Noll. 2001. "Is U.S. Science Policy at Risk? Trends in Federal Support for R&D" Washington, DC: Brookings Institution Press.
Cohen, W. 2002. "Thoughts and Questions on Science Parks." Presented at the National Science Foundation Science Parks Indicators Workshop. University of North Carolina at Greensboro.
Cohen, W., R. Florida, and R. Goe. 1994. *University-Industry Research Centers in the United States*. Pittsburgh, PA: Carnegie-Mellon University.
Cohen, W., R. Nelson, and J. Walsh. 2000. "Protecting Their Intellectual Assets: Appropriability Conditions and Why U.S. Manufacturing Firms Patent (or Not)." NBER Working Paper 7552. Cambridge, MA: National Bureau of Economic Research.

Cohen, W. M., A. Goto, A. Nagata, R. R. Nelson, and J. P. Walsh. 2002. "R&D Spillovers, Patents and the Incentives to Innovate in Japan and the United States." *Research Policy* 1425:1-19.

Colatat, Phech Georgeta Vidican, and Richard K. Lester. 2009."Innovation Systems in the Solar Photovoltaic Industry: The Role of Public Research Institutions." Massachusetts Institute of Technology Industrial Performance Center, Working Paper Series, MIT-IPC-09-007, June.

Combs, Kathryn L., and Albert N. Link. 2003. "Innovation Policy in Search of an Economic Paradigm: The Case of Research Partnerships in the United States." *Technology Analysis & Strategic Management* 15(2).

The Commission of Experts on Innovation. 2011. Research, Innovation and Technological Performance in Germany Report 2011, p. 130. February.

Computer History Museum. 2011. "Interview with Chun-yen Chang, *Taiwanese IT Pioneers: Chun-yen Chang.*" recorded February 16, 2011. p. 11.

Conant, Jennet. 2002. *Tuxedo Park.* New York: Simon & Shuster.

Conant, Jennet. 2005. 109 East *Palace :* Robert Oppenheimer and the secret city of Los Alamos. New York: Simon & Shuster.

Confederation of Indian Industry and Boston Consulting Group. 2005. "Manufacturing Innovation: A Senior Executive Survey."

Congressional Budget Office. 2005. "Corporate Income Tax Rates: International Comparison." November.

Congressional Budget Office. 2006. "Research and Development in the Pharmaceutical Industry." October. Pp. 3-4

Congressional Budget Office. 2009. "Pharmaceutical R&D and the Evolving Market for Prescription Drugs." October 26.

Cooley, Ed. The Washington Post. 2006. "Chinese to Develop Sciences, Technology." February 10. P. A16.

Cooper, Helene and M. Landler. 2011. "U.S. Shifts Focus to Press China for Market Access." New York Times. January 18. Pp. A1.

Council of Canadian Academies. 2009. Innovation and Business Strategy: Why Canada Falls Short. Report by Expert Panel on Business Innovation.

Council on Competitiveness. 2005. *Innovate America: Thriving in a World of Challenge and Change.* Washington, DC: Council on Competitiveness.

Council of Economic Advisors. 1995. Economic Report to the President. Washington, DC: Government Printing Office.

Council of Economic Advisers. 2001. Economic Report of the President, Washington, DC: Government Printing Office.

Council on Government Relations. 2000. *Technology Transfer in U.S. Research Universities: Dispelling Common Myths.* Washington, DC: Council on Government Relations.

Courant, Paul, James Duderstadt, and Edie Goldenberg. 2010. "Needed: A National Strategy to Preserve Public Universities." *The Chronicle of Higher Education,* Jan. 3.,

Cowen, Tyler. 2011. *The Great Stagnation: How America Ate All The Low-Hanging Fruit of Modern History, Got Sick, and Will (Eventually) Feel Better.* New York: Dutton.

Crafts, N. F. R. 1995. "The Golden Age of Economic Growth in Western Europe, 1950-1973." Economic History Review 48(3):429-447.

Crutsinger, Martin. 2010. "U.S. Challenges Chinese Wind-Power Subsidies." Associated Press article published in Seattle Times, Dec. 22.

Curtis, Keith. 2012. Testimony before House Committee on Appropriations. March 22.

Czarnitzki, Dirk, and Niall O'Byrnes. 2007. "Innovation and the Impact on Productivity in Flanders." *Tijdschrift voor Economie en Management* 52(2).

Dahlman, Carl J., and Jean Eric Aubert. 2001. China and the Knowledge Economy: Seizing the 21st Century. Washington, DC: The World Bank.

Dahlman, Carl, and Anuja Utz. 2005. India and the Knowledge Economy: Leveraging Strengths and Opportunities. Washington, DC: The World Bank .

Dahlman, Carl,. 2011. The World Under Pressure: How China and India are Influencing the Global Economy and Environment. Palo Alto: Stanford UP.The Daily. 2010. "Spending on Research and Development." Statistics Canada, December 9

Dalton, Matthew. 2011. "EU Finds China Gives Aid to Huawei, ZTE." Wall Street Journal, Feb. 3.

Daneke, Gregory A. 1998. "Beyond Schumpeter: Non-linear Economics and the Evolution of the U.S. Innovation System." Journal of Socio-economics 27(1):97-117.

Darby, Michael, Lynne G. Zucker, and Andrew J. Wang. 2002. Program Design and Firm Success in the Advanced Technology Program: Project Structure and Innovation Outcomes. NISTIR 6943. Gaithersburg, MD: National Institute of Standards and Technology.

Das, Gurcharan. 2006. "The India Model." *Foreign Affairs* 85(4).

Dasgupta, P., and P. David. 1994. "Toward a New Economics of Science." *Research Policy* 23:487-521.

Datamonitor. 2010. "Achillion Pharmaceuticals Receives Phase I SBIR Grant from NIH." March 19.

David, P. A. 1985. "Clio and the Economics of QWERTY." *American Economic Review* 75(2):332-337.

Davidsson, Per. 1996. "Methodological Concerns in the Estimation of Job Creation in Different Firm Size Classes." Working Paper, Jönköping International Business School.

Davis, Bob. 2011. "U.S. Targets State Firms, Eyeing China" Wall Street Journal, October26,. Accessed at: http://online.wsj.com/article/SB10001424052970203752604576648040995983406.html

Davis, Megan. 2012. "Rusnano, US fund to invest $760 mln in pharma venture." Reuters. March 6.

Davis, Steven, John Haltiwanger, and Scott Schuh. 1993. "Small Business and Job Creation: Dissecting the Myth and Reassessing the Facts." NBER Working Paper No. 4492. Cambridge, MA: National Bureau of Economic Research.

Debackere, Koenraad, and Reinhilde Veugelers. 2005. "The Role of Academic Technology Transfer Organizations in Improving Industry Science Links." *Research Policy* 34(3):321-342.

Debackere, Koenraad, and Wolfgang Glänzel. 2004. "Using a Bibliometric Approach to Support Research Policy Making: The Case of the Flemish BOF-key." Scientometrics 59(2).

Defense Advanced Research Projects Agency. 2003. *DARPA Over The Years*. Arlington, VA: Defense Advanced Research Projects Agency. October 27.

Defense Advanced Research Projects Agency. 2005. *DARPA—Bridging the Gap, Powered by Ideas*. Arlington, VA: Defense Advanced Research Projects. February

de Jonquieres, Guy. 2004. "China and India Cannot Fill the World's Skills Gap." Financial Times July 12.

de Jonquieres, Guy. 2004. "To Innovate, China Needs More than Standards." Financial Times July 12.

de Jonquieres, Guy. 2006. "China's Curious Marriage of Convenience." Financial Times July 19.

De la Mothe, J., and Gilles Paquet. 1998. "National Innovation Systems, 'Real Economies' and Instituted Processes." Small Business Economics 11:101-111.

Delgado, Mercedes, Michael E. Porter and Scott Stern. 2011. "Clusters, Convergence, and Economic Performance." March 11. [Viewed at http://www.isc.hbs.edu/pdf/DPS_ClustersPerformance_08-20-10.pdf on 3/29/2012]

Dempsey, Paul. 2011. "Foundry Overcapacity – Yes, It Could Happen." Tech Design Forum, June 20.

Deng, Xiaoping, General Secretary of the Communist Party of China Central Committee. 1978. Address at the First National Science Congress.

Department of Commerce. 2011. Press Release, "Commerce Secretary John Bryson Lays Out Vision for Department of Commerce." December 15.

Department of Commerce. 2012. "The Competitive and Innovative Capacity of the United States." January. Access at http://www.commerce.gov/americacompetes

Department of Defense. 1997. "SEMATECH 1987-1997: A Final Report to the Department of Defense." Defense Science Board Task Force on Semiconductor Dependency." Feb. 21.

Department of Defense. 2010. "Quadrennial Defense Review Report. February.

Department of Energy. 2010. "The Recovery Act: Transforming America's Transportation Sector—Batteries and Electric Vehicles." July 14. [Viewed at http://www.whitehouse.gov/files/documents/Battery-and-Electric-Vehicle-Report-FINAL.pdf on 3/29/2012]

Department of Energy. 2010. "Penn State to Lead Philadelphia-based team that will pioneer new energy-efficient building designs." Aug.24. [Viewed at http://www1.eere.energy.gov/buildings/building_america/news_detail.html?news_id=16259 on 3/29/2012]

Department of Energy. 2011. "DOE Announces $27 Million to Reduce Costs of Solar Energy Projects, Streamline Permitting and Installation." Press Release. June 1.

Department of Energy. 2011. *U.S.-China Clean Energy Cooperation: A Progress Report by the U.S. Department of Energy*. January.

Department of Electronic & Information Technology, Ministry of Communications & Information Technology, Government of India. 2011? Special Manpower Development Programme in the Area of VLSI Design and Related Software. [Viewed at http://www.mit.gov.in/content/special-manpower-development-programme#MainContent on 3/29/2012 "Last updated July 4, 2011"]

Department of Justice Press Release. 2009. "Attorney General Holder and HHS Secretary Sebelius Announce New Interagency Health Care Fraud Prevention and Enforcement Action Team." May 21.

Department of Science & Technology press release, Ministry of Science & Technology. 2009. "New Millennium Indian Technology Leadership Initiative Scheme." Feb. 27.

De Proft, A. 2006. Presentation at National Academies symposium on "Synergies in Regional and National Policies in the Global Economy." Leuven, Belgium. September.

Devereaux, M. P., R. Griffith, and A. Klemm. 2002. "Corporate Income Tax Reforms and International Tax Competition." *Economic Policy* vol. 35(17). October.

DeVol, Ross and Perry Wong. 2010. "Jobs for America: Investments and Policies for Economic Growth and Competitiveness." Milken Institute, January 26.

DeVol, Ross C. et. al. 2009. "Manufacturing 2.0: A More Prosperous California." Milken Institute, June.

DeVol, Ross Armen Bedsoussian, and Bejamin Yeo. 2011.*The Global Biomedical Industry: Preserving U.S. Leadership.* Milken Institute.

Dewey & LeBoeuf. 2009 "Maintaining America's Competitive Edge: Government Policies Affecting Semiconductor Industry R&D and Manufacturing Activity." [a white paper prepared for the SIA by Dewey & LeBoeuf, March . accessed at http://www.choosetocompete.org/downloads/Competitiveness_White_Paper.pdf]

DIW Berlin [authors: Marius Clemens, Dieter Schumacher]. 2010. "Germany is Well Positioned for International Trade with Research-Intensive Goods." DIW Berlin Weekly Report, No. 11/2010, Viewed here http://www.diw.de/documents/publikationen/73/diw_01.c.353970.de/diw_wr_2010-11.pdf.

DiGregorio, D., and Shane, S. 2003. "Why do some universities generate more start-ups than others?" *Research Policy*, 32(2), 209-227.

Dirks, S and M. Keeling. 2009. "A Vision of Smarter Cities: How Cities Can Lead the Way into a Prosperous and Sustainable Future." IBM Global Business Services.

Dixon, Robert K, Elizabeth McGowan, Ganna Onysko, and Richard M. Scheerb. 2010, "US energy conservation and efficiency policies: Challenges and opportunities." *Energy Policy* 38(11):6398–6408

Doloreux, David. 2004. "Regional Innovation Systems in Canada: A Comparative Study." Regional Studies 38(5):479-492.

Dong-A Ilbo. 2008. "Pharmaceutical Giant to Expand Korea Operations." February 18.

Dong-A Ilbo. 2009. "War Declared on Drug Makers' Rebates to Doctors." July 31.

Dries, Ilse, Peer van Humbeek, and Jan Larosse. 2005. *Linking Innovation Policy and Sustainable Development in Flanders*. Paris, France: Organization for Economic Co-operation and Development.

Dudas, J. 2005. "Statement of the Honorable Jon W. Dudas Deputy Under Secretary of Commerce for Intellectual Property and Director of the U.S. Patent and Trademark Office before the Subcommittee on Intellectual Property, Committee on the Judiciary." U.S. Senate. <http://judiciary.senate.gov>.

Deutsche Press Agentur. 2008. "Compulsory Thai Licensing of AIDS Drug Sets Precedent." July 29.

Duranton, Gilles, Philippe Martin, Thierry Mayer, and Florian Mayneris. 2010. The economics of clusters. Lessons from the French experience. Oxford: Oxford University Press.

Eaton, Jonathan, Eva Gutierrez, and Samuel Kortum. 1998. "European Technology Policy." NBER Working Papers 6827.

Economic Development Agency. 2006. Measuring Broadband's Economic Impact, National Technical Assistance, Training, Research, and Evaluation Project #99-07-13829, February.

Economic Development Agency. 2011. "U.S. EDA Announces Registry to Connect Industry Clusters Across the Country." October 6.

Economist Intelligence Unit. 2008. "Will US Inspections Help Improve the Safety of Chinese Drugs?" April 15.

Economist Intelligence Unit. 2009. "GSK and Fiocruz to Develop and Product Vaccines." September 14.

The Economic Times. 2009. "DuPont India Growing by Leaps and Bounds Despite Slowdown." April 5.
The Economic Times. 2010. "DuPont to Invest $100 million to step-up R&D base.". October 5.
The Economic Times. 2011. "India Will Be 3rd Biggest Carmaker: Diane Gulyas." September 4.
The Economic Times. 2012. "Merck and Company Firms Up Plan for Emerging Markets." February 12.
The Economist. 2005. "Competing Through Innovation." December 17.
The Economist. 2009. "Up, Up and Huawei: China has Made Huge Strikes in Network Equipment." September 24.
The Economist. 2012. "Demography: China's Achilles Heel." April 21.
The Economist. 2012. "The Rise of State Capitalism". January 21.
Edler, J., and S. Kuhlmann. 2005. "Towards One System? The European Research Area Initiative, the Integration of Research Systems and the Changing Leeway of National Policies." Technikfolgenabschätzung: Theorie und Praxis 1(4):59-68.
Edler, Jakob and Luke Gerghiou, 2007. "Public procurement and innovation – Resurrecting the demand side." *Research Policy*. 36, 9, 949-963. EE Times. 2006. "Chinese Province Pays to Get 300-mm Wafer Fab." June 28.
EE Times Eastern Europe. 2008. "ESilicon to expand Romanian Chip Design Chip Operation." November 13.
Ehlen, Mark A. 1999. *Economic Impacts of Flow-Control Machining Technologies: Early Applications in the Automobile Industry*. NISTIR 6373. Gaithersburg, MD: National Institute of Standards and Technology.
Ehrenberg, R. 2002. *Tuition Rising: why colleges cost so much?*. Cambridge, MA: Harvard University Press.
Ehrlich, Everett. 2011. A Study of the Economic Impact of GLOBALFOUNDRIES. June.
Eickelpasch, Alexander, and Michael Fritsch. 2005. "Contests for Cooperation: A New Approach in German Innovation Policy." Research Policy 34:1269-1282.
Einhorn, Bruce. 2005. "A Creativity Lab for Taiwan." *BusinessWeek*, May 16.
Electronic News. 2006. "SMIC Gets $3B Nod from Chain's Wuhan Government." May 22.
Electronics World. 2010. "Flexible Graphene Memristors." December 9.
Eluvangal, Sreejiraj. 2010. "Renewable Energy Goal Quadrupled." DNA Money, December 30.
Endquist, Charles, ed. 1997. *Systems of Innovation: Technologies, Institutions, and Organizations*. London, UK: Pinter.
Energy Information Administration. 2009 "State Energy Consumption Estimates: 1960 through 2007." Tables 8-12.
Engardio, Pete and Arlene Weintraub. 2008. "Outsourcing the Drug Industry." BusinessWeek, September 4.

Engardio, Pete. 2006. "The Future of Outsourcing: How it's Transforming Whole Industries and Changing the Way We Work." BusinessWeek, Jan. 30.

Engardio, Pete. 2008. "Who's Who in Chinese American Life Sciences " Bloomberg BusinessWeek. September

Engardio, Pete. 2008. "Chinese Scientists Build Big Pharma Back Home." BusinessWeek, Sept. ~~154~~.

Engardio, Pete. 2009. "Singapore's One North." BusinessWeek, June 1.

Engardio, Pete. 2009. "Barcelona's Big Bet on Innovation." BusinessWeek. June 8.

Engardio, Pete. 2009. "Can the Future be Built in America? Inside the U.S. Manufacturing Crisis." BusinessWeek: Sept. 10.

Engardio, Pete. 2009. "Innovation Goes Downtown." BusinessWeek, November 19.

EOS Gallup Europe. 2004. Entrepreneurship. Flash Eurobarometer 146. January. Accessed at <http://ec.europa.eu/enterprise/enterprise_policy/survey/eurobarometer146en.pdf>.

Eppinger, Steven and Anil R. Chitkara. 2006. "The New Practice of Global Product Management." MIT Sloan Management Review, 47(4) Summer, Pp. 22-30

Ernst, Dieter. 2011. "China's Innovation Policy is a Wake-Up Call for America". *Asia-Pacific Issues.* No. 100.

European Commission. 2003. Innovation in Candidate Countries: Strengthening Industrial Performance. Luxembourg: Office for Official Publications of the European Communities, May. European Commission. 2003. "Investing in Research: An Action Plan for Europe 2003." Luxembourg: Office for Official Publications of the European Communities.

European Commission. 2003. Third European Report on Science and Technology Indicators 2003.

European Photovoltaic Industry Association. 2011. Global Market Outlook for Photovoltaics Until 2015, April.

European Union Chamber of Commerce in China. 2010. European Business in China Position Paper 2009/2010, executive summary.

Evaluate Pharma. 2011. "Cost of New Drug Development Remains High." March 10.

Evans, D., and Jovanovic, B. 1989. "An Estimated Model of Entrepreneurial Choice under Liquidity Constraints." *Journal of Political Economy* 97:808-827.

Evans, Sir Harold, Buckland Gail Lefer David -2004. *They Made America*. New York: Little, Brown and Company.

Ewing Marion Kauffman Foundation 2007. The 2007 State New Economy Index. http://sites.kauffman.org/pdf/2007_State_Index.pdf

Executive Office of the President. 2009. "A Strategy for American Innovation: Driving Towards Sustainable Growth and Quality Jobs." National Economic Council Office of Science and Technology Policy, September.

Executive Yuan. 2009. "National Science and Technology Development Plan (2009-12). (http://web1.nsc.gov.tw/public/Attachment/91214167571.PDF)

Executive Yuan. 2011. R.O.C. (Taiwan): Council for Economic Planning and Development, Economic Planning Council. Taiwan Statistical Data Book 2011, July.

Export-Import Bank of the United States. 2010. Report to the US Congress on Export Credit Competition and the Export-Import Bank of the United States. June.

Ezell, Stephen. 2011. Fighting Innovation Mercantilism, *Issues in Science and Technology*, Winter.

Ezell, Stephen. 2011. "Understanding the Importance of Export Credit Financing to U.S. Competitiveness." Washington, DC: ITIF, June.

Ezell, Steven and Robert D. Atkinson. 2011. "The Case for a National Manufacturing Strategy." April. Washington, DC: ITIF.

Faems, Dries, Bart Van Looy, and Koenraad Debackere. 2005. "Inter-organizational Collaboration and Innovation: Toward a Portfolio Approach." *Journal of Product Innovation Management* 22(3):238-250.

Faiola, Anthony. 2010. "Germany Seizes on Big Business in China." Washington Post, September 18.

Fairlie, Robert W. 2011. "Kauffman Index of Entrepreneurial Activity, 1996-2010." Kansas City, MO: Ewing Marion Kauffman Foundation, March.

Fan, W., and White, M. J. 2002. "Personal Bankruptcy and the Level of Entrepreneurial Activity." NBER Working Paper, Series 9340. Boston, MA: National Bureau of Economic Research.

Fangerberg, Jan. 2002. *Technology, Growth, and Competitiveness: Selected Essays*. Cheltenham, UK: Edward Elgar.

Federal Communications Commission. 2009. *Connecting America: The National Broadband Plan*. Washington, DC: Federal Communications Commission.

Federal Ministry of Economics and Technology and Federal Ministry for the Environment, Nature Conservation and Nuclear Safety. 2010. Energy Concept for an Environmentally Sound, Reliable and Affordable Energy Supply, Sept. 28.

Federal Ministry of Education and Research. 2001 Knowledge Creates Markets: Action Scheme of the German Government, March.

Federal Ministry of Education and Research. 2009. Research and Innovation for Germany: Results and Outlook.

Federal Ministry of Education and Research. 2010. ICT Strategy of the German Federal Government: Digital Germany 2015, November.

Federal Ministry of Education and Research. 2010. Ideas. Innovation. Prosperity. High-Tech Strategy 2020 for Germany. Innovation Policy Framework Division.

Federal Ministry of Education and Research. 2011. "Germany and the United States Increase Their Cooperation." March 24.

Federal Register Notice. 2004. "2004 WTO Dispute Settlement Proceeding Regarding China: Value Added Tax on Integrated Circuits." April 21.

Federal Trade Commission. 2003. *To Promote Innovation: The Proper Balance of Patent and Competition Law Policy*. Washington, DC: U.S. Government Printing Office.

Federal Trade Commission. 2010. Pay-for-Delay: How Drug Company Pay-Offs Cost Consumers Billions. January.

Federation of the Indian Chamber of Commerce and Industry. 2007. "FICCI Survey on Emerging Skill Shortages in Indian Industry." New Delhi : FICCI, Federation House. July [Viewed here http://www.ficci-hen.com/Skill_Shortage_Survey_Final_1_.pdf on 3/30/2012]

Feigenbaum, Evan and Adam Segal.2003. China's techno-Warriors: National Security and Strategic Competition from the Nuclear Age to the Information Age, Palo Alto: Stanford University Press.

Feldman, Maryann and Albert N. Link. 2001. "Innovation Policy in the Knowledge-based Economy*." Economics of Science, Technology and Innovation*. 23. Boston, MA: Kluwer Academic Press.

Feldman, Maryann P., Albert N. Link, and Donald S. Siegel. 2002. *The Economics of Science and Technology: An Overview of Initiatives to Foster Innovation, Entrepreneurship, and Economic Growth*. Boston, MA: Kluwer Academic Press.

Feldman, Maryann, Irwin Feller, Janet Bercovitz, and Richard Burton. 2002. "Equity and the Technology Transfer Strategies of American Research Universities*." Management Science* 48(1):105-121.

Feller, Irwin. 1997. "Technology Transfer from Universities." In John Smart, ed. *Higher Education: Handbook of Theory and Research*. Vol. XII. New York: Agathon Press.

Feller, Irwin. 2004. *A Comparative Analysis of the Processes and Organizational Strategies Engaged in by Research Universities Participating in Industry-University Research Relationships*. Final report submitted to the University of California Industry-University Cooperative Research Program. Agreement No. M-447646-19927-3.

Ferguson, R. and C. Olofsson. 2004. "Science Parks and the Development of NTBFs: Location, Survival and Growth." *Journal of Technology Transfer* 29(1): 5-17.

Field, A. J. 2003. "The Most Technologically Progressive Decade of the Century." *American Economic Review* (September):1406.

Pilling, David. 2005. "World Leader in Patents Focuses on Incremental Innovations." The Financial Times, October 12. [Accessed at: http://www.ft.com/intl/cms/s/1/083945e0-3b22-11da-a2fe-00000e2511c8.html#axzz1qc0WiG9E. Fierce Biotech. 2011. "The World's Biggest R&D Spenders." March 8.

Fierce Pharma. 2010. "Just How Big is the Counterfeit-Drug Problem?" September 13.

The Financial Express. 2008. "India Inside Intel Chips." September 25.

The Financial Express. 2010. "MNC R&D Centers Generate $40bn in savings: Study." July 18.Financial Times. 2009. J&J Wants Deals with Rivals to Share Risk." October 25.

Financial Times. 2010. "Big Drug Groups Urged to Buy in Test Products." January 31.

Financial Times. 2011. "China's Rail Disaster." July 27.

Financial Times. 2012. "More with less." May 19.

Finnish Science and Technology Information Service and Statistics Finland. 2011.Research and Development 2010, October 27. See here http://tilastokeskus.fi/til/tkke/index_en.html

First Solar, Inc. 2007. "First Solar Announces 100MW Manufacturing Plant Expansion in Malaysia." News Release, January 25.

First Solar, Inc. 2011. *Annual Report 2010*. February. p. 6-10.

Fischbach, Amy. 2009. "Engineering Shortage Puts Green Economy and Smart Grid at Risk." Transmission and Distribution World, April 21.

Fitzpatrick, Ryan, Josh Freed, and Mieke Eoyang. 2011. "Fighting for Innovation: How DoD Can Advance Clean Energy Technology... And Why It Has To." Washington, DC: The Third Way, June.

Flamm, K. 1996. Mismanaged Trade: Strategic Policy and the Semiconductor Industry. Washington, DC: The Brookings Institution.

Flamm, K. 2003. "Microelectronics Innovation: Understanding Moore's Law and Semiconductor Price Trends." *International Journal of Technology, Policy, and Management* 3(2).

Flamm, K. 2003. "The New Economy in Historical Perspective: Evolution of Digital Technology." In *New Economy Handbook*. St. Louis, MO: Academic Press.

Flamm, K. 2003. "SEMATECH Revisited: Assessing Consortium Impacts on Semiconductor Industry R&D." In National Research Council. *Securing the Future: Regional and National Programs to Support the Semiconductor Industry*. Charles W. Wessner, ed. Washington, DC: The National Academies Press.

Florida, Richard, Tim Gulden, and Charlotta Mellander. 2007. "The Rise of the Mega-Region." October.

Foerst, Anne. 2005. *God in the Machine*. New York: Penguin Books.

Fonfria, Antonio, Carlos Diaz de la Guardia, and Isabel Alvarez. 2002. "The Role of Technology and Competitiveness Policies: A Technology Gap Approach." *Journal of Interdisciplinary Economics* 13(1-2-3):223-241.

Fong, Glenn R. 1998. "Follower at the Frontier: International Competition and Japanese Industrial Policy." *International Studies Quarterly*. 42(2):339-366.

Fong, Glenn R. 2001. "ARPA Does Windows: The Defense Underpinning of the PC Revolution." *Business and Politics* 3(3).

Fong, Glenn R. "Breaking New Ground, Breaking the Rules—Strategic Reorientation in U.S. Industrial Policy." International Security 25:2 pp 152.

Food and Drug Law Institute. 2006. "Increased Scrutiny of Investor Communications by Federal Regulators." January/February.

Food and Drug Law Journal. 2008. "Looking Abroad: Clinical Drug Trials." p. 673.

Foray, Dominique, and Patrick Llerena. 1996. "Information Structure and Coordination in Technology Policy: A Theoretical Model and Two Case Studies." *Journal of Evolutionary Economics* 6(2):157-173.

Forbes. 2009. "Officials: Pfizer to Pay Record $2.3 B Penalty." September 23.

Foresight Science & Technology. 2010. "Regional Overviews: Asian Industry Overview (China, Japan, Korea)." July 27.Fox News. 2009. "Obama's Stem Cell Policy Hasn't Reversed Legislative Restrictions." March 14.

Fraunhoffer-Gesellschaft. 2009. Annual Report 2009: With Renewed Energy Fraunhoffer-Gesellschaft: Berlin

Freear and Jeff E. Sohl "Angles on Angels and Venture Capital: Financing Entrepreneurial Ventures" in *Financing Economic Development in the 21st Century*, 2nd Edition, Z. Kotval and S. White, eds., M.E. Sharpe, Inc: NY

Freeman, C. 1987. *Technology Policy and Economic Performance: Lessons from Japan*. London : Frances Pinter.Freeman, Christopher. 1988. "Japan a New National Innovation System." in G. Dosi, et al, *Technology and Economy Theory*. London: Pinter

Freeman, Will. 2010. "The Big Engine That Can: China's High-Speed Rail Project." *Gavekal Dragonomics China Insight Note, May 24, 2010,*

Freedman, Ron. 2011. "Re-Thinking Canada's BERD Gap." The Impact Group, January.

Friedman, Thomas. 2005. *The World Is Flat: A Brief History of the 21st Century*. New York:-Farrar, Straus and Giroux.Frosch, Dan. 2011. "New Mexico's Bet on Space Tourism Hits a Snag." New York Times, Feb. 23. Pp A16.

Fuchs, Erica R. H. 2010. "Rethinking the Role of the State in Technology Development: DARPA and the Case for Embedded Network Governance." *Research Policy* 39(9): 1133-1147

Fuchs, Erica R. H. and Rondolph Kirchain. 2010. "Design for Location? The Impact of Manufacturing Off-Shore on Technology Competitiveness in the Optoelectronics Industry." *Management Science*, 56(12), pp. 2323-2349.

Fukugawa, N. 2006. "Science Parks in Japan and Their Value-Added Contributions to New Technology-based Firms." *International Journal of Industrial Organization* 24(2):381-400

Fukuyama, Francis. 1992. *The End of History and the Last Man*, New York: The Free Press.

Furman, Jeffrey L., Michael E. Porter, and Scott Stern. 2002. "The Determinants of National Innovative Capacity." *Research Policy* 31(6):899-933.

Furong , Zhou and Zhang Zhao. 2010. "Suzhou Industrial Park Faces Challenges on Path to Change." China Daily, March 16.

Gallaher, M. P., A. N. Link, and J. E. Petrusa. 2006. *Innovation in the U.S. Service Sector*. London, UK: Routledge.

Gansler, Jacques. 1995. Defense Conversion: Transforming the Arsenal of Democracy, Cambridge, MA: MIT Press.

Gansler, Jacques. 2011. Democracy's Arsenal, Creating a 21st Century Defense Industry. Cambridge MA: MIT Press.

Gansler, Jacques. 2011. "Solving the Nation's Security Affordability Problem." in *Issues in Science and Technology*, XXVII(4).

Gartner, John and Clint Wheelock, 2009 "Lithium Ion Batteries for Plug-in Hybrid and Battery Electric Vehicles: Market Analysis and Forecasts." executive summary, Pike Research.

Gaule, Patrick. 2011."Return Migration: Evidence From Academic Statistics." [National Bureau of Economic Research}.-May 31 [view PDF online]

GBI Research. 2010. Future of Global Advanced Batteries Market Outlook to 2020: Opportunity Analysis in Electronics and Transportation, January.

"GE Plans to Build Largest US Solar Factory in Colorado, Expand Solar Innovation in New York and Deliver Lighter, Larger, More Efficient Thin Film Panels." 2011. GE Press Release, October 13. [Accessed at: http://www.rttnews.com/1733617/ge-to-build-largest-us-solar-panel-factory-in-colorado-to-add-455-jobs.aspx

Geiger, R. 1986. *To Advance Knowledge: The Growth of American universities*. New York: Oxford University Press.

Geiger, R. 1993. *Research and Relevant Knowledge: American research universities since World War II*. New York: Oxford University Press.

Geithner, Timothy. 2010. Joint Press Availability with Secretary of the Treasury Timothy Geithner in Beijing, China. May 25.

Geoghegan-Quinn, Máire. 2011. "Innovation for stronger regions: opportunities in FP7". Committee of the Regions" Brussels, July 14.

George, Gerard, and Ganesh N. Prabhu. 2003. "Developmental Financial Institutions as Technology Policy Instruments: Implications for Innovation and Entrepreneurship in Emerging Economies." *Research Policy* 32(1):89-108.

Gereffi, Gary, Vivek Wadhwa, Ben Rissing, and Ryan Owen. 2008. "Getting the Numbers Right: International Engineering Education in the United States, China, and India." *Journal of Engineering Education*, Vol. 97(1):13-25.

German Federal Ministry of Education and Research. Innovation Policy Framework Division. 2010. Ideas. Innovation. Prosperity: High Tech Strategy 2020 for Germany. Bonn, Berlin: BMBF. [Viewed here http://www.bmbf.de/pub/hts_2020_en.pdf]

Gertner, Jon. 2011. "Does America Need Manufacturing?" The New York Times Magazine, August 24.

Gibb, M. J. 1985. *Science Parks and Innovation Centres: Their Economic and Social Impact*. Amsterdam, The Netherlands: Elsevier.

Gillen, Andrew et al. 2011. "Net Tuition and Net Price Trends in the United States (2000-2009), Washington, DC: Center for College Affordability. November.

Gilman, Douglas, .2010 . "The New Geography of Global Innovation."Goldman Sachs Global Markets Institute, September 24. Accessed at: http://www.innovationmanagement.se/wp-content/uploads/2010/10/The-new-geography-of-global-innovation.pdf.

Gittell, Ross, Jeffrey Sohl, and Edinaldo Tebaldi. 2010. "Is there a Sweet Spot for U.S. Metropolitan Areas? Exploring the Growth in Employment and Wages in U.S. Entrepreneurship and Technology Centers in Metropolitan Areas over the last Business Cycle, 1991 To 2007, *Frontiers of Entrepreneurship Research*: Vol. 30(15). Article 13.

Goldfarb, Brent and Magnus Henrekson. 2003. "Bottom-up versus top-down policies towards the commercialization of university intellectual property." *Research Policy:* 32. Pp. 639–658

Goldstein, H. A., and M. I. Luger. 1990. "Science/Technology Parks and Regional Development Theory." *Economic Development Quarterly* 4(1):64-78.

Goldstein, H. A., and M. I. Luger. 1992. "University-based Research Parks as a Rural Development Strategy." *Policy Studies Journal* 20(2):249-263.

Goodenough , J. B. and M.S. Whittingham. 1977. Solid State Chemistry of Energy Conversion and Storage :A Symposium. American Chemical Society, Advances in chemistry Series 163.

Goodwin, James C., et al. 1999. Technology Transition. Arlington, VA: Defense Advanced Research Projects Agency.

Gonzalez, Heather, John F. Sargent, and Patricia Moloney Figliola. 2010. "America COMPETES Reauthorization Act of 2010 (H.R. 5116) and the America COMPETES Act (P. L. 110-69): Selected Policy Issues." Congressional Research Service, July 28. [http://www.ift.org/public-policy-and-regulations/~/media/Public%20Policy/0728AmericaCompetesAct.pdf]

Government Accountablity Office. 2011. "Reforms are needed to minimize the risks and costs to current program". GAO-11-26.

Government of India. 2011. "A Triad of Policies to drive a National Agenda for ICTE. October.

Government of India Planning Commission. 2006."Report of the Steering Committee on Science and Technology for Eleventh Five Year Plan (2007-2012)." December[02/02/2010 – view here http://erawatch.jrc.ec.europa.eu/erawatch/opencms/information/country_pages/in/policydocument/policydoc_mig_0002

Government of the People's Republic of China. 2004. *Anticompetitive Practices of Multinational Companies and Countermeasures.* Administration of Industry and Commerce, Office of Antimonopoly, Fair Trade Bureau, State Administration of Industry and Commerce. May.

Government of the People's Republic of China. 2005. *Article 10: Exemptions of Monopoly Agreements.* Anti-Monopoly Law of the People's Republic of China. Revised July 27.

Government of the People's Republic of China, Ministry of Information Industry. 2006. "Outline of the 11th Five-Year Plan and Medium-and-Long-Term Plan for 2020 for Science and Technology Development in the Information Industry." Xin Bu Ke, No. 309, August 29.

Government of the People's Republic of China, National Development and Reform Commission. 2006. The 11th Five-Year Plan. March 19. <http://english.gov.cn/2006-07/26/content_346731.htm>.

Graham, Hugh Davis and Nancy A. Diamond. 1997. *The Rise of American Research Universities: Elites and Challengers in the Postwar Era*, Baltimore: Johns Hopkins UP.

Grande, Edgar. 2001. "The Erosion of State Capacity and European Innovation Policy: A Comparison of German and EU Information Technology Policies." *Research Policy* 30(6):905-921.

Grau, Thilo, Molin Huo, and Karsten Neuhoff. 2011. Survey of Photovoltaic Industry and Policy in Germany and China, Climate Policy Initiative Report, DIW Berlin and Tsinghua University, March.

Grayson, L. 1993. *Science Parks: An Experiment in High-Technology Transfer*. London, UK: The British Library Board.

Green World Investor. 2010. "Reasons Behind Malaysia's Surprising Success in Solar Industry Beating Larger Rivals USA and Japan". October 26.

Greenstone, Michael and Adam Looney. 2011. "Building America's Job Skills with Effective Workforce Programs: A Training Strategy to Raise Wages and Increase Work Opportunities." Washington, DC: Brookings Institution, September.

Griffing, Bruce. 2001. "Between Invention and Innovation, Mapping the Funding for Early-Stage Technologies." Presentation at Carnegie Conference Center. Washington, DC. January 25.

Griliches, Z. 1993. "The Search for R&D Spillovers." *Scandinavian Journal of Economics* 94(S): S29-S47.

Grimaldi, Rosa Martin Kinney, Donald S. Siegel, and Mike Wright. 2011. "30 years after Bayh–Dole: Reassessing academic entrepreneurship" *Research Policy* 40(8):1045-1057.

Grindley, Peter, David Mowery, and Brian Silverman. 1994. "SEMATECH and Collaborative Research: Lessons in the Design of High Technology Consortia." *Journal of Policy Analysis and Management* 13(4):723-758.

Grove, A. 2001. *Swimming Across*. New York: Warner Books.

Grove, Andy. 2010. "How to Make an American Job Before it is Too Late." Bloomberg BusinessWeek, July 1.

Gruber, Martin and Tim Studt. 2010. "2011 Global R&D Funding Forecast: The Globalization of R&D." R&D Magazine, Dec. 15.

GTM Research. 2011. "U.S. Solar Energy Trade Assessment 2011: Trade Flows and Domestic Content for Solar Energy-Related Goods and Services in the United States." prepared for Solar Energy Industries Association, August.

Gu, Shulin and Lundvall Bengt-Åke. 2006. "Policy learning as a key process in the transformation of the ChineseInnovation Systems." In Asian Innovation Systems in Transition. Edward Elgar Publishing Ltd.The Guardian. 2011. "China plans to make a million electric vehicles a year by 2015." February 18.

GUIRR. 20006. "Re-Engineering the Partnership: Summit of the University-Industry Congress." Meeting of 25 April 2006, Washington, DC.

Gulbranson, Christine A. and David B. Audretsch. 2008. "Proof of concept centers: accelerating the commercialization of university innovation." *Journal of Technology Transfer.* 33:249–258.

Guo Ban Han No. 30. 2006. Letter from the General Office of the State Council on Approving the Formulation of the Rules for Implementation of the Several Supporting Policies for Implementation of the Outline of the National Medium and Long-term Plan for Development of Science and Technology. Gazette of the State Council. Issue No. 17, Serial No. 1196, June 20.

Gupta, Anil K. and Haiyan Wang. 2009. Getting China and India Right : Strategies for Leveraging the World's Fastest Growing Economies for Global Advantage, San Francisco, Calif.: Jossey-Bass.

Gupta, Anil K. and Haiyan Wang. 2011. "Chinese Innovation is a Paper Tiger." Wall Street Journal, July 28.

Guy, I. 1996. "A Look at Aston Science Park." *Technovation* 16(5):217-218.

Guy, I. 1996. "New Ventures on an Ancient Campus." *Technovation* 16(6):269-270.

Hackett, S. M., and D. M. Dilts. 2004. "A Systematic Review of Business Incubation Research." *Journal of Technology Transfer* 29(1):55-82.

Hall, Bronwyn H. 2002. "The Assessment: Technology Policy." *Oxford Review of Economic Policy* 18(1):1-9.

Hall, Bronwyn H 2005 "Exploring the Patent Explosion." *Journal of Technology Transfer* 30 (1/2):35-48.

Hall, Bronwyn H.., A. N. Link, and J. T. Scott. 2001. "Barriers Inhibiting Industry from Partnering with Universities : Evidence from the Advanced Technology Program." *Journal of Technology Transfer* 26(1-2):87-98.

Hall, Bronwyn H.., A. N. Link, and J. T. Scott. 2003. "Universities as Research Partners." *Review of Economics and Statistics* 85(2):485-491.

Hall, Bronwyn H.., A. Jaffe, and M. Trajtenberg. 2005. "Market Value and Patent Citations." *RAND Journal of Economics* 36(1):16-38.

Hall, Bronwyn H.. and Beethika Khan. 2003. "Adoption of New Technology." NBER Working Paper 9730.

Hall, Bronwyn H.., C. Helmers, M. Rogers, and V. Sena. 2012. "The Choice between Formal and Informal Intellectual Property: A Literature Review". NBER Working Paper No. 17983. April.

Hall, Bronwyn H.., and R. H. Ziedonis. 2001. "The Patent Paradox Revisited: An Empirical Study of Patenting in the U.S. Semiconductor Industry. 1979-1995." *RAND Journal of Economics* 32(1):101-128.

Halpern, Mark. 2006. "The Trouble with the Turing Test." *The New Atlantis* 11(Winter):42-63.

Haltiwanger, John and CJ Krizan. 1999. "Small Business and Job Creation in the United States: The Role of New and Young Businesses." In *Are Small Firms Important? Their Role and Impact,* Zoltan Acs, ed. Dordrecht: Kluwer.

Hane, G. 1999. "Comparing University-Industry Linkages in the United States and Japan." In L. Branscomb, F. Kodama, and R. Florida, eds. *Industrializing Knowledge*. Cambridge, MA: The MIT Press.

Hamilton, Robert V., McNeely, Connie L. and Perry, Wayne D., Natural Sciences Doctoral Attainment By Foreign Students at U.S. Universities (February 2, 2012). Available at SSRN: http://ssrn.com/abstract=1999816 or http://dx.doi.org/10.2139/ssrn.1999816

Hansson, F., K. Husted, and J. Vestergaard. 2005. "Second Generation Science Parks: From Structural Holes Jockeys to Social Capital Catalysts of the Knowledge Society." *Technovation* 25(9):1039-1049.

Harper, Sarah. 2006. "Addressing the Implications of Global Aging." Journal of Population Research, Vol. 23(2).

Harris, Gardiner. 2011. "Federal Research Center Will Help Develop Medicines." *New York Times*, January 22.

Hart, David. 1998. *Forged Consensus*. Princeton, NJ: Princeton University Press.

Hashimoto, T. 1999. "The Hesitant Relationship Reconsidered: University-Industry Cooperation in Postwar Japan." In Lewis M. Branscomb, Fumio Kodama, and Richard Florida, eds. *Industrializing Knowledge*: University-Industry Linkage in Japan and the United States. Cambridge, MA: The MIT Press.

Hassan, Mohamed H.A. 2005. Small Things and Big Changes in the Developing World." *Science* 309: 65-66

Hayashi, Fumio, and Edward C. Prescott. 2002. "The 1990s in Japan: A Lost Decade." *Review of Economic Dynamics* 5(1):206-235.

Headd, Brian. 2010. *An Analysis of Small Business and Jobs*. U.S. Small Business Administration: Office of Advocacy. March.

Hegedus, Steven S. and Antonio Luque. 2003. "Status, Trends, Challenges and the Bright Future of Solar Electricity from Photovoltaics." chapter in Antonio Luque and Steven Hegedus, editors, *Handbook of Photovoltaic Science and Engineering*, Chichester, England: John Wiley & Sons.

Heller, M., and R. Eisenberg. 1998. "Can Patents Deter Innovation? The Anticommons in Biomedical Research." *Science* 280:698.

Helper, Susan and Howard Wial. 2011. "Accelerating Advanced Manufacturing with New Research Centers." Brookings-Rockefeller Project on State and Metropolitan Innovation, February.

Susan Helper, Timothy Krueger, and Howard Wial, 2012. "Why Does Manufacturing Matter? Which Manufacturing Matters? A Policy Framework." Washington, DC: Brookings, February.

Henderson, J. V. 1986. "The Efficiency of Resource Usage and City Size." *Journal of Urban Economics* 19(1):47-70.

Henderson, Jennifer A., and John J. Smith. 2002. "Academia, Industry, and the Bayh-Dole Act: An Implied Duty to Commercialize." Center for the Integration of Medicine and Innovative Technology. October.

Henderson, R., L. Orsenigo, and G. P. Pisano. 1999. "The Pharmaceutical Industry and the Revolution in Molecular Biology: Interactions among Scientific, Institutional, and Organizational Change." In D. C. Mowery and R. R. Nelson, eds. *Sources of Industrial Leadership*. Cambridge, UK: Cambridge University Press.

Henderson, R., M. Trajtenberg, and A. Jaffe. 1998. "Universities as a Source of Commercial Technology: A Detailed Analysis of University Patenting, 1965-1988." *Review of Economics and Statistics* 80(1):119-127.

Hennessy, J. L., and D. A. Patterson 2002. *Computer Architecture: A Quantitative Approach*. 3rd edition. San Francisco, CA: Morgan Kaufmann Publishers Inc.

Hess, Charlotte and Elinor Ostrom, eds. 2007. Understanding Knowledge as a Commons. Cambridge MA: MIT Press.

Hicks, D., T. Breitzman, D. Olivastro, and K. Hamilton. 2001. "The Changing Composition of Innovative Activity in the U.S.—A Portrait Based on Patent Analysis." *Research Policy* 30(4):681-704.

Hilpert, U., and B. Ruffieux. 1991. "Innovation, Politics and Regional Development: Technology Parks and Regional Participation in High-Technology in France and West Germany." In U. Hilpert, ed. *Regional Innovation and Decentralization: High-Technology Industry and Government Policy*. London, UK: Routledge.

Himmelberg, C., and B. C. Petersen. 1994. "R&D and Internal Finance: A Panel Study of Small Firms in High-Tech Industries." *Review of Economics and Statistics* 76(1):38-51.

The Hindu. 2008. "Intel India Team Lofts a Sixer" September 21.

Hirschler, Ben. 2011. "Analysis: Big Pharma Strips Down Broken R&D Engine." Reuters. May 11.

Holly, Krisztina "Z". 2010. "The Full Potential of University Research, A Model for Cultivating New Technologies and Innovation Ecosystems." *Science Progress*, June.

Holtz-Eakin, D., D. Joulfian, and H. Rosen. 1994. "Striking it Out: Entrepreneurial Survival and Liquidity Constraints." *Journal of Political Economy* 102(1):53-75.

Hong, Jiang. 2011. "State-owned Enterprises Research Project Press Release Conference & Academic Seminar Successfully Held in Beijing." Unirule Institute of Economics.

Hoshi, Takeo, and Anil Kashyap. 2004. "Japan's Financial Crisis and Economic Stagnation :." The Journal of Economic Perspectives 18(1):3-26Accessed at: http://www.jstor.org/stable/3216873

Hounshell, David A. 1984. *From the American System to Mass Production, 1800-1932: The Development of Manufacturing Technology in the United States*, Baltimore, Maryland, USA: Johns Hopkins University Press.

Hourihan, Matt. 2012. "R&D in the FY 2013 Budget." AAAS. April 26,

House Ways and Means Committee. 2008. "Promoting U.S. Worker Competitiveness in a Globalized Economy." June 14, 2007. Serial No. 110-47, Washington, DC: USGPO. Access at http://www.gpo.gov/fdsys/pkg/CHRG-110hhrg43113/pdf/CHRG-110hhrg43113.pdf

Houseman, Susan. 2011. "Offshoring Bias in U.S. Manufacturing." Journal of Economic Perspectives 25: 111-132.

Howell, Thomas. 1988. Steel and the State; Government Intervention and Steel's Structural Crisis. New York: Westview Press.

Howell, Thomas. 2003. "Competing Programs: Government Support for Microelectronics." In National Research Council. *Securing the Future: Regional and National Programs to Support the Semiconductor Industry*. Charles W. Wessner, ed. Washington, DC: The National Academies Press.

Howell, Thomas, Brent Bartlett, and Warren Davis. 1992. *Creating Advantage: Semiconductors and Government Industrial Policy in the 1990s*. Semiconductor Industry Association and Dewey Ballentine.

Howell, Thomas R., William A. Noellert, Janet H. McLaughlin, and Alan Wm. Wolff. 1988. *The Microelectronics Race : The Impact of Government Policy on International Competition.* Boulder, Colorado and London: Westview Press.

Howell, Thomas, et al. 2003. *China's Emerging Semiconductor Industry*. San Jose, CA: Semiconductor Industry Association, October.

Howell, Thomas, William A. Noellert, Gregory Hume, and Alan Wm. Wolff. 2010. China's Promotion of the Renewable Electric Power Equipment Industry: Hydro, Wind, Solar, Biomass, Dewey & LeBoeuf LLP prepared for National Foreign Trade Council, March.

Hu, Jintao. General Secretary of the Communist Party of China Central Committee. 2005. Keynote Speech, November 27.

Hu, Jintao. General Secretary of the Communist Party of China Central Committee. 2007. Speech to the Seventeenth Communist Party of China National Conference, October.

Hu, Jintao. President of China. 2009. Speech to United Nations General Assembly, September 23.

Hu, Zhijian. 2006. "IPR Policies In China: Challenges and Directions." Presentation at Industrial Innovation in China. Levin Institute Conference. July 24-26.

Huang, Alice S. and Chris Y. H. Tan. 2010. "Achieving Scientific Eminence Within Asia." *Science*, Vol. 329:1471-2.

Huang, Can, Celeste Amorim, Mark Spinoglio, Borges Gouveia and Augusto Medina. 2004. "Organization, Programme and Structure: An Analysis of the Chinese Innovation Policy Framework." *R&D Management* 34, 4. http://www.globelicsacademy.net/pdf/CanHuang_paper.pdf

Huang, Yasheng, and Tarun Khanna. 2003."Can India Overtake China?" *Foreign Policy*. July-August.

Huawei . 2011. Press Release. "Huawei Receives Innovation Awards for Contribution to CDMA Development." June 17.

Huddleson, Lillian, and Vicki Daitch. 2002. *True Genius—The Life and Science of John Bardeen*. Washington, DC: Joseph Henry Press.

Hughes, Kent. 2005. *Building the Next American Century: The Past and Future of American Economic Competitiveness*. Washington, DC: Woodrow Wilson Center Press. Chapter 14.

Hughes, Kent H. 2005. "Facing the Global Competitiveness Challenge." Issues in Science and Technology XXI(4):72-78.

Hughes, Kent H and Lynn Sha, eds. 2006. *Funding the Foundation: Basic Science at the Crossroads.* Washington, DC: Woodrow Wilson Center.

Hugin. 2008. "ESilicon Accelerates Expansion to Europe." October 28.

Hulsink, W., H. Bouwman, and T. Elfring. 2007. "Silicon Valley in the Polder? Entrepreneurial Dynamics, Virtuous Clusters and Vicious Firms in the Netherlands and Flanders." ERIM Report Series Research in Management (ERS-2007-048-ORG).

Hundley, Richard O. 1999. *Past Revolutions, Future Transformations: What Can the History of Revolutions in Military Affairs Tell Us About Transforming the U.S. Military*. Rand Corporation, National Research Institute.

Hurt, David M. 1998."Antitrust and Technological Innovation." *Issues in Science and Technology*. Winter.

Iansati, Marco, and Roy Levien. 2005. *The Keystone Advantage*. Cambridge, MA: Harvard Business School Press.

IBN Live. 2012. "Union Budget 2012: Full Text of Pranab Mukherjee's Speech". March 16.

IC Insights. 2010 "Taiwan to Pass Japan as Largest Source of IC Wafer Fab Capacity." Research Bulletin, November 11.

ICIS.com. 2010. Rajith Gopinathan, "Pharmaceutical Companies Seek Biotech Acquisitions to Boost Drug Pipelines." February 12.

The Independent on Sunday. 2007. "Patent Battles Could Savage Drug Giant." March 18.

India Business Insight. 2008. "DuPont plans to Double Manpower in India, " March 28.

India Business Insight. 2011. "DuPont Bets on Helmet, Vest Maker (Who Use its Products Made Under the Kevlar Brand)." April 13.

India Business Insight. 2011. "DuPont India to Recruit 800 Scientists in Two Years." February 21.

India Business Line. 2008. "Developing Technology to Meet Market Needs is DuPont's Priority." January 19.
India Business Line. 2009. "DuPont Adds Seed Research Centers." September 22.
India Semiconductor Association. 2011. Study on Semiconductor Design, Embedded Software and Services Industry , prepared by Ernst & Young, April.
Indian Express. 2012. "$1 bn India Innovation Fund by July, January 17.
Indianapolis Business Journal. 2008. "Generic Meds Don't Come Cheap." December 15.
Indianapolis Business Journal. "R&D Spending Soars Above Value of New Drugs." July 5.
Indo-Asian News Service. 2007. "National Mission to Make India Global Nano Hub." November 5.
Indo-Asian News Source. 2011. "Electronics Development Fund to Promote innovation Soon—Official.". February 21.
Industrial College of the Armed Forces. 2010. Electronics 2010, Industry Study Final Report, National Defense University, Spring .
Industrial Research Institute, Inc. 2001. "Industry-University Intellectual Property." Position Paper. External Research Directors Network. April.
Industry Canada. 2001. Achieving Excellence: Investing in People, Knowledge and Opportunity—Canada's Innovation Strategy. Ottawa, ON : Government of Canada.
Industry Canada. 2007. Mobilizing Science and Technology to Canada's Advantage. Ottawa, ON : Government of Canada]
Industry Canada. 2009. Foreign Affairs and International Trade Canada, and Statistics Canada, "Survey of Innovation and Business Strategy."
Inkpen, Andrew C. and Wang Pien. 2006. "An Examination of Collaboration and Knowledge Transfer: China-Singapore Suzhou Industrial Park." *Journal of Management Studies* 43(4):779-811.
Inside Higher Education. 2011. "The Sinking States." Jan. 24.
Inside U.S.-China Trade. 2006. "Industry Worried China Backing out of Commitment to Join GPA." September 27.
Institute of International Education. 2011. "International Student Enrollments Rose Modestly in 2009/10, Led by Strong Increase in Students from China." Press Release <http://www.iie.org/en>.
International Association of Science Parks (IASP). 2000. <http://www.iaspworld.org/information/definitions.php>.
International Monetary Fund. 2011. "Changing Patterns of Global Trade." June 15, [http://www.imf.org/external/np/pp/eng/2011/061511.pdf
International Monetary Fund. 2011.World Economic and Financial Surveys. World Economic Outlook Database (WEO): Slowing Growth, Rising Risk. September. [http://www.imf.org/external/pubs/ft/weo/2011/02/index.htm].

International Telecommunications Union. 2005. "ITU Internet Reports 2005: The Internet of Things, Executive Summary <http://www.itu.int/osg/spu/publications/internetofthings/InternetofThings_summary.pdf

Institute of Medicine. 2012."Evolution of Translational Omics: Lessons Learned and the Path Forward." Report Brief. March.

Investor's Business Daily. 2008. "Sequenom Raises Bar in Prenatal Test Field." December 16.

Irish Times. "Cancer, Diabetes, Dementia and Cystic Fibrosis: Having the Genome Has Not Meant an End to These Afflictions." February 25.

.Irwin, Douglas A. 1996. "Trade Politics and the Semiconductor Industry." in Anne O. Kreuger, editor, *The Political Economy of American Trade Policy*, Chicago: University of Chicago Press.

Ishikawa, Rikio. "Current Status of Lithium-Ion Production in Japan." Central Research Institute of Electric Power Industry, Tokyo.

Jacob, Merle Mats Lundqvist, and Hans Hellsmark. 2003. "Entrepreneurial Transformations in the Swedish University System: The Case of Chalmers University of Technology." *Research Policy*, 32(9):1555-1568.

Jacob, Klaus and Martin Janicke. 2004. Lead Markets for Environmental Innovations, ZEW Economic Studies, 4(1):29-46. Heidelberg: Physica-Verlag.

Jacobs, Tom. 2002. "Biotech Follows Dot.com Boom and Bust." *Nature Biotechnology* 20(10):973.

Jaffe, A. B. 1989. "Real Effects of Academic Research." American Economic Review 79(5): 957-970.

Jaffe, A. B. 1997. *Economic Analysis of Research Spillovers: Implications for the Advanced Technology Program*. NIST GCR 97-708. Gaithersburg, MD: National Institute of Standards and Technology.

Jaffe, A. B. 1998. "The Importance of 'Spillovers' in the Policy Mission of the ATP." *Journal of Technology Transfer* 23(1):11-19.

Jaffe, A. B. 2000. "The U.S. Patent System in Transition: Policy Innovation and the Innovation Process." *Research Policy* 29:531-557.

Jaffe, A. B., and J. Lerner. 2006. *Innovation and Its Discontents: How Our Broken Patent System Is Endangering Innovation and Progress, and What to Do About It*. Princeton, NJ: Princeton University Press.

Jaffe, A. B., J. Lerner, and S. Stern, eds. 2003. *Innovation Policy and the Economy*. Volume 3. Cambridge, MA: The MIT Press.

Jaffe, A. B., M. Trajtenberg, and R. Henderson. 1993. "Geographic Localization of Knowledge Spillovers as Evidenced by Patent Citations." *Quarterly Journal of Economics* 108(3):577-598.

Japan Patent Office. 2003. *Reports on Technology Trend and Patent Application: Life Science*. (in Japanese).

Jaruzelski, Barry and Kevin Dehoff. 2010. "How the Top Innovators Keep Winning." Booz & Co. 61.Winter (http://www.booz.com/media/file/sb61_10408-R.pdf).

Jasanoff, Sheila, ed. 1997. Comparative Science and Technology Policy. Elgar Reference Collection. International Library of Comparative Pubic Policy. Volume 5. Cheltenham, UK: Edward Elgar.

Jensen, J. Bradford. 2011. "Global Trade in Services: Fear, Facts, and Offshoring" Peterson Institute for International Economics

Johansson, Börje, Charlie Karlsson, Mikaela Backman and Pia Juusola. 2007. "The Lisbon Agenda from 2000 to 2010." CESIS Working Paper No., 106,

Johnson, Toni. 2010. "Health Care Costs and U.S. Competiveness". Washington, DC: Council on Foreign Relations. March.

Jordan, Miriam. 2011 "U.S. to Assist Immigrant Job Creators." Wall Street Journal. August 3 2

Jordon, Steve. 2011. "Tobacco Money Gives Nebraska an Economic, Research Lifeline." Omaha World-Herald, February 3.

Jorgenson, D. W. 2001. "Information Technology and the U.S. Economy." *American Economic Review* 91(1) March.

Jorgenson, D. W. 2001. "U.S. Economic Growth in the Information Age." Issues in Science and Technology Fall.

Jorgenson, D. W. 2005. Productivity: Information Technology and the American Growth Resurgence. Cambridge MA: MIT Press.

Jorgenson, D.W., and Kevin Stiroh. 2002. "Raising the Speed Limit: Economic Growth in the Information Age." In National Research Council. *Measuring and Sustaining the New Economy*. Dale Jorgenson and Charles Wessner, eds. Washington, DC: The National Academies Press.

Jorgenson, D.W., Kevin Stiroh and Mun S. Ho. 2005. Productivity, Volume 3, Information Technology and the American Growth Resurgence. Cambridge MA: MIT Press.Jorgenson, D. W., and K. Motohashi. 2003. "The Role of Information Technology in the Economy: Comparison between Japan and the United States." Prepared for RIETI/KEIO Conference on Japanese Economy: Leading East Asia in the 21st Century? Keio University. May 30.

Joy, William. 2000. "Why the Future Does Not Need Us." Wired 8.04. April.

Jun-ichi Abe, "Small Business Finance & Support for Startups in Japan (Case of NLFC)." National Life Finance Corporation, December 2004

Kaelble, Steve. 2004. "Good Neighbors: Indiana's Certified-Technology-park Program." Indiana Business Magazine July 1.

Kapur, Devesh. 2003. "Indian Diaspora as a Strategic Asset." Economic and Political Weekly 38(5):445-448. ````

Kapur, Devesh and John McHale. 2005. Give us Your Best and Brightest : the global hunt for talent and its impact on the developing world, Washington, DC: Center for Global Development.

Kasturi, Charu Sudan Hindustan Times. 2011. "More Autonomy, New Programmes for IITs." Jan. 16. Accessed at: http://www.hindustantimes.com/India-news/NewDelhi/More-autonomy-new-programmes-for-IITs/Article1-650939.aspx

Kats, Greg. 2010. Greening Our Built World, Costs, Benefits, and Strategies, Washington, DC: Island Press.

Kats, Greg, Aaron Menkin, Jeremy Dommu and Matthew DeBold. 2012. "Energy Efficiency Financing - Models And Strategies." Capital E For The Energy Foundation, March.

Kawamoto, Takuji. 2004. "What Do the Purified Solow Residuals Tell Us about Japan's Lost Decade?" Bank of Japan IMES Discussion Paper Series. No. 2004-E-5. Tokyo, Japan: Bank of Japan.

Kelderman, Eric. 2008. "Look Out Below! American's Infrastructure is Crumbling." Stateline.org, Pew Research Center, January 22.

Kelley, Charles et al. 2004. "High-Technology Manufacturing and U.S. Competitiveness." TR-136-OSTP, prepared for the Office of Science and Technology Policy, RAND Corp., March.

Kelley, Donna J., Niels Bosma, Jóse Ernesto Amorós. 2011. "Global Entrepreneurship Monitor 2010 Global Report." Global Entrepreneurship Research Association.

Kenney, Martin, ed. 2000. Understanding Silicon Valley: The Anatomy of an Entrepreneurial Region. Stanford, CA: Stanford University Press.

Kerr, Drew. 2010. "GlobalFoundries Seeks New Incentive Money From State to Expand Operations." PostStar, March 26.

Kiese, Matthias. 2008. "Cluster Approaches to Local Economic Development: Conceptual Remarks and Case Studies from Lower Saxony, Germany." in Uwe Blien and Gunther Maier, eds., *The Economics of Regional Clusters: Networks, Technology and Policy (New Horizons in Regional Science)*., Cheltenham: Edward Elgar Publishing, p. 269-303

Kim, Linsu. 1997. *Imitation to Innovation; The Dynamics of Korea's Technological Learning*. Boston: Harvard Business School Press. pp. 192-213, 234-243.

Kim, Yong-June. 2006. "A Korean Perspective on China's Innovation System." Presentation at Industrial Innovation in China. Levin Institute Conference. July 24-26.

Kin, Peter and Hari Polavarapu. 2011. "Solar Photovoltaic Industry : 2011 Outlook—FIT Cuts in Key Markets Point to Over-Supply." Deutsche Bank, January 5.

Kinast, Juliane, Christian Reiermann, and Michael Sauga. 2007. "Labor Paradox in Germany: Where have the Skilled Workers Gone?", Spiegel Online, June 22.

Klepper, Stephern. 2011. "Nano-economics, Spinoffs, and the Wealth of Regions", *Small Business Economics*. 37: 141-154

Kneller, R. 2003. "University-Industry Cooperation and Technology Transfer in Japan Compared with the U.S.: Another Reason for Japan's Economic Malaise?" University of Pennsylvania Journal of International Economic Law 24:329-449.

Koizumi, Kei. 2007. "Historical Trends in Federal R&D." In AAAS Report XXXII: Research and Development FY2008. AAAS Publication Number 07-1A. Washington, DC: American Association for the Advancement of Science~~~

Kolakowski, Nicholas. 2009. "Intel Investing $7 Billion in Manufacturing Facilities." eWeek, February 10.

Kondo, Masayuki. 2002. "Kyutenkaishihajimeta Nippon no Daigakuhatsubencha no Genjou to Kadai" ("The Current State and Issues of Rapidly Increasing University Spin-offs in Japan"), *Venture Review*, No. 3, 101 - 107.

Kondo, Masayuki. 2003. "Chinese Model to Create High-Tech Start-Ups from Universities and Research Institutes." In M. von Zedtwitz, G. Haour, T. Khalil, and L. Lefebvre, eds. *Management of Technology: Growth through Business, Innovation and Entrepreneurship.* Oxford, UK: Pergamon Press.

Kondo, Masayuki. 2004. "Policy Innovation in Science and Technology in Japan—from S&T Policy to Innovation Policy." (In Japanese.) *Journal of Science Policy and Research Management* 19(3/4):132-140.

Kondo, Masayuki. 2004. "University spin-offs in Japan." Asia Pacific Tech Monitor. March-April 2004. Pp. 37-43. Asian and Pacific Centre for Transfer of Technology, ESCAP, UN.

Kondo, Masayuki. 2005. "Spin-offs from Public Research Institutes as Domestic Technology Transfer Means—The Case of RIKEN and Riken Industrial Group." (In Japanese.) Development Engineering 11:31-41.

Kondo, Masayuki. 2006. "University-Industry Partnerships in Japan." Presentation at the conference, 21st Century Innovation Systems for the United States and Japan: Lessons from a Decade of Change. Tokyo, Japan. The National Academies, NISTEP of Japan, and The Institute of Innovation Research of Hitotsubashi University. http://www.nistep.go.jp/IC/ic060110/pdf/5-2.pdf

Koopman, Robert, William Powers, Zhi Wang and Shang-Jin Wei. 2010. "Give Credit Where Credit Is Due: Tracing Value Added in Global Production Chains." NBER Working Paper No. 16426, September.

Koopman, Robert, Zhi Wang and Shang-Jin Wei. 2008. "How Much of China's Exports is Really Made in China? Estimating Domestic Content in Exports When Processing Trade is Pervasive". NBER Working Paper No. 14109. June.

Koopman, Robert, Zhi Wang and Shang-Jin Wei. 2009. "A World Factory in Global Production Chains: Estimating Imported Value Added in Chinese Exports." Centre for Economic Policy Research Discussion Paper No. 7430, September

The Korea Herald. 2004. "Novartis Stays Ahead with New Ideas: Country Head Says Dedication." March 31.

The Korea Herald. 2010. "Korea Emerging as Global Trial Hub." May 26.

The Korea Herald. 2011. "Korea Set to Approve World's First Stem Cell Drug." June 24.

Korea Times. 2007. "US Drug Maker to Have Headquarters in Korea." July 9.

Korea Times. 2009. "Cleanup Drive to Sweep Pharm Industry." March 31.

Korea Times. 2009. "ROK Firms Plan Tamiflu Generics Production." Korea Times September 9.

Korea Times. 2009. "Tamiflu Generics Protection Planned." September 9.

Korea Times. 2012. "Korea's ETRI: World Top Agency in Patents." April 4.

Kortum, S., and J. Lerner. 1999. "What is Behind the Recent Surge in Patenting?" *Research Policy* 28:1-22.

Koschatzky, Knut. 2003. "The Regionalization of Innovation Policy: New Options for Regional Change?" In G. Fuchs and Phil Shapira, eds. Rethinking Regional Innovation: Path Dependency or Regional Breakthrough? London, UK: Kluwer.

Koskenlinna , Markus. 2003."Additionality and Tekes." Impact Analysis, November. 25.

Kota, Sridhar. "Technology Development and Manufacturing Competitiveness", Presentation to NIST, Extreme Manufacturing workshop, January 11, 2011.

Kovacic, William C. and Carl Shapiro. 2000. "Antitrust Policy: A Century of Economic Thinking." *Journal of Economic Perspectives 14*.

Krishnan, Rishikesha T. 2010. *From Jugaad to Systematic Innovation: The Challenge for India*. Indian Institute of Management, Bangalore.

Krugman, Paul. 1991. *Geography and Trade*. Cambridge, MA: MIT Press.

Krugman, Paul. 1994. "Competitiveness: A Dangerous Obsession." Foreign Affairs March/April.

Kuhlmann, Stephan and Jakob Edler. 2003. "Scenarios of Technology and Innovation Policies in Europe: Investigating Future Governance—Group of 3." Technological Forecasting & Social Change 70.

Kruse, Wilfried. 2003. "Lifelong Learning in Germany –Financing and Innovation: Skill Development, Education Networks, Support Structures." Berlin: Federal Ministry of Education and Research (BMBF). Access at http://www.bmbf.de/pub/lifelong_learning_oecd_2003.pdf.

Kumar, Deepak. 1995. Science and the Raj: 1857-1905. New York: Oxford University Press.

Kyodo. 2007. "Chemicals Flow Unchecked from China to Drug Market." November 1.

Lacey , Stephen. 2011"How China Dominates Solar Power: Huge Loans from the Chinese Development Bank are Helping Chinese Solar Companies Push American Solar Firms Out of the Market." The Guardian. September 12.

Lall, Sanjaya. 2002. "Linking FDI and Technology Development for Capacity Building and Strategic Competitiveness." Transnational Corporations 11(3):39-88.

Lambe, Patrick. 2002. "The Engineer's Dilemma: Innovation in Singapore." Straits Knowledge.

Lane, Amy. 2011. "Snyder Budget: The Era of the Tax Credit is Over." Crain's Detroit Business, February.17

Lanjouw, J. O., and I. Cockburn. 2000. "Do Patents Matter? Empirical Evidence after GATT." NBER Working Paper No. 7495. January.

Lanjouw, J. O., and J. Lerner 1997. "The Enforcement of Intellectual Property Rights: A Survey of the Empirical Literature." NBER Working Paper No. W6296. Available at <http://ssrn.com/abstract=226053>.

Laredo, Philippe, and Philippe Mustar, eds. 2001. *Research and Innovation Policies in the New Global Economy: An International Perspective.* Cheltenham, UK: Edward Elgar.

Larosse, J. 2004. 49 Towards a 'Third Generation' Innovation Policy in Flanders: Policy Profile of the Flemish Innovation System. Brussels, Belgium: IWT.

Lakshmanan, T.R. 2011. "The broader economic consequences of transport infrastructure investments." *Journal of Transport Geography* 19(1).

Laursen, Keld and Ammon Salter. 2005. "The fruits of intellectual production: economic and scientific specialization among the OECD countries." *Cambridge Journal of Economics* 29(2):289-308.

Lazowska, Edward D., and David Paterson. 2005. "An Endless Frontier Postponed." Science 308(5723):757. May 6.

Leachman, Robert C. and Chien H. Leachman. 2004. "Globalization of Semiconductors: do real men have fabs, or virtual fabs?." in Martin Kenney and Richard L. Florida (eds.), *Locating Global Advantage: Industry Dynamics in the International Economy,* Palo Alto, Calif.: Stanford University Press.

Lebra, Takie Sugiyama. 1971. "The Social Mechanism of Guilt and Shame: The Japanese Case." Anthropological Quarterly 44(4):241-255.

Lechner, Michael, Ruth Miquel, and Conny Wunsch. 2011. "Long-Run Effects of Public Sector Sponsored Training in West Germany." *Journal of the European Economic Association* 9(4).

LEDinside. 2009. "Ranking of LED Chip Manufacturers in China—Report on China's LED Epitaxy Industry."

Lee, Doohee. 2010. Regional Innovation Activity: The Role of Regional Innovation Systems in Korea." KIET Occasional Paper No. 78. February.

Leger, Donna Leinwand. 2011. "End of Shuttle Program Slams Space Coast Economy, USA Today, July 5.

Lembke, Johan. 2002. *Competition for Technological Leadership: EU Policy for High Technology.* Cheltenham, UK: Edward Elgar.

Lemola, Tarmo. 2002. "Convergence of National Science and Technology Policies: The Case of Finland." *Research Policy* 31(8-9):1481-1490.

Leonard, Jeromy. 2008. "The Tide Is Turning: An Update on Structural Cost Pressures Facing U.S. Manufacturer." The Manufacturing Institute and Manufacturers Alliance/MAPI, November 2008

Lerner, J. 1995. "Patenting in the Shadow of Competitors." *Journal of Law and Economics* 38(Oct):463-495.

Lerner, J. 1999. "Public Venture Capital." In National Research Council. *The Small Business Innovation Program: Challenges and Opportunities.* Charles W. Wessner, ed. Washington, DC: National Academy Press.

Lerner, J. 2002. "Patent Protection and Innovation Over 150 Years." NBER Working Paper No. 8977–View here-http://inno.snu.ac.kr/wayboard/db/SES/file/Lesner_Patent%20Protection.pdf

Lerner, J., and J. Tirole. 2004. "Efficiency of Patent Pools." *American Economic Review* 94(3):691-711.

Leslie, Stuart, and Robert Kargon. 2006. "Exporting MIT." *Osiris* 21:110-130.

Levin, R. C., A. K. Klevorick, R. R. Nelson, and S. G. Winter. 1987. "Appropriating the Returns to Industrial R&D." *Brookings Papers on Economic Activity* 783-820.

Levitt, Rachelle, ed. 1987. *The University/Real Estate Connection: Research Parks and Other Ventures*. Washington, DC: Urban Land Institute.

Lewenstein, Bruce V. 1992. "The Meaning of 'Public Understanding of Science' in the United States After World War II." *Public Understanding of Science* 1 (1):45-68.

Lewis, David A, Elsie Harper-Anderson, and Lawrence A. Molnar. 2011. "Incubating Success; Incubation Practices that lead to Successful Ventures." Washington, DC: Economic Development Administration. Access at http://www.edaincubatortool.org/pdf/Master%20Report_FINALDownloadPDF.pdf

Lewis, James A. 2005. *Waiting for Sputnik: Basic Research and Strategic Competition*. Washington, DC: Center for Strategic and International Studies.

Lewis, Tracy and Dennis Yao. 2001. "Innovation, Knowledge Flow, and Worker Mobility." Working Paper. *Wharton School of Business.*

Leyden, D. P., A. N. Link, and D. S. Siegel. 2008. "A Theoretical and Empirical Analysis of the Decision to Locate on a University Research Park." *IEEE Transactions on Engineering Management* 55(1):23-28.

Li, Jianjun. 2006. "The Development and Opening of Tianjin Binhai: New Area & China's Biotechnical Innovations." Presentation at Industrial Innovation in China. Levin Institute Conference. July 24-26.

Licklider, J. C. R. 1960. "Man-Computer Symbiosis." *IRE Transactions on Human Factors in Electronics*. March.

Liljemark, Thomas. 2005 "Innovation Policy in Canada: Strategies and Realities." Swedish Institute for Growth Policy Studies, A2004:24

Lim, Linette. 2011. "P&G Invests $250 million in Innovation Centre." The Business Times, January 27.

Lin, Otto. 1998. "Science and Technology Policy and its Influence on the Economic Development of Taiwan." In Henry S. Rowen, ed. *Behind East Asian Growth: The Political and Social Foundations of Prosperity*. New York: Routledge.

Lindell, Rebecca. 2010. "Canadian R&D Spending Continues Downward Spiral: StatsCan." Postmedia News, December.9

Lindelöf, P., and H. Löfsten. 2003. "Science Park Location and New Technology-Based Firms in Sweden: Implications for Strategy and Performance." *Small Business Economics* 20(3):245-258.

Lindelöf, P. and H. Löfsten. 2004. "Proximity as a Resource Base for Competitive Advantage: University-Industry Links for Technology Transfer." *Journal of Technology Transfer* 29(3-4): 311-326.

Lindh, T., and Ohlsson, H. 1996. "Self-Employment and Windfall Gains: Evidence from the Swedish Lottery." *Economic Journal* 106:1515-1526.

Linebaugh, Kate. 2011, "GE to Build Solar-Panel Plant in Colorado, Hire 355 People." Wall Street Journal, October 13.

Linebaugh, Kate. 2011. "GE Makes Big Bet on Software Development." The Wall Street Journal, November 17.

Link, A. N. 1981. "Basic Research and Productivity Increase in Manufacturing: Some Additional Evidence." *American Economic Review* 71(5):1111-1112.

Link, A. N. 1981. Research and Development Activity in U.S. Manufacturing. New York: Praeger.

Link, A. N. 1995. *A Generosity of Spirit: The Early History of the Research Triangle Park*. Research Triangle Park, NC: University of North Carolina Press for the Research Triangle Park Foundation.

Link, A. N. 2002. *From Seed to Harvest: The Growth of the Research Triangle Park*. Research Triangle Park, NC: University of North Carolina Press for the Research Triangle Park Foundation.

Link, A. N., and D. S. Siegel. 2003. *Technological Change and Economic Performance*. London, UK: Routledge.

Link, A. N., and J. T. Scott. 1998. *Public Accountability: Evaluating Technology-Based Institutions*. Norwell, MA: Kluwer Academic Publishers.

Link, A. N., and J. T. Scott. 2001. "Public/Private Partnerships: Stimulating Competition in a Dynamic Market." *International Journal of Industrial Organization* 19(5):763-794.

Link, A. N., and J. T. Scott. 2003. "The Growth of Research Triangle Park." *Small Business Economics* 20(2):167-175.

Link, A. N., and J. T. Scott. 2003. "U.S. Science Parks: The Diffusion of an Innovation and Its Effects on the Academic Mission of Universities." *International Journal of Industrial Organization* 21(9):1323-1356.

Link, A. N., and J. T. Scott. 2005. "Opening the Ivory Tower's Door: An Analysis of the Determinants of the Formation of U.S. University Spin-Off Companies." *Research Policy* 34(7):1106-1112.

Link, A. N., and J. T. Scott. 2006. "U.S. University Research Parks." *Journal of Productivity Analysis* 25(1):43-55.

Link, A. N., and J. T. Scott. 2007. "The Economics of University Research Parks." *Oxford Review of Economic Policy* 23(4):661-674.

Link, A. N., and K. R. Link. 2003. "On the Growth of U.S. Science Parks." *Journal of Technology Transfer* 28(1):81-85.

Lipton, Eric and John M. Broder. 2011." In Rush to Assist Solar Company, U.S. Missed Signs." New York Times, September 22.

Litan, Robert E., and Lesa Mitchell. 2008. "Should Universities be Agents of Economic Development?" Astra Briefs 7(7-8).

Litan, Robert E. and Lesa Mitchell. 2010. "A Faster Path from Lab to Market." in *Harvard Business Review*, January/February.

Litan, Robert E., Lesa Mitchell, E. J. Reedy, 2007 "Commercializing University Innovations: Alternative Approaches." National Bureau of Economic Research working paper Available at SSRN: http://ssrn.com/abstract=976005 or http://dx.doi.org/10.2139/ssrn.976005

Loach, Rod, Dan Galves, Patrick Nolan. 2008. "Electric Cars: Plugged In: Batteries Must be Included." Deutsche Bank Securities Inc., June 9.

Lobban, P. and A. Kaiser. 1973. "Enzymatic End-to-End Joining of DNA Molecules." *Journal of Molecular Biology* 78(3).

Locke, Gary. 2010. "Upcoming Clean Energy Trade Mission to China and Indonesia." Briefing at the Washington Foreign Press Center: May 12.

Löfsten, Hans, and Peter Lindelöf. 2002. "Science Parks and the Growth of New Technology-based Firms—Academic-industry Links, Innovation and Markets." Research Policy 31(6):859-876.

Long, Guogiang. 2005. "China's Policies on FDI: Review and Evaluation." In Theodore H. Moran, Edward M. Graham, and Magnus Blomström, eds. Does Foreign Direct Investment Promote Development? Washington, DC: Institute for International Economics.

Lott, Melissa C. 2011. "Solyndra—Illumunating Energy Funding Flaws?", *Scientific American*, September 27.

Lowe, Marcy Saori Tokuoka, Tali Trigg and Gary Gereffi. 2010.Lithium-ion Batteries for Electric Vehicles: The U.S. Value Chain, Center on Globalization, Governance & Competitiveness, Duke University, October 5.

Lowell, B. Lindsay Hal Salzman, Hamutal Bernstein, and Everett Henderson. 2009. "Steady as She Goes? Three Generations of Students Through the Science and Engineering Pipeline." Paper presented at Annual Meeting of the Association for Public Policy Analysis and Management. Washington, DC October [Viewed at http://policy.rutgers.edu/faculty/salzman/steadyasshegoes.pdf

Lowell, B. Lindsay. 2011. "Immigration and the Science and Engineering Workforce: Failing Pipelines, Restrictive Visas, and the "Best and Brightest". Testimony before the House Judiciary Committee. October 5.

Lu, Jody. 2011. "Who is Making Junk Patents?" China Daily, March 6.

Lucas, Jr., Robert E. 1988. "On the Mechanics of Economic Development." *Journal of Monetary Economics* 22. pp. 3-42.

Luger, Michael I. and Harvey A. Goldstein, *Technology in the Garden.* 1991. Chapel Hill: University of North Carolina Press, pg. 34.

Luger, M. I. 2001. "Introduction: Information Technology and Regional Economic Development." *Journal of Comparative Policy Analysis*: Research & Practice.

Luger, M. I. 2007. "Smart Places for Smart People: Using Cluster-based Planning in the 21st Century." In *Creating Enterprise: Igniting Innovation through Business-University-Government Networks*. Cheltenham, UK: Edward Elgar.

Luger, M. I., and H. A. Goldstein. 2006. *Research Parks Redux: The Changing Landscape of the Garden.* Washington, DC: U.S. Economic Development Administration.

Lundstörm, A., and L. Stevenson. 2001. *Entrepreneurship Policy for the Future.* Stockholm, Sweden: FSF.

Lux Research. 2009. "The Slow Dawn of Grid Parity : State of the Market Report."

Lundvall, Bengt-Åke ed. 1992 National Innovation Systems: Towards a Theory of Innovation and Interactive Learning, London: Pinter.

Lyne, Jack. 2000. "IBM's Cutting-Edge $2.5 billion Fab Reaps $500 Million in NY Incentives." Site Selection. November.

Lynn, Leonard. 2006. "Collaborative Advantage and China's Evolving Position in the Global Technology System." Presentation at Industrial Innovation in China. Levin Institute Conference. July 24-26.

Lyon, John. 2009. "Beebe Signs Tobacco Tax Hike Into Law." Arkansas New. February 17.

M2 Presswire. "Rensselaer Polytechnic Institute Appoints Cyberinfrastructure Expert James Myers to Lead the Computational Center for Nanotechnology Innovations." August 30.

Macher, Jeffrey, David Mowery, and David Hodges. 1999. "Semiconductors." In National Research Council. *U.S. Industry in 2000: Studies in Competitive Performance.* David C. Mowery, ed. Washington, DC: National Academy Press.

Maddison, Angus, and Donald Johnston. 2001. *The World Economy: A Millennial Perspective.* Paris, France: Organization for Economic Co-operation and Development.

Malecki, E. J. 1991. *Technology and Economic Development.* New York: John Wiley.

Mani, Sunil. 2004. "Government, Innovation and Technology Policy: An International Comparative Analysis." *International Journal of Technology and Globalization* 1(1).

Manners, David. 2010. "Top 25 Fabless Companies." Electronics Weekly, January 19.

Mansfield, E. 1986. "Patents and Innovation: An Empirical Study." *Management Science* 32:173-181.

Mansfield, E. 1996. *Estimating Social and Private Returns from Innovations Based on the Advanced Technology Program: Problems and Opportunities.* GCR 99-780. Gaithersburg, MD: National Institute of Standards and Technology.

Mansfield, E. 1996. "How Fast Does New Industrial Technology Leak Out?" *Journal of Industrial Economics* 34(2):217-224.

The Manufacturing Institute. 2009. "The Facts About Modern Manufacturing-8th Edition." 8th ed. Gaithersburg MD: NIST.

Manyika, James et. al., 2011. Growth and Renewal in the United States: Retooling America's Economic Engine, McKinsey Global Institute, February.

Markoff, John. 2011. "China Has Homemade Supercomputer Gain." The New York Times, October 28. Pp A5.

Marshall, Alfred. 1920. *Principles of Economics,* London: Macmillan.

Marshall, Andrew. 1993. "Some Thoughts on Military Revolutions-Second Version." ONA Memorandum for the Record. August 23.

Martin, Holly B. 2011."Miracle Material: Two-Dimensional Graphene May Lead to Faster Electronics, Stronger Spacecraft and Much More." National Science Foundation. May 19.

Mashelker, R.A. 2008. *Technology in Society,* Vol,30/3-4, Pp 299-308 (Annexure 3). April.

Mashelker, R.A. 2009. Technonationalism to Technoglobalism *Journal of India & Global Affairs*, 90-97. (Annexure 4).

Mashelker, R.A. 2012. *Reinventing India.* Pune: Sahyadri Publications.

Mashelkar, R.A. and C.K. Prahalad. 2010. "Innovation's Holy Grail". *Harvard Business Review.* July.

Matthews, John A. 2010. "The Hsinchu Model: Collective Efficiency, Increasing Returns and Higher-Order Capabilities in the Hsinchu Science-Based Industry Park, Taiwan". Keynote Address, Chinese Society for Management of Technology, 20th Anniversary Conference, Tsinghua University, Hsinchu, Taiwan. December 10.

Matthews, John A. and Dong-Sung Cho. 2000. *Tiger Technology: The Creation of a Semiconductor Industry in East Asia.* Cambridge: Cambridge University Press.

Mazuzan, George. 1988. *The National Science Foundation: A Brief History (1950-1985).* NSF 88-16. Arlington, VA: The National Science Foundation.

McGrath, Dylan. 2009. "China's Fab Utilization Sinks to 43%, says iSuppli." EE Times, April 20.

McGregor, James. "China's Drive for Indigenous Innovation: A Web of Industrial Policies". U.S. Chamber of Commerce, Global Intellectual Property Center, APCO Worldwide.

McKibben, William. 2003. Enough: Staying Human in an Engineered Age. New York: Times Books

McKinsey & Company. 2009. "The Economic Impact of the Achievement Gap in America's Schools." April.

McMillan, Carolyn. 2011. "Regents Scrutinize Fiscal Crisis." UC Newsroom, March 16.

McDonald, Neil. 2012. "Euro Commisioner Visits US". *Federal Technology Watch.* Vol. 10(4).

McPherson, Peter, David Shulenburger, Howard Gobstein, and Christine Keller. 2009. "Competitiveness of Public Research Universities & the Consequences for the Country: Recommendations for Change." A NASULGC Discussion Paper Working Draft Association of Public and

Land-Grant Universities, March [Viewed here http://www.aplu.org/NetCommunity/Document.Doc?id=1561 .
McPherson Peter, Howard J. Gobstein, and David E. Schulenburger. 2010. "Forging a Foundation for the Future: Keeping Public Research Universities Strong." Discussion Paper at April 2010 Regional Meetings on Research Universities. Washington, DC: Association of Public and Land-Grant Universities.
Megginson, William L. 2004. "Towards a Global Model of Venture Capital?" *Journal of Applied and Corporate Finance* 16(1), Winter.
Merges, R. P. 1996. "A Comparative Look at Intellectual Property Rights and the Software Industry." In David C. Mowery, ed. 1996 *The international computer software industry : a comparative study of industry evolution and structure*. New York: Oxford University Press.
Merges, R. P. 1999. "As Many as Six Impossible Patents before Breakfast: Property Rights for Business Concepts and Patent System Reform." *Berkeley Technology Law Journal*, 14(2):577
Merges, R. P., and R. R. Nelson. 1990. "On the Complex Economics of Patent Scope." *Columbia Law Review* 90:839-916.
Merrill, Peter R. 2010. "Corporate Tax Policy for the 21st Century." *National Tax Journal*. December. 63 (4, Part 1), 623–634
Merton, R. K. 1968. "The Matthew Effect in Science." *Science* 159(3810):56-63.
METI (Small and Medium Enterprise Agency, Ministry of Economy, Trade and Industry). 2003. *The 2003 White Paper on Small and Medium Enterprises in Japan*. Tokyo: Japan Small Business Research Institute.
METI. 2005. *The 2005 White Paper on Small and Medium Enterprises in Japan*. Tokyo: Japan Small Business Research Institute.
MEXT. 2004. *Annual Report on Science and Technology Promotion Measures 2004: White Paper on Science and Technology*. (Heisei 16 nendo Kagaku-Gijutu Sinkou-ni-kansuru Nenji-houkoku). MEXT.
Meyer-Krahmer, Frieder. 2001. "The German Innovation System." Pp. 205-252 in P. Larédo and P. Mustar, eds. *Research and Innovation Policies in the New Global Economy: An International Comparative Analysis*. Cheltenham, UK: Edward Elgar.
Meyer-Krahmer, Frieder. 2001. "Industrial Innovation and Sustainability: Conflicts and Coherence." Pp. 177-195 in Daniele Archibugi and Bengt-Ake Lundvall, eds. *The Globalizing Learning Economy*. New York: Oxford University Press.
Michigan Lawyers Weekly. 2011. "Gene Sequence Patents are Being Questioned." June 27.
Middleton, Andrew, and Steven Bowns. With Keith Hartley and James Reid. 2006. "The Effect of Defense R&D on Military Equipment Quality." *Defense and Peace Economics* 17(2):117-139.

Miller Center of Public Affairs. University of Virginia. 2010. *Well Within Reach: America's New Transportation Agenda,* David R. Goode National Transportation Policy Conference. October 4.

Miller, Roger, and Marcel Cote. 1987. *Growing the Next Silicon Valley: A Guide for Successful Regional Planning..* Lexington, MA : Lexington Books.

Mills, K, E. Reynolds, and A Reamer. 2008. *Clusters and Competitiveness: A New Federal Role for Stimulating Regional Economie*s. Washington, DC: Brookings Institution.

Meng, Jiao. 2009. "Photovoltaic Bubble Shattered in China." *China Development Gateway*, Aug. 28, chinagate.cn

Ministry of Economic Affairs. 2011. "Multinational Innovative R&D Center in Taiwan." Access at http://investtaiwan.nat.gov.tw/matter/show_eng.jsp?ID=433.

Ministry of Economy, Trade and Industry. 2006. New Economic Growth Strategies 2006. Tokyo, Japan: Research Institute of Economy, Trade and Industry. (In Japanese).

Ministry of Economy, Trade and Industry. 2010. "100 Actions to Launch Japan's New Growth Strategy. Tokyo, Japan: Research Institute of Economy, Trade and Industry. August.

Ministry of Economy, Trade and Industry. 2010. "Next-Generation Vehicle Plan 2010 (Outline)." Tokyo, Japan: Research Institute of Economy, Trade and Industry.

Ministry of Economy, Trade and Industry. 2010. "The Industrial Structure Vision 2010." Tokyo, Japan: Research Institute of Economy, Trade and Industry. June.

Government of the People's Republic of China, Ministry of Information Industry. 2006. "Outline of the 11th Five-Year Plan and Medium-and-Long-Term Plan for 2020 for Science and Technology Development in the Information Industry." Xin Bu Ke No. 309. August 29

Ministry of Trade and Industry [Singapore] 2006. *Science & Technology Plan 2010 : Sustaining Innovation-Driven Growth*, Government of Singapore, February.

Mints, Paula. 2011. Navigant Consulting, "Reality Check: The Changing World of PV Manufacturing." October 5. Viewed here http://www.renewableenergyworld.com/rea/news/article/2011/10/reality-check-the-changing-world-of-pv-manufacturing.

Mitra, Raja. 2006. "India's Potential as a Global R&D Power." In Magnus Karlsson, ed. *The Internationalization of Corporate R&D*. Östersund: Swedish Institute for Growth Policy Studies.

Miyata, Shinpei. 1983. *A Free Paradise for Scientists*. (In Japanese.) Bungeishunju.

Mlive.com. 2009. "GM taps LG Chem, Compact Power to Supply Batteries for Volt Plug-in." January 12.

Modern Healthcare. 2007. "This Problem Made in China." October 22.

Modern Healthcare. 2009. "J&J Wins $1.67 Billion Lawsuit Against Abbott." June 30.
Modern Healthcare. 2009. "Making Them Pay." October 12.
Modern Healthcare. 2009. "Obama Administration Plans to Take More Regulatory Approach on Healthcare Mergers." June 24.
Mody, Ashok, and Carl Dahlman. 1992. "Performance and Potential of Information Technology: An International Perspective." *World Development* 20(12):1703-1719.
Moore, Glen. "Solar Energy and the Reagan Administration." 1982. Mini Brief Number MB81265, Science Policy Research Division, Congressional Research Service, the Library of Congress.
Moore, Gordon. 2003. "The SEMATECH Contribution." In National Research Council. *Securing the Future: Regional and National Programs to Support the Semiconductor Industry.* Charles W. Wessner, ed. Washington, DC: The National Academies Press.
Moris, Francisco. 2005. "The U.S. Research and Experimentation Tax Credit in the 1990s." NSF InfoBrief, NSF 05-316 . July
Morris, P. R. 1990. *A History of The World Semiconductor Industry*. Stevenage, UK: Peter Peregrinus Ltd.
Morrison, Wayne M. 2010. "China-U.S. Trade Issues" Washington, DC: Congressional Research Service: September 30th [Accessed at: http://www.fas.org/sgp/crs/row/RL33536.pdf
Morrow, Daniel S. 2003. Dr. J. Craig Venter-Oral History. Computer World Honors Program. April 21. Available at <http://cwheroes.org/archives/histories/venter>.
Moser, P. 2001. "How Do Patent Laws Influence Innovation? Evidence from Nineteenth Century World Fairs." University of California at Berkeley: working paper.
Motohashi, Kazuyuki. 2005. "University-industry Collaborations in Japan: The Role of New Technology-based Firms in Transforming the National Innovation System." *Research Policy* 34:583-594.
Motohashi, Kazuyuki, and Xiao Yun. 2007. "China's Innovation System Reform and Growing Industry and Science Linkages." *Research Policy* 36:1251-1260.
Mowery, D. C., and A. A. Ziedonis. 2002. "Academic Patent Quality and Quantity Before and After the Bayh-Dole Act in the United States." *Research Policy* 31(3):399-418.
Mowery, D. C., and B. N. Sampat. 2005. "The Bayh-Dole Act of 1980 and University-Industry Technology Transfer: A Model for Other OECD Governments?" *Journal of Technology Transfer* 30(1/2):115-127.
Mowery, D. C., and B. N. Sampat. 2005. "Universities in national innovation systems". Oxford Handbook of Innovation
Mowery, D. C., R. Nelson, B. Sampat, and A. Ziedonis. 2001. "The Growth of Patenting and Licensing by U.S. Universities: An Assessment of the Effects of the Bayh-Dole Act of 1980." *Research Policy* 30(1): 99-119

Mowery, D. C., R. Nelson, B. Sampat, and A. Ziedonis. 2003. *The Ivory Tower and Industrial Innovation.* Stanford, CA: Stanford University Press.

Mrinalini, N. and Sandhya Wakdikar. 2008. "Foreign R&D Centres in India: Is There any Positive Impact?", *Current Science, 94(4): 452-458.*

Mu, R., Z. Ren, H. Song and F. Chen. 2010. "Innovative Development and Innovation Capacity-Building in China." International Journal of Technology Management, Vol. 51(2-4):427-452.

Mui, Ada C. 2010. "Productive ageing in China: a human capital perspective." *China Journal of Social Work* 3(2-3).

Muro, Mark and Bruce Katz. 2010. "The New 'Cluster Moment': How Regional Innovation Clusters Can Foster the Next Economy." Brookings Institution Metropolitan Policy Program, September.

Mufson, Steven. 2009. "Asian Nations Could Outpace U.S. in Developing Clean Energy." Washington Post, July 16.

Mukul, Akshaya. 2011. "Govt Plans 50 Centres of Excellence for Science & Tech." The Times of India, January 17.

Muro, Mark and S. Rahman. 2010. "Budget 2011: Industry Clusters as a Paradigm for Job Growth." The Avenue, a blog of The New Republic, February 2.

Murphy, L. M., and P. L. Edwards. 2003. *Bridging the Valley of Death: Transitioning from Public to Private Sector Financing.* Golden, CO: National Renewable Energy Laboratory. May.

Mustar, Phillipe, and Phillipe Laredo. 2002. "Innovation and Research Policy in France (1980-2000) or the Disappearance of the Colbertist State." *Research Policy* 31:55-72.

Nadiri I. 1993. Innovations and Technological Spillovers. NBER Working Paper No. 4423. Boston, MA: National Bureau of Economic Research.

Nagaoka, Sadao. 2005. "How Does Priority Rule Work? Evidence from the Patent Examination Records in Japan." A Paper Presented for Patent Statistics and Innovation Research Workshop, November 25. Research Center for Advanced Science and Technology, University of Tokyo.

Nagaoka, Sadao. 2006. "Reform of Patent System in Japan and Challenges." Presentation at the conference, "21st Century Innovation Systems for the United States and Japan: Lessons from a Decade of Change." Tokyo, Japan. The National Academies, NISTEP of Japan, and The Institute of Innovation Research of Hitotsubashi University. http://www.nap.edu/openbook.php?record_id=12194&page=153

Nagaoka, Sadao. 2007. "Assessing the R&D Management of Firms by Patent Citation: Evidence from the U.S. Patents." *Journal of Economics & Management Strategy* 16(1).

Nagaoka, Sadao, Naotoshi Tsukada, and Tomoyuki Shimbo. 2006. "The Emergence and Structure of Essential Patents of Standards: Lessons from Three IT Standards." IIR Working Paper WP#06-08. *Institute of Innovation Research*, Hitotsubashi University.

Nakamura, K., Y. Okada, and A. Tohei. 2006. "Does the Public Sector Make a Significant Contribution to Biomedical Research in Japan? A Detailed Analysis of Government and University Patenting, 1991-2002." Discussion Paper Series. Competition Policy Research Center. Fair Trade Commission of Japan. CPDP25-E.

Nakayama, Yasuo, Mitsuaki Hosono, Nobuya Fukugawa, and Masayuki Kondo. 2005. University-Industry Cooperation: Joint Research and Contract Research. (In Japanese.) NISTEP Research Material. No. 119. Tokyo, Japan: National Institute of Science and Technology Policy, Ministry of Education, Culture, Sports, Science and Technology.

Nanotechnology Law and Business. 2006. "Top Ten Regions for Nanotech Start-ups" September. p. 383

Nanotechnology Now. 2010. "Researchers at Rensselaer Polytechnic Institute Develop New Method for Mass Producing Graphene." June 23.

NASSCOM Research. 2007. Whitepaper. "Tracing China's IT Software and Services Industry Evolution." New Delhi, India : National Association of Software and Service Companies August. International Youth Centre, Teen Murti Marg, Chanakya Puri. August. [Viewed here http://s3.amazonaws.com/zanran_storage/www.business-standard.com/ContentPages/2809607.pdf]

NASSCOM Research . 2011. "Indian IT-BPO Industry." Viewed here http://www.nasscom.org/indian-itbpo-industry.

National Academy of Engineering. 2004. *The Engineer of 2020: Visions of Engineering in the New Century.* Washington, DC: The National Academies Press.

National Academy of Engineering. 2008. *Grand Challenges for Engineering,* Washington, DC: The National Academies Press

National Academy of Engineering, 2010. Lifelong Learning Imperative in Engineering, Summary of a Workshop, D. Dutta, Rapporteur, Washington, DC: The National Academies Press.

National Academy of Sciences/National Academy of Engineering/ Institute of Medicine. 2007. *Rising Above the Gathering Storm: Energizing and Employing America for a Brighter Economic Future.* Washington, DC: The National Academies Press.

National Academy of Sciences/National Academy of Engineering/ Institute of Medicine. 2010. *Rising Above the Gathering Storm, Revisited: Rapidly Approaching Category 5.* Washington, DC: The National Academies Press.

National Academy of Sciences. 2009. *America's Energy Future, Technology and Transformation,* Washington, DC: The National Academies Press.

National Applied Research Laboratories. 2011. "Report I: Six Emerging Industries – Recreating Prosperity."

The National Association of Manufacturers, the Manufacturing Institute, and Deloitte & Touche. 2003. "Keeping America Competitive: How a Talent Shortage Threatens U.S. Manufacturing." April 21.

National Center for Science and Engineering Statistics. 2010. *National Patterns of R&D Resources: 2008 Data Update*, Detailed Statistical Tables, NSF 10-314 (March 2010).

The National Economic Council. 2011. "*A Strategy for American Innovation, Securing our Economic Growth and Prosperity.*" Washington, DC: The White House. February.National Foreign Trade Council. 2010. "China's Promotion of the Renewable Electric Power Equipment Industry: Hydro, Wind, Solar, Biomass." prepared by Dewey & LeBoeuf LLP, March.

National Foreign Trade Council. 2011. "Promoting Cross-Border Data Flows: Priorities for the Business Community." November 3.

National Innovation Council. 2010. "Towards a More Inclusive and Innovative India" Strategy Paper. Office of Advisor to the Prime Minister, Public Information, Innovation & Infrastructures. March

National Institute of Science and Technology Policy and Mitsubishi Research Institute. 2005. Government S&T Budget Analysis during the First and Second S&T Basic Plans. (Dai-ikki oyobi dai-niki Kagaku-Gijutu-Kihonkeikaku kikanchuu no kenkyuu-kaihatu-soushi no naiyou-bunseki). NISTEP Report No. 84. March.

National Intelligence Council. 2008. *Global Trends 2025: A Transformed World.*Washington, DC: U.S. Government Printing Office. Accessed at http://www.dni.gov/nic/PDF_2025/2025_Global_Trends_Final_Report.pdf

National Manufacturing Competitiveness Council. Government of India. 2006. "The National Strategy for Manufacturing." New Delhi, India. March. Viewed here on 4/4/2012 http://www.nmcc.nic.in/nmcc_stratergy.aspx

National Nuclear Security Administration. 2007. "DOE/NNSA to Dedicate Half Billion Dollar Microsystems Engineering Sciences Complex at Sandia." News Release. August 20.

National Research Council of Canada. 2012. "NRC-Industrial Research Assistance Program." Power Point presentation, March 2010 Data Modified: 2012-03-26. Viewed here on 4/4/2012 http://www.nrc-cnrc.gc.ca/eng/ibp/irap.html

National Research Council. 1982. *Scientific Communication and National Security.* Washington, DC: National Academy Press.

National Research Council. 1987. *Balancing the National Interest: U.S. National Security Export Controls and Global Economic Competition.* Washington, DC: National Academy Press.

National Research Council. 1996. *Conflict and Cooperation in National Competition for High-technology Industry.* Washington, DC: National Academy Press.

National Research Council. 1999. *The Advanced Technology Program: Challenges and Opportunities.* Charles W. Wessner, ed. Washington, DC: National Academy Press.

National Research Council. 1999. Defense Manufacturing in 2010 and Beyond, Washington, DC: National Academy Press.

National Research Council. 1999. *Funding a Revolution: Government Support for Computing Research.* Washington, DC: National Academy Press.

National Research Council. 1999. *Industry-Laboratory Partnerships: A Review of the Sandia Science and Technology Park Initiative.* Charles W. Wessner, ed. Washington, DC: National Academy Press.

National Research Council. 1999. *New Vistas in Transatlantic Science and Technology Cooperation.* Charles W. Wessner, ed. Washington, DC: National Academy Press.

National Research Council. 1999. *The Small Business Innovation Research Program: Challenges and Opportunities.* Charles W. Wessner, ed. Washington, DC: National Academy Press.

National Research Council. 2000. *The Small Business Innovation Research Program: A Review of the Department of Defense Fast Track Initiative.* Charles W. Wessner, ed. Washington, DC: National Academy Press.

National Research Council. 2000. *U.S. Industry in 2000: Studies in Competitive Performance.* David C. Mowery, ed. Washington, DC: National Academy Press.

National Research Council. 2001. *The Advanced Technology Program: Assessing Outcomes.* Charles W. Wessner, ed. Washington, DC: National Academy Press.

National Research Council. 2001. *Building a Workforce for the Information Economy.* Washington, DC: National Academy Press.

National Research Council. 2001. *Capitalizing on New Needs and New Opportunities: Government-Industry Partnerships in Biotechnology and Information Technologies.* Charles W. Wessner, ed. Washington, DC: National Academy Press.

National Research Council. 2001. *Energy Research at DOE: Was It Worth It? Energy Efficiency and Fossil Energy Research 1998 to 2000.* Washington, DC: National Academy Press.

National Research Council. 2001. *A Review of the New Initiatives at the NASA Ames Research Center.* Charles W. Wessner, ed. Washington, DC: National Academy Press.

National Research Council. 2001. *Trends in Federal Support of Research and Graduate Education.* Stephen A. Merrill, ed. Washington, DC: National Academy Press.

National Research Council. 2003. *Government-Industry Partnerships for the Development of New Technologies: Summary Report.* Charles W. Wessner, ed. Washington, DC: The National Academies Press.

National Research Council. 2003. *Securing the Future: Regional and National Programs to Support the Semiconductor Industry.* Charles W. Wessner, ed. Washington, DC: The National Academies Press.

National Research Council. 2004. *A Patent System for the 21st Century.* Washington, DC: The National Academies Press.

National Research Council. 2004. *Productivity and Cyclicality in Semiconductors: Trends, Implications, and Questions.* Dale W. Jorgenson and Charles W. Wessner, ed. Washington, DC: The National Academies Press.

National Research Council. 2004. *The Small Business Innovation Research Program: Program Diversity and Assessment Challenges.* Charles W. Wessner, ed. Washington, DC: The National Academies Press.

National Research Council. 2005. *Getting Up to Speed: The Future of Superconducting.* Susan L. Graham, Marc Snir, and Cynthia A. Patterson, eds. Washington, DC: The National Academies Press.

National Research Council. 2005. *Large-Scale Biomedical Science* Washington, DC: The National Academies Press. p. 35.

National Research Council. 2005. *Policy Implications of International Graduate Students and Post-doctoral Scholars in the United States.* Washington, DC: The National Academies Press.

National Research Council. 2007. *Enhancing Productivity Growth in the Information Age: Measuring and Sustaining the New Economy.* Dale W. Jorgenson and Charles W. Wessner, eds. Washington, DC: The National Academies Press.

National Research Council. 2007. *India's Changing Innovation System: Achievements, Challenges, and Opportunities for Cooperation.* Charles W. Wessner and Sujai J. Shivakumar, eds. Washington, DC: The National Academies Press.

National Research Council. 2007. *Innovation Inducement Prices at the National Science Foundation.* Washington, DC: The National Academies Press.

National Research Council. 2007. *Innovation Policies for the 21st Century.* Charles W. Wessner, ed. Washington, DC: The National Academies Press.

National Research Council. 2007. *SBIR and the Phase III Challenge of Commercialization.* Charles W. Wessner, ed. Washington, DC: The National Academies Press.

National Research Council. 2008. *An Assessment of the SBIR Program.* Charles W. Wessner, ed. Washington, DC: The National Academies Press.

National Research Council. 2008. *An Assessment of the SBIR Program at the Department of Energy.* Charles W. Wessner, ed. Washington, DC: The National Academies Press.

National Research Council. 2008. *An Assessment of the SBIR Program at the National Science Foundation.* Charles W. Wessner, ed. Washington, DC: The National Academies Press.

National Research Council. 2008. *Innovative Flanders: Innovation Policies for the 21st Century.* Charles W. Wessner, ed. Washington, DC: The National Academies Press.

National Research Council. 2008. *Innovation in Global Industries: U.S. Firms Competing in a New World.* Jeffrey T. Macher and David C Mowery, eds. Washington, DC: The National Academies Press.

National Research Council. 2009. *21st Century Innovation Systems for Japan and the United States: Lessons from a Decade of Change.* Sadao Nagaoka, Masayuki Kondo, Kenneth Flamm, and Charles Wessner, eds. Washington, DC: The National Academies Press.

National Research Council. 2009. *An Assessment of the SBIR Program at the Department of Defense.* Charles W. Wessner, ed. Washington, DC: The National Academies Press.

National Research Council. 2009. *An Assessment of the SBIR Program at the National Institutes of Health.* Charles W. Wessner, ed. Washington, DC: The National Academies Press.

National Research Council. 2009. *An Assessment of the SBIR Program at the National Aeronautics and Space Administration.* Charles W. Wessner, ed. Washington, DC: The National Academies Press.

National Research Council. 2009. *Revisiting the Department of Defense SBIR Fast Track Initiative.* Charles W. Wessner, ed. Washington, DC: The National Academies Press.

National Research Council. 2009. *Understanding Research, Science, and Technology Parks: Global Best Practices.* Charles W. Wessner, ed. Washington, DC: The National Academies Press.

National Research Council. 2010. *Choosing Our Nation's Fiscal Future.* Washington, DC: The National Academies Press.

National Research Council 2010. *The Dragon and the Elephant, Understanding the Development of Innovation Capacity in China and India.* S. Merrill ed., Washington, DC: The National Academies Press.

National Research Council. 2010. *Managing University Intellectual Property in the Public Interest*, Stephen A. Merrill and Ann-Marie Mazza, editors, Washington, DC: The National Academies Press.

National Research Council. 2010. *S&T Strategies of Six Countries: Implications for the United States.* Washington, DC: National Academy Press.

National Research Council. 2011. *The Future of Photovoltaics Manufacturing in the United States.* Charles W. Wessner, Rapporteur. Washington, DC: The National Academies Press.

National Research Council. 2011. *Growing Innovation Clusters for American Prosperity.* Charles W. Wessner, Rapporteur. Washington, DC : The National Academies Press.

National Research Council. 2011. Research Training in the Biomedical, Behavioral and Clinical Research Sciences. Washington, DC.; The National Academies Press.

National Research Council. 2012. Breaking Through: Ten Strategic Actions to Leverage Our Research Universities for the Future of America. Washington, DC: The National Academies Press.

National Research Council. Forthcoming. *Building the U.S. Battery Industry for Electric-Drive Vehicles: Progress, Challenges, and Opportunities.* Charles W. Wessner, Rapporteur. Washington, DC: The National Academies Press.

National Research Council. Forthcoming. *Clustering for 21st Century Prosperity.* Charles W. Wessner, Rapporteur. Washington, DC: The National Academies Press.

National Research Council of Canada. 2008. "Impact Evaluation of the NRC Industrial Research Assistance Program (NRC-IRAP)." Executive Summary. [Viewed here on 4/4/2012 http://www.nrc-cnrc.gc.ca/eng/evaluation/evaluation-irap.html]

National Research Foundation. 2008. (c2007) "National Framework for Innovation and Enterprise." Prime Minister's Office, Republic of Singapore. Last Updated on 23 September 2011. [Viewed here on 4/4/2012 http://www.nrf.gov.sg/nrf/otherProgrammes.aspx?id=1206]

National Science Board. 2004. *Science and Engineering Indicators 2004.* Arlington, VA: National Science Foundation.

National Science Board. 2006. *Science and Engineering Indicators 2006.* Arlington, VA: National Science Foundation.

National Science Board. 2008. *Science and Engineering Indicators 2008.* Arlington, VA: National Science Foundation.

National Science Foundation. 2004. *Science and Engineering Doctorate Awards*: 2003. NSF 05-300. Arlington, VA: National Science Foundation.

National Science Foundation. 2006. "Where has the Money Gone? Declining Industrial Support of Academic R&D." InfoBrief, NSF 06-328 September.

National Science Foundation, Division of Science Resources Statistics. 2010. *Key Science and Engineering Indicators.*. National Science Foundation: Arlington, VA

National Science Foundation, Division of Science Resources Statistics. 2010. *National Patterns of R&D Resources: 2008 Data Update.* NSF 10-314. Arlington, VA. Available at http://www.nsf.gov/statistics/nsf10314/

Nature. 2001."Initial Sequencing and Analysis of the Human Genome." February 15.

Needham, Joseph, andLing Wang. *1954-1986:. Science and Civilization in China.* Volume V. Cambridge: Cambridge University Press.

Nelson, Richard R. 1993. *National Innovation Systems: A Comparative Analysis.* New York: Oxford University Press.

Nelson, Richard R. 2005. *Technology, Institutions, and Economic Growth.* Cambridge MA: Harvard University Press.

Nelson, R. R., and G. Wright. 1992. "The Rise and Fall of American Technological Leadership: The Postwar Era in Historical Perspective." *Journal of Economic Literature*, Vol. 30(4).

Nelson, R. R., and K. Nelson. 2002. "Technology, Institutions, and Innovation Systems." *Research Policy* 31:265-272.

Nelson, R. R., and N. Rosenberg. 1993. "Technical Innovation and National Systems." In *National Innovation Systems: A Comparative Analysis.* Richard R. Nelson, ed. Oxford, UK: Oxford University Press.

Nelson, R. R., and S. G. Winter. 1982. *An Evolutionary Theory of Economic Change.* Cambridge, MA: Belknap Press of Harvard University Press.

Neuffer, John. 2010 "China: Intellectual Property Infringement, Indigenous Innovation Policies, and Frameworks for Measuring the Effects on the U.S. Economy." Testimony to the United States International Trade Commission Investigation No. 332-514 June 15. USITC 4199 (amended), November [Viewed here on 4/4/2012 http://www.usitc.gov/publications/332/pub4199.pdf]

Neumark, David, Brandon Wall, and Junfu Zhang. "Do Small Businesses Create More Jobs? New Evidence for the United States from the National Establishment Time Series." *The Review of Economics and Statistics.* February. Vol . 93. No. 1. Pp 16-29.

New Energy and Industrial Technology Department Organization. 2008. "2008 Roadmap for the Development of Next Generation Automotive Battery Technology." Ministry of Economy, Trade, and Industry, Republic of Japan.

New Mexico GovernorSusana Martinez' Office and New Mexico Economic Development Department. 2009. Technology 21: Innovation and Technology in the 21st Century Creating Better Jobs for New Mexicans: A Science and Technology Roadmap for New Mexico's Future, January. [Viewed here on 4/4/2012 http://www.gonm.biz/uploads/files/plan.pdf]

New York Times. 2008. "China Didn't Check Drug Suppliers, Files Show." February 16.

New York Times. 2011. "Federal Research Center Will Help Develop Medicines." January 12.

New York Times. 2012. "J&J Fined $1.2 Billion in Drug Case." April 11.

Nishimura, Kiyohiko, G., Takanobu Nakajima, and Kozo Kiyota. 2003. "Does the Natural Selection Mechanism Still Work in Severe Recessions? Examination of the Japanese Economy in the 1990s." *Journal of Economic Behavior and Organization* 58:53-78.

Odagari, H. 1999. "University-Industry Collaboration in Japan: Facts and Interpretations." In Lewis M. Branscomb, Fumio Kodama, and Richard Florida, eds. *Industrializing Knowledge: University-Industry Linkage in Japan and the United States.* Cambridge, MA: The MIT Press.

Odagiri, H., and A. Goto. 1993. "The Japanese System of Innovation: Past, Present and Future." In R. R. Nelson, ed. *National Systems of Innovation.* Oxford, UK: Oxford University Press.

Odagiri, H., and A. Goto. 1996. *Technology and Industrial Development in Japan: Building Capabilities by Learning, Innovation and Public Policy.* Oxford, UK: Oxford University Press.

Odilon Antonio Marcuzzo do Canto, 2007 "Incentives to Support Innovation in the Private Sector: The Brazilian Experience." Brazil Innovation Agency. [Viewed at http://idbdocs.iadb.org/wsdocs/getdocument.aspx?docnum=976025 on March 7/2012]

Office of the US Trade Representative. 2009. National Trade Estimates Report on Foreign Trade Barriers. p. 316-317.

"Opinions on the Implementation of Decisions on Expanding Domestic Demand and Promoting Economic Growth and Further Strengthening Supervision of Tendering and Bidding Projects." 2009. Circular 1361, May 27.

Organisation for Economic Co-operation and Development. 1999. *Boosting Innovation: The Cluster Approach.* Paris, France: Organisation for Economic Co-operation and Development.

Organisation for Economic Co-operation and Development. 1999. *Managing National Innovation Systems.* Paris, France: Organisation for Economic Co-operation and Development.

Organisation for Economic Co-operation and Development. 2001. *The New Economy: Beyond the Hype—The OECD Growth Project.* Paris, France: Organisation for Economic Co-operation and Development.

Organisation for Economic Co-operation and Development. 2001. *Social Sciences and Innovation.* Washington, DC: Organisation for Economic Co-operation and Development.

Organisation for Economic Co-operation and Development. "OECD Guidelines on Corporate Governance of State Owned Enterprises." Paris, France: Organisation for Economic Co-operation and Development.

Organisation for Economic Co-operation and Development. 2007. *Micro-policies for Growth and Productivity.* DSTI/IND(2004)7. Paris, France: Organisation for Economic Co-operation and Development. October.

Organisation for Economic Co-operation and Development. 2007. *Competitive Regional Clusters: National Policy Approaches*, Paris: Organisation for Economic Co-operation and Development.

Organisation for Economic Co-Operation and Development. 2008. *OECD Reviews of Innovation Policy: China.* Paris, France: Organisation for Economic Co-operation and Development.

Organisation for Economic Co-operation and Development. 2009. *OECD Science, Technology and Industry Scoreboard 2009.* Organisation for Economic Co-operation and Development.

Organisation for Economic Co-operation and Development. 2010. *OECD Science, Technology and Industry Outlook 2010.* Organisation for Economic Co-operation and Development.

Organisation for Economic Co-operation and Development. 2011. *Main Science and Technology Indicators.* Paris, France: Organisation for Economic Co-operation and Development. [Viewed here on 4/4/2012 http://www.oecd-ilibrary.org/science-and-technology/main-science-and-technology-indicators/volume-2011/issue-2_msti-v2011-2-en-fr;jsessionid=7j4ea99mv2ce6.epsilon

Organisation for Economic Co-operation and Development. 2011. *Financing High Growth Firms, The Role of Angel Investors.* Organisation for Economic Co-operation and Development. Pp. 96.

Organisation for Economic Co-operation and Development. 2011. *OECD Science, Technology and Industry Scoreboard 2011.* Organisation for Economic Co-operation and Development.

Orszag, Peter, and Thomas Kane. 2003. "Funding Restrictions at Public Universities: Effects and Policy Implications." Washington, DC : Brookings Institution Working Paper. September.

Ostrom, Elinor. 2005. Understanding Institutional Diversity, Princeton, N. J.: Princeton University Press.

Ostry, Sylvia and Richard R. Nelson. *Techno-Nationalism and Techno-Globalism: Conflict and Cooperation.* Washington, DC: Brookings Institution. 1995.

Oughton, Christine. 1997. "Competitiveness in the 1990s." The Economic Journal 107(444): 1486-1503.

Oughton, Christine, Mikel Landabaso, and Kevin Morgan. 2002. "The Regional Innovation Paradox: Innovation Policy and Industrial Policy." The Journal of Technology Transfer 27(1):97-110.

Owens, William, with Edward Offley. 2001. Lifting the Fog of War. Baltimore, MD: Johns Hopkins University Press.

Park, W. G., and J. C. Ginarte. 1997. "Intellectual Property Rights and Economic Growth." *Contemporary Economic Policy XV*(July):51-61.

Packan, Paul A. 1999. "Pushing the Limits: Integrated Circuits Run Into Limits Due to Transistors." *Science*, September 24, 285(5476): 2079-2081 [DOI:10.1126/science.285.5436.2079]

Patel, P., and K. Pavitt. 1994. "National Innovation Systems: Why They are Important and How They Might Be Compared?" *Economic Change and Industrial Innovation*. Vol. 3: 77-95

Patil, Pandit G. 2008."Developments in Lithium-Ion Battery Technology in the Peoples Republic of China." Argonne National Laboratories, ANL/ESD/08-1, January. [Viewed here on 4/5/2012 http://www.ipd.anl.gov/anlpubs/2008/02/60978.pdf

Pavitt, K. 2001. "Public Policies to Support Basic Research: What Can the Rest of the World Learn from US Theory and Practic? (And What They Should Not Learn)." *Industrial and Corporate Change* 10(3):761-799.

Pearce, Robert. 2011. China and the Multinationals: International Business and the Entry of China into the Global Economy. Aldershot: Edward Elgar Publishing.

Pearson, David. 2009. "France Backs Battery-Charging Network for Cars." Wall Street Journal, October 1.

Peck, M. J., R. C. Levin, and Akira Goto. 1988. "Picking Losers: Public Policy Toward Declining Industries in Japan." In J. B. Shoven, ed. *Government Policy Toward Industry in the United States and Japan*. Cambridge, UK: Cambridge University Press. Pp. 165-239.

Peek, Joe, and Eric S. Rosengren. 2005. "Unnatural Selection: Perverse Incentives and the Misallocation of Credit in Japan." *The American Economic Review* 95(4):-1144-1166.

Pele, Anne-Francoise. 2007. "Freescale Eases Out of Crolles2 Alliance." *EE Times Europe*, June 26.

Pele, Anne-Francoise. 2010. "Mentor Joins Nano 2012 R&D Alliance." EE Times, March 16.

Penrose, E. 1951. The Economics of the International Patent System. Baltimore, MD: Johns Hopkins University Press.

People's Daily Online. 2010. "Strategic Emerging Industries Likely to Contribute 8% of China's GDP by 2015." October 19.

People's Daily Online. 2010. "China to Sell 1 Million New-Energy Cars Annually by 2015." November 23.

Perez, Carlota. 2002. *Technological Revolutions and Financial Capital: The dynamics of bubbles and golden ages*. Cheltenham, UK ; Northampton, MA, USA: Edward Elgar.

Perlin, John. 1999. *From Space to Earth: The Story of Solar Electricity*, Ann Arbor: Aatec Publications.

Pew Charitable Trusts. 2010. "Who's Winning the Clean Energy Race? Growth, Competition and Opportunity in the World's Largest Economies." G-20 Clean Energy Factbook [viewed here on 4/5/2012 http://www.pewtrusts.org/uploadedFiles/wwwpewtrustsorg/Reports/Global_warming/G-20%20Report.pdf

Pham, Nam d. (NDP Consulting Group). 2010. "The Impact of Innovation and the Role of Intellectual Property Rights on U.S. Productivity, Competitiveness, Jobs, Wages, and Exports.", April. Washington, DC : Global Intellectual Property Center, U.S. Chamber of Commerce. [Viewed here on 4/5/2012 http://www.theglobalipcenter.com/reports/impact-innovation-and-role-ip-rights-us-productivity-competitiveness-jobs-wages-and-exports]

Phan, Phillip H., and Donald S. Siegel. 2006. "The Effectiveness of University Technology Transfer: Lessons Learned from Qualitative and Quantitative Research in the U.S. and U.K." *Rensselaer Working Papers in Economics* 0609. Troy, NY: Rensselaer Polytechnic Institute Department of Economics.

Phan, Phillip H., Donald S. Siegel, and Mike Wright. 2005. "Science Parks and Incubators: Observations, Synthesis and Future Research." *Journal of Business Venturing* 20(2):165-182.

Pharma Times. 2008. "U.S. Congress Warned of 'Gathering Storm' at FDA." February 8.

Pharma Times. 2009. "NIH Award US $117 Million to Rare Disease Consortia." October 7.

Pharma Times. 2009. "NIH to Create Development Pipeline for Rare, Neglected Diseases." May 25.

Pharma Times. 2009. "U.S. $23 Million Grant Makes Cincinnati University CTSA Members" April 8.

Pharma Times. 2011."Obama Seeks $135B Drug Price Cuts Over 10 Years." September 23.

Pharma Times. 2011. "Pfizer Steps Up Campaign in Fight Against Counterfeit Drugs." September 30.
Pharma Times. 2011. "U.S. CBO Doubles Estimated Savings from Pay-for-Delay Ban." November 13.
Pharma Times. 2011. "U.S. Govt Drug Research Agency 'To Start Work in October,'" January 25.
Pharma Times. 2011. "U.S. Pharma's 'Record' 460 Drugs in Development for Rare Diseases." February 28.
Pharma Times. 2011. "U.S. Public Research 'Responsible for Many Major New Drugs,'" February 18.
Pharma Times. 2012. "Counterfeit Avastin Seized in the US." February 16.
Pharma Times. 2012. "Fitch Looks at Implications of FDA Biosimilar Guidance." February 13.
Pharma Times. 2012. "Pharma Blasts Obama Budget, FDA to Get $4.49 Billion." February 14.
Pharma Times. 2012. "Teva Fast Out of the Blocks to Sell Generic Seroquel." March 27.
Phillimore, J. 1999. "Beyond the Linear View of Innovation in Science Park Evaluation: An Analysis of Western Australian Technology Park." *Technovation* 19(11):673-680.
Phillips, Charles, Laura Tyson, and Robert Wolf. 2010. "The U.S. Needs an Infrastructure Bank." Wall Street Journal, January 15.
Piekkola, Hannu. 2006. "Knowledge and Innovation Subsidies as Engines of for Growth—The Competitiveness of Finnish Regions." *Research Institute of the Finnish Economy* (ETLA), Sarja B 216 Series, Helsinki: Taloustieto Oy. Viewed here on 4/5/2012 http://www.etla.fi/files/1453_B216_22_01_06.pdf.
Pike Research. 2010. "Asian Manufacturers Will Lead the $8 Billion Market for Electric Vehicle Batteries." June 1.
Pilat, Dirk and Anita Wölfl. 2005. "Measuring the Interaction Between Manufacturing and Services." OECD STI Working Paper, DSTI/DOC, May 31.
Pisano, Gary and Willy C. Shih. 2009. "Restoring American Competitiveness." *Harvard Business Review*, Paper Series, July-August Pp. 1-15
Pisano, Gary and Willy C. Shih. 2012. "Does America Really Need Manufacturing." *Harvard Business Review* 90, no. 3
Pitelis, Christos, Roger Sugden, and James R. Wilson, eds. 2006. *Clusters and Globalisation: The Development of Urban and Regional Economies*, Cheltenham: Edward Elgar Publishing,Plastic Electronics. 2011. "Nanomaterials firms turn to Asia for Commercial Opportunities" April 15.
Poh, Lim Chuan. 2010. "Singapore Betting on Biomedical Science." *Issues in Science and Technology*, Spring. 26(3)
Polenske, Karen, Nicolas Rockler, et al. 2004. Closing the Competitive Gap: A Retrospective Analysis of the ATP 2mm Project. NIST GCR 03-856. Gaithersburg, MD: National Institute of Standards and Technology.

PolymerOhio. 2008. "Strength of Workforce." September 23. Accessed at: http://www.polymerohio.org/index.php?option=com_content&view=article&id=70&Itemid=87

Porter, M. E., 1998. "Clusters and the New Economics of Competition." *Harvard Business Review* 76(6):77-90.

Porter, M. E., and M. Sakakibara. 2004. "Competition in Japan." *The Journal of Economic Perspectives* 18(1).

Porter, Michael. 1990. The Competitive Advantage of Nations, New York: The Free Press.

Porter, Michael E. 2005. South Carolina Competitiveness Initiative: A Strategic Plan for South Carolina. [South Carolina] : South Carolina Council on Competitiveness.

Porter, Michael . 2007. "Clusters and Economic Policy: Aligning Public Policy with the New Economics of Competition." ISC White Paper, Harvard Business School, November.

Porter, Michael E. 2009. Revised. "Clusters and Economic Policy: Aligning Public Policy with the New Economics of Competition." *Institute for Strategy and Competitiveness* White Paper.

Porter, Michael E. 2011. "South Carolina Competitiveness: State and Cluster Economic Performance." Harvard Business School, February 26.

Porter, Michael E and Ian W. Rivkin. 2012. "Prosperity at Risk." Harvard Business School. January.

Posen, Adam. 1998. Restoring Japan's Economic Growth. Washington, DC: Peterson Institute for International Economics.

Posen, Adam S. 2001. "Japan." In Benn Steil, David G. Victor, and Richard R. Nelson, eds. *Technological Innovation and Economic Performance*. Princeton, NJ: Princeton University Press.

Posner, Richard A. 1979."The Chicago School of Antitrust Analysis." *University of Pennsylvania Law Review* 127.

Powell, Buonassisi, et al. 2012. Crystalline Silicon Photovoltaics, a Cost Analysis Framework, Energy and Environmental Science 5, 5874. February.

Powell, Jeanne, and Francisco Moris. 2002. *Different Timelines for Different Technologies*. NISTIR 6917. Gaithersburg, MD: National Institute of Standards and Technology.

Powell, Walter W. Kelley A. Packalen, and Kjersten Bunker Whittington. 2012. "Organizational and Institutional Genesis: The Emergence of High-Tech Clusters in the Life Sciences." In John Padgett, Walter W. Powell, eds, *The Emergence Of Organization And Markets*, Princeton: Princeton University Press, Chapter 13.

Prabju, Jadeep C., Andreas B Eisengerich, Rajesh K. Chandy, and Gerard J. Tellis. 2010. "Patterns in the Global Location of R&D Centres by the World's Largest Firms: The Role of India and China". Paper presented at Druid Summer Conference. June.

Prahalad, C. K., 2005. The Fortune at the Bottom of the Pyramid: Eradicating Poverty Through Profits. Wharton School Publishing.

President's Council of Advisors on Science and Technology. 2004. "Sustaining the Nation's Innovation Ecosystems." Washington, DC: Executive Office of the President. January.

President's Council of Advisors on Science and Technology. 2004. "Sustaining the Nation's Innovation Ecosystems: Information Technology Manufacturing and Competitiveness". [Washington, DC : President's Council of Advisors on Science and Technology.

President's Information Technology Advisory Committee. 2005. "Cybersecurity: A Crisis of Prioritization." Report to the President. Washington, DC: President's Information Technology Advisory Committee. February.

President's Council of Advisors on Science and Technology. 2011. Report to the President on Ensuring American Leadership in Advanced Manufacturing.[Washington, DC] : Executive Office of the President, June. [Viewed here on 4/5/2012 http://www.whitehouse.gov/sites/default/files/microsites/ostp/pcast-advanced-manufacturing-june2011.pdf

President's Council of Advisors on Science and Technology. 2012. "Report to the President And Congress on the Fourth Assessment of the National Nanotechnology Initiative." Washington, DC: The White House. April.

Prestowitz, Clyde. 1989. Trading Places: How We Allowed Japan to Take the Lead. New York: Basic Books.

PricewaterhouseCoopers. 2006. "China's Impact on the Semiconductor Industry: 2005 Update." New York: PricewaterhouseCoopers.

PricewaterhouseCoopers. 2010. Global Reach: China's Impact on the Semiconductor Industry 2010 Update, November p.14.

PricewaterhouseCoopers/National Venture Capital Association. 2011. The MoneyTreeTM Report: Overview of Venture Capital Investments Third Quarter 2011.PRO INNO Europe. 2010. European Commission, ZIM, the Central Programme for SMEs (Zentrales Innovationsprogramm Mittelstand), INNO-Partnering Forum, Document ID: IPF 11-005.

Pro Inno Europe. 2011. *Innovation Union Scoreboard 2010: The Innovation Union's Performance Scoreboard for Research and Innovation*, Feb. 1. [Viewed here on 4/5/2012 http://ec.europa.eu/research/innovation-union/pdf/iu-scoreboard-2010_en.pdf]

Procassini, Andrew A. 1995. Competitors in Alliance: Industry Associations, Global Rivalries, and Business-government Relations, Westport, CT: Quorum Books.

Purvis, G. 2002. "Moving into the Real World." EDN July 1.

PV Magazine. 2011 "Korea Expected to Pick up the PV Pace." January 12.

PV Magazine. 2011. "Korea: Construction of 100 MW Thin Film Plant Underway." April 18.

Qiang, Christine Zhen-Wei. 2009. "Broadband Infrastructure Investment in Stimulus Packages: Relevance for Developing Countries." Washington, DC: World Bank. [Viewed here on 4/5/2012 http://siteresources.worldbank.org/EXTINFORMATIONANDCOMMUNICATIONANDTECHNOLOGIES/Resources/282822-1208273252769/Broadband_Investment_in_Stimulus_Packages.pdf]

R & D Magazine . 2010. "R&D 100 Awards: 2010 Winners." July.

Radjou, Nav, Jaideep Prabhu and Simone Ahuja. 2012. *Jugaad Innovation: Think Frugal, Be Flexible, Generate Breakthrough Growth*, San Francisco: Jossey-Bass.

Raduchel, William. 2006. "The End of Stovepiping." In National Research Council. *The Telecommunications Challenge: Changing Technologies and Evolving Policies.* Charles W. Wessner, ed. Washington, DC: The National Academies Press.

RAND Corporation. 2005. *Health Information Technology: Can HIT Lower Costs and Improve Quality?*, (http://www.rand.org/pubs/research_briefs/RB9136/index1.html).

Reich, Eugene. 2010. "US Congress passes strategic science bill." *Nature*, December 22.

Reid, T. R. 2004. *The United States of Europe: The New Superpower and the End of American Supremacy*. New York: Penguin Press.

Reiko, Yamada. 2001. "University Reform in the Post-massification Era in Japan: Analysis of Government Education Policy for the 21st Century." *Higher Education Policy* 14(4):277-291.

Reuters. 2006. "China Sees No Quick End to Economic Boom." February 21.

Reuters. 2010., "China Electric Vehicles to Hit 1 Million by 2020: Report". October 16.

Reuters. 2011. "China Eases Government Procurement Rules After U.S. Pressure." June 29.

Reuters. 2011. "Drug R&D Spending Fell in 2010, and Heading Lower." June 26.

Rich, Ben R. and Janos Leo. 1994.. Skunk Works : a Personal Memoir of My Years at Lockheed Boston : Little Brown.

Richburg, Keith B. 2011. "Are China's High-Speed Trains Heading Off the Rails?" Washington Post, April 23

Ridley, Tony, Lee Yee-Cheong, and Calestous Juma. 2006. "Infrastructure, Innovation, and Development." *International Journal of Technology and Globalisation,* 2(3-4):268-278.

Roberts, Dexter and Pete Engardio. 2009. "China's Economy: Behind All the Hype." BusinessWeek, October. 23 22 [Viewed here on 4/6/2012 http://www.businessweek.com/magazine/content/09_44/b4153036870077.htm].

Roberts, Edward P. and Charles Eesley, Entrepreneurial Impact; the Role of MIT, Kauffman Foundation Report (2009). Access at http://web.mit.edu/dc/policy/MIT-impact-full-report.pdf

Rodrik, Dani and Arvind Subramanian. 2004. "From 'Hindu Growth' to Productivity Surge: The Mystery of the Indian Growth Transition." NBER Working Paper 10376.

Rohatyn, Felix. 2010. "The Case for an Infrastructure Bank", *Wall Street Journal*, September 15.

RolandBerger Strategy Consultants. 2011. "Global Study on the Development of the Automotive Li-ion Battery Market." Press Release, September 6.

Rolfstam, M, W. Phillips, and E. Bakker 2011. "Public procurement and the diffusion of innovations: exploring the role of institutions and institutional coordination." *International Journal of Public Sector Management*, 24 (5).

Romanainen, Jari. 2001. "The Cluster Approach in Finnish Technology Policy." Pp. 377-388 in Edward M. Bergman, Pim den Hertog, and David Charles, Svend Remoe, eds. Innovative Clusters: Drivers of National Innovation Systems. OECD Proceedings. Washington, DC: Organisation for Economic Co-operation and Development.

Romer, Paul. 1990. "Endogenous Technological Change." *Journal of Political Economy* 98:72-102.

Romer, Paul. 1993. "Idea Gaps and Object Gaps in Economic Development." *Journal of Monetary Economics* 32(3): 543-573.

Rosen, Howard F. 2008. "Designing a National Strategy for Responding to Economic Dislocation." Testimony before the Subcommittee on Investigation and Oversight House Science and Technology Committee. June 24.

Rosen, Harold F. 2008. "Strengthening Trade Adjustment Assistance." Policy Brief 08-02. Peterson Institute for International Economics. Access at http://www.iie.com/publications/pb/pb08-2.pdf.

Rosen, Jeffrey. 2011. "Universal Service Fund Reform: Expanding Broadband Internet Access in the United States." *Issues In Technology Innovation*, Number 8, April.

Rosenberg, N., and R. Nelson. 1994. "American Universities and Technical Advance in Industry." *Research Policy* 23:325-348.

Rosenzweig, R. and B. Turlington. 1982. The *Research Universities and Their Patrons.* Berkeley, CA: University of California Press.

Roth, Aleda, et. Al. 2010. "Global Manufacturing Competitiveness Index." Deloitte Touche Tohmatsu and U.S. Council on Competitiveness, June. [Viewed here on 4/6/2012 http://www.compete.org/images/uploads/File/PDF%20Files/2010_Global_Manufacturing_Competitiveness_Index_FINAL.pdf

Rothaermel, F. T., and M. C. Thursby. 2005. "Incubator Firm Failure or Graduation? The Role of University Linkages." *Research Policy* 34(7):1076-1090.

Rothaermel, Frank T. and Marie Thursby. 2005. "University–incubator firm knowledge flows: assessing their impact on incubator firm performance." *Research Policy* 34(3):305–320.

Rowen, Henry S., and A. Maria Toyoda. 2002. "From Kiretsu to Start-ups: Japan's Push for High Tech Entrepreneurship." Asia-Pacific Research Center Working Paper. Stanford, CA.

Ruegg, Rosalie, and Irwin Feller. 2003. *A Toolkit for Evaluating Public R&D Investment: Models, Methods, and Findings from ATP's First Decade.* NIST GCR 03-857. Gaithersburg, MD: National Institute of Standards and Technology.

Rulison, Larry. 2009. "GlobalFoundries Board Approves Malta Fab Go-Ahead." *Albany Times Union,* March 20.~~~

Russell, Alene. 2008. "Dedicated Funding for Higher Education: Alternatives for Tough Economic Times." American Association of State Colleges and Universities, Higher Education Policy Brief, December.

Ruttan, Vernon W. 2001. Technology, Growth, and Development: An Institutional Design Perspective, New York: Oxford University Press.

Ruttan, Vernon W. 2002. *Technology, Growth and Development: An Induced Innovation Perspective.* Oxford, UK: Oxford University Press.

Ruttan, Vernon W. 2006. *Is War Necessary for Economic Growth, Military Procurement and Technology Development?* New York: Oxford University Press.

Ruttan, Vernon W. 2006. "Will Government Programs Spur the Next Breakthrough?" *Issues in Science and Technology* Winter. 66(2);55-61.

Rutten, Roel, and Frans Boekema. 2005. "Innovation, Policy and Economic Growth: Theory and Cases." European Planning Studies 13(8):1131-1136.

Rycroft, Robert W., and Don E. Kash. 1999. "Innovation Policy for Complex Technologies: U.S. Technology Policy Must be Revamped to Deal with Accelerating Technological and Organizational Complexity*." Issues in Science and Technology* Fall. 16(1): 73-79.

Saguy, Sam. 2011. "Paradigm shifts in academia and the food industry required to meet innovation challenges." *Trends in Food Science & Technology* 22(9):467–475.

Saito, Ken. 1987. Research on a New Concern: RIKEN Industrial Group. (In Japanese.) Jichosha

Sakakibara, M., and L. Branstetter. 2001. "Do Stronger Patents Induce More Innovation? Evidence from the 1988 Japanese Patent law Reforms." *Rand Journal of Economics* 32:77-100.

Sakakibara, M., and L. Branstetter. 2002. *Measuring the Impact of ATP-Funded Research Consortia on Research Productivity of Participating Firms, A Framework Using Both U.S. and Japanese Data.* NIST GCR 02-830. Gaithersburg, MD: National Institute of Standards and Technology.

Sakamoto, Kozo, and Masayuki Kondo. 2004. "The Analysis of University-Industry Research Collaborations by Time Series and Corporate Characteristics." (In Japanese.) *Development Engineering* 10:11-26.

Sallet, J, J. Masterman, and E. Paisley. 2009. "The Geography of Innovation." Washington, DC: Center for American Progress. [See here the sponsoring organization Science Progress http://scienceprogress.org/2009/09/the-geography-of-innovation/

Sandia Labs news release. 2010. "Report: Sandia Science & Technology Park Fuels Economy With Jobs, Tax Revenue, Spending." Aug 3. [Viewed here on 4/6/2012 https://share.sandia.gov/news/resources/news_releases/report-sandia-science-technology-park-fuels-economy-with-jobs-tax-revenue-spending/

Sapolsky, Harvey M. 1990. *Science and the Navy—The History of the Office of Naval Research*. Princeton, NJ: Princeton University Press.

Sarfraz, Mian A.. 2011. "University's involvement in technology business incubation: what theory and practice tell us?" *International Journal of Entrepreneurship and Innovation Management*, Vol. 13(2):113-121

Saxenian, AnnaLee. 1994. *Regional Advantage: Culture and Competition in Silicon Valley and Route 128*. Cambridge, MA: Harvard University Press.

Saxenian, AnnaLee. 1999., *Silicon Valley's New Immigrant Entrepreneurs*, San Francisco: Public Policy Institute of California.

Schaefer, Ulrich. 2010. "2010-2013 Semiconductor Market Forecast." WSTS European Chapter, EECA-ESIA & WSTS, June 8 [See this URL for the slides http://www.slideshare.net/StephanCadene/20102013-semiconductor-market-forecast-seizing-the-economic-amp-political-momentum-in-europe-for-key-enabling-technologies

Schaffer, Teresita. 2002. "Building a New Partnership with India." *Washington Quarterly* 25(2):31-44.

Scherer, F. M. 2001. "U.S. Government Programs to Advance Technology." *Revue d'Economie Industrielle* 0(94):69-88.

Schmitz, Hardy Rudolph. 2009. "100 Years of Innovation from Adlershof: Dawns, Damage, and Determination." Wista-Management GMBH, Sept. 9. [Viewed here on 4/7/2012 http://www.adlershof.de/fileadmin/web/ansprechpartner/netzwerke/internationales/events/Hardy_Schmitz_-_Adlershof_100_years_of_innovation_speech.pdf]

Schwab, Klaus, editor. 2011. World Economic Forum. 2011. The Global Competitiveness Report 2011-2012. [Viewed here on 4/7/2012 http://www3.weforum.org/docs/WEF_GCR_Report_2011-12.pdf]

Schurman, Rachel and Dennis Kelso. 2003. *Engineering Trouble: Biotechnology and its Discontents*. Berkeley and Los Angeles: University of California Press.

Science. 2001. "The Sequence of the Human Genome." February 16.

Science. 2011. Stem Court Ruling a Decisive Win for NIH." July 27.

Science Committee of the Council for Science and Technology of the Ministry of Education, Culture, Sports, Science and Technology. 2005. Sciences Policy to Support Diversity in Research. (Kenkyuu-no-tayousei wo-sasaeru Gakujutu-seisaku). October 13.

Science Insider. 2012. Panel Calls for Closer Oversight of Biomark Tests." March 23.

Science, Technology, and Innovation Council2008. State of the Nation 2008 [Viewed here on 4/7/2012 http://www.stic-csti.ca/eic/site/stic-csti.nsf/vwapj/08-141_IC_SOTN_EN_Final_no_trans2.pdf/$FILE/08-141_IC_SOTN_EN_Final_no_trans2.pdf]

Scotchmer, Suzanne. 1996. "Protecting Early Innovators: Should Second-Generation Products Be Patentable?" *Rand Journal of Economics* 27:322-331.

Scotchmer, Suzanne. 2004. Innovation and Incentives. Cambridge, MA: The MIT Press.

Segal, Adam. 2010. "China's Innovation Wall: Beijing's Push for Homegrown Technology." *Foreign Affairs,* Sept. 28.

Semiconductor Industry Association "Nanoelectronics Research Initiative: A Model Government-Industry Partnership Promoting Basic Research." Access at http://www.sia-online.org/clientuploads/One%20Pagers/Nanoelectronics_SRC_FINAL.pdf

Semiconductor Industry Association (SIA). 2009. *Maintaining America's Competitive Edge: Government Policies Affecting Semiconductor Industry R&D and Manufacturing Activity*, March. Accessed at http://www.sia-online.org/galleries/default-file/Competitiveness_White_Paper.pdf.

Shane, S. 2004. *Academic Entrepreneurship*. Cheltenham, UK: Edward Elgar Publishing.

Shang, Yong. 2006. "Innovation: New National Strategy of China." Presentation at Industrial Innovation in China. Levin Institute Conference. July 24-26.

Shanghai Municipal Government. 2004. Notice of the Shanghai Municipal Government Regarding Distributing the Outline of Shanghai's Intellectual Property Strategy (2004-2010). September 14.

Shapira, Philip. 2001. "US manufacturing extension partnerships: technology policy reinvented?" *Research Policy*, 30(6): 977–992.

Shapira, Philip. 2010. Building capabilities for innovation in SMEs: a cross-country comparison of technology extension policies and programmes." *International Journal of Innovation and Regional Development* 3(3-4).

Shapira, Philip and Jan Youtie, 2010. "The Innovation System and Innovation Policy in the United States." Chap. 2 in Rainer Frietsch and Magrot Schüller eds. *Competing for Global Innovation Leadership: Innovation Systems and Policies in the USA, EU, and Asia*, Fraunhofer IRB Verlag, Stuttgart.

Shapira, Philip, Jan Youtie, and Luciano Kay. 2011. "Building Capabilities for Innovation in SMEs: A Cross-Country Comparison of Technology Extension Policies and Programs" *International Journal of Innovation and Regional Development,* 3-4: 254-272.

Shapiro, Carl. 2001. "Navigating the Patent Thicket: Cross License, Patent Pools and Standard-Setting." In Adam Jaffe, Joshua Lerner, and Scott Stern, eds. *Innovation Policy and the Economy*. Cambridge, MA: The MIT Press.

Sharpe, Andrew. 2006. "Lessons for Canada from the International Productivity Experience." *Centre for the Study of Living Standards*, Research Report 2006-02.

Shearmur, R., and D. Doloreux. 2000. "Science Parks: Actors or Reactors? Canadian Science Parks in their Urban Context." *Environment and Planning* 32(6):1065-1082.

Sheehan, Jerry, and Andrew Wyckoff. 2003. "Targeting R&D: Economic and Policy Implications of Increasing R&D Spending." DSTI/DOC(2003)8. Paris, France: Organisation for Economic Co-operation and Development.

Sherk, James, 2010. "Technology Explains Drop in Manufacturing Jobs." Backgrounder #2476, Heritage Foundation, October 12.

Shenzhen Daily. 2005. "Nation May Introduce Antimonopoly Law." December 30.

Sherwin, Martin, and Kai Bird. 2005. *American Prometheus: The Triumph and Tragedy of J. Robert Oppenheimer*. New York: Alfred A. Knopf.

Shi, Yigong and Yi Rao, "China's Research Culture." Science, Vol. 329(5996). Sept. 3, 2010.

Shin, Roy W. 1997. "Interactions of Science and Technology Policies in Creating a Competitive Industry: Korea's Electronics Industry." *Global Economic Review* 26(4):3-19.

Shoven, John. 2011. *Demography and the Economy*, Chicago: University of Chicago Press.

Siegel, Donald S., David Waldman, and Albert Link. 2004. "Toward a Model of the Effective Transfer of Scientific Knowledge from Academicians to Practitioners: Qualitative Evidence from the Commercialization of University Technologies." *Journal of Engineering and Technology Management* 21(1-2):115-142.

Siegel, Donald S., P. Westhead, and M. Wright. 2003. "Assessing the Impact of Science Parks on Research Productivity: Exploratory Firm-Level Evidence from the United Kingdom." *International Journal of Industrial Organization* 21(9):1357-1369.

Simon, Denis Fred and Cong Cao. 2009. China's Emerging Technological Edge: Assessing the Role of High-End Talent, Cambridge, MA: Cambridge University Press.

Simon, Denis Fred, Cong Cao, and Richard P. Suttmeier. 2007. "The Evolution of Business China's New Science and Technology Strategy: Implications for Foreign Firms." *China Currents*, Vol. 6(2). Spring. [http://www.chinacurrents.com/spring_2007/cc_simon.htm]

SinoCast . 2007. "Eli Lilly Asia VC Fund Settles in Shanghai." November 16.

SinoCast . 2008. "Eli Lilly Opens China R&D Headquarters in Shanghai." October 17.

SinoCast. 2010. "Foreign Giants Dominate China Pharmaceutical Market." November 5.

SinoCast . 2011." Tenacent, Intel to jointly set up Research Center." April 13.

Sirkin, Harold L, Michael Zinser, and Douglas Hohner. 2011. *Made in America, Again: Why Manufacturing Will Return to the U.S.* Boston Consulting Group, August .

Skolnikoff, Eugene B. 1993. "Knowledge Without Borders? Internationalization of the Research Universities." *Daedalus* 122(4):225-252.

Slaughter, Matthew J. 2010. *How Piracy in China Costs U.S. Jobs.* Tuck School of Business at Dartmouth and NBER. September.

SmartGridNews. 2009. "$2.4 Billion Going to Accelerate Advanced Battery and EV Manufacturing." August 5.

Smith, Adam. 1776. An Iquiry Into the Nature and Causes of the Wealth of Nations.

Smith, Brian D. 2011. *The Future of Pharma: Evolutionary Threats and Opportunities*. Farakaan, U.K.: Gower Publishing Limited.

Smith, Kathlin. 1998. *The Role of Scientists in Normalizing U.S.-China Relations: 1965-1979*. Washington, DC : Council on Library and Information Resources [Viewed here on 4/7/2012 http://china-us.uoregon.edu/pdf/Smith's%20NYAS%20article.pdf]

Smith, Tobin, Josh Trapani, Anthony Decrappeo and David Kennedy. 2011."Reforming Regulation of Research Universities" in *Issues in Science and Technology*. Summer.

Smits, Ruud, and Stefan Kuhlmann. 2004. "The Rise of Systemic Instruments in Innovation Policy." *International Journal of Foresight and Innovation Policy* 1(1/2):4-32.

Soete, Luc G., and Bastiann J. ter Weel. 1999. "Innovation, Knowledge Creation and Technology Policy: The Case of the Netherlands." *De Economist* 147(3). September.

Sofouli, E., and N. S. Vonortas. 2007. "S&T Parks and Business Incubators in Middle-Sized Countries: The Case of Greece." *Journal of Technology Transfer* 32(5):525-544.

Sohl, Jeff E. 2007. "The Organization of the Informal Venture Capital Market." in *Handbook of Research on Venture Capital,* Hans Landström, editor. Northampton MA: Edward Elgar

Solar Energy Industry Association. 2011. U.S. Solar Market Insight 2nd Quarter 2011: Executive Summary.

Solar Energy Industry Association and GTM Research. 2010. "U.S. Solar Market Insight: 2010 Year in Review." Executive Summary.

Solow, Robert. 1956. "A Contribution to the Theory of Economic Growth." *Quarterly Journal of Economics*, 70 (1): 65–94.

Solow, Robert. 1957. "Technical Change and the Aggregate Production Function." *The Review of Economics and Statistics,* 39 (3): 312-320.

Solow, Robert. 2000. *Growth Theory: An Exposition.* New York: Oxford University Press. 2nd edition.

Solow, Robert. 2000."Toward a Macroeconomics of the Medium Run." *Journal of Economic Perspectives* 14(1):151-158.

Solvell, Örjan, Göran Lindqvist, and Christian Ketels. 2003. "The Cluster Initiative Greenbook". Stockholm: Ivory Tower.

Special Chem Coatings and Inks. 2008. " DuPont opens High-Performance Coatings R&D Center in Russia." July 28.

Spence, Michael. 1974. *Market Signaling: Informational Transfer in Hiring and Related Processes*. Cambridge, MA: Harvard University Press.

Spence, Michael. 2011. "Globalization and Unemployment: The Downside of Integrating Markets." Foreign Affairs. July/August.

Spence, Michael and Sandile Hlatshwayo. 2011. "The Evolving Structure of the American Economy and the Employment Challenge." Council on Foreign Relations, Center for Geoeconomic Studies.

Spencer, William, and T. E. Seidel. 2004. "International Technology Roadmaps: The U.S. Semiconductor Experience." In National Research Council. *Productivity and Cyclicality in Semiconductors: Trends, Implications, and Questions*. Dale W. Jorgenson and Charles W. Wessner, eds. Washington, DC: The National Academies Press.

Spencer, W. J., L. Wilson, and R. Doering. 2005 "The Semiconductor Technology Roadmap." Future Fab International 18. [See this http://www.future-fab.com/documents.asp?d_ID=3004]

Spiegel, Eric. 2011. "Make Permanent the Research and Experimentation Tax Credit", The Atlantic, July 12.

Spigel, Saul. 2005. "University of Albany Nanotechnology Program, OLR Research Report, 2005-R-0146, February. 9.

Springut, Micah, et al. 2011. "China's Program for Science and Technology Modernization: Implications for American Competitiveness." *U.S.-China Economic and Security Review Commission*. January 2011.

SRI International. 2009. Making an Impact: Assessing the Benefits of Ohio's Investment in Technology-Based Economic Development Programs. September.

Stafford, Sean. 2009. Why the Garden Club Couldn't Save Youngstown: The Transformation of the Rust Belt. Cambridge MA: Harvard University Publishing.

Stanford University. 1999. *Inventions, Patents and Licensing: Research Policy Handbook*. Document 5.1. [Current version: July 15, 1999[see here http://rph.stanford.edu/5-1.html]

State Council. of The Peoples' Republic of China. 2006. *The National Medium- and Long-term Program on Scientific and Technological Development (2006-2020): An Outline*. February 9.

Statistics Bureau, Ministry of Internal Affairs and Communications. 2005. *Report on the Survey of Research and Development 2004*. Ministry of Internal Affairs and Communications.

Stephan, P. 2001. "Educational Implication of University-Industry Technology Transfer." Journal of Technology Transfer 26:199-205.

Sternberg, R. 1990. "The Impact of Innovation Centres on Small Technology-based Firms: The Example of the Federal Republic of Germany." Small Business Economics 2(2):105-118.

Stevens, Ashley J., Jonathan J. Jensen, Katrine Wyller, Patrick C. Kilgore, Sabarni Chatterjee, and Mark L. Rohrbaugh. 2011. "The Role of Public-Sector Research in the Discovery of Drugs and Vaccines." *The New England Journal of Medicine*, February 9. 364(6):535-541.

Stiglitz, Joseph. 1999. "Knowledge as a Global Public Good" in I. Kaul, I. Grunberg, and M. Stern, eds. International Cooperation in the 21st Century. New York: UNDP.

Stiglitz, Joseph. 2001. "Bankruptcy Law; Basic Economic Principles." in Stijn Claessens et al. eds., *The Resolution of Financial Distress, An International Perspective on the Design of Bankruptcy Laws*, Washington, DC: The World Bank.

Stiglitz, Joseph. 2005. "The Ethical Economist." *Foreign Affairs*. Council on Foreign Affairs. November/December. Accessed at http://www.foreignaffairs.com/articles/61208/joseph-e-stiglitz/the-ethical-economist

Stockinger, Dennis, Rolf Sternberg and Matthias Kiese. 2009. "Cluster Policy in Co-Ordinated vs. Liberal Market Economies: A Tale of Two High-Tech States".

Stokes, Donald E. 1997. *Pasteur's Quadrant: Basic Science and Technological Innovation.* Washington, DC: Brookings Institution Press.

The Street. 2011. United States and South Korea Invest in Lithium Battery Technology." December 22.

Strout, E. 2005. "Gift of a Book Was a Key to Intel Founder's Big Donation to City College of New York." *Chronicle of Higher Education.* December 2. P. A27.

Su, Yun-Shan, and Ling-Chun Hung. 2008. "Spontaneous vs. Policy-driven: The Origin and Evolution of the Biotechnology Cluster." *Technological Forecast and Social Change* 76(5):608-619.

Sun, Steven and Garry Evans. 2010. "Emerging Strategic Industries: Aggressive Growth Plans Targets." HSBC Global Research, Oct 19). [Viewed here on 4/7/2012 http://www.research.hsbc.com/midas/Res/RDV?p=pdf&key=lg0uISbcyh&n=280786.PDF.

Sun Mountain Capital. 2011. "New Mexico Private Equity Investment Program: Overview and 2010 Review." June.

Suttmeier, Richard P., Cong Cao, and Denis Fred Simon. 2006. "China's Innovation Challenge and the Remaking of the Chinese Academy of Sciences.*" Innovations,* Summer.

Sutton, Stephanie. 2011. "The Status of Pharma R&D." BioPharm. July 5.

Swann, G. M. P. 1998. "Towards a Model of Clustering in High-Technology Industries." In G. M. P. Swann, M. Prevezer, and D. Stout, eds. *The Dynamics of Industrial Clustering*. Oxford, UK: Oxford University Press.

Taiwan Economic News. 2008."Welldone Ventures into Production of Lithium Battery." July 15.

Taiwan Economic News. 2011. "Taiwan Alliance, Set Up to Develop Extended-Range Electric Buses." May 4.

Taiwan Economic News. 2011. "Taiwan's Investment in Lithium Batteries to Exceed NT$50 B. in 2012." March 22.

Taipei Times. 2004. "Cabinet Inks Deal with Israeli Fund" October 19.

Taipei Times. 2008. "Taiwan Spearheads Lithium-Battery Module Effort." May 15.

Taipei Times. 2009. "Fund to Invest in Venture Capital Firms" March 19.

Taipei Times. 2009. "ITRI's Battery Technology Wins Oscar of Invention." October 17.

Talele, Chitram J. 2003. "Science and Technology Policy in Germany, India and Pakistan." *Indian Journal of Economics and Business* 2(1):87-100.

Tanaka, Nobua. 2005. Presentation at the International Forum on Technology Foresight and National Innovation Strategies. Seoul, Republic of Korea. November 4.

Tassey, G. 1997. *The Economics of R&D Policy*. Westport, CT: Quorum Books.

Tassey, G. 2002. R&D and Long-Term Competitiveness: Manufacturing's Central Role in a Knowledge-Based Economy, NIST Planning Report 02-2, February.pp. 31-40.

Tassey, G. 2004. "Policy Issues for R&D Investment in a Knowledge-based Economy." *Journal of Technology Transfer* 29:153-185.

Tassey, G. 2007. *The Innovation Imperative*. Cheltenham, UK: Edward Elgar.

Tassey, Gregory. 2007. "Tax incentives for innovation: time to restructure the R&E tax credit." *The Journal of Technology Transfer* 32(6): 605-615.

Tassey, Gregory. 2010. "Rationales and Mechanisms for Revitalizing US Manufacturing R&D Strategies." *Journal of Technology Transfer*-35 (3): 283-333

Tejaswi, Mini Joseph and Sujit John. 2010. "IBM is India's second largest pvt sector employer." Times of India, Aug. 18.

The Telegraph Online. 2012. "India Uses Arm-Twist Rule for Cancer Drug." March 13.

Teubal, Morris. 2002. "What Is the Systems Perspective to Innovation and Technology Policy and How Can We Apply It to Developing and Newly Industrialized Economies?" *Journal of Evolutionary Economics* 12(1-2):233-257.

Thai Press Reports. 2008. "Commerce Ministry Asks Council of State for Opinion on Legality of Compulsory Licensing of Cancer Drug." August 28.

Thai Press Reports. 2011. "LG Chem Builds World's Largest Electric-Car Battery Plant." April 7.

Thierer, Adam D. 2002. "Solving the Broadband Paradox." Issues in Science and Technology Spring.

Thelin, J. 2004. *A History of American Higher Education*. Baltimore, MD: Johns Hopkins University Press.

Thomas, J. R., and W. H. Schacht. 2008. "Patent Reform in the 110th Congress: Innovation Issues." Report for Congress. Washington, DC: Congressional Research Service. Order Code RL33996.

Thompson, Roger. 2011. "Why Manufacturing Matters". Harvard Business School. March 28. Access at http://hbswk.hbs.edu/item/6664.html.

Thomson Reuters Pharma. 2010. "The Ones to Watch: A Pharma Matters Report." July-September.

Thomson Reuters News & Insight. 2012. "FTC Loses Bid to Block Pay-for-Delay Drug Settlements." April 25.

Thursby, J., and M. Thursby. 2002. "Who is Selling the Ivory Tower? Sources of Growth in University Licensing." *Management Science* 48:90-104.

Thursby, J., and M. Thursby. 2004. "Industry Perspectives on Licensing University Technologies: Sources and Problems." AUTM Journal P. 2000. [See here http://www.provendis.info/fileadmin/info/pdfs/1255.pdf]

Thursby, J., and M. Thursby. 2006. "Here or There? Report to the Government-University-Industry Research Roundtable." Washington, DC: The National Academies Press. 2006.

Tierney, Sean. 2012. "High-speed rail, the knowledge economy, and the next growth wave." *Journal of Transport Geography* 22:285-287

The Times of India. 2012. "DuPont Opens World-Class Ballistics Facility in City." April 14.

The Times of London. 2008. "Drug Companies Trigger European Ire for Holding Back Supplies of Cheap Medicine." November 28.

The Times of London. 2011. "Why the Gene Revolution Has Been Postponed — It Costs $1bn to Develop a New Drug, so Don't Expect Personalized Treatments, But the Genome Project is Still Worthwhile." August 25.

Timmerman, Luke. 2011. Gates Foundation Makes First Equity Investment in Biotech Startup, Liquidia Technologies." xconomy.com.

Tong-hyung, Kim. 2009. "5% of GDP Set Aside for Science Research." Korea Times, Dec. 12.

Turner, James. 2006. "The Next Innovation Revolution, Laying the Groundwork for the United States." *Innovations*, Spring.

Tyson, Laura D'Andrea. 1992. Who's Bashing Whom? Trade Conflict in High Technology Industries, Washington, DC.: Institute for International Economics.

United Nations Environment Programme. 2010. Overview of the Republic of Korea's National Strategy for Green Growth, April. [Viewed here on 4/7/2012, http://www.unep.org/PDF/PressReleases/201004_unep_national_strategy.pdf].

US-China Business Council. 2011. Issues Brief: China's Domestic Innovation and Government Procurement Policies, [Updated: March 2011].

U.S.-China Joint Commission on Commerce and Trade (JCCT). 2005. *Outcomes on Major U.S. Trade Concerns.* Washington, DC: The Office of the United States Trade Representative.

U.S. Congress. 1988. *Omnibus Trade and Competitiveness Act of 1988* (P.L. 100-418, codified in 15 U.S.C. 278n.) and later amended by the *American Technology Preeminence Act of 1991* (P.L. 102-245, codified in 15 U.S.C. 3701).

U.S. Department of Commerce. 2008. *Innovation Measurement: Tracking the State of Innovation in the American Economy,* Report to the Secretary. Washington, DC: U.S. Department of Commerce. January. Available at <http://www.innovationmetrics.gov/Innovation%20Measurement%2001-08.pdf>.

U.S. Department of Defense. 1987. *Report of Defense Science Board Task Force on Defense Semiconductor Dependency.* Washington, DC, Office of the Under Secretary of Defense for Acquisition.

U.S. Department of State. 2011. "Background Note: Taiwan." Bureau of East Asian and Pacific Affairs. July 7. [Viewed here on 4/7/2012 http://purl.access.gpo.gov/GPO/LPS33822]

U.S. International Trade Commission. 2010. China: Intellectual Property Infringement, Indigenous Innovation Policies, and Frameworks for Measuring the Effects on the U.S. Economy, Investigation No. 332-514, USITC Publication 4199 (amended), November.

U.S. General Accounting Office. 2002. *Export Controls: Rapid Advances in China's Semiconductor Industry Underscore Need for Fundamental U.S. Policy Review.* GAO-020620. Washington, DC: U.S. General Accounting Office. April.

U.S. Small Business Administration. 2004. "Small Business by the Numbers." Office of Advocacy. Washington, DC: U.S. Small Business Administration. June.

Vaidyanathan, G. 2008. "Technology Parks in a Developing Country: The Case of India." *Journal of Technology Transfer* 33(3):285-299.

Van Atta, Richard. 2004. "Energy and Climate Change Research and the DARPA Model." Presentation to the Washington Roundtable on Science and Public Policy. November 3.

Van Atta, Richard. 2008. "Fifty Years of Innovation and Discovery." In *DARPA: 50 Years of Bridging the Gap.* Arlington, VA: Defense Advanced Research Projects Agency. April.

Van Atta, Richard, et al. 1991. *DARPA Technical Accomplishments.* Volumes I-V. Alexandria, VA: Institute for Defense Analysis.

Van Atta, Richard, et al. 1991. *DARPA Technological Accomplishments: An Historical Review of Selected DARPA Projects.* Alexandria, VA: Institute for Defense Analysis.

Van Atta, Richard, and Michael Lippitz. 2003. *Transformation and Transition: DARPA's Role in Fostering an Emerging Revolution in Military Affairs.* Volume 1: Overall Assessment. Alexandria, VA: Institute for Defense Analysis.

Van Looy, Bart, K. Debackere, and T. Magerman. 2005. *Assessing Academic Patent Activity: The Case of Flanders.* Leuven, Belgium: SOOS.

Van Looy, Bart, Marina Ranga, Julie Callaert, Koenraad Debackere, and Edwin Zimmermann. 2004. "Combining Entrepreneurial and Scientific Performance in Academia: Towards a Compounded and Reciprocal Matthew-effect?" *Research Policy* 33(3):425-441.

Vedovello, C. 1997. "Science Parks and University-Industry Interaction: Geographical Proximity between the Agents as a Driving Force." *Technovation* 17(9):491-502.

Vervliet, Greta. 2006. *Science, Technology, and Innovation.* Brussels, Belgium: Ministry of Flanders, Science and Innovation Administration.

Vest, C. 2005. "Industry, Philanthropy and Universities-The Roles and Influences of the Private Sector in Higher Education." 2005 Clark Kerr Lecture. University of California at Berkeley. September 13.

Vest, Charles M. 2011. "Chancellor's Colloquium." University of California, Davis, November 30, p. 8. [see link to video presentation here http://chancellor.ucdavis.edu/initiatives/colloquium/

Vest, Charles M. 2011. "Engineers: The Next Generation." President's Address. National Academy of Engineering Annual Meeting. October 16.

Veugelers, Reinhilde, Jan Larosse, Michele Cincera, Donald Carchon, and Roger Kalenga-Mpala. 2004. "R&D Activities of the Business Sector in Flanders: Results of the R&D Surveys in the Context of the 3% Target." IWT-Studies (46). Brussels, Belgium.

Vogel, Gretchen. 2006. "A German Ivy League Takes Shape." Science Magazine, Oct. 13. Accessed at: news.sciencemag.org/sciencenow/2006/10/13.01.html

Wadwha, Vivek et. al. 2008."The Globalization of Innovation: Pharmaceuticals: Can India and China Cure the Global Pharmaceutical Market?" Duke University Pratt School and Engineering and Harvard Labor and Work Life Program. [See these links http://ssrn.com/abstract=1143472 or http://dx.doi.org/10.2139/ssrn.1143472]

Wadhwa, Vivek, AnnaLee Saxenian, Richard Freeman, and Alex Salkever. 2009. "Losing the World's Best and Brightest: America's New Immigrant Entrepreneurs." Ewing Marion Kauffman Foundation, March http://www.kauffman.org/uploadedFiles/ResearchAndPolicy/Losing_the_World's_Best_and_Brightest.pdf

Wadhwa, Vivek, and Gary Gereffi. 2005. *Framing the Engineering Outsourcing Debate.* Durham, NC: Duke University.

Wadhwa, Vivek, Ben Rissing, AnnaLee Saxenian, Gary Gereffi. 2007 "Education, Entrepreneurship and Immigration: America's New Immigrant Entrepreneurs, Part II." Duke University Pratt School of Engineering, U.S. Berkeley School of Information, Ewing Marion Kauffman Foundation. June 11.

Wadhwa, Vivek, Ben Rissing, Gary Gereffi, John Trumpbour, and Pete Engardio. 2008. "The Globalization of Innovation: Pharmaceuticals." Duke Pratt School of Engineering, Kauffman Foundation, Harvard Law School Labor and Worklife Program, June.

Wadhwa, Vivek. 2010 "Chinese and Indian Entrepreneurs Are Eating America's Lunch." *Foreign Policy*, December 28.

Wagner, Caroline S. 2005. "Network structure, self-organization and the growth of international collaboration in Science." *Research Policy* 34(10) 1608-1618

Wagner, Caroline S. 2008. *The New Invisible College: Science for Development*, Washington, DC: Brookings.

Wagner, Caroline S. 2011. "The Shifting Landscape of Science." *Issues in Science and Technology*, Volume XXVII, No. 1, Fall.

Waits, Mary Jo. 2000. "The Added Value of the Industry Cluster Approach to Economic Analysis, Strategy Development, and Service Delivery." *Economic Development Quarterly* 14 :1 35-50.

Walcott, Susan M. 2003. *Chinese Science and Technology Industrial Parks*. Burlington, VT: Ashgate Publishing.

Wald, Matthew L. 2011. "Maker of Silicon Wafers Wins Millions in U.S. Loan Support." New York Times, June 17.

Waldrop, M. Mitchell. 1992. *Complexity: The Emerging Science at the Edge of Order and Chaos*. New York: Simon & Schuster.

Waldrop, M. Mitchell. 2001. *The Dream Machine*. New York: Viking Press.

Wall Street Journal. *2012*. "Car Battery Start-ups Fizzle." May 31.

Wall Street Journal. "U.S.'s Afghan Headache: $400-a-Gallon Gasoline." December 6.

Walsh, J., A. Arora, and W. Cohen. 2003. "Research Tool Patent and Licensing and Biomedical Innovation." In *Patents in the Knowledge-Based Economy*. W. Cohen and S. Merrill, eds. Washington, DC: The National Academies Press.

Walsh, J. P., and W. M. Cohen. 2004. "Does the Golden Goose Travel? A Comparative Analysis of the Influence of Public Research on Industrial R&D in the U.S. and Japan." Mimeo.

Wang, Changyong. 2005. "IPR Sails Against Current Stream." Caijing Magazine October 17. Available online at <http://caijing.hexun.com>.

Wang, Liwei. 1993. "The Chinese Traditions Inimical to the Patent Law." *Northwestern Journal of International Law and* Commerce Fall.

Warsh, David. 2006. *Knowledge and the Wealth of Nations*. New York: W. W. Norton.

Weaver, Courtney. 2011. "Welcome to Russia's Silicon Valley." Financial Times, August 21.

Wellman-Laback, Olivier and Youwen Zhou. 2010. "The U.S. Orphan Drug Act: Rare Disease Research Stimulator or Commercial Opportunity?" *Health Policy*. May.

Wen, Jiang, and Shinichi Kobayashi. 2001. "Exploring Collaborative R&D Network: Some New Evidence in Japan." *Research Policy* 30:1309-1319.

Wennberg, Karl and Gören Lindqvist, "How Do Entrepreneurs in Clusters Contribute to Economic Growth?" SSE/EFI Working Paper Series in Business Administration No. 2008:3

Wessner, Charles W. 2005. "Entrepreneurship and the Innovation Ecosystem Policy: Lessons From the United States." In David B. Audretsch, Heike Grimm, and Charles W. Wessner, eds. *Local Heroes in the Global Village: Globalization and the New Entrepreneurship Policies*. New York: Springer.

West, Darrell M. 2011. "Creating a 'Brain Gain' for U.S. Employers: The Role of Immigration." Brookings Policy Brief Series #178, Brookings Institution, January.

Westhead, P. 1995. "New Owner-Managed Businesses in Rural and Urban Areas in Great Britian: A Matched Pairs Comparison." *Regional Studies* 29(4):367-380.

Westhead, P. 1997. "R&D 'Inputs' and 'Outputs' of Technology-based firms Located On and Off Science Parks." *R&D Management* 27(1):45-61. Kingdom. London, UK: HMSO.

Westhead, P., and D. Storey. 1994. *An Assessment of Firms Located On and Off Science Parks in the United States*. London: HMSO

Westhead, P., and D. Storey. 1997. "Financial Constraints on the Growth of High-Technology Small Firms in the U.K." *Applied Financial Economics* 7(2):197-201.

Westhead, P., D. J. Storey, and M. Cowling. 1995. "An Exploratory Analysis of the Factors Associated with the Survival of Independent High-Technology Firms in Great Britain." In F. Chittenden, M. Robertson, and I. Marshall, eds. *Small Firms: Partnerships for Growth*. London, UK: Paul Chapman.

Westhead, P., and M. Cowling. 1995. "Employment Change in Independent Owner-Managed High-Technology Firms in Great Britain." *Small Business Economics* 7(2):111-140.

Westhead, P., and S. Batstone. 1998. "Independent Technology-based Firms: The Perceived Benefits of a Science Park Location." Urban Studies 35(12):2197-2219.

White House. 1987. "Increase in the Rate of Duty for Certain Articles from Japan." Proclamation 5631. April 17

The White House 2009. "Remarks by President Obama." Prague: April 5

The White House. 2011. "Remarks by President and Obama and President Hu in a Roundtable with American and Chinese Business Leaders." Office of the Press Secretary, January 19.

The White House. 2011. "U.S.-China Joint Statement." Paragraph 27, Office of the Press Secretary." January. 19.
The White House. 2011. Fact Sheet: Obama's Plan to Win the Future." Office of the Press Secretary, January. 25.
White House Press Release. 2011."President Obama and Skills for America's Future Partners Announce Initiatives Critical to Improving Manufacturing Workforce." June 8.
Wiener, Norbert. 1948. *Cybernetics or Control and Communication in the Animal and the Machine.* Cambridge, MA: The MIT Press.
Williamson, John. 1989. "What Washington Means by Policy Reform." in John Williamson, editor, *Latin American Readjustment: How Much has Happened.* Washington, DC: Institute for International Economics.
Wilson, Patrick 2011. Director of Government Affairs, Semiconductor Industry Association, "Maintaining US Leadership in Semiconductors." AAAS Annual Meeting.
Wista-Management GMBH. 2011."The Economic Significance of Adlershof: Impact on Added Value, Employment, and Tax Revenues in Berlin." study by the German Institute for Economic Research commissioned by Wista-Management.
Wolff, Alan Wm. 2011. "China's Indigenous Innovation Policy." Testimony before the U.S. China Economic and Security Review Commission, Washington, DC, May 4.
Wolff, Alan Wm. 2012. "America's Trade Policy and the Future of U.S. Trade Negotiations". Testimony before the House Ways and Means Committee, Washington, DC, February 29.
Wong, Joseph. 2011. Betting *on Biotech: Innovation and the Limits of Asia's Developmental State*, Ithica, NY: Cornell University Press.
Wong, Siew Ying. 2010. "Biomedical Manufacturing Output Grew to S$21b in 2009." channelnewsasia.com, March 17.
Woodhouse, Michael et. al. 2011. "An Economic Analysis of Photovoltaics Versus Traditional Energy Sources: *Where Are We Now and Where Might We Be in the Near Future.* NREL/CP ;; 6A20-50714 Retrieved April 10, 2012 from http://permanent.access.gpo.gov/gpo12708/50714.pdf
Woodward, Douglas. 2005. "Porter's Cluster Strategy Versus Industrial Targeting." University of South Carolina, presentation at ICIT Workshop, July 1.
Woodward,Douglas P, Joseph C. Von Nessen and Veronica Watson. 2011. "The Economic Impact of South Carolina's Automotive Cluster." Darla Moore School of Business, University of South Carolina. Study prepared for South Carolina Automotive Council. January http://www.moore.sc.edu/UserFiles/moore/Documents/rev1_19.pdf
World Bank. 2004. *Innovation Systems: World Bank Support of Science and Technology Development.* Vinod Kumar Goel, ed. Washington, DC: World Bank.

World Bank. 2006. *The Environment for Innovation in India.* South Asia Private Sector Development and Finance Unit. Washington, DC: World Bank.

World Bank International Finance Corporation. 2006. *Doing Business in 2006: Creating Jobs.* Washington, DC: International Bank for Reconstruction and Development.

World Bank. 2007. *Unleashing India's Innovation: Toward Sustainable and Inclusive Growth,* Mark A. Dutz, editor, The International Bank for Reconstruction and Development.

World Bank. 2012. *China 2030.* Washington, DC: The World Bank.

World Bank. 2012. Supporting Report 2: China Grows Through Technological Convergence and Innovation. Washington, DC: World Bank. Pp 177-178.

World Trade Organization Working Party on the Accession of China. 2001. Report of the Working Party on the Accession of China. WT/MN(01)/3. November 10.

World Trade Organization. 2011. "Trade Growth to Ease in 2011 But Despite 2010 Record Surge, Crisis Hangover Persists." WTO Press/628, April 7. Viewed here http://www.wto.org/english/news_e/pres11_e/pr628_e.htm

Wu, Ching. 2006. "China to Build 30 New Science and Technology Parks." SciDev.net. April 19.

Xing, Yuqing and Neal Detert. 2010. "How the iPhone Widens the United States Trade Deficit with the People's Republic of China." ADBI Working Paper 257, Asian Development Bank Institute, December. Retrieved from http://www.adbi.org/files/2010.12.14.wp257.iphone.widens.us.trade.deficit.prc.pdf

Xinhua. 2011. "China 2010 International Patent Filings up 56.2%." China Daily, Feb. 2. Retrieved from http://www.chinadaily.com.cn/bizchina/2011-02/10/content_11975479.htm

Xinhua. 2007. "Innovation tops Hu Jintao's Economic Agenda." October 15. Viewed here http://news.xinhuanet.com/english/2007-10/15/content_6883390.htm

Xinhua. 2011. "China to Boost IC Sector as 'State Strategy'." April 16. Viewed here http://news.xinhuanet.com/english2010/sci/2011-04/16/c_13831924.htm

Xinhua. 2011. "Financing agency boosts Brazil's innovation, productivity." March 6.

Xue, Lan. 2006. "Universities in China's National Innovation System." prepared for the UNESCO Forum on Higher Education, Research, and Knowledge. Retrieved from http://portal.unesco.org/education/en/files/51614/11634233445XueLan-EN.pdf/XueLan-EN.pdf

Xue, Lan. 2007. "China's Innovation Policy in the Context of national Innovation System Reform." Tsinghua University, August 27.

Yan, Cathy. 2011. "Road-Building Rage to Leave U.S. in Dust." Wall Street Journal, Jan. 18. Viewed here http://blogs.wsj.com/chinarealtime/2011/01/18/road-building-rage-to-leave-us-in-dust/

Yan, Dai. 2004. "Anti-Monopoly Legislation on the Way." China Daily June 18. Viewed here on 4/6/2012 http://www.chinadaily.com.cn/english/doc/2004-06/18/content_340471.htm

Yang, Lei, ed. 2006. "Chinese WAPI Delegation Quits Prague Meeting." Xinhua. June 8. Available at <http://news.xinhuanet.com>.

Yasuda, T. 2005. "Seisakukinyuu no riyou" (The Utilization of Public Finance). In Nippon no shinki kaigyou kigyou (Startup Enterprises in Japan). K. Kustuna and T. Yasuda, ed. Hakutousha.

Yonhap News Agency. 2007. "Pfizer Pharmaceutical Company to Invest 300m Dollars in South Korea by 2012." June 14.

Yonhap News Agency. 2010. "S. Korea Aims to Become Dominant Producer of Rechargeable Batteries in 2020." July 11. Viewed here on 4/6/2012 http://english.yonhapnews.co.kr/business/2010/07/10/38/0501000000AEN20100710002600320F.HTML

Yoshida, Fujio. 1967. "Preparation of Legal System for Capital Liberalization (Part 3)." Panel Discussion in Zaikei Shoho. July 17.

Yoshida, Junko. 2006. "Grenoble Lure: Un-French R&D." EE Times. June 12.

Young, John. 2007. Info Memo for Secretary of Defense Robert M. Gates. DoD Science and Technology Program. August 24.

Yourkin, Peter. 2008. "Making the Market: Howe the American Pharmaceutical Industry Transformed itself During the 1940s." University of California at Berkeley. November.

Yukio, Sato. 2001. "The Structure and Perspective of Science and Technology Policy in Japan." In Phillipe Laredo and Phillipe Mustar, eds. *Research and Innovation Policies in the New Global Economy: An International Comparative Analysis.* Cheltenham, UK: Edward Elgar.

Zachary, G. Pascal. 1999. *Endless Frontier: Vannevar Bush, Engineer of the American Century.* Cambridge, MA: The MIT Press.

Zeigler, Nicholas J. 1997. *Governing Ideas: Strategies for Innovation in France and Germany.* Ithaca, NY: Cornell University Press.

Zemin, Jiang, General Secretary of the Communist Party of China Central Committee. 1999. Keynote Speech at the National Technological Innovation Conference, August 23.

Zemsky, Robert and James J. Duderstadt. 2004. "Reinventing the Reserarch University: An American Perspective. In *Revinventing the Research University,* Luc E. Weber and James J. Duderstadt, eds, London: Economica.

Ziedonis, R. H., and BRONWYN H. Hall. 2001. "The Effects of Strengthening Patent Rights on Firms Engaged in Cumulative Innovation: Insights from the Semiconductor Industry." In Gary Libecap, ed., *Entrepreneurial Inputs and Outcomes: New Studies of Entrepreneurship in the United States.* Volume 13 of Advances in the Study of Entrepreneurship, Innovation, and Economic Growth. Amsterdam, The Netherlands: Elsevier Science.

Zhang, Chunlin, Douglas Zhihua Zeng, William Peter Mako, and James Seward. 2009. *Promoting Enterprise-Led Innovation in China*, Washington, DC: The International Bank for Reconstruction and Development/The World Bank Retrieved April 6, 2012 from http://siteresources.worldbank.org/CHINAEXTN/Resources/318949-1242182077395/peic_full_report.pdf.

Liang Z, and Lan Xue 2010 "The evolution of China's IPR system and its impact on the patenting behaviours and strategies of multinationals in China." Volume 51 of International Journal of Technology Management. 51 (2-4) pp.-469-496.Zhou, Evey Y. and Bob Stembridge."Patented in China: The Present and Future State of Innovation in China." Thompson Reuters, 2010. . Viewed here on 4/6/2012 http://www.ipeg.eu/blog/wp-content/uploads/Patented-in-China-The-Present-and-Future-State-of-Innovation-in-China-Eve-Y.-Zhou-Bob-Stembridge.pdf.

Zycher, Benjamin, Joseph DiMasi, and Christopher-Paul Milne. 2008. "The Truth About Drug Innovation: Thirty –Five Summary Case Histories on Private Sector Contributions to Pharmaceutical Science. *Medical Progress Report 6.*